Springer Tracts in Modern Physics

Volume 281

Springer Tracts in Modern Physics provides comprehensive and critical reviews of topics of current interest in physics. The following fields are emphasized:

- Elementary Particle Physics
- Condensed Matter Physics
- Light Matter Interaction
- Atomic and Molecular Physics
- Complex Systems
- Fundamental Astrophysics

Suitable reviews of other fields can also be accepted. The Editors encourage prospective authors to correspond with them in advance of submitting a manuscript. For reviews of topics belonging to the above mentioned fields, they should address the responsible Editor as listed in "Contact the Editors".

More information about this series at http://www.springer.com/series/426

Ion I. Geru

Time-Reversal Symmetry

Seven Time-Reversal Operators for Spin Containing Systems

Springer

Ion I. Geru
Institute of Chemistry
Academy of Sciences of Moldova
Chisinau, Moldova

ISSN 0081-3869 ISSN 1615-0430 (electronic)
Springer Tracts in Modern Physics
ISBN 978-3-030-01209-0 ISBN 978-3-030-01210-6 (eBook)
https://doi.org/10.1007/978-3-030-01210-6

Library of Congress Control Number: 2018955912

This Springer imprint is published by the registered company Springer Nature Switzerland AG
The registered company address is: Gewerbestrasse 11, 6330 Cham, Switzerland

Foreword

Time-reversal symmetry is of fundamental importance to physics but remains to puzzle physicists until the present day. My first step in research concerned the consequences of time-reversal symmetry for dynamical systems, leading me to observe that while recognized as important, time-reversal symmetry had not received the central attention it deserved and as a consequence, the corresponding theory was not always developed to the required level. I was most fortunate to be able to stimulate further research, in particular, through organizing a *Workshop on Time-reversal Symmetry in Dynamical Systems* at the University of Warwick in 1996 and writing an accompanying survey article with John Roberts.[1]

Ion Geru is a well-established physicist and I am happy to have been approached to write a foreword for his book which addresses a similar deficiency of the literature on time-reversal symmetry in quantum spin systems as I observed before in the context of dynamical systems.

Ion Geru's book is a very welcome addition to the literature, which counts few attempts to focus in comparative detail on time-reversal symmetry. The first five chapters provide an excellent accessible exposition of the classical theory of time-reversal symmetry in classical and quantum systems, and the last chapter summarizes the importance of time-reversal symmetry violation in a variety of contexts. The remaining chapters are more specialized, and Chaps. 8–9 address a particular deficiency in the theory, re-enforcing the point that time-reversal remains a topic that deserves further scrutiny and dedicated exploration.

[1] J.S.W. Lamb and J.A.G. Roberts, Time-reversal symmetry in dynamical systems: A survey, *Physica D* **112** (1998), 1–39.

I warmly recommend this book as a primer for all interested to learn about the state of the art in time-reversal symmetry. Personally, it has revealed to me some unexplored directions for future research.

London, UK

Jeroen S.W. Lamb
Professor of Applied Mathematics
Imperial College London

Preface

The symmetry of time reversal is one of the most important discrete symmetries in the natural sciences. In classical mechanics, Newton's differential equations of motion are invariant with respect to the change of the variable t by $-t$. The equations of classical electrodynamics are also invariant with respect to time reversal if, along with the substitution of t by $-t$, invert the directions of the axial vectors of the magnetic induction **B** and the magnetic field strength **H** (respectively, the axial magnetization vector **M**) and the polar current density vector **j**.

In Chap. 1, the symmetry of the time reversal in classical mechanics and in classical and relativistic electrodynamics is discussed, as well as the dynamic reversibility and statistical irreversibility of time, and the reversibility of fluctuations in closed systems as a consequence of the Onsager relations and Poincaré recurrence theorem.

In Chap. 2, the symmetry of the time reversal in quantum mechanics and in quantum electrodynamics is considered.

In quantum mechanics, Wigner introduced, in 1932, the time-reversal operator on the basis of the invariance of the Schrödinger equation with respect to the substitution of t by $-t$, which is accompanied by the operation of complex conjugation (and by a certain unitary transformation, if the spin is also taken into account). Unlike linear operators describing spatial symmetry, this operator is an antiunitary (antilinear and unitary) operator.

After a brief presentation of the fundamentals of quantum mechanics and the properties of antilinear operators, the properties and various applications of the Wigner antiunitary time-reversal operator are discussed. There are given the criteria (a), (b), and (c), found for the first time by Wigner, concerning the absence or presence of an additional degeneracy of the energy levels due to the time-reversal symmetry without taking into account and taking into account the spin of the particles forming the quantum system. We also consider the corepresentations of symmetry groups introduced by Wigner, connected with the presence of the time-reversal symmetry.

A geometric interpretation of the time-reversal transformation and the Kramers theorem is given and various ways are given for constructing the matrix of the unitary operator **U** entering into the time inversion operator $\mathbf{T} = \mathbf{UK}$, where **K** is the complex conjugation operator for systems with integer or half-integer spin $S \geq 1/2$. The matrix of the unitary operator **U** is determined in the basis of 32 point symmetry groups.

Further, the Herring criteria (a), (b), and (c), established by him in 1937 and relating to the cases of the absence or presence of an additional degeneracy of the energy bands in crystals due to the time-reversal symmetry, are discussed.

After Wigner introduced the time-reversal operator and after establishing its properties, determining Wigner criteria (a), (b), (c) for atomic and molecular systems and Herring criteria (a), (b), (c) for crystals, it might be considered that the theory of groups applied to the time-reversal operator is complete. Indeed, these results are included in the tables of irreducible representations of 32 point symmetry groups and are found for a number of groups of the wave vector. However, as shown in this book (Chaps. 8 and 9), it turned out that the properties of the time-reversal operator were not fully studied.

There are also discussed in Chap. 2 the non-conventional time-reversal symmetry, selection rules due to the time-reversal symmetry, the detailed balance principle and the time-reversal symmetry, application of the time-reversal operator to the dynamic matrix in the classical theory of lattice vibrations, as well as the time-reversal symmetry in the theory of quantized field and the *CPT* theorem.

Chapter 3 is devoted to magnetic symmetry point groups. Problems of the invariant spin arrangement and of such called admissible magnetic point groups of non-Kramers systems are discussed. It is presented the well-known method for constructing point magnetic symmetry groups using a cyclic group of the second order, containing the identity transformation and the time-reversal operator. On the basis of this method, different authors found 58 nontrivial black-and-white groups of magnetic symmetry. We have shown that this result is valid only for the case of non-Kramers systems. In the case of systems with Kramers degeneracy of energy levels, the extension of 32 point symmetry groups to magnetic symmetry groups must be performed not with the help of a cyclic group of the second order, but using a cyclic group of the fourth order, constructed on the basis of the time-reversal operator. With such an extension, the point groups of magnetic symmetry are not two colors, but four colors and the number of nontrivial groups is not equal to 58, but to four. Magnetic symmetry groups were also constructed for the more general case when instead of a fourth-order cyclic group containing four consecutive anti-rotations by 90° of a square with differently colored neighboring vertices (around an axis passing through the center of the square perpendicular to its plane), a noncyclic group of the eighth order, formed by all symmetry elements of a square with colored vertices, is used. In this case, the number of groups of magnetic symmetry is also four, and these groups are also four-colored, but they naturally have a different structure.

In Chap. 4, anomalous properties of trihomonuclear Kramers clusters due to their four-color symmetry are studied. It is shown that some anomaly of the magnetic properties of trihomonuclear Kramers clusters is due to the presence of structural distortions of these clusters, which are a consequence of the existence of time-reversal symmetry. This theoretical conclusion is confirmed by experimental data.

Chapter 5 contains the results of a study of the time-reversal symmetry in quantum systems with a quasienergy spectrum. The behaviors of nonstationary states of quantum systems under time-reversal operator, the time-reversal invariance of the Schrödinger equation for the Green function, the quasienergy spectrum, and the Brillouin zone in the quasienergy space are considered. It is predicted the existence of quasienergy doublets due to non-commuting time translation and time-reversal operators. The doublet structure of the quasienergy levels has no Kramers origin, since it also exists in the case of systems formed from spinless particles.

Chapter 6 proposes an original method of converting the antiferromagnetic-type exchange interaction into a ferromagnetic one and vice versa in magnetic dimer clusters. The possibility of such a transformation is based on the fact that the wave function of the hole in the unfilled electron shell of the 3*d*- or 4*f*-ion is a time-reversed wave function of the electron. If the Hamiltonian of the exchange interaction of the magnetic homonuclear dimer is acted upon by the time-reversal operator in only one of the dimer ions, then the invariance with respect to the time reversal will be violated. If we restore the violated invariance not by performing the time-reversal transformation for the second ion, but by changing the sign of the exchange interaction constant, then an inversion of the dimer spin levels occurs. For example, for the dimer $Cu^{2+} - Cu^{2+}$ this is equivalent to a transition from the hole wave function in an unfilled 3*d*-shell to the wave function of an electron in the same shell. Thus, in order for a homonuclear dimer $3d^n - 3d^n$ or $4f^m - 4f^m$ with the exchange interaction of an antiferromagnetic type to be converted into a heteronuclear dimer with a ferromagnetic exchange, it is necessary to replace one of the dimer ions by a time-reversed ion, that is, to replace electrons in this ion by holes (or holes by electrons). On the one hand, this method is of interest, because the number of dimeric coordination compounds with antiferromagnetic exchange interaction is much higher than the number of compounds with ferromagnetic exchange and it is possible to increase the number of these latter compounds on account of the first ones. On the other hand, an increase in the number of compounds with a ferromagnetic-type exchange interaction is promising in practical terms in connection with the widespread practical application of ferromagnetic substances, in particular, for recording and storing information.

In the last, fifth paragraph of this chapter, an experimental confirmation of the method proposed by us is given.

Chapter 7 discusses the possibility of the existence or absence of the effect of instability of spin populations of an equilateral trimeric magnetic cluster containing antiferromagnetic coupled half-integer spins, with respect to a weak structural deformation. The ground state of such a cluster is formed by two Kramers doublets. It is shown that this instability is forbidden because the contributions to the magnetic moments of the paramagnetic ions localized at each of the vertices of the triangle due to the spin wave functions φ_1 and φ_3 of the first and second Kramers doublets are compensated by the contributions due to the spin wave functions $\varphi_2 = \mathbf{T}\varphi_1$ and $\varphi_4 = \mathbf{T}\varphi_3$ of the same Kramers doublets, where $\mathbf{T}$ is the time-reversal operator. It is shown that the effect of instability of spin populations in the presence of weak structural deformation can exist in the case of a time-reversal symmetry breaking. In particular, this takes place in the presence of a constant electric field, since in this case invariants containing spin projection operators in the first degree can be constructed in the form of products with components of the electric field strength vector. Such a Hamiltonian for the interaction of the spins of a magnetic cluster with a constant electric field is not invariant with respect to time reversal.

According to the novelty of the results, Chaps. 8 and 9 occupy a special place among other chapters of the book. In the Preface it was already noted that after the establishment of Wigner criteria (a), (b), (c) and Herring criteria (a), (b), (c), in atomic and molecular physics, as well as in solid state physics, it is only necessary to apply these criteria to solving specific problems. It would seem that no new group-theoretical results can be obtained, but we still succeeded in obtaining some new theoretical results. They concern, first of all, the nontrivial factorization of the Wigner time-reversal operator, that is, in the representation of the operator **T** as the product of two or three operators, each of which commutes separately with the corresponding Hamiltonian of the quantum system. To prove the existence of a nontrivial factorization of the operator **T**, previously unknown non-Abelian groups of 8th and 16th orders containing the operator **T** (for the Kramers systems) and Abelian groups of 8th and 16th orders containing the operator **T** (for non-Kramers systems) were constructed and studied.

In Chap. 8, symmetry groups of the 8th and 16th orders were constructed, containing the time-reversal operator **T** related to systems with half-integer angular momentum and it was shown that they are non-Abelian groups. Each of these groups contains a cyclic subgroup of the fourth order, constructed on the basis of the operator **T**. In the case of non-Kramers systems, the corresponding groups of 8th and 16th orders are Abelian. The structures of non-Abelian and, respectively, Abelian groups are different, but these structures do not change when the angular momentum changes.

On the basis of the group-theoretic approach, it is proved that there are six new time-reversal operators, along with the antiunitary operator **T**. Among the new operators, the three time-reversal operators are antiunitary, and three are unitary. A characteristic feature of the new time-reversal operators is that under their action, the signs do not change for all three angular momentum projection operators, but only for two or only for one of them. As a result, such operators can be called operators of incomplete time reversal, in contrast to the Wigner operator, which in this context is an operator of the full-time reversal.

In Chap. 9, based on the results obtained in Chap. 8, a nontrivial factorization of the operator of the complete time reversal **T** as a product of two or three operators of incomplete time reversal is carried out. It is proved that when the time-reversal symmetry decreases in systems with Kramers degeneracy of energy levels, a violation of Kramers theorem occurs, with the exception of one case when the degree of lowering the time-reversal symmetry turns out to be insufficiently high to remove the Kramers degeneracy.

The commutation and anticommutation relations between operators of incomplete time reversal and between these operators and the **T** operator are found. It is shown that these relations are different for Kramers and non-Kramers systems.

It is proved that in the two-boson representation of operators of the angular momentum J, the sum of the boson occupation numbers can be negative for integer J. This corresponds to the anti-boson states, which can be obtained from boson states under the action of the time-reversal operator. It is found that for a special type of time-reversal symmetry breaking, shifts and broadening of the energy levels occur in Zeeman systems, which depend on the value of the angular momentum J (Sect. 9.6).

The six possible types of T-symmetry reducing to $T_{1(yz)}$-, $T_{2(zx)}$-, $T_{2(xy)}$-, $T_{1(x)}$-, $T_{1(y)}$- or $T_{1(z)}$-symmetries, discussed in Chap. 8, can be restored by introducing six types of corresponding meta-particles. The types of meta-particles depend on spin values of initial particles of systems with broken T-symmetry. The meta-particles corresponding to spin-1/2 and spin-1 particles are presented in Sect. 9.7 and Appendix E. Existence of six types of meta-particles theoretically means the availability of six metaworlds in the Universe.

In Chap. 10, experimental data related to time-reversal symmetry violation in meson systems, atomic nuclei, atoms and molecules, and high-T_c superconductors are discussed. It is also discussed here the enhancement of quantum transport at time-reversal symmetry breaking (Sect. 10.5) and the implication of T violation processes in the phenomenological unidirectionality of time observed in the Universe (Sect. 5.6).

In the last Sect. 10.7 of Chap. 10, we propose a simple experimental method for testing the time-reversal symmetry breaking in paramagnetic molecules and ions in gases, liquids, and solids based on EPR spectroscopy. Here are given the experimental results obtained by this method (called by us the method of virtual time reversal) for EPR spectra of free radicals in DPPH and EPR spectra with hyperfine structure for 0.001 M aqueous solution of manganese chloride, $MnCl_2$. The EPR spectra detected in the natural flow of time and during virtual time reversal coincide in shape, which indicates the presence of the time-reversal symmetry. In this case, a shift in the resonant frequency was detected. This shift is much greater than the frequency shift found by other methods and interpreted as a consequence of the time-reversal symmetry breaking (the presence of a constant electric dipole moment in protons, electrons, nuclei, atoms, molecules, etc.). Apparently, such a shift can be related to the non-coincidence of sweep rates of the magnetic field in the forward and reverse directions during detecting the EPR spectrum.

Generally, the time-reversal symmetry violation remains the most mysterious symmetry violation, of which origin is not yet fully understood.

Chisinau, Moldova

Ion I. Geru

Contents

Chapter 1
Time Reversal in Classical and Relativistic Physics

The symmetry of time reversal is one of the basic symmetries considered in the natural sciences. It occurs in many physical dynamic systems, in particular, in classical and relativistic mechanics and electrodynamics. These consider the time conception, time translation invariance, and time-reversal symmetry. It was shown that the symmetry under time translation, which is a manifestation of the time homogeneity, is stipulated by the law of conservation of total energy of a closed system. This is proved in the Lagrange formalism of classical mechanics, as well as on the basis of Nöether's theorem in the case of Einstein's special relativity.

In this chapter it is presented the proof of Painlevé theorem. The time-reversal symmetry in classical and relativistic mechanics and electrodynamics is discussed in the absence and presence of external magnetic field. The reversibility, including microscopical reversibility of fluctuations in closed systems, and irreversibility, as well as the Poincaré recurrence, are analyzed.

1.1 The Time Conception and Time Translation Invariance

Time is a dimension in which events can be ordered from the past to the future and a measure of the duration of events and the intervals between them [1]. Time is a part of the fundamental structure of the Universe. In classical physics, time is a dimension that does not depend on events. In this dimension, events occur in a certain sequence. Time was historically closely related to space, the two merging into space-time in the special relativity and general relativity of Einstein.

The nature and role of time variable in physics are different from those of space variables. Our perception about time is influenced by the irreversibility of our every-day life. Therefore it is of special importance to make a distinction between the motion equations and the actual motion as far as the time variable is concerned [2].

I. I. Geru, *Time-Reversal Symmetry*, Springer Tracts in Modern Physics 281,
https://doi.org/10.1007/978-3-030-01210-6_1

The time variable is distinguished also in other ways. Namely, the roles of space and time variables in Newtonian dynamics and in special relativity are quite different. In the first case, space and time variables are separated, while in special relativity the Lorentz transformations interrelating time and space variables make an invariant distinction between timelike and spacelike quantities. These distinctions mean that not only the time variables must be treated differently from the space variables, but also that the consequences of symmetry considerations in these two cases can be different. Specifically, it may be shown that consequences of time reversal are different from the space inversion [2].

In order to elucidate the physical aspect of the time variable nature, it is convenient to consider the analogy with spatial variables, as far as they are applicable. Among the features of the space that appear to be simply a question of convenience, the location of coordinate system would be included, because we assume that the space is homogeneous. Similarly, the orientation of the coordinate axes is assumed to be arbitrary, because we believe the space to be isotopic. Both of those assumptions have important physical consequences: leading to the laws of conservation of the total linear momentum and the total angular momentum for isolated systems. In analogy with these conservation laws, the assumption that the time is uniform leads to the law of conservation of total energy for isolated systems. All these fundamental conservation laws, which are consequences of the homogeneity and isotropy of the space, and uniformity of the time, can be strongly proved in the Lagrange formalism of classical mechanics [3–5].

Let us prove that the time uniformity leads to the law of conservation of the total energy of a conservative system. We begin with the Lagrangian[1] of the system from n particles, written in generalized coordinates (q_k and $\dot{q}_k = dq_k/dt$ are generalized coordinates and generalized velocities, respectively; $k = 1, 2, \ldots, s$), which has the form

$$\mathcal{L}(q_k, \dot{q}_k, t) = T(q_k, \dot{q}_k) - U(q_k, t), \tag{1.1}$$

where $T(q_k, \dot{q}_k)$ and $U(q_k, t)$ are kinetic and potential energy, respectively. For a closed system or a system situated in a stationary field of force, Lagrangian does not depend explicitly on time ($\partial\mathcal{L}/\partial t = 0$). In this case the time by virtue of its homogeneity can not be included explicitly in Lagrangian, since homogeneity means equivalence of all time moments. That is why the replacement of one time moment by another with the replacement of coordinates and velocities of particles must not change mechanical properties of the system.

In the case of stationary bonds and stationary potential forces the potential energy U from (1.1) does not depend on time, but the kinetic energy T is determined by the expression [5]

$$T = \frac{1}{2}\sum_{k=1}^{s}\sum_{l=1}^{s}\sum_{i=1}^{n} m_i \frac{\partial x_i}{\partial q_k}\frac{\partial x_i}{\partial q_l}\dot{q}_k\dot{q}_l, \tag{1.2}$$

[1] Lagrangian $\mathcal{L}$ is a functional, so (strongly speaking) its dependence on generalized coordinates q_k, generalized velocities $\dot{q}_k$ and time t should be notated as $\mathcal{L}[q_k, \dot{q}_k, t]$.

where $x_i = x_i(q_1, q_2, \ldots, q_s)$ is the coordinate of particle with mass m_i and n is the total number of particles.

The derivative with respect to time of Lagrange function is

$$\frac{\mathrm{d}\mathfrak{L}}{\mathrm{d}t} = \sum_{k=1}^{s} \left(\frac{\partial \mathfrak{L}}{\partial q_k} \frac{\partial q_k}{\partial t} + \frac{\partial \mathfrak{L}}{\partial \dot{q}_k} \frac{\partial \dot{q}_k}{\partial t} \right), \tag{1.3}$$

which for a conservative system with regard to Euler–Lagrange equations

$$\frac{\mathrm{d}}{\mathrm{d}t} \left(\frac{\partial \mathfrak{L}}{\partial \dot{q}_k} \right) = \frac{\partial \mathfrak{L}}{\partial q_k}, \quad (k = 1, 2, \ldots, s) \tag{1.4}$$

leads to the form

$$\frac{\mathrm{d}}{\mathrm{d}t} \left(\sum_{k=1}^{s} \dot{q}_k \frac{\partial \mathfrak{L}}{\partial \dot{q}_k} - \mathfrak{L} \right) = 0. \tag{1.5}$$

Since only kinetic energy T depends on $\dot{q}_k$, we replace $\frac{\partial \mathfrak{L}}{\partial \dot{q}_k}$ by $\frac{\partial T}{\partial \dot{q}_k}$ in (1.5), and after that we obtain

$$\frac{\mathrm{d}}{\mathrm{d}t} \left(\sum_{k=1}^{s} \dot{q}_k \frac{\partial T}{\partial \dot{q}_k} - \mathfrak{L} \right) = 0. \tag{1.6}$$

The expression in brackets from (1.6) is equal to a constant which we denote by H

$$H = \sum_{k=1}^{s} \dot{q}_k \frac{\partial T}{\partial \dot{q}_k} - \mathfrak{L}. \tag{1.7}$$

Now we can apply Euler theorem in accord to which a homogeneous function of variables $q_1, q_2, \ldots, q_s$ of degree m satisfies the equation

$$\sum_{i=1}^{s} \frac{\partial f}{\partial q_i} q_i = m\, f(q_1, q_2, \ldots, q_s). \tag{1.8}$$

Comparing (1.8) with the sum from (1.7), we find that this sum is the doubled kinetic energy (the degree of homogeneity of the function $T(q_k, \dot{q}_k)$ is $m = 2$)

$$H = 2T - \mathfrak{L} = \text{const}. \tag{1.9}$$

Taking into account (1.1), the formula (1.9) takes the form

$$H = T + U = \text{const}. \tag{1.10}$$

Thus H is the Hamiltonian function and (1.10) is the law of conservation of total energy of conservative system as consequence of the time uniformity.

The general theory of conservation laws was developed on the base of the Nöether theorem [6] and continuous groups of symmetry Lee [4, 7–9]. The Nöether theory asserts that if the Lagrange function for a physical system is not affected by changes in the coordinate system used to describe it, then the corresponding conservation law will be satisfied. For example, if the Lagrangian is not dependent on the time, then energy is conserved. Below we briefly outline the theory of conservation laws, following [10].

Let us consider the Euler–Lagrange equation

$$\frac{\delta \mathcal{L}}{\delta u^\alpha} = 0, \quad (\alpha = 1, \ldots, m) \tag{1.11}$$

which admits the group G of Lie–Bäcklund transformation [11, 12]. Here

$$\mathcal{L} = \mathcal{L}(x, u, u_{(1)}, \ldots, u_{(p)}),\ x = (x^1, \ldots, x^n),\ u = (u^1, \ldots, u^m),$$
$$u_{(1)} = \left\{ u^\alpha_{i_1 \ldots i_s} \mid \alpha = 1, \ldots, m;\ i_1 \ldots i_s = 1, \ldots, n \right\}, s = 1, 2, \ldots.$$

The operator $\frac{\delta}{\delta u^\alpha}$ in (1.11) is

$$\frac{\delta}{\delta u^\alpha} = \frac{\partial}{\partial u^\alpha} + \sum_{s \geq 1} (-1)^s D_{i_1} \ldots D_{i_s} \cdot \frac{\partial}{\partial u^\alpha_{i_1 \ldots i_s}}, \tag{1.12}$$

where

$$D_i = \frac{\partial}{\partial x^i} + u^\alpha_i \frac{\partial}{\partial u^\alpha} + \sum_{s \geq 1} u^\alpha_{i j_1 \ldots j_s} \cdot \frac{\partial}{\partial u^\alpha_{j_1 \ldots j_s}}. \tag{1.13}$$

Here and further the summation is done on repeating indexes.

The Lie–Bäcklund infinitesimal operators of the continuous group G are

$$X = \xi^i \frac{\partial}{\partial x^i} + \eta^\alpha \frac{\partial}{\partial u^\alpha} + \sum_{s \geq 1} \zeta^\alpha_{i_1 \ldots i_s} \cdot \frac{\partial}{\partial u^\alpha_{i_1 \ldots i_s}}, \tag{1.14}$$

where $\xi^i = \xi^i(x, u, u_{(1)}, \ldots)$, $\eta^\alpha = \eta^\alpha(x, u, u_{(1)}, \ldots)$ are analytical functions of finite numbers of variables, and

$$\zeta^\alpha_{i_1 \ldots i_s} = D_{i_1} \ldots D_{i_s} \left(\eta^\alpha - \xi^j u^\alpha_j \right) + \xi^j u^\alpha_{j i_1 \ldots i_s}. \tag{1.15}$$

Besides the operator X, its equivalent canonical Lie–Bäcklund operator is

$$\overline{X} = X - \xi^i D_i = \overline{\eta}^\alpha \frac{\partial}{\partial u^\alpha} + \cdots, \qquad \overline{\eta}^\alpha = \eta^\alpha - \xi^j u^\alpha_j. \tag{1.16}$$

The system (1.11) is said to have a conservation law if there exists a vector $C = (C^1, \ldots, C^n)$ satisfying the condition

$$D_i C^i = 0 \tag{1.17}$$

for any solution of the system.

One of the possible algorithms for constructing conserved vectors is provided by Nöether's theorem [6, 10]: if the action (which is a functional) is invariant with respect to G, then the vector

$$C^i = N^i \mathfrak{L}, \tag{1.18}$$

where

$$N^i = \xi^i + \overline{\eta}^\alpha \left(\frac{\partial}{\partial u_i^\alpha} + \sum_{s \geq 1} (-1)^s D_{j_1} \ldots D_{j_s} \frac{\partial}{\partial u^\alpha_{ij_1 \ldots j_s}} \right) + \tag{1.19}$$
$$+ \sum_{r \geq 1} D_{k_1} \ldots D_{k_r} \left(\overline{\eta}^\alpha \right) \left(\frac{\partial}{\partial u_{ik_1 \ldots k_r}} + \sum_{s \geq 1} (-1)^s D_{j_1} \ldots D_{j_s} \frac{\partial}{\partial u^\alpha_{ik_1 \ldots k_r j_1 \ldots j_s}} \right)$$

and C^i from (1.18) satisfies the equation (1.17). The infinitesimal criterion of invariance of the action is [13]

$$X\mathfrak{L} + \mathfrak{L} D_i \xi^i = 0. \tag{1.20}$$

If X is admitted by the system (1.11), and C is any conserved vector, then the vector

$$P = \overline{X}(C) \tag{1.21}$$

also satisfies the conservation law (1.17).

Theorem *Suppose the operators X, X_1, X_2 correspond (in accordance with (1.17)) to the conserved vectors C, C_1, C_2, and*

$$X_2 = adX(X_1) \equiv [X, X_1]. \tag{1.22}$$

Then the vectors $\overline{X}C_1$ and C_2 define equivalent conservation laws.

This theorem is proved in [10].

To illustrate the application of the general theory of conservation laws, let obtain the law of conservation of energy in relativistic mechanics. Let at $t' = t = 0$ the origin of the coordinate system x', y', z', associated with the moving particle of rest mass m, coincides with the origin of the coordinate system with respect to which this particle moves with the speed $\mathbf{v}$ ($v \leq c$, c is the speed of light in vacuum). According to the special theory of relativity, the transformation of coordinates from x, y, z, t to x', y', z', t' (Lorentz transformation) leaves invariant the quantity $x^2 + y^2 + z^2 - c^2t^2$. Among infinitesimal operators of Lorentz group, which is one of Lee groups, we choose the operator

$$X_0 = \frac{\partial}{\partial t}. \tag{1.23}$$

The Lagrangian of a free particle in special relativity is

$$\mathfrak{L} = -mc^2\sqrt{1-\beta^2}, \qquad \beta = v/c. \tag{1.24}$$

Applying the Nöether theorem to X_0 from (1.23) with the Lagrangian (1.24) we can obtain the corresponding conservation law. Since for the Lorentz group

$$X = \xi(t, \mathbf{x})\frac{\partial}{\partial t} + \eta^\alpha(t, \mathbf{x})\frac{\partial}{\partial x^\alpha}, \qquad (\alpha = 1, 2, 3) \tag{1.25}$$

we obtain a conserved quantity of the form [13]

$$T = \xi\mathfrak{L} + (\eta^\alpha - \xi v^\alpha)\frac{\partial \mathfrak{L}}{\partial v^\alpha}, \tag{1.26}$$

where

$$\frac{\partial \mathfrak{L}}{\partial v^\alpha} = \frac{mv^\alpha}{\sqrt{1-\beta^2}}. \tag{1.27}$$

For $X_0 = \frac{\partial}{\partial t}$ we have $\xi = 1,\ \eta^1 = \eta^2 = \eta^3 = 0$. Equation (1.26) yields

$$T = \mathfrak{L} - v^\alpha\frac{\partial \mathfrak{L}}{\partial v^\alpha} = -mc^2\sqrt{1-\beta^2} - \frac{m|\mathbf{v}|^2}{\sqrt{1-\beta^2}} = -\frac{mc^2}{\sqrt{1-\beta^2}}. \tag{1.28}$$

Substituting T by $-E$ ($T = -E$) we get

$$E = \frac{mc^2}{\sqrt{1-\beta^2}}. \tag{1.29}$$

This formula is the law of energy conservation in the special relativity of a free particle moving with a velocity $\mathbf{v}$.

1.2 Kinematically Admissible Transformations and Time Reversal

In the Sect. 1.1 it was mentioned the distinction between equations of movement and the movement itself at time reversal. The movement equations of classical mechanics, containing a second derivative with respect to time do not change at time reversal. In contrast to this, at macroscopic scale the real movement corresponds to only one direction course of time (from the past to the future). Thus transformation of time reversal is based on properties of forces or interchanges, which determine the dynamic behavior of the system. Since the movement properties studied in kinematics are independent of those studied in dynamics, we may introduce admissible time reversal transformations, which must be formulated in terms of kinematics. For this we shall require that the equation of movement must leave invariant under the symmetry transformation in conditions when all forces or interactions are absent.

Thus, for a mass point in classical mechanics, kinematically admissible transformations of position and time kinematic variables from $\boldsymbol{r}$ to $\boldsymbol{\xi}$ and from t to τ, respectively, will reduce the free-particle equation

$$\frac{\mathrm{d}^2\mathbf{r}}{\mathrm{d}t^2} = 0 \tag{1.30}$$

to the form

$$\frac{\mathrm{d}^2\boldsymbol{\xi}}{\mathrm{d}\tau^2} = 0. \tag{1.31}$$

If we take $\tau \equiv t$, the form of an admissible space transformation is found to be

$$x_i \to \xi_i = \sum_j a_{ij}x_j + b_i + c_i t,\ i = 1, 2, 3, \tag{1.32}$$

where x_i is the ith component of $\mathbf{r}$ ($\mathbf{r} = x_1\mathbf{e}_1 + x_2\mathbf{e}_2 + x_3\mathbf{e}_3$), a_{ij}, b_i and c_i are real numbers, and the matrix of a_{ij} is nonsingular ($\det a \neq 0$) [2].

The kinematically admissible transformations of the time variable may be identified by noting that if $\tau = \tau(t)$, then [14]

$$\frac{\mathrm{d}^2\mathbf{r}}{\mathrm{d}t^2} = \frac{\mathrm{d}^2\mathbf{r}}{\mathrm{d}\tau^2}\left(\frac{\mathrm{d}\tau}{\mathrm{d}t}\right)^2 + \frac{\mathrm{d}\mathbf{r}}{\mathrm{d}\tau}\cdot\frac{\mathrm{d}^2\tau}{\mathrm{d}t^2}. \tag{1.33}$$

The conditions of (1.30) and (1.31), with $\boldsymbol{\xi} \equiv \mathbf{r}$, lead to

$$\frac{\mathrm{d}^2\tau}{\mathrm{d}t^2} = 0. \tag{1.34}$$

In this case, admissible transformations are of the form

$$\tau = At + B. \tag{1.35}$$

We can see that (1.35) includes both time translations ($A = 1$) and time reversal ($A = -1,\ B = 0$). Thus, we find that conceptions of uniformity and reversibility of time are built into the kinematics. Below we will examine some consequences of time-reversal invariance of the motion equations of classical mechanics.

The basic measure of time is taken to be a constant time interval (for example, the constant period of oscillations of a clockwork drive based on Mössbauer effect). It is known that the unit of time is the second, defined as the duration of 9 192 631 770 periods of the oscillations of the electromagnetic radiation associated with the allowed magneto-dipole quantum transition between the two neighboring hyperfine levels of the ground state of the ^{133}Cs atom. This method of determining the time unit is less exact than the method based on Mössbauer effect. Because the basic measure of the time is taken to be a constant time interval, only the time interval is defined and the choice of the time origin, $t = 0$, is a matter of convenience. Once the choice of an origin, $t = 0$, for the time variable has been made, the assignment of a sign (positive for a t later than $t = 0$ and negative for a t earlier than $t = 0$) appears to be a matter of convention. The "time-reversed" variable $t' = -t$ appears to have equal standing, because the time interval $\Delta t = t_2 - t_1$ can be expressed in terms of t': $\Delta t' = t'_2 - t'_1$.

However, can be seen that

$$\Delta t' = -\Delta t, \tag{1.36}$$

which leads to some differences in the form of equations describing the motion. For example, if a particle goes from the position $\mathbf{r}_1$ to the position $\mathbf{r}_2$ during the interval $\Delta t = t_2 - t_1$, then its average velocity is $\mathbf{v} = (\mathbf{r}_2 - \mathbf{r}_1)/\Delta t$. If we use t' instead the time-reversed system, then $(\mathbf{r}_2 - \mathbf{r}_1)/\Delta t' = -\mathbf{v}$.

These relations can be written also by introducing the time-reversal operator $\mathbf{T}$

$$\mathbf{T} : t \to t' = -t, \tag{1.37}$$

where $\mathbf{T}$ is an improper transformation (time reversal) similar to inversion $\mathbf{P}$. For the velocity we have

$$\mathbf{T} : \mathbf{v} \to \mathbf{v}' = -\mathbf{v}, \tag{1.38}$$

and the transformation may be said to reverse the velocity. For this reason the transformation $\mathbf{T}$ can be called also motion reversal. The term "motion reversal" is convenient as a supplementary concept related only to the transformation of variables describing motion, such as velocity, momentum, angular momentum, and so forth.

A reference system using t as a variable can be called "standard system" and the one using t' as a variable, "time-reversal" system. Since the $\mathbf{T}$ transformation of one system to another reverses velocities, it might appear that the two reference systems are physically distinguishable. This situation is in a contradiction with the evident assertion that the difference between the systems is merely a matter of convention. In fact, the reversal of velocities does not distinguish the reference systems, because the sign of velocity vector itself is merely a matter of convention.

In spite of fact that the difference between "standard" and "time-reversed" systems is merely a matter of convention, there are any physical implications of the transformation $\mathbf{T}$, which arise from the dynamics, not the kinematics. The equations of motion (which are solved by methods of dynamics) for all physical systems are invariant under the operator $\mathbf{T}$. Thus, the acceptance of a convention led to an assumption of time-reversal invariance (T-invariance) despite the apparent irreversibility of nature.

In conclusion, we emphasize once again that **T** is an improper transformation by analogy with the spatial inversion **P**. In the last case, the violated parity symmetry can be restored by introduction of charge conjugation operation and, respectively, by introduction the notions of antiparticles, antinucleus, antiatoms, antimolecules and antiworld.

By analogy with this, the time-reversal symmetry can be broken by any known and unknown now interactions. In this case the restoration of improper symmetry will lead to conceptions of meta-particles. This problem will be discussed more detailed in the Chap. 9.

1.3 Time-Reversal Symmetry in Dynamical Systems

In the Hamiltonian formalism of classical mechanics the equations

$$\dot{q}_k = \frac{\partial H}{\partial p_k},\ \dot{p}_k = -\frac{\partial H}{\partial q_k},\ (k = 1, 2, \ldots, s), \tag{1.39}$$

where s is the number of generalized coordinates (momenta), are invariant under the reversal of the time direction $t \to -t$. Under such transformation the positions and momenta are transformed according to

$$q_k \to q_k,\ p_k = m_k \frac{\mathrm{d}q_k}{\mathrm{d}t} \to m_k \frac{\mathrm{d}q_k}{\mathrm{d}(-t)} = -p_k. \tag{1.40}$$

Thus

$$\frac{\mathrm{d}q_k}{\mathrm{d}(-t)} = \frac{\partial H}{\partial(-p_k)},\ \frac{\mathrm{d}(-p_k)}{\mathrm{d}(-t)} = -\frac{\partial H}{\partial q_k}, \tag{1.41}$$

$$\dot{q}_k = \frac{\partial H}{\partial p_k},\ \ \dot{p}_k = -\frac{\partial H}{\partial q_k}.$$

The classical notion of time-reversal symmetry as discussed above is directly related to symmetry properties of the Hamilton function:

$$H(q_k, p_k) = H(q_k, -p_k). \tag{1.42}$$

Namely, if the Hamiltonian satisfies (1.39), then the equations of motion (1.41) are invariant under the transformation

$$\mathbf{T} : (q_k, p_k, t) \to (q_k, -p_k, -t) \tag{1.43}$$

In turn, it means that if $(\mathbf{q}(t), \mathbf{p}(t)) \equiv (q_1(t), q_2(t), \ldots, q_s(t); p_1(t), p_2(t), \ldots, p_s(t))$ is a trajectory in the phase space describing a possible motion of the system with the initial position and momentum $(\mathbf{q}_0, \mathbf{p}_0)$, the same is $(\mathbf{q}(-t), -\mathbf{p}(-t))$ with the initial position $(\mathbf{q}_0, -\mathbf{p}_0)$. It means that if there is a trajectory $\mathbf{q}(t)$, then there exists a trajectory $\mathbf{q}(-t)$.

The same result we shall obtain, if we start from the motion equations of Newton mechanics in Cartesian coordinates:

$$m_i \frac{\mathrm{d}^2\mathbf{r}_i}{\mathrm{d}t^2} = \mathbf{F}_i(\mathbf{r}_1, \mathbf{r}_2, \ldots, \mathbf{r}_n; t), \tag{1.44}$$

where $\mathbf{F}_i$ is the force acting on the ith mass point. The designation of the force as a three-vector implies that it transforms in the same way as $\mathbf{r}_i$ under rotations of coordinate axes. Thus, if the only vectors determining $\mathbf{F}_i$ are $\mathbf{r}_1, \mathbf{r}_2, \ldots, \mathbf{r}_n$, it must be a function of these variables. In this case, a rotation applied to $\mathbf{r}_i$ leaves (1.44) invariant and they are now consistent with the assumption of space isotropy. A particular consequence of this symmetry is conservation of the total angular momentum. However, if $\mathbf{F}_i$ depends on any external fixed vectors, as in the case when any external forces are acting, (1.44) will no longer be invariant, because the fixed vectors are not rotated. In this case the total angular momentum will change with time in a manner determined by the torque associated with external forces [2].

Since (1.44) involve the second time derivative, they are invariant under time-reversal transformation $t \to t' = -t$, if $\mathbf{F}_i$ are independent on the sign of t. Even assuming that $\mathbf{F}_i$ does not depend explicitly on time or on the particle velocity, this is not a trivial requirement due to the possible involvement of internal parameters such as spin variables and pseudoscalars. As long as the internal properties of particles and the three-dimensional configuration space character are not dependent on time, there are no explicit time dependences of the forces acting within the isolate system. Therefore (1.44) remain invariant under time translations because of uniformity of the time variable. If the forces are conservative, i.e., they can be expressed as a gradient of the potential, which does not depend explicitly on time, the energy will be conserved.

The conception about the time-reversal symmetry in classical physics may be demonstrated by the following: mental experiment. Let $\mathbf{r}(t_0), \mathbf{v}(t_0)$ be the position and velocity of a classical particle subjected to some statistical forces at $t = t_0$ and allow it to proceed undisturbed for a time t_1, when its position and velocity will become $\mathbf{r}(t_0 + t_1), \mathbf{v}(t_0 + t_1)$. Let at $t = t_0 + t_1$ another identical particle start off at $\mathbf{r}(t_0 + t_1) \equiv \mathbf{r}'$ with a velocity $-\mathbf{v}(t_0 + t_1) \equiv \mathbf{v}'$. Then at later time $t_0 + 2t_1$, we will find that the new position and velocity are $\mathbf{r}(t_0), -\mathbf{v}(t_0)$. We can see that in the case of motion in a force field derivable from a potential, the basic laws (equation of motion) are invariant under the time reversal. On the other hand, if consider the motion of a charged particle in a magnetic field, there is no time-reversal symmetry [15].

As it can be seen on Fig. 1.1a for time-reversed orbits in the case when the equations of motion are invariant under the time reversal, the two trajectories (1 and 1′)

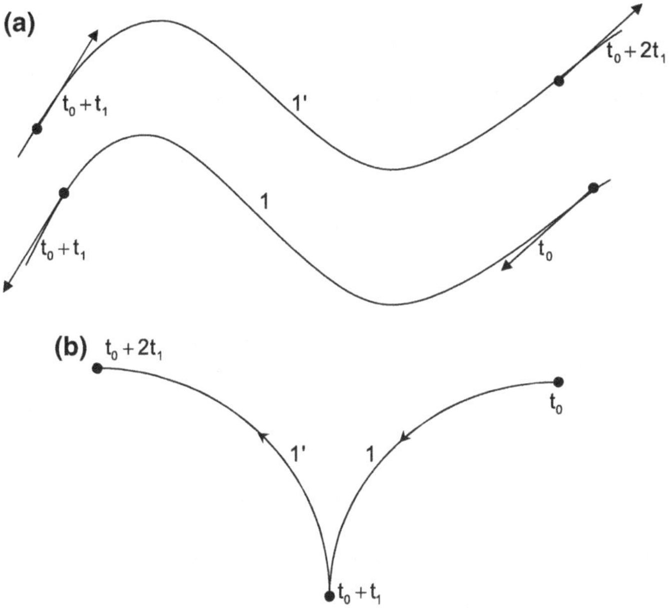

Fig. 1.1 Particle trajectories corresponding to forward and backward directions of the time: **a** in absence of a magnetic field, **b** the charged particle in a magnetic field

really coincide (they are shown for clarity). The tangential arrows indicate the velocities. On the Fig. 1.1b it is shown the trajectory of the negative charged particle ($q < 0$) in a magnetic field, when its direction is normal to the page (oriented to the rider); 1 and $1'$ correspond to the forward and backward the time flow. It should be noted that in this case the orbit $1'$ coincides with 1 if we also change the sign of the charge.

To summarize, the classical equations of motion are said to be invariant under time reversal operator T if one can separate all motions in two subsets with a one-to-one correspondence between them. Let $q_1(t)$ and $p_1(t)$ be the canonical variables of the motion belonging to the subset 1, and $q_{1'}(t)$ and $p_{1'}(t)$ - the canonical variables of the subset $1'$. The above-mentioned one-to-one correspondence is

$$q_1(t_0 + t) = q_{1'}(t_0 - t),$$

$$p_1(t_0 + t) = -p_{1'}(t_0 - t). \tag{1.45}$$

In the case of the motion in a magnetic field, the correspondence (1.45) does not exist. Instead one has

$$q_1(\mathbf{H}, t_0 + t) = q_{1'}(-\mathbf{H}, t_0 - t),$$

$$p_1(\mathbf{H}, t_0 + t) = -p_{1'}(-\mathbf{H}, t_0 - t), \tag{1.46}$$

where $\mathbf{H}$ is the vector of the magnetic field strength.

1.4 Painlevé Theorem

In 1904 P. Painlevé had formulated and proved the theorem about the relative motion of an isolate system of particles that are governed by the conservative forces acting between particles in conditions when the initial velocities vanish [16].

Theorem *If the relative motion of an isolated system of particles is governed by the action of conservative forces and if the initial velocities (at $t = t_0$) of all particles vanish, then the system cannot resume the configuration again with the same relative positions as the initial position, but with a different general orientation in space.*

The proof of this theorem, which differs from the proof presented in the original Painlevé paper, will be carried out according to [2].

The Painlevé theorem proof may be summarized as follows. Let $V(\mathbf{r}_{12}, \mathbf{r}_{13}, \ldots, \mathbf{r}_{\alpha\beta}, \ldots)$ be the potential generating the conservative forces between particles, where $\mathbf{r}_{ij} = \mathbf{r}_i - \mathbf{r}_j$ is the vector distance between the ith and jth particles. The potential energy V is a scalar function assumed to depend only on the $\mathbf{r}_{ij}$. The equations of motion (1.44) take the form

$$m_i \frac{\mathrm{d}^2\mathbf{r}_i}{\mathrm{d}t^2} = -\nabla_i V, \; i = 1, \ldots, n. \tag{1.47}$$

These equations are invariant under the time-reversal operator $\mathbf{T}$ changing t by $t' = -t$.

At the initial time $t_0 = 0$ the particle positions are $\mathbf{r}_i^{(0)}$ and respectively all velocities vanish

$$\left(\frac{\mathrm{d}\mathbf{r}_i}{\mathrm{d}t}\right)_{t=0} = 0. \tag{1.48}$$

In this case the initial conditions are invariant under $\mathbf{T}$, since

$$\left(\frac{\mathrm{d}\mathbf{r}_i}{\mathrm{d}t}\right)_{t=0} = -\left(\frac{\mathrm{d}\mathbf{r}_i}{\mathrm{d}t'}\right)_{t'=0} = 0 \tag{1.49}$$

Hence the motion is invariant under $\mathbf{T}$. Thus the solution $\mathbf{r}_i(t)$ of (1.47) and (1.48) satisfy the relationship

$$\mathbf{r}_i(-t) = \mathbf{r}_i(t). \tag{1.50}$$

Let us suppose the contrary of the theorem, namely, suppose that at some time $t_1 > 0$ the system does return to its initial relative configuration, but with a reorientation in space. Then, because positions $\mathbf{r}_i^{(1)}$ at $t = t_1$ differ from the $\mathbf{r}_i^{(0)}$ by a common rotation R, that is the same for all particles, the distance vectors $\mathbf{r}_{ij}$ can be presented as

$$\mathbf{r}_{ij}(t_1) = R[\mathbf{r}_{ij}(0)], \tag{1.51}$$

where R is a three-dimensional orthogonal matrix (rotation transformation) operating on the vector $\mathbf{r}_{ij}$, that is the same for all i, j. The potential energy is a scalar function of $\mathbf{r}_{ij}$ and, therefore, has the same value at $t = t_1$ as at $t = 0$. Taking into account the conservation energy law, the kinetic energy also has the same value at $t = t_1$ and $t = 0$. But the kinetic energy vanishes at $t = 0$ and therefore it must vanish again at $t = t_1$. Since the kinetic energy is a positive definite function of the velocity, each velocity of particles of the system must vanish:

$$\left(\frac{\mathrm{d}\mathbf{r}_i}{\mathrm{d}t}\right)_{t=t_1} = 0. \tag{1.52}$$

The conditions at $t = t_1$ are identical to those at $t = t_0$ excepting the initial positions that are rotated from $\mathbf{r}_i^{(0)}$ to $\mathbf{r}_i^{(1)}$. Therefore, if $t = t_1$ rather than $t = 0$ is taken as the initial point, then previous arguments made relative to t may be repeated without change, but starting from the new positions $\mathbf{r}_i^{(1)}$. Thus, if we use a reference system rotated by R with respect to the original one, the motion from $t = t_1$ to $t = 2t_1$ in the new system will be identical to the motion from $t = 0$ and $t = t_1$ in the old reference system. Since the magnitude r_{ij} of the vector $\mathbf{r}_{ij}$ is independent on the orientation of the reference frame, we find after n repetitions of this argument that r_{ij} are periodic functions of t with the period t_1:

$$r_{ij}(t + nt_1) = r_{ij}(t). \tag{1.53}$$

We may also consider t_1 as the initial time by introducing a translated time variable

$$\tau = t - t_1 \tag{1.54}$$

to write the equations of motion and initial conditions. In the rotated frame the equations, written for position vectors $\boldsymbol{\rho}_i = R[\mathbf{r}_i]$ in terms of τ, will be identical to those given for $\mathbf{r}_i$ in terms of t in the original frame. Therefore, the motion will again be invariant under reversal of the translated time variable

$$\boldsymbol{\rho}_i(-\tau) = \boldsymbol{\rho}_i(\tau). \tag{1.55}$$

We can see from (1.54) that the solution $\mathbf{r}_i(t)$ satisfies the condition

$$\mathbf{r}_i(t_1 + \tau) = \mathbf{r}_i(t_1 - \tau) \tag{1.56}$$

in analogy with (1.50) and, therefore,

$$r_{ij}(t_1 + \tau) = r_{ij}(t_1 - \tau). \tag{1.57}$$

If we consider a third time variable,

$$\tau' = t - \frac{1}{2}t_1 = \tau + \frac{1}{2}t_1, \tag{1.58}$$

then from (1.57) we have

$$r_{ij}\left(t_1 + \tau' - \frac{1}{2}t_1\right) = r_{ij}\left(t_1 - \tau' + \frac{1}{2}t_1\right) \tag{1.59}$$

or

$$r_{ij}\left(\frac{1}{2}t_1 + \tau'\right) = r_{ij}\left(\frac{3}{2}t_1 - \tau'\right). \tag{1.60}$$

On the other hand, according to (1.53), the periodicity condition is

$$r_{ij}\left(\frac{1}{2}t_1 + \tau'\right) = r_{ij}\left(\frac{1}{2}t_1 - \tau'\right) \tag{1.61}$$

for arbitrary τ'. Thus $r_{ij}(\frac{1}{2}t_1 + \tau')$ is an even function of τ', and its time derivative must be an old function vanishing at $\tau' = 0$:

$$\left(\frac{\mathrm{d}r_{ij}}{\mathrm{d}t}\right)_{t=\frac{t_1}{2}} = \left[\frac{\mathrm{d}}{\mathrm{d}\tau'}r_{ij}\left(\frac{1}{2}t_1 + \tau'\right)\right]_{\tau'=0} = 0. \tag{1.62}$$

By presenting $\mathbf{r}_{ij}$ in the form

$$r_{ij} = (\mathbf{r}_{ij} \cdot \mathbf{r}_{ij})^{\frac{1}{2}}, \tag{1.63}$$

we obtain

$$\frac{\mathrm{d}r_{ij}}{\mathrm{d}t} = \mathbf{r}_{ij} \cdot \widehat{\mathbf{r}_{ij}}, \tag{1.64}$$

where $\widehat{\mathbf{r}_{ij}}$ is the unit vector in the direction $\mathbf{r}_{ij}$. As a consequence, the motion of the system may be described as a rigid body rotation. However, since total angular momentum is conserved and it vanished at $t = 0$, the system cannot be rotating at $t = \frac{1}{2}t_1$. Therefore,

$$\left(\frac{\mathrm{d}\mathbf{r}_i}{\mathrm{d}t}\right)_{t=\frac{t_1}{2}} = 0. \tag{1.65}$$

Thus at $t = \frac{1}{2}t_1$ the velocities satisfy the same initial conditions as those at $t = 0$ and the arguments, that we made using $t = t_1$ as an initial point, may be repeated by using $t = \frac{1}{2}t_1$. Particularly, the time-reversal argument for the variable t' leads to the equation

$$\mathbf{r}_i\left(\frac{1}{2}t_1 + \tau'\right) = \mathbf{r}_i\left(\frac{1}{2}t_1 - \tau'\right) \tag{1.66}$$

that is equivalent to (1.56). For $\tau' = \frac{1}{2}t_1$ the system has the same position at $t = t_1$ as it had at $t = 0$. The only "rotated" set of positions $\mathbf{r}_i^{(1)}$ that the system can attain are identical to the initial positions $\mathbf{r}_i^{(0)}$. This completes the proof of the Painlevé theorem.

1.5 Time-Reversal Symmetry in Classical Electrodynamics

For beginning, let us consider a charged particle of the mass m and charge q in an homogeneous electric field of strength $\mathbf{E}$. In this case the equation of motion in vectorial form is

$$m\frac{\mathrm{d}^2\mathbf{r}}{\mathrm{d}t^2} = q\mathbf{E}(\mathbf{r}). \tag{1.67}$$

This equation is invariant under the time-reversal transformation $t \to t' = -t$, because the left-hand side contains the second time derivative and right one is time independent. Thus at the time-reversal transformation we have: $t \to -t$, $\mathbf{r} \to \mathbf{r}$, $\mathbf{E} \to \mathbf{E}$ (m and q are scalars and, along with $\mathbf{r}$ and $\mathbf{E}$, do not change at time-reversal). However, this property does not hold for magnetic forces (see Fig. 1.1b). In this case the equation of motion includes the first-order time derivative

$$m\frac{\mathrm{d}^2\mathbf{r}}{\mathrm{d}t^2} = \frac{q}{c}\cdot\frac{\mathrm{d}\mathbf{r}}{\mathrm{d}t} \times \mathbf{B}(\mathbf{r}), \tag{1.68}$$

where $\mathbf{B}(\mathbf{r})$ is the vector of magnetic induction, q is the particle charge and c is the light velocity.

The left-hand side of this equation is invariant under $t \to -t$, while the second term in the right-hand side changes its sign. For example, the sense of the circular motion of a charged particle in a constant magnetic field is determined by the charge of the particle, not by the initial conditions. If we make the replacement $\mathbf{B} \to -\mathbf{B}$ as well as $t \to -t$ then the motion equation (1.68) becomes invariant under the time-reversal operation. It means that the time-reversed motion is physically admissible in the reversed magnetic field. One can see that a system that interacts with external fields does not possess the invariance of time reversal in itself. However, if one includes the charges and currents producing external fields in the definition of "the system", then the invariance of the Hamiltonian under time-reversal operator may be restored. If the motion of all charges in a closed system is reversed, then charge densities and currents are transformed according to

$$\rho \to \rho, \mathbf{j} \to -\mathbf{j}, \tag{1.69}$$

while the electric and magnetic fields produced by these charges and currents are transformed according to

$$\mathbf{E} \to \mathbf{E}, \mathbf{B} \to -\mathbf{B}. \tag{1.70}$$

These are the basic transformation equations of the electromagnetic field under time-reversal operator, which are combined with (1.67) and (1.68) to show that electromagnetic phenomena are time-reversed.

The presented above argumentation concerning restoration of the time-reversal symmetry by extension of the definition of "the system" has also a connection with the invariance of classical Maxwell equations under time-reversal operator. In a Gauss system, the Maxwell equations for macroscopic electromagnetic field in the matter are

$$\mathrm{rot}\,\mathbf{E} = -\frac{1}{c}\frac{\partial \mathbf{B}}{\partial t}, \quad \mathrm{div}\,\mathbf{B} = 0,$$

$$\mathrm{rot}\,\mathbf{H} = \frac{4\pi}{c}\mathbf{j} + \frac{1}{c}\frac{\partial \mathbf{D}}{\partial t}, \quad \mathrm{div}\,\mathbf{D} = 4\pi\rho, \tag{1.71}$$

where **E** and **H** are vectors of strengths of electric and magnetic fields, while **D** and **B** are vectors of inductions of electric and magnetic fields determined by relationships

$$\mathbf{D} = \mathbf{E} + 4\pi\mathbf{P}, \ \ \mathbf{H} = \mathbf{B} - 4\pi\mathbf{M}. \tag{1.72}$$

Here **P** and **M** are depending on **E** and **B** vectors of electrical polarization (**P**) and magnetization (**M**) of the matter, **j** is the vector of the density of the conduction current and ρ is the density of electric charges.

The fundamental Maxwell equations (1.71) do not yet constitute a complete system of equations of the electromagnetic field. They must be supplemented by correlations named material equations.

The material equations are most simple in the case of sufficiently weak electromagnetic fields, comparatively weakly changing in space and time. In this case, for an isotropic medium, that do not contain ferroelectrics and ferromagnetics, the material equations have the form

$$\mathbf{D} = \varepsilon\mathbf{E}, \ \ \mathbf{B} = \mu\mathbf{H}, \ \ \mathbf{j} = \sigma(\mathbf{E} + \mathbf{E}^*), \tag{1.73}$$

where ε and μ are electric and magnetic permeabilities, σ is the electrical conductivity and $\mathbf{E}^*$ is the strength of the field of foreign forces, conditioned by chemical and thermal processes.

The Maxwell equations are not symmetrically relative to electric and magnetic fields. This is conditioned by the fact that in the nature there exist electric charges, but there are no magnetic charges (magnetic monopoles predicted theoretically by Dirac are not yet discovered experimentally).

However, in a neutral homogeneous and not conducting medium, where $\rho = 0$ and $\mathbf{j} = 0$, Maxwell equations (1.71) have almost a symmetrical form. In this case **E** is connected with $\frac{\partial \mathbf{B}}{\partial t}$, like **B** with $\frac{\partial \mathbf{E}}{\partial t}$, excepting the fact that signs in front of these derivatives do not coincide. Different signs in front of $\frac{\partial \mathbf{B}}{\partial t}$ and $\frac{\partial \mathbf{E}}{\partial t}$ mean that lines of the vortex electric field induced by change of the field **B** form a leftspiral system with the vector $\frac{\partial \mathbf{B}}{\partial t}$, while lines of magnetic field induced by change of **D** form a rightspiral system with the vector $\frac{\partial \mathbf{D}}{\partial t}$.

Note that in the first pair of Maxwell equations from (1.71) only principal characteristics of electromagnetic field **E** and **B** enter, while in the second pair of equations auxiliary vectors **D** and **H** appear. Moreover, Maxwell equations can be written in such a way that the vectors **D** and **H** do not enter into these equations. For this, it is necessary to change **D** and **H** in the second pair of equations (1.71) by respective vectors from (1.72):

$$\operatorname{rot}\mathbf{E} = -\frac{1}{c}\frac{\partial \mathbf{B}}{\partial t},\ \operatorname{div}\mathbf{B} = 0,$$

$$\operatorname{rot}\ \mathbf{B} = \frac{4\pi}{c}\left(\mathbf{j} + c\operatorname{rot}\mathbf{M} + \frac{\partial \mathbf{P}}{\partial t}\right) + \frac{1}{c}\frac{\partial \mathbf{E}}{\partial t},\ \operatorname{div}(\mathbf{E} + 4\pi\mathbf{P}) = 4\pi\rho. \quad (1.74)$$

In the absence of polarizable and magnetizable mediums Maxwell equations for the electromagnetic field in vacuum can be presented in the form [5]:

$$\operatorname{rot}\mathbf{E} = -\frac{1}{c}\frac{\partial \mathbf{B}}{\partial t},\ \operatorname{div}\mathbf{B} = 0,$$

$$\operatorname{rot}\mathbf{B} = \frac{4\pi}{c}\mathbf{j} + \frac{1}{c}\frac{\partial \mathbf{E}}{\partial t},\quad \operatorname{div}\mathbf{E} = 0. \quad (1.75)$$

Since in the Gauss system for a field in vacuum **H** coincides with **B** (see form (1.72) at $\mathbf{M} = 0$), in this case one may change **B** by **H** in (1.75). However, since **B** is a fundamental characteristic of the magnetic field and **H** is an auxiliary vector, we will not do it.

In (1.71) **E**, **D**, **P** and **j** are polar vectors, while **B**, **H** and **M** are axial vectors. The first and third equations from (1.71) contain the first time derivative in the right-hand side (third equation contains also the vector **j**). Therefore, the Maxwell equations (1.71) will be invariant under the time-reversal, if along with the transformation $t \to -t$ perform transformations $\mathbf{B} \to -\mathbf{B}$, $\mathbf{H} \to -\mathbf{H}$ (correspondingly $\mathbf{M} \to -\mathbf{M}$) and $\mathbf{j} \to -\mathbf{j}$.

Thus, we can see that in classic electrodynamics the Lorentz equation (1.68) is invariant under time reversal, if in parallel with the transformation $t \to -t$ change the direction of the vector **B** to the opposite. The Maxwell equations (1.71) are invariant under time reversal, if in parallel with the transformation $t \to -t$ change to the opposite directions of axial vectors **B** and **H** (axial vector **M**, respectively), and that of the polar vector **j**.

If Maxwell equations are written in the form (1.74), then the transformation $\mathbf{H} \to -\mathbf{H}$ is absent, since these equations do not contain **H** in this case.

1.6 Time-Reversal Symmetry in Relativistic Electrodynamics

According to the special theory of relativity of Einstein [17], the four-dimension coordinates of one and the same event in different inertial systems are expressed with each other linearly and, moreover, the square of the interval

$$S_{AB}^2 = c^2(t_A - t_B)^2 - (x_A - x_B)^2 - (y_A - y_B)^2 - (z_A - z_B)^2 \tag{1.76}$$

between any two events is the same in all frames of reference. There the positive signature is used, when the time part of the square of the interval S_{AB}^2 enters with the sign plus, and the space part with the sign minus. The negative signature is used no less frequently under the choice of which the time part of the square of interval enter S_{AB}^2 with the sign minus and the space part, with the sign plus. We separate in all inertial frames of references the set of all frames, of which origins in some moment of time coincide, and take this moment as the beginning of time reference in all these inertial frames of reference. It is clear that the event with coordinates $(0, 0, 0, 0)$ in one of the inertial frame of reference has the same coordinates in every other inertial frame. That is why the dependence between the coordinates of some event in different inertial frames of reference is not only linear, but homogeneous as well. We denote by $a(K, K')$ the matrix realizing the transition from coordinates in the inertial frame of references K to coordinates in the frame of reference K', moving uniformly relative to K. Then

$$x_i' = \sum_j a_{ij}(K'K)x_j, \tag{1.77}$$

where $i, j = 0, 1, 2, 3$, $x_0 = ct$, $x_1 = x$, $x_2 = y$, $x_3 = z$.

If in the frame K'' coordinates of the same event are denoted by $x_l''(l = 0, 1, 2, 3)$, then one may write

$$x_l'' = \sum_j a_{lj}(K''K)x_j, \; x_l'' = \sum_i a_{li}(K''K')x_i'. \tag{1.78}$$

From (1.78) and (1.77) it follows that

$$a_{lj}(K''K) = \sum_i a_{li}(K''K')a_{ij}(K'K). \tag{1.79}$$

The relationship (1.79) shows that matrices a form a continuous group relative to matrix multiplication, in addition

$$a(K''K) = a(K''K')a(K'K). \tag{1.80}$$

This group is called the complete Lorentz group $\mathfrak{L}$ [7]. The group contains as a subgroup the so-called proper Lorentz group $\mathfrak{L}_1$, which is obtained, if from all inertial frames of references only left (or only right) ones are separated.

Let us consider an important particular case when axes of K' are parallel to axes of K and it is moving relative to the last one with the speed v, directed parallel to the axis X. In this case, from the explicit form of the matrix a, which may be found by algebraic transformations beginning with the invariance of the interval relative to transitions from one inertial frame of reference to another [18], the known Lorentz transformations follows:

$$x' = \frac{x - vt}{\sqrt{1-\beta^2}}, \quad y' = y, \quad z' = z, \quad t' = \frac{t - \frac{\beta}{c}x}{\sqrt{1-\beta^2}}, \tag{1.81}$$

where $\beta = v/c$.

If there are two inertial frames of reference, the frame K and the frame K' moving relative to it with the speed v and values of fields $\mathbf{E}$ and $\mathbf{B}$ in some space-time point of frame K are known, then values of fields $\mathbf{E}'$ and $\mathbf{B}'$ in the same space-time point in frame K' are determined by formulas

$$\begin{aligned} E'_x &= E_x, & B'_x &= B_x, \\ E'_y &= \frac{E_y - vB_z}{\sqrt{1-\beta^2}}, & B'_y &= \frac{B_y + \frac{\beta}{c}E_z}{\sqrt{1-\beta^2}}, \\ E'_z &= \frac{E_z + vB_y}{\sqrt{1-\beta^2}}, & B'_z &= \frac{B_z - \frac{\beta}{c}E_y}{\sqrt{1-\beta^2}}, \end{aligned} \tag{1.82}$$

where it is supposed that coordinate axes X and X' are directed along the vector of speed $\mathbf{v}$, the axis Y' is parallel to the axis Y and the axis Z' is parallel to the axis Z.

It is easy to see that the system of equation (1.82) is invariant under the time-reversal transformation. For this, we rewrite this system of equations in a vector form

$$\begin{aligned} \mathbf{E}'_\parallel &= \mathbf{E}_\parallel, & \mathbf{B}'_\parallel &= \mathbf{B}_\parallel, \\ \mathbf{E}'_\perp &= \frac{\mathbf{E}_\perp + [\mathbf{vB}]}{\sqrt{1-\beta^2}}, & \mathbf{B}'_\perp &= \frac{\mathbf{B}_\perp + [\mathbf{vE}]/c^2}{\sqrt{1-\beta^2}}, \end{aligned} \tag{1.83}$$

where by symbols $\parallel$ and $\perp$ are marked the longitudinal and transversal components of electric and magnetic fields, relative to the vector $\mathbf{v}$.

According to (1.83), in the case of alternating in time electric and magnetic fields the longitudinal components of electric fields $\mathbf{E}_\parallel$ and $\mathbf{E}'_\parallel$ do not change at the transformation $\mathrm{T}(t \to t' = -t)$, while for the vector $\mathbf{E}'_\perp$ to leave invariant it is necessary that side by side with the change of direction of the speed vector to the opposite ($\mathbf{v} \to -\mathbf{v}$), the change of the vector $\mathbf{B}$ to the opposite ($\mathbf{B} \to -\mathbf{B}$) should take place as well. Moreover, the vectors $\mathbf{B}_\parallel$ and $\mathbf{B}_\perp$ (as well as $\mathbf{B}'_\parallel$ and $\mathbf{B}'_\perp$) change their directions to opposite, so that the system of equation (1.83) as a whole leaves invariant. This means an invariance with respect to the time reversal of laws of transformation of electric and magnetic fields under transition from one frame of reference to another.

Now, let us consider Maxwell equations for the electromagnetic field in vacuum in the absence of charges ($\rho = 0,\ \mathbf{j} = 0$), obtained from (1.75) at $\mathbf{j} = 0$, which have the form

$$\begin{aligned} \operatorname{rot}\mathbf{E} &= -\tfrac{\partial \mathbf{B}}{\partial t},\ \operatorname{div}\mathbf{B} = 0, \\ \operatorname{rot}\mathbf{B} &= \tfrac{\partial \mathbf{E}}{\partial t},\quad \operatorname{div}\mathbf{E} = 0. \end{aligned} \tag{1.84}$$

Here and below on we use the system of units, in which the light speed $c = 1$. Further the problem of time-reversal symmetry in Relativistic Electrodynamics will be presented according to the monograph [19].

We introduce vector and scalar potentials $\mathbf{A}$ and φ of the electromagnetic field and express vectors $\mathbf{E}$ and $\mathbf{B}$ as follows:

$$\mathbf{E} = -\frac{\partial \mathbf{A}}{\partial t} - \nabla\varphi,\quad \mathbf{B} = \operatorname{rot}\mathbf{A}. \tag{1.85}$$

In this case the first two equations (1.84) turn into identities. However, the relations (1.85) do not define potentials uniquely, but only to within a gauge transformation [19]

$$\mathbf{A} \to \mathbf{A} + \nabla\chi,\quad \varphi \to \varphi - \frac{\partial \chi}{\partial t}, \tag{1.86}$$

where χ is an arbitrary function of coordinates and time.

Obviously, all physical relations, containing vectors of the field, are invariant relative to caliber transformations.

The lack of uniqueness in the choice of potentials allows to impose to them a supplementary condition (Lorentz condition)

$$\operatorname{div}\mathbf{A} + \frac{\partial \varphi}{\partial t} = 0. \tag{1.87}$$

The second pair of Maxwell equations in (1.84) at substitution (1.85) with regard to (1.87) takes the form of wave equations for the potentials

$$\Box\mathbf{A} = 0,\quad \Box\varphi = 0, \tag{1.88}$$

where $\Box$ is D'Alembert operator

$$\Box = \nabla - \frac{\partial^2}{\partial t^2} \tag{1.89}$$

The Lorentz condition (1.87) does not determine the calibration of potentials $\mathbf{A}$ and φ completely. We can see that Lorentz condition will be satisfied if potentials $\mathbf{A}$ and φ are subjected to the gauge conversion (1.86), in which the function χ satisfies the wave equations

$$\Box\chi = 0. \tag{1.90}$$

In order to ensure the relativistic invariance of Maxwell equations (1.84), it is sufficient to introduce a four dimensional vector of electromagnetic field constructed on the basis of potentials $\mathbf{A}$ and φ with components

$$\begin{aligned} A_\mu &= (\mathbf{A})_\mu, \ \mu = 1, 2, 3, \\ A_4 &= i\varphi. \end{aligned} \tag{1.91}$$

Under the Lorentz transformations these components behave as components of the four dimensional radius vector $x_\mu(\mathbf{r}, it)$. Then the equations for potentials take the form

$$\Box A_\mu = 0, \tag{1.92}$$

and the Lorentz condition will be written in the form

$$\frac{\partial A_\mu}{\partial x_\mu} = 0. \tag{1.93}$$

The gauge transformation (1.86) may be written in the form

$$A_\mu \to A_\mu + \frac{\partial \chi}{\partial x_\mu}. \tag{1.94}$$

The formula (1.94) conserves its form at Lorentz transformations, if χ is a four dimensional scalar.

In the 4-vector writing, equation (1.85) determine a four dimensional antisymmetric tensor of the electromagnetic field, $F_{\mu\nu}$

$$F_{\mu\nu} = \frac{\partial A_\nu}{\partial x_\mu} - \frac{\partial A_\mu}{\partial x_\nu}, \tag{1.95}$$

joining the vectors of electromagnetic fields

$$F_{jk} = \varepsilon_{jkl} B_l, \ F_{4k} = i E_k, \tag{1.96}$$

$(i, j, k = 1, 2, 3, \ \varepsilon_{jkl}$ is the unit antisymmetric tensor).

Relations (1.96) permit to write Maxwell equations (1.84) in the form

$$\begin{aligned} &\frac{\partial F_{\mu\nu}}{\partial x_\lambda} + \frac{\partial F_{\nu\lambda}}{\partial x_\mu} + \frac{\partial F_{\lambda\mu}}{\partial x_\nu} = 0, \\ &\frac{\partial F_{\mu\nu}}{\partial x_\nu} = 0, \end{aligned} \tag{1.97}$$

where the summation is done by twice encountered index ν from 1 to 4.

For electromagnetic field in the presence of charges, the equation (1.92) and the second one from the equation (1.97) become non-homogeneous

$$\Box A_\mu = -j_\mu, \tag{1.98}$$

$$\frac{\partial F_{\mu\nu}}{\partial x_\nu} = -j_\mu, \tag{1.99}$$

where j_μ is a 4-vector of electric current density.

Returning to matrices a, and performing the transition from one inertial reference frame to another in formulas (1.77)–(1.81), one can observe that the proper Lorentz group is the totality of all transformations of the complete Lorentz group, which contain the identity transformation I and infinitesimal transformations

$$x'_\mu = x_\mu + \sum_\nu \varepsilon_{\mu\nu} x_\nu. \tag{1.100}$$

In other words, these transformations are realized by the matrix

$$a = I + \varepsilon, \tag{1.101}$$

where ε is an infinitesimal antisymmetric matrix

$$\varepsilon_{\mu\nu} = -\varepsilon_{\mu\nu}, \; |\varepsilon_{\mu\nu}| \ll 1. \tag{1.102}$$

Thus, every transformation of the proper Lorentz group may be obtained from the identity one by continuous changing the elements of matrix a, for which

$$\det \alpha = 1, \; a_{44} > 1. \tag{1.103}$$

At the transition from one inertial frame of reference to another, determined by the transformation of the proper Lorentz group, the four dimensional vector potential of electromagnetic field 4-vector is transformed like x, that is

$$A'(x') = aA(x) \tag{1.104}$$

at $x' = ax$, where a is a transformation of the proper Lorentz group.

It is easy to see from (1.95), that the tensor of electromagnetic field $F_{\mu\nu}$, which is a tensor of second rank, is transformed in the following way:

$$F'(x') = aF(x)\tilde{a}, \tag{1.105}$$

where $\tilde{a}$ is a matrix transposed to matrix a.

Every transformation of the complete Lorentz group (not entering the proper group) differs from a proper transformation by the sign of det a or a_{44} (or of both these quantities). It may be obtained as a totality of the proper transformation, the discrete transformation of space inversion $x' = a_P x$ and the time-reversal transformation $x' = a_T x$, where

$$a_P = \begin{pmatrix} -1 & 0 & 0 & 0 \\ 0 & -1 & 0 & 0 \\ 0 & 0 & -1 & 0 \\ 0 & 0 & 0 & 1 \end{pmatrix}, \quad a_T = \begin{pmatrix} 1 & 0 & 0 & 0 \\ 0 & 1 & 0 & 0 \\ 0 & 0 & 1 & 0 \\ 0 & 0 & 0 & -1 \end{pmatrix}, \tag{1.106}$$

$\det(a_P) = -1$, $\det(a_T) = -1$.

Each of these transformations (together with the identity transformation) forms a group, which is a subgroup of the complete Lorentz group. These subgroups are formed by matrices a_P and I or a_T and I. Matrices a_P and a_T satisfy the relationship

$$a_P^2 = I, \quad a_T^2 = I. \tag{1.107}$$

At transformations of the complete Lorentz group, different 4-vectors may be transformed in different ways, depending on their behavior at the space inversion **P** and the time reversal **T**, of which matrices of transformations have the form

$$\mathbf{P} = \gamma_P a_P, \mathbf{T} = \gamma_T a_T. \tag{1.108}$$

Since $\mathbf{P}^2 = I$ and $\mathbf{T}^2 = I$, then $\gamma_P^2 = 1$, $\gamma_T^2 = 1$, whence it follows that

$$\gamma_P = \pm 1, \gamma_T = \pm 1. \tag{1.109}$$

A four-dimensional vector is called polar at $\gamma_P = 1$ and axial at $\gamma_P = -1$. At $\gamma_T = -1$, the vector is called t-even and at $\gamma_T = 1$ such a vector is called t-odd [19].

Determining the quantities γ_P and γ_T for the vector of potential of the electromagnetic field can not be done on the basis of equations (1.92) and (1.93) for potentials, because these equations admit any values for γ_P and γ_T. However, this can be done on the basis of non-homogeneous equation (1.98), in which the density of current j_μ is a polar even vector. It is easy to see this, considering the connection of the charge q with the time component of the density of current $j_4 = i\rho$

$$q = \int \rho \mathrm{d}^3 x. \tag{1.110}$$

Since the charge is an invariant quantity and the volume $\int d^3x$ is not changed at space inversion and time reversal, then $j_4 \to j_4$ like at $P-$, as well as at $T-$ transformation. On the other hand, the space components of the current density, which are determined by formula

$$j_k = \rho v_k = \rho \frac{\mathrm{d}x_k}{\mathrm{d}t}, \ (k = 1, 2, 3), \tag{1.111}$$

change the sign at space inversion ($x_k \to -x_k, t \to t$) and time reversal ($x_k \to x_k, t \to -t$): $j_k \to -j_k$.

From (1.98), it follows that the 4-potential A_μ is a polar even vector

$$\gamma_P = 1, \gamma_T = -1. \tag{1.112}$$

From this and (1.95) it follows the transformation of components of the electromagnetic field at space inversion and time reversal:

$$\begin{aligned} \mathbf{P}E_j &= -E_j, \ \mathbf{P}B_j = B_j, \\ \mathbf{T}E_j &= E_j, \ \mathbf{T}B_j = -B_j. \end{aligned} \tag{1.113}$$

Components D_j and H_j of auxiliary vectors, which characterize the electromagnetic field will be transformed under the action of operators **P** and **T** in a similar way.

We see that although in relativistic electrodynamics the space and time are not independent, but in a single four dimensional Minkovsky space operation of time reversal is realized in fact in the same way, as in the case of the classical electrodynamics. It consists in the reversion of the direction of movement of all electrically charged particles, that leads to the change of sign of space components of the polar even vector of current density, the components of polar t-even four dimensional vector potential $A_\mu(\mathbf{A}, i\varphi)$ and, as a consequence, to the reversion of the direction of the three-dimensional vector of magnetic induction **B**. More cardinal changes of conceptions at time-reversal symmetry transformation, which take place in quantum mechanics and quantum electrodynamics, will be considered in the next chapter.

1.7 Dynamical Reversibility and Statistical Irreversibility of Time

Let a system evolving forward in time at some initial condition up to a time t. At t the evolution is stopped and a time-reversal transformation is performed. Further, the system is allowed to evolve once again for a time interval t. The system, obviously, will return to its original starting point in the phase space. In accordance with mechanics and microcanonical ensemble, there does not exist a preferred direction of the time flow.

The motion of a N-particle system is determined by the $3N$ initial positions of particles and $3N$ initial their velocities, when the forces acting on the particles are known. If we consider n possible one-particle initial states for each of the particles, the number of initial states for the system will be n^N. Therefore, as the number of particles, N, increases, a very rapid increase in the number of available states occurs and, consequently, an increase in the amount of information necessary for determining the motion. This rapid increase in the complexity of the information required to determine the state of the system has a strong influence on its behavior under time reversal, regardless of whether the dynamics is T-invariant or not [2].

Thus, the irreversibility of the motion from an ordered state to a disordered state of a complex system is associated with the irreversibility of the initial conditions. Although in a T-invariant system the motion of each particle is reversible, the probability of finding the precise initial conditions required to attain the reversed motion is exponentially small.

There is an apparent contradiction which consists in the fact that the entropy of a system increases inevitably with time, while the behavior of the atoms and molecules is determined by reversible (T-invariant) equations of motion. Such a situation is known as "Locshmidt paradox", because the most important application of T-invariance in classical physics was made by Locshmidt [20]. He shown that Boltzmann proof of the H theorem [21] leads to a paradox, because in the proof of this theorem T-invariant equations were used to derive result that violated T-invariance. Considering elastic collisions between molecules of the perfect gas by means of the distribution of velocities $f(\mathbf{v}, t)$, after making a plausible assumption about the separation of velocity and spatial distributions, Boltzmann obtained the relation

$$\frac{\partial H}{\partial t} \leqslant 0, \tag{1.114}$$

where

$$H = \int \mathrm{d}^3\mathbf{v} f(\mathbf{v}, t) ln f(\mathbf{v}, t). \tag{1.115}$$

If $f(\mathbf{v}, t)$ from (1.115) is replaced by $f(\mathbf{v}_0)$, the Maxwell–Boltzmann distribution for the perfect gas, $f_0(\mathbf{v})$, then $H = H_0$ and the following relation take place

$$S_0 = -kH_0, \tag{1.116}$$

where S_0 is the equilibrium value of the entropy of the gas per unit volume and k is the Boltzmann constant. This suggests that H should be considered similarly related to the non-equilibrium value of entropy and that (1.114), which implies the inevitable increase of the entropy toward equilibrium, is a "proof" of the second law of thermodynamics for the perfect gas [2].

The Locshmidt observation was that, in view of the T-invariance of the dynamics underlying the motion of molecules, with every distribution $f(\mathbf{v}, t)$, it must be associated with another possible distribution $f_1(\mathbf{v}, t) = f(-\mathbf{v}, -t)$ obtained by time reversal of the motion. But the quantity H_1 defined in (1.115) with f replaced by f_1 must satisfy

$$\frac{\partial H_1}{\partial t} \geqslant 0 \tag{1.117}$$

in contradiction with the second law of thermodynamics [2].

A first solution of the Locshmidt paradox was proposed by Gibbs [22], who gave an explanation based on the concepts of fine-grained and course-grained structure of the phase space. However, to date, many works have been written that explain the paradoxical situation, when a system that has the symmetry of time reversal on a microscopic scale breaks this symmetry in its macroscopic behavior. A popular solution to the Loschmidt paradox is to consider that, despite the reversibility of the equations of motion, not all solutions must possess the complete symmetry of time reversal. A different view is presented in [23]. For a historical reference see [24]. An extensive bibliography on the topic of time-reversal symmetry in dynamical systems is presented in [25]. In this context, it should be noted that the concepts of reversibility and irreversibility for a long time contributed to the lack of clarity in classical thermodynamics. Particularly, Clausius defines a reversible process as a slowly varying process (called quasi-static transformation), the successive states of which differ by infinitesimals from the equilibrium state of the system.

Alternatively, the Kelvin concept of reversibility includes the ability of the system to restore its initial state from the final state of the system. Planck introduced several concepts of reversibility. The main of these concepts is complete reversibility, which includes the restoring capacity of the initial state of the dynamic system and, at the same time, the restoration of the environment to its original state. Unlike the concept of the reversibility of Clausius, the concepts of reversibility of Kelvin and Planck do not require that the system accurately return the original trajectory in reverse order [26]. Another concept of thermodynamic reversibility that includes restoration of the system state in an adiabatic process was introduced by Caratheodory [27, 28]. All these notions of reversibility were used during a long time discussion of Loschmidt's paradox.

From our point of view, here is no paradox. According to the definition of the symmetry transformation, it is not required that such a transformation was really realized, it is sufficient for it to be realized in principle. For example, in order to state that some crystal possesses a center of space inversion it is not necessary to really change all atoms situated in some positions in the crystal lattice by atoms in a equivalent positions. For this it is sufficient only to make such a change in principle. Therefore from the presence of time-reversal invariance of dynamical equations on microscopical level it does not follow at all the reverse of direction of motion of particles. The law of entropy (the second law of thermodynamics) in the form of the Boltzmann H-theorem (in the proof of which, although time-reversal invariance of the equations is used, but the supposition about the real reverse of direction of particle motion is not contained) testifies only the fact that ongoing processes are not reversible, and so there exists the "arrow of time".

The above-presented subtle distinctions between reversibility and irreversibility, as opposite conceptions are often unrecognized in the thermodynamic literature. An exceptions to this fact are [26, 29, 30], which give an excellent description of the relationship between irreversibility, the second law of thermodynamics and the time arrow [30]. The arrow of time is one of "most perplexing enigmas" in physics [31–36]. The idea that the second law of thermodynamics provides the physical basis of the arrow of time was voiced by many authors [32, 37, 38]. However, a convincing argument of this assertion has never been given [30, 33, 35]. An approach in which the irreversible dynamics of non-equilibrium thermodynamical systems is modeled by reversible dynamical systems, is given in papers [39–41]. A general discussion related to reversibility/irreversibility and arrows of the time in physics and chemistry is given in the works of Prigogine [42, 43].

It is necessary to note that in statistical thermodynamics, the time arrow is viewed as a consequence of a high system dimension and randomness. However, since in statistical thermodynamics there is no absolute certainty that entropy increases in each dynamic process, the direction of time determined by the increase in entropy has only statistical and not absolute certainty. Therefore, it cannot be deduced from statistical thermodynamics that the flow of time has a single direction [26].

In a global aspect, if the laws are invariant under the time reversal, for every physically admissible process there is a time-reversed one, which is also admissible. Initial conditions then forbids on average a large class of local processes. For instance, a low entropy initial condition for the Universe can explain the second thermodynamical principle [44]. An example would be the Universe undergoing a big crunch accompanied by a reduction of the total entropy [45] (see, however, [46, 47]). If the classical state of the Universe at maximum expansion is not a time-reversal invariant, then we must expect a contracting epoch to be substantially different from the expanding one. Thus, there will be no time-reversed epoch of expansion of the Universe [44]. The fact that we observe entropy increasing everywhere rather than decreasing could be explained, if we clearly understood the peculiarities of events in low and high entropy epochs [32, 44, 48, 49].

1.8 Reversibility of Fluctuations in the Closed Systems and Onsager's Relationships

At small deviations from the equilibrium state, not only in equilibrium processes in closed system, it is possible to describe, beginning with very general considerations, firstly started by Onsager [50]. We shall characterize the state of a closed system by the some macroscopic parameters x_i, which are functions of time. By x_i it is to understand the difference between the values of thermodynamic quantities in a given, not in equilibrium, state and in equilibrium state.

At small values of x_i all quantities which characterize the state of the system and the speed of its change may be expanded into a series on powers of x_i, keeping in these series only the first terms [51]

$$\frac{\mathrm{d}x_i}{\mathrm{d}t} = \sum_k \alpha_{ik} x_k, \tag{1.118}$$

$$S = S_0 - \frac{1}{2} \sum_{i,k} \beta_{ik} x_i x_k, \tag{1.119}$$

$$\frac{\mathrm{d}S}{\mathrm{d}t} = -\sum_{i,k} \frac{\mathrm{d}x_i}{\mathrm{d}t} \cdot x_k, \tag{1.120}$$

where S is the entropy, α_{ik} and β_{ik} are kinetic coefficients.

Formula (1.118) shows that all processes near the state of equilibrium are slow. The entropy of the system not in state of equilibrium is given by a quadratic form, moreover, from the condition (1.119) it follows that $\beta_{ik} = \beta_{ki}$. Formula (1.120) describes the production of entropy, during the unit of time, which is a small value. Formulas (1.118)–(1.120) may be applied only to changes of state of the system due to limited time t ($\tau_{\text{micro}} \ll t \ll \tau_{\text{macro}}$), where τ_{micro} is the time of microscopic scale characterizing changes of the state of the microscopic part of system, while τ_{macro} is the time of macroscopic scale, after which in the system it is established the state of complete statistical equilibrium.

We denote, according to [51], by I_i' the flow corresponding to the macroscopic parameter x_i, $I_i' = \frac{\mathrm{d}x_i}{\mathrm{d}t}$, and by X_i' the so-called thermodynamic force

$$X_i' = \frac{\partial S}{\partial x_i} = -\sum_k \beta_{ik} x_k. \tag{1.121}$$

Then the relations (1.118)–(1.120) may be rewritten in the form

$$I_i' = \sum_i \alpha_{ik} x_k = -\sum_{j,k} \alpha_{ij} \beta_{jk}^{-1} X_k' = \sum_k \gamma_{ik} X_k', \; \gamma_{ik} = -\sum_j \alpha_{ij} \beta_{jk}^{-1}, \tag{1.122}$$

$$S = S_0 + \frac{1}{2} \sum_i X_i' x_i, \tag{1.123}$$

$$\frac{\mathrm{d}S}{\mathrm{d}t} = \sum_i I_i' X_i'. \tag{1.124}$$

According to Onsager hypothesis, the macroscopic non-equilibrium state near equilibrium may be considered as some fluctuation. The change in time of the state of a macroscopic non-equilibrium system and that of a microscopic system the fluctuation is described by common laws. Thus, the relation between flows and forces or, in other words, the macroscopic law

$$I_i = \sum_k L_{ik} X_k, \tag{1.125}$$

may be equally applied to non-equilibrium processes and to processes of fluctuation relation. Macroscopic flows I_i and forces X_i are obtained, averaging that I_i' and X_i', and coefficients β_{ik} and γ_{ik} coincides.

Onsager proved [50] that $L_{ik} = L_{ki}$. Note that it is possible to come to the same result, if the hypothesis of Gibbs local distribution is used, instead of Onsager hypothesis.

The symmetry of kinetic coefficients $L_{ik} = L_{ki}$ has a profound sense. It shows that if, for example, in the system two parameters are changed, then two flows I_1 and

I_2 appear. In this case, the force X_1 gives contribution to the flow I_2, and the force X_2 gives the same contribution to the flow I_1.

Onsager relations may be obtained on the basis of the general theory of fluctuations as well [51]. According to the principle of microscopic reversibility, fluctuations in a closed system are reversed in time, so that for the correlation function $\langle x_l(t)x_k(t+\tau)\rangle$ one may write

$$\langle x_l(t)x_k(t+\tau)\rangle = \langle x_l(t)x_k(t-\tau)\rangle \,. \tag{1.126}$$

On the other hand, changing the origin of time reference in the right part of the equality (1.126), we obtain

$$\langle x_l(t)x_k(t+\tau)\rangle = \langle x_l(t+\tau)x_k(t)\rangle \,. \tag{1.127}$$

Here symbol $\langle\cdots\rangle$ means the average over ensemble. Averaging once more over time, we have

$$\overline{\langle x_l(t)x_k(t+\tau)\rangle} = \overline{\langle x_l(t+\tau)x_k(t)\rangle}. \tag{1.128}$$

Both averages are independent and equivalent. After subtracting from both parts of the equality (1.128) of $\langle x_l(t)\overline{x_k(t)}$, dividing by τ and passing to limit $\tau \to 0$, we have

$$\left\langle x_l(t)\frac{\overline{\mathrm{d}x_k(t)}}{\mathrm{d}t}\right\rangle = \left\langle x_k(t)\frac{\overline{\mathrm{d}x_l(t)}}{\mathrm{d}t}\right\rangle. \tag{1.129}$$

On the basis of Onsager hypothesis for fluctuations, as well as for macroscopic processes, it takes place the relation

$$I_i(t) = \frac{\overline{\mathrm{d}x_i(t)}}{\mathrm{d}t} = \sum_k L_{ik}X_k(t). \tag{1.130}$$

Using (1.130), the relation (1.129) leads to the equality

$$\left\langle x_l(t), \sum_i L_{ki}X_i(t)\right\rangle = \left\langle x_k(t), \sum_i L_{li}X_i(t)\right\rangle. \tag{1.131}$$

Since it is possible to prove that $\langle x_l X_i\rangle = \delta_{li}$ and $\langle x_k X_i\rangle = \delta_{ik}$ [51], then from (1.131) it follows

$$\sum_i L_{ki}\delta_{li} = \sum_i L_{li}\delta_{ki}, \tag{1.132}$$

whence we find $L_{kl} = L_{lk}$.

At this proof of the symmetry principle of kinetic coefficients there is used only the principle of the microscopic reversibility in time of the fluctuation in a closed system and the Onsager hypothesis. However, the sense of kinetic coefficients at this proof is not revealed.

When the Onsager principle was deduced, it was not supposed in fact that the system is not in a magnetic field and is not rotated. But if the system is in a magnetic field **H**, then at change of the sign of time $\tau \rightarrow -\tau$ there takes place the relation

$$\mathbf{H} \rightarrow -\mathbf{H}. \tag{1.133}$$

In this case, in order to fulfill the principle of the microscopic reversibility of fluctuations, the Lorentz's force does not change the sign. Therefore at time-reversal in the presence of a magnetic field the following relationship takes place:

$$\frac{dx(\tau)}{d\tau} \rightarrow \frac{dx(-\tau)}{d\tau}. \tag{1.134}$$

The same refers to the angular velocity of the motion of charged particles in a magnetic field. By repeating the previous reasoning it is easy to obtain the Onsager principle for non-equilibrium processes in the magnetic field at non-large deviations of the equilibrium state

$$L_{kl}(\mathbf{H}) = L_{lk}(-\mathbf{H}). \tag{1.135}$$

This equality always takes place, if two parameters x_k and x_l are such that at time reversal one of them changes its sign and the other does not. The relation (1.135) is the Casimir-Onsager reciprocity relation.

1.9 Poincaré Recurrence Theorem

In the Sect. 1.7 we have shown that the assumption of molecular chaos allows us to prove the Boltzmann H-theorem, a theorem that predicts an increase in entropy until equilibrium is reached. However, the Boltzmann assumption of molecular chaos remains unproved.

Poincaré proved the so-called recurrence theorem, which established that if the flow of a dynamical system preserves its volume and has only bounded orbits, then for each open set there exist orbits that intersect the set infinitely often [52].

Theorem *A system having a finite amount of energy and confined to a finite spatial volume will, after a sufficiently long time, return to an arbitrary small neighborhood of its initial state.*

Since the proof of the recurrence theorem has been given by Poincaré, it was discussed by different authors (see, for example, [53]).

In [26] a system-theoretic foundation for thermodynamics is presented. Specifically, using a state space formulation, it was developed a nonlinear compartmental dynamical system model characterized by energy conservation laws, that is consistent with basic thermodynamic principles. It was established the existence of a unique, continuously differentiable global entropy function for a large-scale dynamical system. Moreover, it was shown the absence of Poincaré recurrence for proposed thermodynamic model.

In fact the last result follows also from estimations made on the basis of Poincaré theorem, which leads to a very long (practically infinite) recurrence time. Really, let us consider an initial state, in which all particles of any gas are initially in a corner of a large box of volume V, after which they occupy all the volume of the box. By Poincaré theorem, after some time the gas particles must eventually return to their initial state in the corner of the box. In order to find their recurrence time, one considers that the box is divided into small cells of volume v. The total number of microstates accessible to the gas varies with the number of particles N like V^N. The number of microstates corresponding to all the particles occupying a single cell of volume v is v^N.

Thus, the probability of finding the system in the microstate is approximately $(v/V)^N$. Even if $v = V/2$, for $N \sim 10^{23}$, the probability is vanishingly small. On the other hand, the Poincaré recurrence time is proportional to the inverse of this probability or $(V/v)^N$. Believing $N \sim 10^{23}$ and $v = V/2$, the required time is

$$t \sim 2^{10^{23}}\,\mathrm{s},$$

which is many orders of magnitude longer than the current age of the Universe (10^{10} years). Thus, although the system will return arbitrarily close to its initial state, the time required for this is nonphysically long and will never be observed. We see that although in accordance with Poincaré's recurrence theorem, the physical process of returning of gas molecules in initial state is formally reversible, in fact this process is irreversible.

Chapter 2
Time Reversal in Quantum Mechanics and Quantized Field Theory

In this chapter we discuss the properties of the time-reversal operator (introduced into quantum mechanics by Wigner in 1932 [68, 222]) for particles without spin, as well as taking into account the spin. There are given Wigner criteria (a), (b), and (c) concerning the absence or presence of an additional degeneracy of energy levels due to the symmetry with respect to time reversal, both in the case when the spin of particles that form the quantum system is taken into account and in the case it is not. Wigner corepresentations of symmetry groups associated with the presence of time-reversal symmetry are also discussed.

Criteria (a), (b) and (c) of Herring for the wave vector group are considered, relating to the absence or presence of an additional degeneracy of the energy bands due to the symmetry of the time reversal.

On the basis of available publications, the nonconventional time-reversal symmetry, the selection rules due to the symmetry of time reversal, the role of the time-reversal symmetry for detailed balance principle and for crystalline lattice dynamics, the time-reversal symmetry in the theory of quantized field, and the CPT theorem are analyzed.

In this chapter, the original results belonging to the author refer to the geometric interpretation of the time reversal and the Kramers theorem (Sect. 2.9), the different ways of determining the unitary part **U** of the time-reversal operator, and the presentation of the operator **U** matrix in the basis of irreducible representations for all of 32 point symmetry groups (Sect. 2.5).

2.1 The Basic Concepts of Quantum Mechanics

In contrast to classic mechanics, the description of physical system states in quantum mechanics has a probabilistic character [3, 54–64]. We can not point out in the moment of time t exact values of generalized coordinates $\xi \equiv \{\xi_1, \ldots, \xi_n\}$,

I. I. Geru, *Time-Reversal Symmetry*, Springer Tracts in Modern Physics 281,
https://doi.org/10.1007/978-3-030-01210-6_2

characterizing the system. However, we can give the probability density of distribution (the density of probability) $\rho(\xi, t)$, on the basis of which to determine the probability of the fact that measuring in time moment t the variable ξ in the given state, we shall obtain its value in the interval $(\xi, \xi + \mathrm{d}\xi)$:

$$\mathrm{d}w(\xi, t) = \rho(\xi, t)\mathrm{d}\xi. \tag{2.1}$$

The states of physical systems are divided into pure and mixed, which are described respectively by a complex quadratically integrable function $\psi(\xi, t)$, called the wave function (as well as amplitude of probability), and a matrix of density. The wave function depends on n generalized coordinates (dynamic variables) $\xi_1, \ldots, \xi_n$ and on time t, which is not a dynamic variable and is considered as a parameter. It determines the density of distribution of dynamic variables ξ:

$$\rho(\xi, t) = |\psi(\xi, t)|^2 . \tag{2.2}$$

The complete probability is normed on identity:

$$\|\psi\|^2 = \int |\psi(\xi, t)|^2 \, \mathrm{d}\xi, \tag{2.3}$$

where integration is carried out on whole domain of definition of the function $\psi(\xi, t)$.

The set of all quadratically integrable complex functions of real variables is a linear unitary function space (Hilbert space $\mathfrak{L}_2$) [63]. It is possible to show that this space is infinite-dimensional, that is there are infinite linear independent vectors in it. The orthogonal set of vectors of the space $\mathfrak{L}_2$ is complete and can not be extended by means of introducing some other vectors of the space. The complete set in $\mathfrak{L}_2$ contain an infinite number of mutually orthogonal basis vectors. Thus, in quantum mechanics it is postulated, that some element (vector of state) of the space $\mathfrak{L}_2$ is associated to every system state. In the Dirac notations the inner product between vectors of states $|\Psi\rangle$ and $|\Phi\rangle$ in this space is introduced by means of the relationship

$$\langle\Psi|\Phi\rangle = \int \Psi^*(\xi, t)\Phi(\xi, t)\mathrm{d}\xi \tag{2.4}$$

where $|\Psi\rangle$ and $|\Phi\rangle$ are any elements in $\mathfrak{L}_2$ and the asterisk denotes the complex conjugation. The definition (2.4) satisfies to all axioms of inner product.

If an orthogonal set of basic vectors in the Hilbert space is denoted by $|\eta\rangle$, where η is a continuous or discrete index, or both of them, then

$$|\Psi\rangle = \int \mathrm{d}\eta \, |\eta\rangle \, \langle\eta|\Psi\rangle , \tag{2.5}$$

where $\langle\eta|\Psi\rangle$ is a complex number ("component" of $|\Psi\rangle$) and

$$\langle\Psi|\Phi\rangle = \int \mathrm{d}\eta \, \langle\Psi|\eta\rangle \, \langle\eta|\Phi\rangle \, . \tag{2.6}$$

Here $\langle\Psi|\eta\rangle$ is complex conjugate, $\langle\eta|\Psi\rangle^*$, of the component so that $\langle\Psi|$, is the Hermitian conjugate to $|\Psi\rangle$.

One of principal statements of quantum mechanics is the principle of superposition of states, which in a simplest form is reduced to two statements

(i) If some system can be in two states, described by vectors of states $|\Psi_1\rangle$ and $|\Psi_2\rangle$, then it can also be in states, which are described by the state vectors, obtained from $|\Psi_1\rangle$ and $|\Psi_2\rangle$ using the linear transformation

$$|\Psi\rangle = c_1 \, |\Psi_1\rangle + c_2 \, |\Psi_2\rangle \, , \tag{2.7}$$

where c_1 and c_2 are any complex numbers, not depending on time.

(ii) If a state vector is multiplied by any nonzero complex number, then a new state vector will correspond to the same state of the system.

It is necessary to mention that the superposition of states in quantum mechanics essentially differs from the superposition of oscillations in classical physics, where the superposition of states leads to new oscillations with a higher or smaller amplitude. Moreover, in the classical theory of oscillations there exists the resting state, in which the amplitude of oscillations is equal to zero everywhere, while in the quantum theory the equality to zero of the wave function in all points of the space corresponds to the absence of a state [61].

In quantum mechanics, the following statements are also postulated:

1. Every physical quantity F is associated to some linear Hermitian operator $\mathbf{F}$, acting in the space $\mathfrak{L}_2$. $\mathbf{F}$ is a Hermitian (or self-adjoint) operator if $\mathbf{F}^+ = \mathbf{F}$, i.e., for any $|\Psi\rangle$ and $|\Phi\rangle$ from $\mathfrak{L}_2$ the following relationship takes place:

$$\langle\Psi|\mathbf{F}|\Phi\rangle = \langle\Phi|\mathbf{F}|\Psi\rangle^* \, . \tag{2.8}$$

The explicit form of principal physical quantities is postulated.

2. A physical quantity F in every quantum-mechanical state may take only those values, which belong to the spectrum of the operator $\mathbf{F}$ that corresponds to it.

If $\rho(F_n)$ is the probability of the fact that in the state $\Psi(\xi, t)$ in the time moment t the physical quantity F has the value F_n, and $\rho(f)$ is the corresponding density of probability for a neighborhood of the point f of the continuous spectrum, then

$$\sum_n \rho(F_n) + \int f\rho(f)\mathrm{d}f = 1, \tag{2.9}$$

and the mean value (mathematical expectation) of the quantity F in the state $\Psi(\xi, t)$ is

$$\langle F\rangle = \sum_n F_n\rho(F_n) + \int f\rho(f)\mathrm{d}f. \tag{2.10}$$

3. The average value of the physical quantity F in the state $\Psi(\xi, t)$ is calculated by the formula

$$\langle F\rangle = \frac{\langle\Psi|\mathbf{F}|\Psi\rangle}{\langle\Psi|\Psi\rangle}. \tag{2.11}$$

If the wave function is normed on identity, then

$$\langle F\rangle = \langle\Psi|\mathbf{F}|\Psi\rangle . \tag{2.12}$$

We see that the dependence of $\langle F\rangle$ on t is determined by the time dependence of the wave function and the operator $\mathbf{F}$.

A totality of points (discrete) and continuous spectra of the operator $\mathbf{F}$ is called a complete spectrum of this operator. In the functional analysis it is proved that the complete spectrum of the Hermitian operator lies on the real axis. The reality of the operator spectrum of every physical quantity is in accordance with the requirement of reality of a result of every its measurement.

4. The given physical system can be in those and only those states, whose wave functions satisfy the differential equation

$$i\hbar\frac{\partial\,|\Psi(\xi, t)\rangle}{\partial t} = \mathbf{H}\,|\Psi(\xi, t)\rangle , \tag{2.13}$$

where $\mathbf{H}$ is the Hamiltonian of the system.

Equation (2.13) is written in the Schrödinger representation and is called the Schrödinger equation. An equivalent statement of the dynamics can be made in the Heisenberg representation, for which explicit time dependence is ascribed to the operators representing the observations. These operators satisfy the dynamic equation

$$i\hbar\frac{\partial\mathbf{F}(t)}{\partial t} = [\mathbf{F}(t), \mathbf{H}] , \tag{2.14}$$

where $[\mathbf{F}(t), \mathbf{H}]$ is a commutator of operators $\mathbf{F}(t)$ and $\mathbf{H}$.

From (2.14) it follows that all operators commuting with the Hamiltonian are unchanged in time not only in the Schrödinger representation, but also in the Heisenberg one. Since at $t = 0$ the operators coincide in these two representations, the form of operators commuting with $\mathbf{H}$ remains unchanged at the transition from the Schrödinger representation to the Heisenberg one.

In the case of mixed states of the quantum system, when wave functions can not be used for their characterization, it is necessary to introduce the density matrix and the corresponding statistical operator[1] ρ, which satisfies the following differential equation:

$$i\hbar\frac{\partial\rho}{\partial t} = [\mathbf{H}, \rho] . \tag{2.15}$$

[1]The statistical operator should not be confused with the probability density, denoted earlier also by ρ.

Sometimes this equation is called the quantum Liouville equation, since it corresponds to the Liouville equation for classical distribution function in statistical physics. The density matrix was introduced for the first time by Landau [65] and von Neumann [66].

The operator equation (2.15) permits to determine the statistical operator for every time moment, if it is known in some initial time moment. If the Hamiltonian does not depend explicitly on time and elements of the density matrix are equal to $\rho_{n'n}(0)$ in the initial moment of time, then from (2.15) it follows

$$\rho_{n'n}(t) = \rho_{n'n}(0) \exp\left\{\frac{i}{\hbar}(E_n - E_{n'})t\right\}, \tag{2.16}$$

where E_n and $E_{n'}$ are eigenvalues of the energy of the system.

One of the basic concepts of quantum mechanics is connected with physical quantities, of which operators commute with the Hamiltonian and, as a consequence, are integrals of motion. In particular, if $\mathcal{T}_\tau$ is the operator of time translation of the value τ, $\mathcal{T}_\tau t = t + \tau$ then

$$\mathcal{T}_\tau \Psi(t) = \Psi(t - \tau). \tag{2.17}$$

The homogeneity of time at such a translation will be expressed by the commutation condition

$$[\mathcal{T}_\tau, \mathbf{H}] = 0. \tag{2.18}$$

In spite of the operator $\mathcal{T}_\tau$ it is convenient to use the infinitesimal operator $\mathbf{I}(t)$, which is defined as a derivative of the operator with respect to a parameter at zero value of this parameter:

$$\mathbf{I}(t) = \frac{\partial}{\partial \tau}\mathcal{T}_\tau\bigg|_{\tau=0}. \tag{2.19}$$

It is easy to obtain the explicit form of the operator $\mathbf{I}(t)$, taking into account that

$$\mathbf{I}(t)\Psi(t) = \frac{\partial}{\partial \tau}\mathcal{T}_\tau \Psi(t)\bigg|_{\tau=0} = \frac{\partial}{\partial \tau}\Psi(t-\tau)\bigg|_{\tau=0} = -\frac{\partial}{\partial t}\Psi(t).$$

Thus

$$\mathbf{I}(t) = -\frac{\partial}{\partial t}. \tag{2.20}$$

The energy conservation law is connected with the commutation of the infinitesimal operator $\mathbf{I}(t)$ with the Hamiltonian. For this reason, an operator having a dimension of energy

$$-i\hbar\mathbf{I}(t) = i\hbar\frac{\partial}{\partial t}$$

is called sometimes the energy operator. Really, an operator of energy is the Hamilton operator, which is a function of the operators of coordinates and impulses.

In contrast to space coordinates, the time is not an operator [61]. However, in this context it is necessary to mention that currently there are many studies, containing frequently contradictory data about the existence of time operator and its properties.

As for kinematical aspect of quantum mechanics, we can note that kinematical relationships are associated with algebra of operators representing observables of the system and usually formulated as commutation relations between these operators.

2.2 Antilinear and Antiunitary Operators

An operator $\mathfrak{L}$, acting in a linear space,[2] is called a linear operator if it satisfies the relationships [67]

$$\begin{aligned} \mathfrak{L}(|\varphi_1\rangle + |\varphi_2\rangle) &= \mathfrak{L}\,|\varphi_1\rangle + \mathfrak{L}\,|\varphi_2\rangle\,, \\ \mathfrak{L}(c\,|\psi\rangle) &= c\mathfrak{L}\,|\psi\rangle\,, \end{aligned} \tag{2.21}$$

where $|\varphi_1\rangle$, $|\varphi_2\rangle$ and $|\psi\rangle$ are arbitrary ket-vectors and c is an arbitrary complex number (c-number).

In contrast to linear operators, which have wide application in quantum mechanics, there exist antilinear operators. The antilinear operator A satisfies the relationships

$$\begin{aligned} A(|\varphi_1\rangle + |\varphi_2\rangle) &= A\,|\varphi_1\rangle + A\,|\varphi_2\rangle\,, \\ A(c\,|\psi\rangle) &= c^* A\,|\psi\rangle\,, \end{aligned} \tag{2.22}$$

where c^* is a complex-conjugate c-number relative to c.

An example of antiunitary operator is the operator of complex conjugation, which action on bra- and ket-vectors is determined by formulas

$$\begin{aligned} K(\alpha\,|\varphi_1\rangle + \beta\,|\varphi_2\rangle) &= \alpha^* K\,|\varphi_1\rangle + \beta^* K\,|\varphi_2\rangle = \alpha^*\,\langle\varphi_1| + \beta^*\,\langle\varphi_2|\,, \\ K^2(\alpha\,|\varphi_1\rangle + \beta\,|\varphi_2\rangle) &= (\alpha^*)^*\,\langle\varphi_1|\,K + (\beta^*)^*\,\langle\varphi_2|\,K = \alpha\,|\varphi_1\rangle + \beta\,|\varphi_2\rangle\,, \\ K^2 &= 1, \quad K = K^{-1}, \end{aligned} \tag{2.23}$$

where α and β are c-numbers.

It is clear that the real Hamiltonian is invariant under the operator K:

$$KHK = H. \tag{2.24}$$

Two antilinear operators A_1 and A_2 are considered equal to each other, if $A_1\,|\psi\rangle = A_2\,|\psi\rangle$ for every ket-vector $|\psi\rangle$. The operator of identical transformation $A = I$ is defined for every ket-vector $|\psi\rangle$ by the equality $I\,|\psi\rangle = |\psi\rangle$.

[2]Further, if it does not involve misunderstandings, we shall write $\mathfrak{L}$ instead of $\mathbf{L}$ (respectively, A instead of $\mathbf{A}$) omitting the symbol of operator.

Now, it is possible to construct the algebra of antilinear operators. We define a sum of two operators $A_1 + A_2$ using their action on ket-vector $|\psi\rangle$

$$(A_1 + A_2)|\psi\rangle = A_1|\psi\rangle + A_2|\psi\rangle \tag{2.25}$$

and a product of two operators A_1A_2 in the form

$$(A_1A_2)|\psi\rangle = A_1(A_2|\psi\rangle). \tag{2.26}$$

From this relationship it follows that we can define different powers of the operator A at $A_1 = A_2 = A$. Also, it is evident that the following relationships are fulfilled:

$$\begin{aligned} (A_1 + A_2)|\psi\rangle &= (A_2 + A_1)|\psi\rangle, \\ [(A_1 + A_2) + A_3]|\psi\rangle &= [A_1 + (A_2 + A_3)]|\psi\rangle, \\ [A_1(A_2 + A_3)|\psi\rangle &= A_1A_2|\psi\rangle + A_1A_3|\psi\rangle. \end{aligned} \tag{2.27}$$

The algebra of n-dimensional antilinear square matrices is the same as algebra of antilinear operators. If two antilinear operators A_1 and A_2 satisfy equations

$$A_1A_2 = A_2A_1 = I, \tag{2.28}$$

where I is the identity transformation, then the operator A_2 (if it exists) is an inverse operator to the operator A_1. In this case

$$A_2 = A_1^{-1}, \qquad A_1 = A_2^{-1}. \tag{2.29}$$

By defining the action of antilinear operators on ket-vectors, we must define how these operators act on bra-vectors. So, we consider a ket-vector

$$|\Phi\rangle = A|\psi\rangle. \tag{2.30}$$

It is possible to find a dot product of this ket-vector by any bra-vector (for example, bra-vector $\langle\chi|$):

$$\langle\chi|\Phi\rangle = \langle\chi|(A|\psi\rangle). \tag{2.31}$$

This dot product depends antilinearly on ket-vector $|\psi\rangle$, since A is an antilinear operator. Therefore the dot product $\langle\chi|\Phi\rangle$ from (2.31) can be considered as a dot product of the ket-vector $|\psi\rangle$ by the bra-vector $\langle\chi|A$. Let us agree that antilinear operators acting on bra-vectors always stand at the right to bra-vectors, in contrast to the case when these operators act on ket-vectors, standing at the left to them. It is easy to see that the following relationship

$$\langle\chi|(A|\psi\rangle) = (\langle\chi|A)|\psi\rangle \tag{2.32}$$

takes place.

Every of two sides of the equation (2.32) may be written in the form $\langle\chi|A|\psi\rangle$ and therefore we shall not use round parentheses further. Consequently, the antilinear operator A can first act on bra-vector $\langle\chi|$ and afterwards the obtained result must be scalarly multiplied on ket-vector $|\psi\rangle$. It may come to the same result, if to act first by the operator A on ket-vector $|\psi\rangle$, and afterwards the obtained new ket-vector multiply scalarly by the bra-vector $\langle\chi|$. We note that the scalar product $\langle\chi|A|\psi\rangle$ is a number, which may be a complex one.

Antilinear operators, in contrast to linear ones, have a very limited application in quantum mechanics, which is related in essence only on the average to problems of time-reversal symmetry. In these problems, the antilinear operators are not applicable by themselves, except the case of combination with some unitary operator U, for which (as it is known) an inverse operator coincides with Hermitian-conjugated operator ($U^{-1} = U^{+},\ U^{+}U = UU^{+} = \mathbf{1}$). The product of an antilinear operator by an unitary operator is called the antiunitary operator. Such is the time-reversal operator, as it will be shown in the Sect. 2.3.

2.3 Wigner Time-Reversal Operator

The antiunitary time-reversal operator was first obtained on the basis of theoretical-group analysis by E. Wigner in 1932 [68, 222], starting from the invariance of the Schrödinger equation under time-reversal transformation. Before determining the explicit form of the time-reversal operator in non-relativistic quantum mechanics, we shall show that this operator (according E. Wigner) can not be linear. So we consider the behavior of a physical system relative to a rotation of the coordinate system. Let the relative arrangement of two coordinate systems be such that a point with the radius-vector $\mathbf{r}(x, y, z)$ in the second (turned) system has the radius-vector $\mathbf{r}'(x', y', z')$. Here the rotation is realized by a real orthogonal three-dimensional matrix $\mathbf{R}$ with the determinant equal to unity. By virtue of space isotropy we can define the function $\mathbf{O}_R\Phi$ as a wave function assigned to the state Φ by the second observer in the turned coordinate system, or as a wave function of the initial state Φ, turned by the help of transformation R and observed by the first observer in the original coordinate system.

If the wave function would be dependent only on space coordinates, the operator $\mathbf{O}_R$ would be a point transformation $\mathbf{P}_R$

$$\mathbf{P}_R\left|\varphi(\mathbf{r}')\right\rangle = |\varphi(\mathbf{r})\rangle\,, \tag{2.33}$$

where $|\varphi(\mathbf{r})\rangle$ and $\left|\varphi(\mathbf{r}')\right\rangle$ are the particle vector states in two reference systems (the function $\mathbf{P}_R\left|\varphi(\mathbf{r}')\right\rangle$ in the point (x', y', z') has the same value as the function $|\varphi(\mathbf{r})\rangle$ has in the point (x, y, z)).

When along with space coordinates the spin coordinates are included, the operator $\mathbf{O}_R$ can not remain a point transformation, since the spin can not be subjected to a dot transformation. Therefore the operator $\mathbf{O}_R$ will be more general than $\mathbf{P}_R$ is.

The operator $\mathbf{O}_R$ satisfies to relationship (2.21) (at $\mathfrak{L} = \mathbf{O}_R$) and hence it is a linear operator. Moreover, this operator leaves unchanged the absolute value of the dot product of two arbitrary state vectors, and consequently it does not change the norming condition of wave functions

$$|\langle \mathbf{O}_R\Psi|\mathbf{O}_R\Phi\rangle| = |\langle \Psi|\mathbf{O}_R^{+}\mathbf{O}_R\Phi\rangle| = |\langle \Psi|\Phi\rangle|, \tag{2.34}$$

$$\mathbf{O}_R^{+}\mathbf{O} = \mathbf{1}.$$

Relationships (2.34) take place, since the dot product of two state vectors is independent of choice of a representation. Thus, $\mathbf{O}_R$ is a linear unitary operator.

Let $|\Phi\rangle$ be an arbitrary state vector, decomposed by a complete set of basic ket-vectors $|\Psi_\varkappa\rangle$, beginning from $|\Psi_1\rangle$:

$$|\Phi\rangle = \sum_{\varkappa} a_\varkappa |\Psi_\varkappa\rangle . \tag{2.35}$$

It may be shown [222] that the following relationships take place:

$$|\langle \Psi_1 + \Psi_\varkappa|\Phi\rangle| = |a_1 + a_\varkappa|, \tag{2.36}$$

$$|\langle \mathbf{O}_R\Psi_1 + \mathbf{O}_R\Psi_\varkappa|a_1\mathbf{O}_R\Psi_1 + a_2'\mathbf{O}_R\Psi_2 + a_3'\mathbf{O}_R\Psi_3 + \cdots\rangle| = |a_1 + a_\varkappa'|, \tag{2.37}$$

where $a_1, a_2', a_3', \ldots$ are the coefficients of decomposition of the ket-vector $\mathbf{O}_R|\Phi\rangle$ by the complete orthogonal system of ket-vectors $\mathbf{O}_R|\Psi_1\rangle, \mathbf{O}_R|\Psi_2\rangle, \mathbf{O}_R|\Psi_3\rangle, \ldots,$ so that

$$\mathbf{O}_R|\Phi\rangle = a_1\mathbf{O}_R|\Psi_1\rangle + \sum_{\varkappa=2,3,\ldots} a_\varkappa'\mathbf{O}_R|\Psi_\varkappa\rangle . \tag{2.38}$$

Since

$$\langle \Psi_1 + \Psi_\varkappa|\Phi\rangle = \langle \mathbf{O}_R\Psi_1 + \mathbf{O}_R\Psi_\varkappa|\mathbf{O}_R\Phi\rangle , \tag{2.39}$$

then $|a_1 + a_\varkappa| = |a_1 + a_\varkappa'|$ according to (2.36) and (2.37), and consequently

$$|a_1 + a_\varkappa|^2 = |a_1 + a_\varkappa'|^2, \tag{2.40}$$

or

$$|a_1|^2 + a_1^*a_\varkappa' + a_1a_\varkappa'^* + |a_\varkappa'|^2 = |a_1|^2 + a_1^*a_\varkappa + a_1a_\varkappa^* + |a_\varkappa|^2. \tag{2.41}$$

The quantity $a_\varkappa'^*$ may be excluded from (2.41) on the basis of the equality $a_\varkappa'a_\varkappa'^* = a_\varkappa a_\varkappa^*$. As a result, we obtain for $a_\varkappa'$ the quadratic equation

$$a_1^*a_\varkappa'^2 - (a_1^*a_\varkappa + a_1a_\varkappa^*)a_\varkappa' + a_1|a_\varkappa|^2 = 0, \tag{2.42}$$

of which two solutions have the form

$$a'_{\varkappa,1} = a_{\varkappa}, \qquad a'_{\varkappa,2} = a^*_{\varkappa} a_1 / a^*_1. \tag{2.43}$$

In the first case ($a'_{\varkappa} = a_{\varkappa}$) for every of ket-vectors $|\Phi\rangle = \sum_{\varkappa} a_{\varkappa} |\Psi_{\varkappa}\rangle$ and $|\Psi\rangle = \sum_{\varkappa} b_{\varkappa} |\Psi_{\varkappa}\rangle$ the relationships

$$\mathbf{O}_R |\Phi\rangle = \sum_{\varkappa} a_{\varkappa} \mathbf{O}_R |\Psi_{\varkappa}\rangle, \qquad \mathbf{O}_R |\Psi\rangle = \sum_{\varkappa} b_{\varkappa} \mathbf{O}_R |\Psi_{\varkappa}\rangle \tag{2.44}$$

take place, so we obtain

$$\mathbf{O}_R(a|\Phi\rangle + b|\Psi\rangle) = \mathbf{O}_R \sum_{\varkappa} (aa_{\varkappa} + bb_{\varkappa}) |\Psi_{\varkappa}\rangle = \tag{2.45}$$

$$= \sum_{\varkappa} (aa_{\varkappa} + bb_{\varkappa}) \mathbf{O}_R |\Psi_{\varkappa}\rangle = a\mathbf{O}_R |\Phi\rangle + b\mathbf{O}_R |\Psi\rangle,$$

and the operator $\mathbf{O}_R$ is linear. Moreover, since there are also valid the relationships

$$\langle \mathbf{O}_R\Phi | \mathbf{O}_R\Psi\rangle = \left\langle \sum_{\varkappa} a_{\varkappa} \mathbf{O}_R \Psi_{\varkappa} \Big| \sum_{\lambda} b_{\lambda} \mathbf{O}_R \Psi_{\lambda} \right\rangle = \tag{2.46}$$

$$= \sum_{\varkappa\lambda} a^*_{\varkappa} b_{\lambda} \left\langle \Psi_{\varkappa} | \mathbf{O}^+_R \mathbf{O}_R \Psi_{\lambda} \right\rangle = \sum_{\varkappa\lambda} a^*_{\varkappa} b_{\lambda} \delta_{\varkappa\lambda} = \sum_{\varkappa} a^*_{\varkappa} b_{\varkappa},$$

$$\langle \Phi | \Psi \rangle = \left\langle \sum_{\varkappa} a_{\varkappa} \Psi_{\varkappa} \Big| \sum_{\lambda} b_{\lambda} \Psi_{\lambda} \right\rangle = \sum_{\varkappa\lambda} a^*_{\varkappa} b_{\lambda} \langle \Psi_{\varkappa} | \Psi_{\lambda} \rangle = \sum_{\varkappa} a^*_{\varkappa} b_{\varkappa},$$

it follows

$$\langle \mathbf{O}_R\Phi | \mathbf{O}_R\Psi \rangle = \langle \Phi | \Psi \rangle, \tag{2.47}$$

and we can conclude that the operator $\mathbf{O}_R$ is not only a linear operator, but also an unitary one.

Thus, the second solution $a'_{\varkappa,2} = a^*_{\varkappa} a_1 / a^*_1$ from (2.43) is not related to linear unitary operators, but possibly it is related to antilinear or antiunitary operators. In order to prove this fact, we identify the system of state vectors $|\Psi_1\rangle, |\Psi_2\rangle, |\Psi_3\rangle, \ldots$ with eigenfunctions of the Hamiltonian. Then $|\Psi_i\rangle$ are stationary states. Also, stationary states are the states $\mathbf{T}|\Psi_i\rangle$, where $\mathbf{T}$ is the time-reversal operator. Thus, state vectors $|\Psi_i\rangle$ and $\mathbf{T}|\Psi_i\rangle$ correspond to the same values of energy.

If the first solution of the equation (2.41) $a'_{\varkappa} = a_{\varkappa}$ (given in (2.42)) should also be applicable to the operator $\mathbf{T}$, then $\mathbf{T}$ should be linear. This leads to a contradiction, from which it follows that the second possibility from (2.42) is realized $\mathbf{T}$ [222]. In order to come to this contradiction, we consider an arbitrary state $|\Phi_0\rangle$ and expand it by stationary states $|\Psi_{\varkappa}\rangle$:

$$|\Phi_0\rangle = \sum_{\varkappa} a_{\varkappa} |\Psi_{\varkappa}\rangle . \tag{2.48}$$

We suppose that **T** is a linear operator:

$$\mathbf{T}|\Phi_0\rangle = \sum_{\varkappa} a_{\varkappa}\mathbf{T}|\Psi_{\varkappa}\rangle . \tag{2.49}$$

Since $\mathbf{T}|\Psi_{\varkappa}\rangle$ is also a stationary state with energy $E_{\varkappa}$, then the time interval t passes to the state $\exp(-iE_{\varkappa}t/\hbar)\mathbf{T}|\Psi_{\varkappa}\rangle$ when it outflows. Thus after the time interval t the state $\mathbf{T}|\Phi_0\rangle$ will become the state

$$\mathbf{T}|\Phi_0\rangle_t = \sum_{\varkappa} a_{\varkappa}e^{-iE_{\varkappa}t/\hbar}\mathbf{T}|\Psi_{\varkappa}\rangle . \tag{2.50}$$

This state must coincide with the state obtained by application of the operator **T** to the state vector

$$|\Phi\rangle_{-t} = \sum_{\varkappa} a_{\varkappa}e^{iE_{\varkappa}t/\hbar}|\Psi_{\varkappa}\rangle . \tag{2.51}$$

If the operator **T** is linear, its action on the vector $|\Phi\rangle_{-t}$ leads to the state

$$\mathbf{T}|\Phi\rangle_{-t} = \sum_{\varkappa} a_{\varkappa}e^{iE_{\varkappa}t/\hbar}\mathbf{T}|\Psi_{\varkappa}\rangle . \tag{2.52}$$

The state vector (2.52) is not a multiple (with a constant factor) to the state vector (2.50), since exponents of powers under the signs of sums in (2.52) and (2.50) have different signs. Consequently, the supposition that **T** is linear, leads to a contradiction. Therefore, the operator **T** corresponds to the second possibility from (2.43) ($a'_{\varkappa,2} = a^*_{\varkappa}a_1/a^*_1$) and the state vector $\mathbf{T}|\Phi_0\rangle$ must be represented in the form

$$\mathbf{T}|\Phi_0\rangle = \frac{a_1}{a_1^*}\left(a_1^*\mathbf{T}|\Psi_1\rangle + a_2^*\mathbf{T}|\Psi_2\rangle + a_3^*\mathbf{T}|\Psi_3\rangle + \cdots\right). \tag{2.53}$$

up to a constant factor. Since the constant factor in the definition of $\mathbf{T}|\Phi_0\rangle$ may be selected arbitrarily, we can select it to be equal to a_1^*/a. Then

$$\mathbf{T}|\Phi_0\rangle = \mathbf{T}\left(\sum_{\varkappa} a_{\varkappa}|\Psi_{\varkappa}\rangle\right) = \sum_{\varkappa} a^*_{\varkappa}\mathbf{T}|\Psi_{\varkappa}\rangle \tag{2.54}$$

and thus the operator **T** is antilinear with respect to the system of basis state vectors $|\Psi_{\varkappa}\rangle$ ($\varkappa = 1, 2, 3, \ldots$).

It is easy to show that the time-reversal operator is also antilinear with respect to every system of wave functions. In particular, if $|\Phi_1\rangle = \sum_{\varkappa} b_{\varkappa}|\Psi_{\varkappa}\rangle$, we have

$$\alpha\left|\Phi_0\right\rangle+\beta\left|\Phi_1\right\rangle=\alpha\sum_{\varkappa}a_{\varkappa}\left|\Psi_{\varkappa}\right\rangle+\beta\sum_{\varkappa}b_{\varkappa}\left|\Psi_{\varkappa}\right\rangle=\sum_{\varkappa}(\alpha a_{\varkappa}+\beta b_{\varkappa})\left|\Psi_{\varkappa}\right\rangle, \tag{2.55}$$

where α and β are arbitrary complex numbers, in general case.

The action of the time-reversal operator $\mathbf{T}$ on the linear superposition of state vectors (2.55) leads to the result

$$\begin{aligned}\mathbf{T}(\alpha\left|\Phi_0\right\rangle+\beta\left|\Phi_1\right\rangle)&=\mathbf{T}\left(\sum_{\varkappa}(\alpha a_{\varkappa}+\beta b_{\varkappa})\left|\Psi_{\varkappa}\right\rangle\right)= \\ &=\sum_{\varkappa}(\alpha a_{\varkappa}+\beta b_{\varkappa})^*\mathbf{T}\left|\Psi_{\varkappa}\right\rangle= \\ &=\alpha^*\sum_{\varkappa}a_{\varkappa}^*\mathbf{T}\left|\Psi_{\varkappa}\right\rangle+\beta^*\sum_{\varkappa}b_{\varkappa}^*\mathbf{T}\left|\Psi_{\varkappa}\right\rangle= \\ &=\alpha^*\mathbf{T}\left|\Phi_0\right\rangle+\beta^*\mathbf{T}\left|\Phi_1\right\rangle.\end{aligned} \tag{2.56}$$

The relationship (2.56) is true for every state vectors $|\Phi_0\rangle$ and $|\Phi_1\rangle$ and every two complex numbers α and β and (as it was mentioned in the Sect. 2.2, formulas (2.22) and (2.23)) it is the definition of the antilinear operator. This follows from the fact that the second possibility from (2.43) is realized for the time-reversal operator, as well as from the normalizing condition, accepted for the state vector $\mathbf{T}\left(\sum_{\varkappa}a_{\varkappa}\left|\Psi_{\varkappa}\right\rangle\right)$ in (2.54).

Besides the fact that the operator $\mathbf{T}$ is antiliniar, it is also antiunitary. Let us prove this property of the operator for spinless particles and for spin containing systems. In the first case the operator $\mathbf{T}$ is reduced to the operator of complex conjugation $\mathbf{K}$. If $|\psi_1\rangle$ and $|\psi_2\rangle$ are two ket-vectors related to a spinless particle, then under the operator $\mathbf{T}=\mathbf{K}$ the inner product $\langle\psi_1\mid\psi_2\rangle$ is transformed as follows

$$\langle\mathbf{T}\Psi_1\mid\mathbf{T}\psi_2\rangle=\int(\mathbf{K}\psi_1(\mathbf{r}))^*\,\mathbf{K}\psi_2(\mathbf{r})\,d^3\mathbf{r}=\int\psi_1(\mathbf{r})\,\psi_2^*(\mathbf{r})\,d^3\mathbf{r}=\langle\psi_2\mid\psi_1\rangle. \tag{2.57}$$

This formula is the definition of antiunitarity of the operator $\mathbf{T}$. We consider now the case of a particle with a spin. Let $|\Psi_1\rangle$ and $|\Psi_2\rangle$ be two arbitrary spinors of rank one for a particle with spin $S=\frac{1}{2}$:

$$|\Psi_1\rangle=\begin{pmatrix}\psi^{(1)}_{+\frac{1}{2}}(\mathbf{r})\\ \psi^{(1)}_{-\frac{1}{2}}(\mathbf{r})\end{pmatrix},\quad|\Psi_2\rangle=\begin{pmatrix}\psi^{(2)}_{+\frac{1}{2}}(\mathbf{r})\\ \psi^{(2)}_{-\frac{1}{2}}(\mathbf{r})\end{pmatrix}, \tag{2.58}$$

where $+\frac{1}{2}$ and $-\frac{1}{2}$ denote the eigenvalues (in the system of units with $\hbar=1$) of the operator S_z. It will be shown in this chapter that the operator $\mathbf{T}$ can be presented as a product of two commuting operators

$$\mathbf{T}=\mathbf{UK}=\mathbf{KU}, \tag{2.59}$$

where **U** is any unitary operator and **K** is the operator of complex conjugation. For a spinless particle $\mathbf{T} = \mathbf{K}$, while for a particle with spin $S = \frac{1}{2}$ the operator **T** is determined by (2.59) and, as it will be shown below, the explicit form of the operator **U** is

$$\mathbf{U} = i\sigma_y \tag{2.60}$$

where σ_y is the imaginary Pauli matrix [63, 222]. So the following transformations take place:

$$\mathbf{T}\,|\Psi_1\rangle = i\sigma_y\mathbf{K}\begin{pmatrix}\psi^{(1)}_{+\frac{1}{2}}(\mathbf{r})\\ \psi^{(1)}_{-\frac{1}{2}}(\mathbf{r})\end{pmatrix} = i\sigma_y\begin{pmatrix}\psi^{(1)^*}_{+\frac{1}{2}}(\mathbf{r})\\ \psi^{(1)^*}_{-\frac{1}{2}}(\mathbf{r})\end{pmatrix} = \begin{pmatrix}\psi^{(1)^*}_{-\frac{1}{2}}(\mathbf{r})\\ -\psi^{(1)^*}_{+\frac{1}{2}}(\mathbf{r})\end{pmatrix}, \tag{2.61}$$

$$\mathbf{T}\,|\Psi_2\rangle = i\sigma_y\mathbf{K}\begin{pmatrix}\psi^{(2)}_{+\frac{1}{2}}(\mathbf{r})\\ \psi^{(2)}_{-\frac{1}{2}}(\mathbf{r})\end{pmatrix} = i\sigma_y\begin{pmatrix}\psi^{(2)}_{+\frac{1}{2}}(\mathbf{r})\\ \psi^{(2)}_{-\frac{1}{2}}(\mathbf{r})\end{pmatrix} = \begin{pmatrix}\psi^{(2)^*}_{-\frac{1}{2}}(\mathbf{r})\\ -\psi^{(2)^*}_{+\frac{1}{2}}(\mathbf{r})\end{pmatrix},$$

$$\langle\mathbf{T}\Psi_1 \mid \mathbf{T}\Psi_2\rangle = \int\left[\psi^{(1)}_{-\frac{1}{2}}(\mathbf{r})\,\psi^{(2)*}_{-\frac{1}{2}}(\mathbf{r}) + \psi^{(1)}_{+\frac{1}{2}}(\mathbf{r})\,\psi^{(2)*}_{+\frac{1}{2}}(\mathbf{r})\right]d^3\mathbf{r} = \langle\Psi_2 \mid \Psi_1\rangle\,.$$

The last formula from (2.61) represents the definition of antiunitarity of the operator **T** for particles with spin, which coincides with an analogical definition for spinless particles. We can easily extend this definition to quantum systems with the spin S. Let $|\Psi_1\rangle$ and $|\Psi_2\rangle$ be two spinors of $2S$ rank

$$|\Psi_1\rangle = \begin{pmatrix}\psi^{(1)}_{S}(\mathbf{r})\\ \psi^{(1)}_{S-1}(\mathbf{r})\\ \vdots\\ \psi^{(1)}_{1-S}(\mathbf{r})\\ \psi^{(1)}_{-S}(\mathbf{r})\end{pmatrix}, \quad |\Psi_2\rangle = \begin{pmatrix}\psi^{(2)}_{S}(\mathbf{r})\\ \psi^{(2)}_{S-1}(\mathbf{r})\\ \vdots\\ \psi^{(2)}_{1-S}(\mathbf{r})\\ \psi^{(2)}_{-S}(\mathbf{r})\end{pmatrix}. \tag{2.62}$$

In this case the last formula from (2.61) is transformed into

$$\langle\mathbf{T}\Psi_1 \mid \mathbf{T}\Psi_2\rangle = \sum_{m_s=-S}^{S}\int \psi^{(1)}_{m_s}(\mathbf{r})\psi^{(2)^*}_{m_s}(\mathbf{r})d^3\mathbf{r} = \langle\Psi_2 \mid \Psi_1\rangle\,. \tag{2.63}$$

The transformation $t \to t' = -t$ of time variable is carried out by the time-reversal operator **T**, which transforms the state vector and observable of any system. The transformed state vector or operator representing an observable will be denoted by the prime of the original state or observable. The transformation of the state vector in Schrödinger representation can be written as

$$\left|\psi_1{}'(t')\right\rangle = \mathbf{T}\psi(t). \tag{2.64}$$

The transformations of the vector operator $\mathbf{r}^{\alpha}$ associated with the position of αth particle and the vector operator $\mathbf{P}^{\beta}$ associated with the momentum of βth particle are

$$\begin{aligned} \mathbf{r}^{\alpha'} &= \mathbf{T}\mathbf{r}^{\alpha}\mathbf{T}^{-1}, \\ \mathbf{P}^{\beta'} &= \mathbf{T}\mathbf{P}^{\beta}\mathbf{T}^{-1}. \end{aligned} \tag{2.65}$$

The commutators of operators x^{α}, $\hat{P}_x^{\beta}$ and $x^{\alpha'}$, $\hat{P}_x^{\beta'}$ satisfy the relationships

$$\left[x^{\alpha'}, \hat{P}_x^{\beta'}\right] = \mathbf{T}\left[x^{\alpha}, \hat{P}_x^{\beta}\right]\mathbf{T}^{-1}. \tag{2.66}$$

In accordance to the classical conditions, motion reversal imposes the requirements

$$\mathbf{r}^{\alpha'} = \mathbf{r}^{\alpha}, \qquad \mathbf{P}^{\beta'} = -\mathbf{P}^{\beta}. \tag{2.67}$$

Thus we arrive at the condition

$$\left[x^{\alpha'}, \hat{P}_x^{\beta'}\right] = -\left[x^{\alpha}, \hat{P}_x^{\beta}\right]. \tag{2.68}$$

Comparing this condition with the commutation relation

$$\left[x^{\alpha}, \hat{P}_x^{\beta}\right] = i\hbar\delta_{\alpha\beta} \tag{2.69}$$

and taking into account (2.66), we obtain [2]

$$\hbar\delta_{\alpha\beta}\mathbf{T}i\mathbf{T}^{-1} = -\hbar\delta_{\alpha\beta}i, \qquad \mathbf{T}i\mathbf{T}^{-1} = -i. \tag{2.70}$$

Hence $\mathbf{T}$ must include the operator of complex conjugation $\mathbf{K}$.The same result can be obtained if instead of commutation relation (2.69) commutation relations between operators y^{α} and $\hat{P}_y^{\beta}$ or z^{α} and $\hat{P}_z^{\beta}$ are used. Moreover, the condition (2.70) can be obtained if instead of commutation relations between operators associated with the position of αth particle and the momentum of βth particle ((2.69) and two other analogical equations) we use the commutation relations between operators of projections of the total angular momentum $\mathbf{J}$, which in a compact form are represented by a single vectorial relationship

$$[\mathbf{J} \times \mathbf{J}] = I\hbar\mathbf{J}. \tag{2.71}$$

Operators $\mathbf{L}$ and $\mathbf{S}$ of orbital and spin momenta satisfy the analogical commutation relations. On the other hand, the motion reversal imposes (in accordance with classical conditions) the requirements

$$\mathbf{J}' = -\mathbf{J}, \quad \mathbf{L}' = -\mathbf{L}, \quad \mathbf{S}' = -\mathbf{S}. \tag{2.72}$$

Taking into account (2.71) and (2.72), we obtain again the formula (2.70):

$$\mathbf{TJT}^{-1} = -\frac{1}{\hbar}(\mathbf{T}i\mathbf{T}^{-1})[\mathbf{TJT}^{-1} \times \mathbf{TJT}^{-1}] = -\frac{1}{\hbar}(\mathbf{T}i\mathbf{T}^{-1})[\mathbf{J} \times \mathbf{J}], \qquad (2.73)$$

$$-\mathbf{J} = -\frac{1}{\hbar}(\mathbf{T}i\mathbf{T}^{-1})i\hbar\mathbf{J}, \quad \mathbf{T}i\mathbf{T}^{-1} = -i.$$

Certainly, the same result will be obtained by replacing **J** by operators **L** or **S**. The conclusion about the fact that operator **T** contains in a multiplicative form the operator of complex conjugation follows also directly from the invariance of Schrödinger equation under time reversal. Let consider firstly a system with the Hamiltonian **H** in absence of external forces. Then the Schrödinger equation

$$i\hbar\frac{\partial\,|\psi\rangle}{\partial t} = \mathbf{H}\,|\psi\rangle \qquad (2.74)$$

must be transformed under **T** to

$$i\hbar\frac{\partial\,|\psi'\rangle}{\partial t'} = \mathbf{H}\,|\psi'\rangle, \qquad (2.75)$$

where $t' = -t$ and $|\psi'\rangle = \mathbf{T}\,|\psi\rangle$. By applying **T** from (2.59) to both sides of (2.74), we find

$$-i\hbar\frac{\partial\mathbf{T}\,|\psi\rangle}{\partial t} = \mathbf{THT}^{-1} \times \mathbf{T}\,|\psi\rangle\,,$$

or

$$i\hbar\frac{\partial\,|\psi'\rangle}{\partial t'} = \mathbf{THT}^{-1}\,|\psi'\rangle, \qquad (2.76)$$

which is consistent with (2.75) if and only if

$$\mathbf{THT}^{-1} = \mathbf{H}, \qquad (2.77)$$

that is, if and only if **H** is invariant under time reversal. The operator **T** belongs to a class of operators, which includes those operators that, when repeated, restore the original state. Wigner calls "involutional" such a class of operators [222]. Since multiplication of the state $\mathbf{T}\,|\psi\rangle$ by a phase factor is the only transformation that leaves the physical state unchanged, we have for involution operator

$$\mathbf{T}^2 = \varepsilon\mathbf{1}, \qquad (2.78)$$

where $\varepsilon = e^{i\phi}$ is a constant phase factor, ϕ is an undetermined phase angle that may be $\phi = 0$, and $\mathbf{1}$ is the identity operator. Wigner showed that $\varepsilon = \pm\mathbf{1}$. The proof of this result is as follows

$$\mathbf{T}^2 = \mathbf{UKUK} = \mathbf{UU}^\star. \tag{2.79}$$

Since **U** is unitary,

$$\mathbf{U}^{-1} = \mathbf{U}^+ = \tilde{\mathbf{U}}^*$$

or

$$\mathbf{U}^* = \tilde{\mathbf{U}}^{-1}. \tag{2.80}$$

From (2.79), (2.80) and (2.78) follows

$$\mathbf{T}^2 = \mathbf{U}\tilde{\mathbf{U}}^{-1} = \varepsilon\mathbf{1} \tag{2.81}$$

and

$$\mathbf{U}\tilde{\mathbf{U}}^{-1}\tilde{\mathbf{U}} = \varepsilon\tilde{\mathbf{U}}, \quad \mathbf{U} = \varepsilon\tilde{\mathbf{U}}. \tag{2.82}$$

The transpose of last equation is

$$\tilde{\mathbf{U}} = \varepsilon\mathbf{U}. \tag{2.83}$$

Substituting (2.83) in (2.82), we obtain

$$\mathbf{U} = \varepsilon^2\mathbf{U},$$

from which $\varepsilon = \pm\mathbf{1}$ and taking into account (2.81),

$$\mathbf{T}^2 = \pm\mathbf{1}. \tag{2.84}$$

Thus, there are two classes of quantum mechanical systems, "even" systems and "odd" ones. The coordinates, total energy and kinetic energy belong to the first class. These quantities either are not connected with time, or contain an even degree of time variable. The velocity, linear and angular momenta are those related to the second class. The operators corresponding to quantities of the first class are commuting with **T**, while those corresponding to quantities of the second class are anti-commuting with **T** [222]. For spinless particles $\mathbf{T} = \mathbf{K}$ and $\mathbf{T}^2 = +1$, while for a system, containing n particles with spin $S = 1/2$, $\mathbf{T}^2 = +\mathbf{1}$ if n is an even number and $\mathbf{T}^2 = -\mathbf{1}$ if n is an odd number.

The fundamental property of the antiunitary time-reversal operator **T** was found on the basis of inner product of two arbitrary states $|\psi_1\rangle$ and $|\psi_2\rangle$ of the spinless particle (2.57) or arbitrary states $|\Psi_1\rangle$ and $|\Psi_2\rangle$ in the case of particles with spin ((2.61) and (2.63)). The fundamental property of the inner product under antiunitary transformation yields directly the important result that for "odd" systems (i.e., such systems, of which observables change the sign at time reversal) the state $\mathbf{T}|\Psi_1\rangle$ is orthogonal to $|\Psi_1\rangle$

$$\langle \mathbf{T}\Psi_1 \mid \Psi_2\rangle = \left\langle \mathbf{T}\Psi_1 \mid \mathbf{T}^2\Psi_1\right\rangle$$

from (2.61) or (2.63) with $|\Psi_2\rangle = \mathbf{T}\,|\Psi_1\rangle$, whence, for $\mathbf{T}^2 = -\mathbf{1}$

$$\langle \mathbf{T}\Psi_1 \mid \Psi_1\rangle = -\langle \mathbf{T}\Psi_1 \mid \Psi_1\rangle = 0. \tag{2.85}$$

Thus for "odd" systems the arbitrary state vector $|\Psi\rangle$is orthogonal to the state vector $\mathbf{T}\,|\Psi\rangle$, i.e., there is a doubly degeneracy of energy levels due to the existence of the time-reversal symmetry. Such a type of degeneracy is called Kramers degeneracy, after the author, who first discovered it [69], though Kramers did not connected this degeneracy of the energy levels with the time-reversal symmetry. As we already mentioned, the antiunitary time-reversal operator was firstly introduced in physics by Wigner and he also showed that in the case of a system with odd number of particles with spin $\frac{1}{2}$ the invariance of the Hamiltonian under time-reversal operator leads to a supplementary degeneracy of energy levels. Questions connected with Kramers degeneracy of energetical levels will be discussed in more detail in the Chaps. 4 and 9 after finding the explicit form of the matrix of unitary operator $\mathbf{U}$ for different particular cases. In the Sect. 2.9, geometrical interpretations of time-reversal transformation and Kramers theorem will be brought out.

Let us find the form of operator $\mathbf{U}$ for a particle with spin $S = \frac{1}{2}$. There are several ways to determine the matrix of operator $\mathbf{U}$.

(i) In the standard representation the spin operator $\mathbf{S}$ takes the form

$$\mathbf{S} = \frac{\hbar}{2}\sigma,$$

$$\sigma_x = \begin{pmatrix} 0 & 1 \\ 1 & 0 \end{pmatrix}, \quad \sigma_y = \begin{pmatrix} 0 & -i \\ i & 0 \end{pmatrix}, \quad \sigma_z = \begin{pmatrix} 1 & 0 \\ 0 & -1 \end{pmatrix}. \tag{2.86}$$

A useful time-reversal transformation for the spin $S = \frac{1}{2}$ follows from the requirement

$$\mathbf{TJT}^{-1} = -\mathbf{J}, \tag{2.87}$$

which occurs not only for the orbital momentum, but also for the spin (see (2.72)). However, with respect to the spin, $\mathbf{T}$ cannot be a complex-conjugate operator, since all pure imaginary Hermitian 2×2-matrices commute with each other. The matrix $\mathbf{U}$ (2.87) must satisfy

$$\begin{aligned} \mathbf{T}\sigma_x\mathbf{T}^{-1} &= \mathbf{UK}\sigma_x\mathbf{KU}^{-1} = \mathbf{U}\sigma_x\mathbf{U}^{-1} = -\sigma_x, \\ \mathbf{T}\sigma_y\mathbf{T}^{-1} &= \mathbf{UK}\sigma_y\mathbf{KU}^{-1} = -\mathbf{U}\sigma_y\mathbf{U}^{-1} = -\sigma_y, \\ \mathbf{T}\sigma_z\mathbf{T}^{-1} &= \mathbf{UK}\sigma_z\mathbf{KU}^{-1} = \mathbf{U}\sigma_z\mathbf{U}^{-1} = -\sigma_z. \end{aligned} \tag{2.88}$$

We see that $\mathbf{U}$ must commute with σ_y and anti-commute with σ_x and σ_z. Since any matrix $\mathbf{U}$ can be represented as a sum of Pauli matrices, one can write

$$\mathbf{U} = \alpha\sigma_x + \beta\sigma_y + \gamma\sigma_z + \delta. \tag{2.89}$$

The first of (2.88) gives $\alpha = \delta = 0$, the second yields $\gamma = 0$, whereas β remains unrestricted by (2.88). However, since $\mathbf{U}$ is unitary, β must have unit modules. Therefore, we can choose $\beta = i$, whereupon the time-reversal operation reads [70]:

$$\mathbf{T} = i\sigma_y \mathbf{K} = e^{i\pi \frac{\sigma_y}{2}} \mathbf{K}. \tag{2.90}$$

The operator $\mathbf{T}$ from (2.90) can be considered as conventional time-reversal operator for a particle with spin $S = \frac{1}{2}$.

(ii) We represent the matrix $\mathbf{U}$ in the form

$$\mathbf{U} = \begin{pmatrix} a & b \\ c & d \end{pmatrix}. \tag{2.91}$$

Taking into account formulas (2.88), written for convenience of calculations in the form

$$\begin{aligned} \mathbf{U}\sigma_x &= -\sigma_x \mathbf{U}, \\ \mathbf{U}\sigma_y &= \sigma_y \mathbf{U}, \\ \mathbf{U}\sigma_z &= -\sigma_z \mathbf{U}, \end{aligned} \tag{2.92}$$

we obtain $d = -a, \quad c = -b$, and $a = 0$, whence

$$\mathbf{U} = b \begin{pmatrix} 0 & 1 \\ -1 & 0 \end{pmatrix}. \tag{2.93}$$

The matrix (2.93) coincides with $i\sigma_y$ up to a real coefficient, which can be considered to be equal to identity since $\mathbf{U}^+\mathbf{U} = \mathbf{U}\mathbf{U}^+ = \mathbf{1}$.

(iii) The wave function of the particle with spin $\frac{1}{2}$ can be presented in the form

$$|\Psi\rangle = \sum_{m_s = \pm\frac{1}{2}} \psi_{m_s}(\mathbf{r}) \, |S, m_s\rangle \, . \tag{2.94}$$

Let operator $\mathbf{T} = \mathbf{UK}$ acts on $|\Psi\rangle$ from (2.94) considering that the operator $\mathbf{U}$ does not act on coordinate part of the wave function $\psi_{m_s}(\mathbf{r})$. Then we obtain:

$$\mathbf{UK}\,|\Psi\rangle = \sum_{m_s = \pm\frac{1}{2}} \psi^*_{m_s}(\mathbf{r}) \mathbf{UK} \, |S, m_s\rangle \, . \tag{2.95}$$

The spin wave functions $\left|\frac{1}{2}, \frac{1}{2}\right\rangle$ and $\left|\frac{1}{2}, -\frac{1}{2}\right\rangle$ are transformed under the irreducible representation $D^{(\frac{1}{2})}$ of the rotation group of symmetry, while the state vectors $\mathbf{UK}\left|\frac{1}{2}, \frac{1}{2}\right\rangle$ and $\mathbf{UK}\left|\frac{1}{2}, -\frac{1}{2}\right\rangle$ are transformed under the representation $D^{(\frac{1}{2})^*}$. The representations $D^{(\frac{1}{2})}$ and $D^{(\frac{1}{2})^*}$ are equivalent, because the characters of elements in these representations are the same real numbers. The matrix $\mathbf{D}^{(\frac{1}{2})}(\alpha, \beta, \gamma)$ defined in the spin space

$L^{(\frac{1}{2})}$ by the basis vectors $\left|\frac{1}{2},\frac{1}{2}\right\rangle$ and $\left|\frac{1}{2},-\frac{1}{2}\right\rangle$ has the form [71]

$$\mathbf{D}^{(\frac{1}{2})}(\alpha,\beta,\gamma)=\begin{pmatrix} e^{-i\frac{\alpha+\gamma}{2}}\cos\frac{\beta}{2} & -e^{-i\frac{\alpha-\gamma}{2}}\sin\frac{\beta}{2} \\ e^{i\frac{\alpha-\gamma}{2}}\sin\frac{\beta}{2} & e^{i\frac{\alpha+\gamma}{2}}\cos\frac{\beta}{2} \end{pmatrix}, \tag{2.96}$$

where α, β,nd γ are Euler angles. Since representations $D^{(\frac{1}{2})}$ and $D^{(\frac{1}{2})^*}$ are equivalent, it is easy to find such a spin space $L_1^{(\frac{1}{2})}$ with basis spinors being a linear combination of basis spinors of the space $L^{(\frac{1}{2})}$, namely $0\cdot\left|\frac{1}{2},\frac{1}{2}\right\rangle-\left|\frac{1}{2},-\frac{1}{2}\right\rangle$ and $\left|\frac{1}{2},\frac{1}{2}\right\rangle+0\cdot\left|\frac{1}{2},-\frac{1}{2}\right\rangle$, in which the matrix $\mathbf{D}^{(\frac{1}{2})^*}(\alpha,\beta,\gamma)$ coincides with $\mathbf{D}^{(\frac{1}{2})}(\alpha,\beta,\gamma)$. In other words, under action of the operator $\mathbf{T}$ on spinors $\left|\frac{1}{2},\frac{1}{2}\right\rangle$ and $\left|\frac{1}{2},-\frac{1}{2}\right\rangle$ such a linear combinations of these spinors arise, at which the matrix $\mathbf{D}^{(\frac{1}{2})^*}(\alpha,\beta,\gamma)$ defined in the spin space $L_1^{(\frac{1}{2})}$ coincides with the matrix $\mathbf{D}^{(\frac{1}{2})}(\alpha,\beta,\gamma)$ defined in the spin space $L^{(\frac{1}{2})}$. It follows directly from here that the matrix of unitary operator $\mathbf{U}$ in the basis of spinors $\left|\frac{1}{2},\frac{1}{2}\right\rangle$ and $\left|\frac{1}{2},-\frac{1}{2}\right\rangle$ has the form

$$\mathbf{U}=\begin{pmatrix} 0 & 1 \\ -1 & 0 \end{pmatrix} \tag{2.97}$$

that coincides with $i\sigma_y$. Thus, the transition from the representation $D^{(\frac{1}{2})}$ to the representation $D^{(\frac{1}{2})^*}$ is equivalent to the transition from the basis $\{\left|\frac{1}{2},\frac{1}{2}\right\rangle,\left|\frac{1}{2},-\frac{1}{2}\right\rangle\}$ to $\{\mathbf{UK}\left|\frac{1}{2},\frac{1}{2}\right\rangle,\mathbf{UK}\left|\frac{1}{2},-\frac{1}{2}\right\rangle\}$. However, these notations need some caution since it is possible to give two different meanings to complex-conjugated quantities $\mathbf{UK}\left|\frac{1}{2},\frac{1}{2}\right\rangle$ and $\mathbf{UK}\left|\frac{1}{2},-\frac{1}{2}\right\rangle$. On the one hand, the spinor unit vectors $\left|\frac{1}{2},\frac{1}{2}\right\rangle=\binom{1}{0}$ and $\left|\frac{1}{2},-\frac{1}{2}\right\rangle=\binom{0}{1}$ are real quantities, therefore $\mathbf{UK}\left|\frac{1}{2},\frac{1}{2}\right\rangle=\mathbf{U}\left|\frac{1}{2},\frac{1}{2}\right\rangle$ and $\mathbf{UK}\left|\frac{1}{2},-\frac{1}{2}\right\rangle=\mathbf{U}\left|\frac{1}{2},-\frac{1}{2}\right\rangle$. On the other hand, spinor unit vectors $\left|\frac{1}{2},\frac{1}{2}\right\rangle$ and $\left|\frac{1}{2},-\frac{1}{2}\right\rangle$ are transformed at rotations by the irreducible representation $D^{(\frac{1}{2})}$. Since these matrices of the representation $D^{(\frac{1}{2})}$ are complex, the spinor unit vectors $\left|\frac{1}{2},\frac{1}{2}\right\rangle$ and $\left|\frac{1}{2},-\frac{1}{2}\right\rangle$ are to be considered as complex quantities. The first interpretation at which the state vector

$$|\Psi\rangle=\begin{pmatrix} \boldsymbol{\psi}_{+\frac{1}{2}}(\mathbf{r}) \\ \boldsymbol{\psi}_{-\frac{1}{2}}(\mathbf{r}) \end{pmatrix}=\psi_{\frac{1}{2}}(\mathbf{r})\begin{pmatrix}1\\0\end{pmatrix}+\psi_{-\frac{1}{2}}(\mathbf{r})\begin{pmatrix}0\\1\end{pmatrix}=\psi_{\frac{1}{2}}(\mathbf{r})\,\zeta_{\frac{1}{2}}(\sigma_z)+\psi_{-\frac{1}{2}}(\mathbf{r})\,\zeta_{-\frac{1}{2}}(\sigma_z) \tag{2.98}$$

is represented as a decomposition by basic spinors

$$\left|\frac{1}{2},\frac{1}{2}\right\rangle\equiv\zeta_{\frac{1}{2}}(\sigma_z), \qquad \left|\frac{1}{2},-\frac{1}{2}\right\rangle\equiv\zeta_{-\frac{1}{2}}(\sigma_z), \tag{2.99}$$

where $\zeta_{\frac{1}{2}}(+1)=1$, $\zeta_{\frac{1}{2}}(-1)=0$, $\zeta_{-\frac{1}{2}}(+1)=0$, $\zeta_{-\frac{1}{2}}(-1)=1$, and correspondingly $\mathbf{K}\zeta_{\frac{1}{2}}(+1)=1$, $\mathbf{K}\zeta_{\frac{1}{2}}(-1)=0$, $\mathbf{K}\zeta_{-\frac{1}{2}}(+1)=0$, $\mathbf{K}\zeta_{-\frac{1}{2}}(-1)=1$ are used at computations of matrix elements and determination of the conditions of orthonormalization of basis spinors.

Particularly, the condition of orthonormalization of basis spinors is written in the form

$$\sum_{\sigma_z} \zeta_\alpha^*(\sigma_z)\zeta_\beta(\sigma_z) = \delta_{\alpha\beta}, \tag{2.100}$$

or

$$(1\ 0)\begin{pmatrix}1\\0\end{pmatrix} = 1,\quad (1\ 0)\begin{pmatrix}0\\1\end{pmatrix} = 0,\quad (0\ 1)\begin{pmatrix}1\\0\end{pmatrix} = 0,\quad (0\ 1)\begin{pmatrix}0\\1\end{pmatrix} = 1, \tag{2.101}$$

which is the same. The second interpretation of quantities $\mathbf{K} \mid \frac{1}{2}, \frac{1}{2}\rangle$ and $\mathbf{K} \mid \frac{1}{2}, -\frac{1}{2}\rangle$ follows from their transformation properties. Let denote the elements of the matrix $\mathbf{D}^{(\frac{1}{2})}(\alpha, \beta, \gamma)$ from (2.96) by a_{11}, a_{12}, a_{21}, and a_{22}. Then we have

$$\mathbf{R}\left|\frac{1}{2}, \frac{1}{2}\right\rangle \equiv a_{11}\left|\frac{1}{2}, \frac{1}{2}\right\rangle + a_{12}\left|\frac{1}{2}, -\frac{1}{2}\right\rangle, \tag{2.102}$$
$$\mathbf{R}\left|\frac{1}{2}, -\frac{1}{2}\right\rangle \equiv a_{21}\left|\frac{1}{2}, \frac{1}{2}\right\rangle + a_{22}\left|\frac{1}{2}, -\frac{1}{2}\right\rangle.$$

where the operator of rotation, described by Euler angles α, β and γ, is denoted by $\mathbf{R}$. Since $\mathbf{R}$ is a real transformation of real coordinate axes O_x, O_y and O_z, the action of operator $\mathbf{UK}$ on expressions (2.102) gives

$$\mathbf{RUK}\left|\frac{1}{2}, \frac{1}{2}\right\rangle \equiv a_{11}^*\mathbf{UK}\left|\frac{1}{2}, \frac{1}{2}\right\rangle + a_{12}^*\mathbf{UK}\left|\frac{1}{2}, -\frac{1}{2}\right\rangle, \tag{2.103}$$
$$\mathbf{RUK}\left|\frac{1}{2}, -\frac{1}{2}\right\rangle \equiv a_{21}^*\mathbf{UK}\left|\frac{1}{2}, \frac{1}{2}\right\rangle + a_{22}^*\mathbf{UK}\left|\frac{1}{2}, -\frac{1}{2}\right\rangle.$$

In accordance to the form (2.96) of matrices of transformations $\mathbf{D}^{(\frac{1}{2})}(\alpha, \beta, \gamma) \equiv \mathbf{D}^{(\frac{1}{2})}(R)$, we have

$$a_{11}^* = a_{22},\quad a_{12}^* = -a_{21},\quad a_{21}^* = -a_{12},\quad a_{22}^* = a_{11}. \tag{2.104}$$

and

$$\mathbf{R}\left(\mathbf{UK}\left|\frac{1}{2}, \frac{1}{2}\right\rangle\right) \equiv a_{11}\left(\mathbf{UK}\left|\frac{1}{2}, -\frac{1}{2}\right\rangle\right) + a_{12}\left(-\mathbf{UK}\left|\frac{1}{2}, -\frac{1}{2}\right\rangle\right), \tag{2.105}$$
$$\mathbf{R}\left(-\mathbf{UK}\left|\frac{1}{2}, \frac{1}{2}\right\rangle\right) \equiv a_{21}\left(\mathbf{UK}\left|\frac{1}{2}, -\frac{1}{2}\right\rangle\right) + a_{22}\left(-\mathbf{UK}\left|\frac{1}{2}, \frac{1}{2}\right\rangle\right).$$

Thus, $\mathbf{UK}\left|\frac{1}{2}, -\frac{1}{2}\right\rangle$ and $-\mathbf{UK}\left|\frac{1}{2}, -\frac{1}{2}\right\rangle$are transformed like $\left|\frac{1}{2}, \frac{1}{2}\right\rangle$and $\left|\frac{1}{2}, -\frac{1}{2}\right\rangle$, that is by representation $D^{(\frac{1}{2})}$. Suppose that $\mathbf{UK}\left|\frac{1}{2}, \frac{1}{2}\right\rangle$ and $\mathbf{UK}\left|\frac{1}{2}, -\frac{1}{2}\right\rangle$must be expressed as a linear combination of unit vectors $\left|\frac{1}{2}, \frac{1}{2}\right\rangle$and $\left|\frac{1}{2}, -\frac{1}{2}\right\rangle$. Then according to Schur's lemma [75, 80] $\mathbf{UK}\left|\frac{1}{2}, -\frac{1}{2}\right\rangle$ and $-\mathbf{UK}\left|\frac{1}{2}, \frac{1}{2}\right\rangle$ are proportional to $\left|\frac{1}{2}, \frac{1}{2}\right\rangle, \left|\frac{1}{2}, -\frac{1}{2}\right\rangle$. We may assume that the arbitrary phase factor is equal to identity and obtain

$$\begin{aligned}\mathbf{UK}\left|\frac{1}{2}, \frac{1}{2}\right\rangle &= -\left|\frac{1}{2}, -\frac{1}{2}\right\rangle,\\ \mathbf{UK}\left|\frac{1}{2}, -\frac{1}{2}\right\rangle &= \left|\frac{1}{2}, \frac{1}{2}\right\rangle,\end{aligned} \tag{2.106}$$

from where it follows again that the matrix of unitary operator $\mathbf{U}$ in spinor basis $\{\left|\frac{1}{2}, \frac{1}{2}\right\rangle, \left|\frac{1}{2}, -\frac{1}{2}\right\rangle\}$ has the form (2.97)

$$\mathbf{U} = \begin{pmatrix} 0 & 1 \\ -1 & 0 \end{pmatrix} = i\sigma_y.$$

Note that in the reversibility space defined by basis vectors $\begin{pmatrix} 1 \\ 0 \end{pmatrix}$ and $\begin{pmatrix} 0 \\ -i \end{pmatrix}$ the time-reversal operator $\mathbf{T}$ for a particle with spin $\frac{1}{2}$ obtains a supplementary factor as a unitary antisymmetric (2×2)-matrix [72].

In conclusion, in parallel with the invariance of the Schrödinger equation (2.74) and of the Hamiltonian $\mathbf{H}$ under time-reversal operator $\mathbf{T}$ (see (2.77)), we discuss also the invariance under operator $\mathbf{T}$ of the Pauli equation

$$\mathbf{H}\left|\varphi\right\rangle = \varepsilon\left|\varphi\right\rangle \tag{2.107}$$

with the Hamiltonian [76]

$$\mathbf{H} = mc^2 + \frac{\mathbf{p}^2}{2m} - \frac{\mathbf{p}^4}{8m^3c^2} + V(\mathbf{r}) + \frac{\hbar}{i}\cdot\frac{1}{4m^2c^2}(\nabla V, \mathbf{p}) + \frac{\hbar}{4m^2c^2}[\nabla V \times \mathbf{p}]\cdot\boldsymbol{\sigma}. \tag{2.108}$$

Here $\mathbf{p}$, m, ε and $V(\mathbf{r})$ are respectively the operator of linear momentum, the mass, the total and the potential electron energy; c is the light velocity in vacuum and $\boldsymbol{\sigma}$ is Pauli operator determined by Pauli matrices σ_x, σ_y and σ_z from (2.86). The state vector $\left|\varphi\right\rangle$ is a two-component spinor, which can be represented in the form of a decomposition by basis spinors by analogy with (2.98). The first three summands in (2.108) represent the energy of the free electron. The main summand mc^2 is the proper energy of electron, $\frac{\mathbf{p}^2}{2m}$ and $\frac{\mathbf{p}^4}{8m^3c^2}$ are kinetic energy in null approach by $\frac{v}{c}$ (v is the electron velocity) and mass-velocity correction in the first approach by $\frac{v^2}{c^2}$. The fifth summand described the relativistic correction constant to the potential, which is called the Darvin correction. The last summand corresponds to the interaction of the spin with orbital momentum of the electron. The presence in (2.108) of terms containing operators $\mathbf{p}^4$, $(\nabla V, \mathbf{p})$ does not violate the invariance of the Hamiltonian $\mathbf{H}$ from (2.108) under time-reversal transformation. It is easy to see that the time-reversed state vector $\left|\mathbf{T}\varphi\right\rangle$, as well as the state vector $\left|\varphi\right\rangle$satisfy the Pauli equation

(2.107) and belongs to the same eigenvalue of energy ε. In this case, as well as in the case of the Schrödinger equation for a particle with spin $\frac{1}{2}$, the relationship $\mathbf{T}^2 = -\mathbf{1}$ takes place and the state vectors $|\varphi\rangle$ and $|\mathbf{T}\varphi\rangle$ not only belong to the same energy level, but are also mutually orthogonal (the Kramers degeneracy of energy levels due to time-reversal symmetry).

2.4 Time-Reversal Operator in High Spin Systems

The time-reversal operator $\mathbf{T}$ and, respectively, the unitary operator $\mathbf{U}$ for systems with high spin S can be obtained in the same way as for a particles with spin $S = \frac{1}{2}$.

(i) Starting from (2.90), we can write for a system formed by N particles with spin $S = \frac{1}{2}$:

$$\mathbf{T} = \mathbf{UK}, \quad \mathbf{U} = \mathbf{U}_{1y}\mathbf{U}_{2y\ldots}\mathbf{U}_{Ny} = e^{i\pi S_y}, \tag{2.109}$$

where S_y is the y-component of the total spin $S = \frac{1}{2}(\sigma_1 + \sigma_2 + \cdots + \sigma_N)/2$ in a system of units with $\hbar = 1$. The square of $\mathbf{T}$ depends on the number of particles according to

$$\mathbf{T}^2 = \begin{cases} +\mathbf{1}, & N \text{ even}, \\ -\mathbf{1}, & N \text{ odd}. \end{cases} \tag{2.110}$$

Decomposition of the exponential operator from (2.109) in series leads to an infinite sum of terms containing increasing powers of the operator S_y:

$$\mathrm{U} = e^{i\pi S_y} = \mathbf{1}\cdot\left(1 - \frac{\pi^2}{2!}S_y^2 + \frac{\pi^4}{4!}S_y^4 - \frac{\pi^6}{6!}S_y^6 + \cdots\right) + i\left(\pi S_y - \frac{\pi^3}{3!}S_y^3 + \frac{\pi^5}{5!}S_y^5 - \cdots\right). \tag{2.111}$$

Since there is a restriction on the degree of spin operators, according to which spin operators having the degree $k > 2S$ are reduced to spin operators of less degrees ($k \leq 2S$), the operator polynomial (2.111) of degree $k \to \infty$ is reduced to the polynomial of degree $2S$. As an example, let consider the spin $S = 1$. In this case the greatest degree of operator S_y in (2.111) is equal to 2 because the following relations between different degrees of the operator S_y from (2.111) take place

$$S_y^{2k} = S_y^2, \quad k \geq 1 \text{ and } S_y^{2k+1} = S_y, \quad k \geq 0. \tag{2.112}$$

Taking into account (2.112), the operator polynomial (2.111) can be transformed to the form

$$\mathbf{U} = \mathbf{1} - 2S_y^2. \tag{2.113}$$

The matrix of the operator $\mathbf{U}$ in the basis of spin functions $|1, 1\rangle$, $|1, 0\rangle$ and $|1, -1\rangle$ has the form

$$\mathbf{U} = \begin{pmatrix} 1 & 0 & 0 \\ 0 & 1 & 0 \\ 0 & 0 & 1 \end{pmatrix} - 2 \cdot \frac{1}{2} \begin{pmatrix} 1 & 0 & -1 \\ 0 & 0 & 0 \\ -1 & 0 & 1 \end{pmatrix} = \begin{pmatrix} 0 & 0 & -1 \\ 0 & 1 & 0 \\ -1 & 0 & 0 \end{pmatrix}. \tag{2.114}$$

We see that **U** is a symmetric matrix. Uniutary matrices are also symmetric for every integer values of the spin or summary angular momentum J. In the case of half-integer J unitary matrices **U** are antisymmetric. In both cases the nonzero matrix elements of unitary matrix **U** are situated on the collateral diagonal of the type indicated in (2.114) for matrix **U** in the case of spin $S = 1$. All other matrix elements of the matrix **U** are equal to zero for every systems with integer or half-integer values of the angular momentum.

(*ii*) At first view it seems that using the formulas of type (2.88) to determine the form of the matrix of unitary operator **U** in the case of great values of angular momentum is weakly efficient. This is due to great dimensions of the operator matrix of the angular momentum and, respectively, great number of unknown matrix elements of the operator **U**, in contrast to spin $S = \frac{1}{2}$, for which the number of matrix elements is equal to 4. However, formulas of type (2.88) may be used to determine matrix **U** at $S > \frac{1}{2}$. as a consequence of the fact that the matrix elements of the spin projection operator are equal to zero, excepting those lying on the main diagonal (S_z) or on the neighboring diagonals (S_x, S_y). It is especially simple done, if it is to use the property of anti-commutation of **T** with operator S_z and at the same time to take into account the unitarity of **U**. This is in accordance with the results obtained in [72], where for finding the explicit form of the operator in the representation, in which S_x and S_z have real matrix elements, and operator S_y, excepting null elements, has pure imaginary matrix elements, the solution of the equations

$$\begin{aligned} \mathbf{U}S_x + S_x\mathbf{U} &= 0, \\ \mathbf{U}S_y - S_y\mathbf{U} &= 0, \\ \mathbf{U}S_z + S_z\mathbf{U} &= 0, \end{aligned} \tag{2.115}$$

is written in the form $\mathbf{U} = e^{i\pi S_y}$. The matrix of operator **U** is symmetric at integer values of S and antisymmetric at half-integer of S.

(*iii*) The wave function of the system with spin S we represent in the form

$$|\Psi\rangle = \sum_{m_s=-S}^{S} \psi_{m_S}(\mathbf{r})\,|S, m_s\rangle\,. \tag{2.116}$$

Let act with the operator $\mathbf{T} = \mathbf{UK}$ on Ψ from (2.116) and take into account that the operator **U** does not act on the coordinate part of the wave function,

$$\mathbf{UK}\,|\Psi\rangle = \sum_{m_s=-S}^{S} \psi^*_{m_s}(\mathbf{r})\mathbf{UK}\,|S, m_s\rangle\,. \tag{2.117}$$

The spin wave functions $|S, S\rangle$, $|S, S-1\rangle$, ..., $|S, 1-S\rangle$ and $|S, -S\rangle$ are transformed under the irreducible representation $D^{(S)}$ of the rotation group, while the state vectors $\mathbf{UK}\,|S, S\rangle$, $\mathbf{UK}\,|S, S-1\rangle$, ..., $\mathbf{UK}\,|S, 1-S\rangle$ and $\mathbf{UK}\,|S, -S\rangle$ are transformed under the representation $D^{(S)^*}$. The representations $D^{(S)}$ and $D^{(S)^*}$ are equivalent by virtue of the same reason discussed in the Sect. 2.3 for the representations $D^{(\frac{1}{2})}$ and $D^{(\frac{1}{2})^*}$. Under action of the operator $\mathbf{T} = \mathbf{UK}$ on spinors $|S, S\rangle$, $|S, S-1\rangle$, ..., $|S, 1-S\rangle$ and $|S, -S\rangle$ such linear combinations of these spinors arise, at which the matrix $\mathbf{D}^{(S)^*}(\alpha, \beta, \gamma)$ defined in the space $L_1^{(S)}$ coincides with the matrix $\mathbf{D}^{(S)}(\alpha, \beta, \gamma)$ defined in the space $L^{(S)}$.

The unitary matrix $\mathbf{U}$ satisfies the relationship

$$\mathbf{D}^{(S)^*}(\alpha, \beta, \gamma) = \mathbf{U}\mathbf{D}^{(S)}(\alpha, \beta, \gamma)\mathbf{U}^{+}, \tag{2.118}$$

which can be obtained taking into account that the time-reversal and space rotation operators are commuting. The relation (2.118) is valid, since the representation $D^{(S)}$ is unitary and has a real character. On the other hand, since $D^{(S)}$ is a irreducible representation of the rotation group, the matrix $\mathbf{U}$ is determined up to a constant factor $c = \pm 1$ [222]. Thus, depending on the value of the spin S the matrix $\mathbf{U}$ is either symmetric at integer value of S, or antisymmetric at half-integer S. This can be seen, starting directly from the form of matrix $\mathbf{U}$.

According to Wigner [222], matrix elements of the matrix $\mathbf{U}$ from (2.118) are[3]

$$U_{nm} = (-1)^{S+m}\delta_{n,-m} = (-1)^{S-n}\delta_{n,-m}. \tag{2.119}$$

All matrix elements of the matrix $\mathbf{U}$ are equal to zero, excepting those which are on collateral diagonal. In turn, these elements are equal to $+1$ and -1 beginning in upper right corner and ending in lower left angle with the number $+1$, if S is an integer number, and -1, if S is half-integer number:

$$\mathbf{U} = \begin{pmatrix} 0 & & & 1 \\ & & -1 & \\ & 1 & & \\ & \cdot^{\cdot^{\cdot}} & & 0 \\ \pm 1 & & & \end{pmatrix}, \tag{2.120}$$

Therefore $\mathbf{U}$ is symmetric for integer S and antisymmetric for half-integer S. This conclusion predicted on the basis of formulas (2.97) and (2.114), is now confirmed.

[3] Formula (2.119) differs from that given in [222], since our spinor basis of the matrix $\mathbf{U}$ differs by a permutation of the basis spinors used in [222].

2.5 Time-Reversal Operator in Symmetry Point Groups

In the presence of crystalline or external electric fields the representations $D^{(S)}$ split into irreducible representations Γ_γ of symmetry point groups, and so change the form of the time-reversal operator. By analogy with (2.118) for the unitary operator $\mathbf{U}$ in the case of rotation group, which can be rewritten in the form

$$\mathbf{T}\mathbf{D}^{(S)}(R) = \mathbf{D}^{(S)}(R)\mathbf{T}, \tag{2.121}$$

where R is the element of the rotation group, one can suppose that in the case of symmetry point group we can also write

$$\mathbf{T}\boldsymbol{\Gamma}_\gamma(g) = \boldsymbol{\Gamma}_\gamma(g)\mathbf{T}, \tag{2.122}$$

where γ enumerates matrices $\boldsymbol{\Gamma}_\gamma(g)$ irreducible representations $\boldsymbol{\Gamma}_\gamma$ of the point group of symmetry and g is the element of this group.

The condition (2.122) means that time reversal and transformations of symmetry of point groups are independent operations. For the operator $\mathbf{U}$ these conditions are reduced to the relationship

$$\mathbf{U}\boldsymbol{\Gamma}^*_\gamma(g) = \boldsymbol{\Gamma}_\gamma(g)\mathbf{U}. \tag{2.123}$$

However, in contrast to representations $D^{(S)}$ of the rotation group, of which basis vectors are spinors and the unitary operator $\mathbf{U}$ is also given in the spinor basis, in the case of representations of point groups of symmetry the matrices of operators $\boldsymbol{\Gamma}_\gamma(G)$ and $\mathbf{U}$ are given in the basises of irreducible representations Γ_γ. There are two ways for determining the operator $\mathbf{U}$.

(i) The wave functions of the many-particle system with the total spin S in the case of spherical symmetry are transformed under time-reversal operator $\mathbf{T} = \mathbf{UK}$ as follows [222]

$$\mathbf{T}\,|S, M\rangle = \mathbf{UK}\,|S, M\rangle = (-1)^{S+M}\,|S, -M\rangle\,, \tag{2.124}$$

where M is the eigenvalue of the operator S_z (in the units system with $\hbar = 1$) and the matrix $\mathbf{U}$ is defined by basis vectors $|S, S\rangle\,, |S, S-1\rangle\,, \ldots, |S, 1-S\rangle\,, |S, -S\rangle$, which form the representation $D^{(S)}$ of the rotation group. At lowering the symmetry basis vectors of the irreducible representations Γ_γ of point symmetry groups can be determined by means of Clebsh–Gordan coefficients for point groups of symmetry [73]. The basis vectors of representations Γ_γ, obtained by such a way, are subjected to the action of the time-reversal operator $\mathbf{T}$. As a result, the matrices $\mathbf{U}$ in the bases of irreducible representations Γ_γ can be obtained.

(ii) We start directly from the basis function-operators of the irreducible representations Γ_γ, expressed in a general case as polynomials of the spin projections operators S_x, $S_y$and S_z. The form of matrices of the operators **U** is found as a result of action of the time-reversal operator $\mathbf{T} = \mathbf{UK}$ on basis function-operators of representations Γ_γ.

Using the both methods (i) and (ii) leads to the same results concerning the matrices of operators **U** for different irreducible representations of the point symmetry groups.

The matrices of unitary operators **U** for ordinary irreducible representations of 32 point groups of symmetry are presented in the Appendix A.

2.6 Wigner Criteria of Energy Levels Degeneracy Due to Time-Reversal Symmetry

In the Sect. 2.4, it was shown that the Schrödinger equation (2.76) for the state vector $|\psi'\rangle$ is consistent with (2.75), if **H** is invariant under time-reversal. Under stationary conditions the Schrödinger equation take the form

$$(\mathbf{H} - E)\,|\psi\rangle = 0. \tag{2.125}$$

At real Hamiltonian when $\mathbf{H}^* = \mathbf{THT}^{-1} = \mathbf{H}$ the state vectors $|\psi\rangle$ and $\mathbf{T}\,|\psi\rangle$ correspond to the same energy level, as well as functions $|\psi\rangle$ and $g\,|\psi\rangle$, where g is one of the elements of the Hamiltonian symmetry group G. Therefore, the time-reversal operator **T** can be considered as a new element of symmetry.

It is easy to show that the operator **T** commutes with all symmetry transformations of the space group G. Since $|\psi\rangle$ and $\mathbf{T}\,|\psi\rangle$ are transformed under complex-conjugated representations D and D^* of this group, we can write

$$\begin{aligned} g\,|\psi_i\rangle &= \sum_j D_{ji}(g)\,|\psi_j\rangle, \\ g(\mathbf{T}\,|\psi_i\rangle) &= \sum_j D^*_{ji}(g)\,(\mathbf{T}\,|\psi_j\rangle)\,. \end{aligned} \tag{2.126}$$

On the other hand,

$$\mathbf{T}g\,|\psi_i\rangle = \mathbf{T}\left(\sum_j D_{ji}(g)\,|\psi_j\rangle\right) = \sum_j D^*_{ji}(g)(\mathbf{T}\,|\psi_j\rangle),$$

that is

$$\mathbf{T}g = g\mathbf{T}. \tag{2.127}$$

The state vectors $|\psi\rangle$ and $\mathbf{T}|\psi\rangle$, satisfying the same (2.125) and corresponding to the same energy level may be linearly independent or linearly dependent. In the first case, two orthonormed state vectors $|\psi\rangle$ and $\mathbf{T}|\psi\rangle$ correspond to one eigenvalue of energy E. In the second case, two state vectors $|\psi\rangle$ and $\mathbf{T}|\psi\rangle$ are expressed each by other using the unitary matrix C:

$$\mathbf{T}|\psi_i\rangle = \sum_j C_{ji}|\psi_j\rangle. \tag{2.128}$$

We can see that in the last case the representations D and D^* are equivalent.

In the case, when the state vectors $|\psi\rangle$ and $\mathbf{T}|\psi\rangle$ are linearly independent, they may be transformed either by equivalent representations or by not equivalent ones. We have seen (see (2.85)) that, for "odd" systems, the arbitrary state vector $|\psi\rangle$ is orthogonal to the state vector $\mathbf{T}|\psi\rangle$, i.e., there is a double degeneracy of energy levels due to existence of time-reversal symmetry[4] (the Kramers theorem). The presence of this degeneracy can be shown in other way. It follows from (2.77) that if $\mathbf{H}|\psi\rangle = E\psi$, then $\mathbf{HT}|\psi\rangle = E\mathbf{T}|\psi\rangle$. Therefore, $|\psi\rangle$ and $\mathbf{T}|\psi\rangle$ belong to one and the same energy level. In order for this to lead to degeneracy, it is necessary to show that these state vectors are linearly independent. We suppose the contrary, that is we consider that

$$\mathbf{T}|\psi\rangle = c|\psi\rangle, \tag{2.129}$$

where c is a constant. Then

$$\mathbf{T}^2|\psi\rangle = \mathbf{T}c|\psi\rangle = c^*\mathbf{T}|\psi\rangle = c^*c|\psi\rangle. \tag{2.130}$$

For systems with an odd number of electrons this leads to a contradiction with $\mathbf{T}^2|\psi\rangle = -|\psi\rangle$ for "odd" systems, since c^*c positive and can not be equal to -1. Thus, the supposition (2.129) is false, and $|\psi\rangle$ and $\mathbf{T}|\psi\rangle$ are linearly independent. Since $\mathbf{T}^2|\psi\rangle = -|\psi\rangle$, the multiplicity of degeneracy of every energy level is even according to the Kramers theorem.

It is easy to show that if the Hamiltonian of the system is invariant under time-reversal, then acting with the transformation $\mathbf{T}$ on some degenerated system of eigen state vectors, we obtain another system of eigen state vectors with the same degree of degeneracy. Let the initial system of state vectors is transformed by the representation of symmetry group of the Hamiltonian. Then the new system will be transformed by the complex-conjugated representation D^*. A question arises, whether the new system of state vectors will be linearly independent of the old one. If the new system of state vectors and the old one will be linearly dependent, then none new degeneracy of energy levels will appear. In the opposite case the time-reversal symmetry leads to the appearance of a new degeneracy of energy levels.

[4]In the absence of a magnetic field.

There are three cases[5] [9, 74–78, 222]:

(a) The representation D is real or can be reduced to a real one with the help of an equivalent transformation of basis vectors.

(b) The representations D and D^* are nonequivalent.

(c) The representations D and D^* are equivalent, but can not be reduced to a pure real form with the help of equivalent transformation of basis vectors.

Further we shall subdivide these cases and consider separately systems with even and odd numbers of electrons. In addition, we shall consider that the Hamiltonian is invariant under time-reversal and shall determine, in what cases this leads to the same supplementary degeneracy of energy levels in comparison with that which follows only from the symmetry of the group G. Let us first consider a system with an even number of electrons.

Case (a). Let state vectors $|\psi_i\rangle$ be chosen in such a way that representation D is real. Then

$$\mathbf{RT}\,|\psi_i\rangle = \sum_j D^*_{ji}(\mathbf{R})\mathbf{T}\,|\psi_j\rangle = \sum_j D_{ji}(\mathbf{R})\mathbf{T}\,|\psi_j\rangle, \tag{2.131}$$

and functions $|\varphi_i\rangle = |\psi_i\rangle + \mathbf{T}\,|\psi_i\rangle$ are also transformed by the representation D of the group G. From (2.110), the case of N even ($\mathbf{T}^2 = +1$), it follows $\mathbf{T}^2\,|\varphi_i\rangle = |\varphi_i\rangle$, so that the system $|\varphi_i\rangle$ passes self into itself under the action of all symmetry transformations, that is under action of transformations of the group G and the time-reversal operator. Thus, there are no symmetry transformations, connecting the system $|\varphi_i\rangle$ with other state vectors and therefore the additional degeneration of energy levels is absent.

Case (b). The state vectors $\mathbf{T}\,|\psi_i\rangle$ are transformed under the representation D^*, which is not equivalent to the representation D, so that $|\psi_i\rangle$ and $\mathbf{T}\,|\psi_i\rangle$ are linearly independent. Thus, representations D and D^* always meet each other simultaneously. Since the Hamiltonian $\mathbf{H}$ is invariant with respect to $\mathbf{T}$, the state vectors $|\psi_i\rangle$ and $\mathbf{T}\,|\psi_i\rangle$ belong to the same energy level, and we have a supplementary degeneracy. If there is a symmetry relative to time-reversal, then such two representations are always degenerated and behave as one double degenerated irreducible representation. For this reason such a pair is joined by brackets and is denoted by an unique symbol.

Case (c). In this case the time-reversal symmetry leads to a supplementary degeneracy of energy levels (doubling of degeneracy). It is possible to show that $\mathbf{T}\,\left|\psi_j\right\rangle$ and $|\psi_i\rangle$ can not be linearly independent for all j, since in the opposite case it would be possible in the vector space $(\left|\psi_1\right\rangle, \ldots, |\psi_n\rangle)$ to form functions $\left|\varphi_j\right\rangle = \left|\psi_j\right\rangle + \mathbf{T}\left|\psi_j\right\rangle$, satisfying the condition $\mathbf{T}\left|\varphi_j\right\rangle = \left|\varphi_j\right\rangle$, leading to the *case (a)*. Thus, at least one of the state vector $\mathbf{T}\left|\psi_j\right\rangle$ is not a linear combination of state vectors $|\psi_i\rangle$. Further, using this state vector, on the basis of standard methods it is possible to determine

[5] The indicated classification of cases (a), (b) and (c) is rather different from that introduced in [78], for which (a) it is related to the case when functions $|\psi\rangle$ and $\mathbf{T}\,|\psi\rangle$ are linearly dependent, (b) it is referred the case when functions $|\psi\rangle$ and $\mathbf{T}\,|\psi\rangle$ are linearly independent and are transformed by nonequivalent representations, and (c) is related to the case when $|\psi\rangle$ and $\mathbf{T}\,|\psi\rangle$ are linearly independent and are transformed by equivalent representations. For ordinary representations both classifications coincide, but for spinor representations the cases (a) and (c) change places.

all the other state vectors of the representation D^*, which is equivalent to D. In such a way we can construct a complete system of basis state vectors orthogonal to state vectors $|\psi_i\rangle$.

Consider now a system with an odd number of electrons. In contrast to the previous case, for which the relation $\mathbf{T}^2 = +\mathbf{1}$ was used for N *even*, now it is necessary to use the relationship $\mathbf{T}^2 = -\mathbf{1}$ (or $\mathbf{T}^2 \left|\psi_j\right\rangle = -\left|\psi_j\right\rangle$) for N *odd*.

Case (a). According to (2.126), the state vectors $|\psi_i\rangle$ and $\mathbf{T}|\psi_i\rangle$ are transformed again under representations D and D^* of the symmetry group G, respectively. As in the proof of the Kramers theorem (see (2.129)), the relationship $\mathbf{T}|\psi_i\rangle = c|\psi_i\rangle$ can not be valid. However, it is the unique possible if $|\psi_i\rangle$ and $\mathbf{T}|\psi_i\rangle$ belong to the same irreducible vector space. Therefore, the representation D enters twice and there is an additional doubling of the degeneracy of energy levels due to time-reversal symmetry.

Case (b). Since we have not used the relationship (2.110), the situation is the same for even, as well as for odd number of electrons. Thus, the state vector $\mathbf{T}|\psi_i\rangle$ is transformed under the representation D^* of the group G, which is not equivalent to the representation D, so that $|\psi_i\rangle$ and $\mathbf{T}|\psi_i\rangle$ are linearly independent and belong to the same energy level and representations D and D^* always meet each other simultaneously. The presence of the time-reversal symmetry leads to a supplementary degeneracy of energy levels as a consequence of the fact that such two representations are always degenerated and behave themselves as one twice degenerated irreducible representation.

Case (c). In this case it is possible to show that for an odd number of electrons it is always possible to choose basis vectors $|\psi_i\rangle$ so that $\mathbf{T}\left|\psi_j\right\rangle$ will be a linear combination of $|\psi_i\rangle$ [68, 75]. Hence, the supplementary degeneracy of electron levels due to the time-reversal symmetry is absent.

Thus, the obtained results for the three cases (a), (b), and (c) are essentially different and also essentially depend on the fact, what is (even or odd) the number of electrons and, in a more general case, the number of particles with spin $\frac{1}{2}$ in the system. These results may be stated shortly in the following form:

1. At an even number of electrons or neglecting the spin:
   (a) There is no supplementary degeneracy.
   (b) The representations D and D^* are realized simultaneously and between them a supplementary degeneracy appears.
   (c) There is a supplementary degeneracy.
2. At an odd number of electrons (with spin):
   (a) There is a supplementary degeneracy.
   (b) Representations D and D^* are realized simultaneously and between them a supplementary degeneracy appears.
   (c) There is no supplementary degeneracy.

There exists a simple method, permitting to determine to what case from the three ones is related the given representation. In particular, this is related to cases (a) and (c), since the case (b) may be established directly by the table of characters. We give without proof the following criteria, which depends only on group characters:

$$\sum_R \chi(R^2) = \begin{cases} h & - \text{ the case } (a), \\ 0 & - \text{ the case } (b), \\ -h & - \text{ the case } (c). \end{cases} \tag{2.132}$$

Here R enumerates all elements of the Hamiltonian symmetry group G, χ is the character of the element R^2 in a considered irreducible representation and h is the order of the group. Such a classification of irreducible representations of the group G taking into account the invariance of the Hamiltonian **H** under time-reversal operator was first fulfilled by Wigner [222]. The results obtained by Wigner were based on the theorem, proved for the first time by Frobenius and Schur [74].

Frobenius–Schur Theorem. *If D is any irreducible representation of a finite group G of the order N and if $\chi_D(g)$ is the character of the element g in this representation D, then*

$$\sum_{g\in G} \chi_D(g^2) = \begin{cases} N, & \text{if } D \text{ is equivalent to the representation realized} \\ & \text{by real matrices;} \\ 0, & \text{if representations } D \text{ and } D^* \text{ are nonequivalent;} \\ -N, & \text{if } D \text{ is equivalent to } D^*, \text{ but is equivalent to none} \\ & \text{representation, constructed entirely from real matrices.} \end{cases} \tag{2.133}$$

Summation in (2.132) is carried on by all elements of the group G. In (2.132), the first, second, and third possibilities from (2.133) are called case (a), (b), and (c), respectively.

The proof of the Frobenius–Schur theorem consists in the fact that

$$\sum_{g\in G} \chi_D(g^2) = \sum_{i,j} \sum_{g\in G} D_{ij}(g) D_{ji}(g) = \sum_{i,j} \sum_{g\in G} D_{ij}(g) D^*_{ij}(g^{-1}). \tag{2.134}$$

If the representations D and D^* are non-equivalent, then the sum from (2.134) vanishes by virtue of the main theorem of representation theory concerning the relation of orthogonality for every two irreducible representations D_q and D_p

$$\sum_{g\in G} D^{(q)}_{ij}(g) D^{(p)^{-1}}_{kl}(g) = \frac{N}{d_p} \delta_{il} \delta_{jk} \delta'_{qp}, \tag{2.135}$$

where d_p is the dimension of the representation D_p. Symbol δ'_{qp} means that $\delta'_{qp} = 0$ if the representation D_q is nonequivalent to D_p and $\delta'_{qp} = 1$ for $D_q \equiv D_p$ [76]. But if representations D and D^* are equivalent, that is

$$\mathbf{D}^* = \mathbf{S}^{-1} \mathbf{D} \mathbf{S}, \tag{2.136}$$

where the operator **S** satisfies relationships[6]

$$\mathbf{SS}^* = \pm\mathbf{1}, \tag{2.137}$$

then according to (2.132) we have again

$$\sum_{g\in G}\chi_D(g^2) = \sum_{i,j}\sum_{g\in G} D_{ij}(g)S^{-1}_{im}D_{mn}(g^{-1})S_{nj} = \frac{N}{d}\sum_{i,j} S^{-1}_{ij}S_{ij} = \tag{2.138}$$

$$= \frac{N}{d}\sum_{i,j} S^*_{ji}S_{ij} = \frac{N}{d}Sp(\mathbf{S}^*\mathbf{S}) = \pm N,$$

where d is the dimension of the representation D. The upper (lower) sign before N corresponds to the upper (lower) sign in (2.137).

2.7 Herring Criteria for Energy Bands Degeneracy Due to Time-Reversal Symmetry

Frobenius–Schur theorem is also the basis for elucidation the possibility of existence of additional degeneracy of state vectors due to time-reversal symmetry in the case of space groups. Such an analysis was first fulfilled by Herring [79]. The nontriviality of applying the Frobenius–Schur theorem to a crystal consists in the fact that the continuous translation group, which is an invariant subgroup of the space group, contains an infinite number of elements, while formulas (2.133) are deduced for finite groups. Every space group G contains as an invariant subgroup the group of translations $\mathcal{T}$, consisting from all translations of the form

$$\mathbf{R} = \sum_{i=1}^{3} n_i\mathbf{t}_i, \tag{2.139}$$

where $\mathbf{t}_1, \mathbf{t}_2, \mathbf{t}_3$, are three main translation vectors, directed along crystallographic axes and n_1, n_2, n_3 a set of integer positive and negative numbers. If the crystallographic model is given by three of its dimensions $N_1\mathbf{t}_1$, $N_2\mathbf{t}_2$ and $N_3\mathbf{t}_3$, where N_1, N_2, N_3 are big numbers, then Born–Carman boundary cyclic conditions can be written in the form

$$\mathbf{R}_{N_100} = \mathbf{R}_{0N_20} = \mathbf{R}_{00N_3} = \mathbf{R}_{000} \tag{2.140}$$
$$(\mathbf{R}_{N_100} = N_1\mathbf{t}_1,\ \mathbf{R}_{0N_20} = N_2\mathbf{t}_2,\ \mathbf{R}_{00N_3} = N_3\mathbf{t}_3,\ \mathbf{R}_{000} = 0).$$

---

[6]Relation $\mathbf{SS}^* = \mathbf{1}$ corresponds to case (a), when representation D is equivalent to the real representation D_r ($D \sim D^* \sim D_r$), while the relation $\mathbf{SS}^* = -\mathbf{1}$ corresponds to the case (c), when a real representation equivalent to D does not exist [76].

On the basis of this, the group of translations can be changed by a finite group, consisting of $N_1N_2N_3$ elements. All the irreducible representations of this group are unidimensional with Block wave functions $\psi_{\mathbf{k}}(\mathbf{r}) = u_{\mathbf{k}}(\mathbf{r})\exp(i\mathbf{kr})$, where $u_{\mathbf{k}}(\mathbf{r})$ is the periodical part of the Block function, as basis vectors:

$$\mathbf{R}\psi_{\mathbf{k}}(\mathbf{r}) = \psi_{\mathbf{k}}(\mathbf{r}+\mathbf{R}) = e^{i\mathbf{kR}}\psi_{\mathbf{k}}(\mathbf{r}). \tag{2.141}$$

Therefore each irreducible representation of the finite Abelian group of $N_1N_2N_3$ dimension, replacing the translation group, is determined by the wave vector **k**. However, this determination is not unique, firstly, because in the **k** space there is a whole lattice (inverse lattice) of vectors

$$\mathbf{K} = \sum_{i=1}^{3} m_i \mathbf{b}_i, \tag{2.142}$$

where $\mathbf{b}_i$ $(i = 1, 2, 3)$ are the basis vectors of the inverse lattice and m_1, m_2, m_3 are integer positive or negative numbers, which satisfy the relationship

$$\exp(i\mathbf{KR}) = 1 \tag{2.143}$$

for all **R**. Since for all **R** the relation

$$\exp(i\mathbf{kR}) = \exp\{i(\mathbf{k}+\mathbf{K})\mathbf{R}\}, \tag{2.144}$$

takes peace, the vectors **k** and $\mathbf{k}+\mathbf{K}$ (where **K** is the vector of the inverse lattice) belong to one and the same irreducible representation of the finite group of N_1, N_2, N_3 dimension. Hence, in order to determine the representations uniquely, it is necessary to restrict **k** by same range, which is an elementary cell of inverse lattice (the reduced Brillouin zone). The second restriction on **k** is imposed by the boundary cyclic conditions (2.140) and is reduced to the fact that wave vector **k** can be represented in the form [9]

$$\mathbf{k} = \sum_{i=1}^{3} \frac{q_i}{N_i} \mathbf{b}_i, \tag{2.145}$$

where q_1, q_2 and q_3 are integer numbers. This formula determines the complete number $N_1N_2N_3$ of different irreducible representations in the reduced Brillouin zone. $N_1N_2N_3$ values of the vector **k** are uniformly distributed across the whole zone. At such approach some uncertainty still remains concerning vectors **k** lying on the surface of the zone. Since the Brillouin zone is an unit cell of inverse lattice, the distance between opposite faces is always equal to the length of a vector of inverse lattice.

On Fig. 2.1 the Brillouin zone of a square lattice with indicated singular points of six types is shown. The wave vector $\mathbf{k}_\Gamma = 0$ corresponds to the center of Brillouin zone. Vectors $\mathbf{k}_Z$ and $\mathbf{k}_{Z'}$ differ by a vector of inverse lattice $\mathbf{K}_1$: $\mathbf{k}_{Z'} = \mathbf{k}_Z + \mathbf{K}_1$.

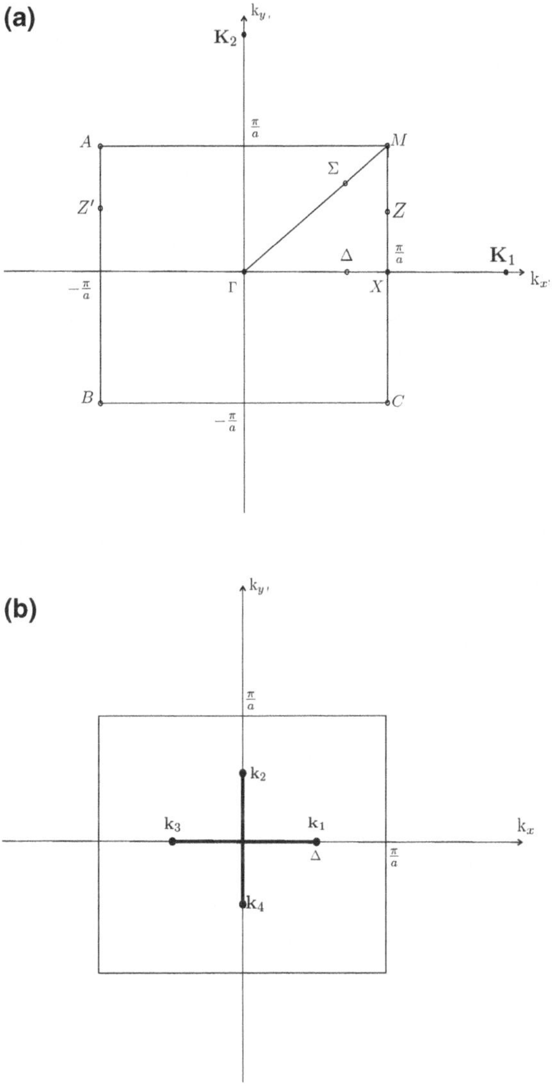

Fig. 2.1 The Brillouin zone of a square lattice with plotted singular points of six types: **a** singular points Γ, Δ, X, Z, Σ, M; **b** star of vector $\mathbf{k}_\Delta$

These vectors determine the same representation. Similarly there is also another representation, which is denoted by one of the four vectors $\mathbf{k}_A$, $\mathbf{k}_B$, $\mathbf{k}_C$ and $\mathbf{k}_M$.

The singular points of the Brillouin zone differ in that for them two or more vectors $\mathbf{k}_i$ are identical. They may be found identical, if they (1) lie on the axis of rotation or on the plane of symmetry, (2) are separated from each other by a vector of inverse lattice and (3) conditions (1) and (2) are fulfilled simultaneously. According to this classification, the points Γ, Δ and Σ are related to the type (1), Z – to the type (2), and X and M – to the type (3).

In order to determine the presence or absence of the additional degeneracy of energy bands of crystals due to time-reversal symmetry it is necessary some information about the space groups of symmetry and groups of the wave vector. From physical point of view, the space group is formed from all the symmetry transformations, as a result of which the crystal is superposed self by itself. This corresponds to the rigorous mathematical definition, according to which the space group is called a discrete subgroup of the real affine group (consisting of various translations and orthogonal transformations of the three-dimensional space), of which translations form its invariant subgroup $\mathcal{T}$ and have the form $\{\boldsymbol{\varepsilon} \mid \mathbf{R}\}$, where $\boldsymbol{\varepsilon}$ is the identity transformation of the space and $\mathbf{R}$ is determined by (2.139). A parallelepiped constructed on the basis vectors t_1, t_2 and t_3 (2.139) is called the unit cell of the crystal. The crystal lattice can be considered as constructed from identical cells. The vertices of parallelepipeds forming each unit cell, which are the ends of the vectors $\mathbf{R}$, are called Bravais nodes. The lattice formed by the Bravais nodes is called the Bravais lattice, which corresponds to the space group of the crystal under consideration. Nodes of this lattice do not necessary coincide with the positions of equilibrium of oscillating particles. Moreover, the number of equilibrium positions of particles in the general case is greater than the number of nodes of Bravais lattice. Let $\{\boldsymbol{\alpha} \mid \mathbf{R}\}$ be an arbitrary element of the space group G ($\boldsymbol{\alpha}$ is an orthogonal transformation). The point transformation $\boldsymbol{\alpha}$ of the element $\{\boldsymbol{\alpha} \mid \mathbf{R}\}$ of the space group form themselves a group, which is called the point group G_0. One of the important property of space groups is that the space group lattice remains invariant relative to all elements of the point group, corresponding to a given space group.

In [76], it is shown that between two wave vectors $\mathbf{k}$ and $\mathbf{k}'$, of which values lie in the first (irreducible) Brillouin zone and which characterize representations $D^{(i)}$ and $D^{(i')}$ of the space group of symmetry G, the following relationship takes place:

$$\mathbf{k}' = \boldsymbol{\alpha}\mathbf{k} + \mathbf{K}. \tag{2.146}$$

Here the superscript i indicates that $D^{(i)}$ is an induced representation obtained on the basis of a representation of the wave vector group $G_{\mathbf{k}}$, the definition of which is given below (similarly for $D^{(i')}$); $\mathbf{K}$ is an arbitrary vector of the inverse lattice. We remark that though vector $\mathbf{k}$ lyes in the Brillouin zone, it may turn out that $\boldsymbol{\alpha}\mathbf{k}$ does not yet lye inside it. So, in order for $\mathbf{k}'$ lyes in the Brillouin zone, the condition $\mathbf{K} \neq 0$ must be fulfilled, since only adding a nonzero vector of the inverse lattice will pass again the vector $\boldsymbol{\alpha}\mathbf{k}$ inside the Brillouin zone. The totality of vectors $\mathbf{k}'$, obtained on the basis of (2.146), when $\boldsymbol{\alpha}$ ranges over all elements of the point group is called the *star* of the vector $\mathbf{k}$. The number of different "rays" of the star is called the order of the star. The same proper value of energy corresponds to all members of star. The order of the star is not greater the h (h is the order of the point group). If the star order equals to h, the star is of a general type. Than the position of the vectors of the star is more symmetric, so its order and the number of its rays is smaller. Particularly, the star of the vector $\mathbf{k} = 0$ consists of only one ray. On Fig. 2.1b the star of the vector $\mathbf{k}_\Delta$ is presented. Δ represents the point of type (1).

Now, it is possible to construct a small group, consisting of elements $\{\boldsymbol{\alpha} \mid \mathbf{R}\}$, for which

$$\alpha\mathbf{k} = \mathbf{k} + \mathbf{K}. \tag{2.147}$$

Here the vector $\mathbf{K}$ is not zero only for vectors $\mathbf{k}$, lying on the surface of the Brillouin zone, and it serves to return vectors $\boldsymbol{\alpha}\mathbf{k}$, if they are out of the Brillouin zone. Such a small group is called the group of wave vector and it is denoted by $G_{\mathbf{k}}$. This group consists of elements $\{\boldsymbol{\alpha} \mid \mathbf{R}\}$ of the space group, the point transformations $\boldsymbol{\alpha}$, which pass star rays of the vector $\mathbf{k}$ into itself (or if $\mathbf{K} \neq 0$, into an equivalent position of a given ray).

The collection of different vectors, which are part of the star of the vector $\mathbf{k}$ corresponds to basis functions of representation $D^{(i)}$ of the space group G. Since this collection includes all the set of vectors $\mathbf{k}$, vectors $\boldsymbol{\alpha}\mathbf{k}$ coincides with one of the vectors $\mathbf{k}_i$ for all elements $g \in G$. Let take any vector $\mathbf{k}_{\mathrm{l}}$ of the star of representation and consider all possible vectors $\boldsymbol{\alpha}\mathbf{k}_{\mathrm{l}}$. All these vectors enter into the star (among them may be also equal). There are only two possibilities [78]:

(i) The set of vectors $\boldsymbol{\alpha}\mathbf{k}_{\mathrm{l}}$ exhaust all stars of the representation $D^{(i)}$. Such a star is called reducible. Each irreducible star is determined by some vector $\mathbf{k}_{\mathrm{l}}$. The other wave vectors are obtained as a result of applying the transformations of the group G. The irreducible star that is characterized by a single vector $\mathbf{k}_{\mathrm{l}}$ is denoted.

(ii) The set $\boldsymbol{\alpha}\mathbf{k}_{\mathrm{l}}$ does not exhaust all the stars. Such a star is called reducible. It is clear that such a reducible star may be decomposed into irreducible stars.

In the study of irreducible representations only irreducible stars are essential. It may be proved that every irreducible representation possesses only an irreducible star [78]. It should be noted that the converse assertion is not true, because in the center of the Brillouin zone a irreducible star consisting of a single vector $\mathbf{k} = 0$ corresponds to every representation, including a reducible one as well. We also mention that if the vector $\mathbf{k}_{\mathrm{l}}$ occupies a general position inside the Brillouin zone, then the star $\{\mathbf{k}_{\mathrm{l}}\}$ is irreducible and the respective representation of the space group G is also irreducible.

In the group $G_{\mathbf{k}}$ of the wave vector, as well as in the space group G, there is an invariant subgroup of translations. The quotient group on this subgroup is isomorphic to the point group, which includes all the rotary elements $\boldsymbol{\alpha} \in G_{0\mathbf{k}}$, which either do not change $\mathbf{k}$ or passes the vector $\mathbf{k}$ into an equivalent vector. The point group $G_{0\mathbf{k}}$ is determined by a vector $\mathbf{k}$ and is a subgroup of the group G_0, which characterizes the crystallographic class. In the case of point $\mathbf{k} = 0$ in the center of the Brillouin zone $G_{0\mathbf{k}} = G_0$ and in this case $G_{\mathbf{k}} = G$. Points groups $G_{0\mathbf{k}}$ for singular points $\Gamma, \Delta, X, Z, \Sigma$ and M on Fig. 2.1a are the groups $4mm$, $4mm$, $2mm$, $4mm$, $4mm$ and $4mm$, respectively (here the international notations for point symmetry groups are used).

The representation $D_{\mathbf{k}}$ of the wave vector group $G_{\mathbf{k}}$, which is also called the small representation, determines uniquely the representation $D^{(i)}$ of the space group with irreducible star $\{\mathbf{k}\}$. Moreover, if the representation $D_{\mathbf{k}}$ is irreducible, then representation $D^{(i)}$ of space group is irreducible as well. Similarly, if the representation $D_{\mathbf{k}}$ is unitary, then it is the same for representation $D^{(i)}$ of a space group.

The problem of construction of irreducible representations of the space group is reduced to the construction of irreducible representations of small groups $G_{\mathbf{k}}$. The admissible irreducible representations of the group $G_{\mathbf{k}}$ are expressed by irreducible representations of the point group $G_{0\mathbf{k}}$ of the wave vector **k**.

In addition, if it is fulfilled at least one of the following two conditions:

(1) vector **k** lies strictly inside the Brillouin zone;

(2) $G_{\mathbf{k}}$ is a symmorphic group (the elements of $G_{\mathbf{k}}$ are only admissible translations), then the number of nonequivalent admissible irreducible representations $D_{\mathbf{k}}$ of the group $G_{\mathbf{k}}$ is equal to the number of nonequivalent irreducible representations $D_{0\mathbf{k}}$ of the point group $G_{0\mathbf{k}}$ of the wave vector **k**.

Between the matrices of representations $D_{\mathbf{k}}$ and $D_{0\mathbf{k}}$ there exists the relation [76]

$$\mathbf{D}_{\mathbf{k}}(\{\boldsymbol{\beta}|\,\mathbf{b}\}) = e^{-i\mathbf{k}\mathbf{b}}\mathbf{D}_{0\mathbf{k}}(\boldsymbol{\beta}), \tag{2.148}$$

where $\{\boldsymbol{\beta}|\,\mathbf{b}\}$ is an arbitrary element of the group $G_{\mathbf{k}}$ ($\boldsymbol{\beta}$ is an orthogonal transformation and $\mathbf{b} = \mathbf{v}_\beta + \mathbf{R}$, $\mathbf{v}_\beta$ is the vector of a non-elementary translation) and $\boldsymbol{\beta}$ in the right side of the (2.148) is the element of the point group $G_{0\mathbf{k}}$. Under the made suppositions (1) and (2) it is possible to prove that matrices $\mathbf{D}_{0\mathbf{k}}(\boldsymbol{\beta})$ and $\mathbf{D}_{\mathbf{k}}(\{\boldsymbol{\beta}|\,\mathbf{b}\})$ really form an irreducible representations of the point group $G_{0\mathbf{k}}$ and the wave vector group $G_{\mathbf{k}}$, respectively. This statement is valid as a consequence of the fact that the relation

$$\mathbf{D}_{\mathbf{k}}(\{\boldsymbol{\beta}|\mathbf{b}\})\mathbf{D}_{\mathbf{k}}(\{\boldsymbol{\beta}'|\mathbf{b}'\}) = \mathbf{D}_{\mathbf{k}}(\{\boldsymbol{\beta}|\mathbf{b}\}\{\boldsymbol{\beta}'|\mathbf{b}'\}) \tag{2.149}$$

is fulfilled for all $\boldsymbol{\beta}, \boldsymbol{\beta}' \in G_{0\mathbf{k}}$ [76].

Since the point symmetry groups have not high orders, using (2.148) it is easy to find irreducible representations of the wave vector group $G_{\mathbf{k}}$, and on its basis to determine irreducible representations of space groups. Difficulties appear only for the non-symmorphic space group in the case when the star vector get into the surface of the Brillouin zone. If in the wave vector group there are non-trivial screw axes or planes of slip planes and the wave vector **k** lies on the frontier of the Brillouin zone, then relations (2.149) are not fulfilled and it is necessary to construct the so-called loaded [7] or that is the same projective [78] representations instead of usual vectorial irreducible representations of the space groups.

The projective representations were first introduced by Schur, which developed the general theory of projective representations and worked out methods of finding the projective representations of finite groups [80] (see also monographs [81, 82]). All irreducible representations (vectorial and projective) of the space group G can be obtained, taking one ray **k** from every star and inducing irreducible representations of group G from all non-equivalent admissible irreducible representations $D_{\mathbf{k}}$ of the wave vector group $G_{\mathbf{k}}$. If m is the dimension of the representation $D_{\mathbf{k}}$ and s is the order of star $\{\mathbf{k}\}$, then the dimension of the representation $D^{(i)}$ of the space group is equal to ms. Herring [79] investigated conditions (2.133), obtained by Frobenius and Schur for finite groups [74], in connection with space groups. In this case, these conditions take the form:

$$\sum_{Q_0} \chi(Q_0^2) = n_{\mathbf{k}} \quad \text{in case (a)}, \tag{2.150}$$
$$\sum_{Q_0} \chi(Q_0^2) = 0 \quad \text{in case (b)},$$
$$\sum_{Q_0} \chi(Q_0^2) = -n_{\mathbf{k}} \text{ in case (c)}.$$

Here it is denoted by Q_0 the element of the space group G, which passes $\mathbf{k}$ in $-\mathbf{k}$, and by $n_{\mathbf{k}}$ – the order of the group of wave vector $\mathbf{k}$. That is why the transformation Q_0^2 does not change the vector $\mathbf{k}$ and it is an element of the wave vector group $G_{\mathbf{k}}$. As χ in (2.150), it means the character of the element Q_0^2 in the irreducible representation of the group $G_{\mathbf{k}}$ (but not G!).

The Herring criteria (2.150) are related to spinless systems, as well as to systems, for which spin is considered, since it is a mathematic property of group representations and does not depend on any its applications in quantum mechanics.

If the space group G contains the inversion $\mathbf{i}$ and $\mathbf{i}$ is the element of the group $G_{\mathbf{k}}$, then Q_0 is an element of the group $G_{\mathbf{k}}$. If $\mathbf{i}$ is also an element of the group G, but does not enter in $G_{\mathbf{k}}$, then Q_0 are elements of $\mathbf{i} \otimes G_{\mathbf{k}}$. That is why the criteria (2.150) are easy to apply, if the table of characters of the group $G_{\mathbf{k}}$ is known. Moreover, it is not necessary to make an implicit summation by all elements of the group, which are represented by identical matrices, but it is sufficient to take one element of each type.

Elliot [83] showed how to use the table of characters of the wave vector group $G_{\mathbf{k}}$ for the elucidation of question about the equivalence of a given irreducible representation of the space group to its complex conjugated representation. If it is to denote by χ and χ' characters of some irreducible representation of the group G and those of representation, which is complex conjugated to it, then between these characters there is the relation

$$\chi'(\{\boldsymbol{\alpha}| \boldsymbol{R}\}) = [\chi(\{\boldsymbol{\alpha}| \boldsymbol{R}\})]^* e^{ik\boldsymbol{R}}. \tag{2.151}$$

By denoting $\overline{\mathbf{k}} = -\mathbf{k} + \mathbf{K}$, one must distinguish three cases depending on the relation between $\mathbf{k}$ and $\overline{\mathbf{k}}$ [76, 83].

Case 1: $\overline{\mathbf{k}} = \mathbf{k}$. Such a relation can occur only at $2\mathbf{k} = \mathbf{K}$, particularly, for $\mathbf{k} = 0$. If there is an additional degeneracy due to time-reversal symmetry (according to Herring, the cases (b) and (c) without spin and the cases (a) and (b) with spin), then the irreducible representation $D_{\mathbf{k}}$ of the wave vector group $G_{\mathbf{k}}$ is not equivalent to the representation $D_{\mathbf{k}}^*$. In this case, every ray of the star is doubly degenerated (additional degeneracy of energy bands).

Case 2: $\overline{\mathbf{k}} \neq \mathbf{k}$, the star $\{\overline{\mathbf{k}}\}$ coincides with the star $\{\mathbf{k}\}$. If there is no additional degeneracy (according to Herring, the case (a) without spin and the case (c) with spin), then representations $D^{(i)}$ and $D^{(i)*}$ of the space group G are given in the same space of the representation.

If there is additional degeneracy due to time-reversal symmetry, then spaces corresponding to representations $D^{(i)}$ and $D^{(i)*}$ are different. Since stars of representations

$D^{(i)}$ and $D^{(i)*}$ coincide, then $D^{(i)*}$ may be obtained from the same ray **k**, as $D^{(i)}$. In this case, firstly $D^{(i)*}$ must be obtained beginning with the ray $\overline{\mathbf{k}}$. Thereupon with the help of transformation of similarity the representation $D^{(i)\prime}$ is obtained. This representation can be already construct beginning with the ray **k**. Functions corresponding to the ray $\overline{\mathbf{k}}$ are basic functions of the representation $D^*_{\mathbf{k}}$ of the wave vector group $G_{\mathbf{k}}$. For matrix $\mathbf{D}'_{\mathbf{k}}$ of the representation $D'_{\mathbf{k}}$ the equality

$$\mathbf{D}'_{\mathbf{k}}(g) = \mathbf{D}'_{\mathbf{k}}(\overline{g}g\overline{g}^{-1}) \quad (g \in G_{\overline{\mathbf{k}}} = G_{\mathbf{k}}) \tag{2.152}$$

takes place [76]. Since $G_{\overline{\mathbf{k}}} = G_{\mathbf{k}}$, then $D'_{\overline{\mathbf{k}}}$ is the representation of the group $G_{\mathbf{k}}$. In this case, the representation $D'_{\mathbf{k}}$ coincides with $D^*_{\mathbf{k}}$, so for the matrix of the representation $D'_{\mathbf{k}}$, from which $D^{(i)*}$ is obtained, we have

$$\mathbf{D}'_{\mathbf{k}}(g) = \mathbf{D}^*_{\mathbf{k}}(\overline{g}g\overline{g}^{-1}) \quad (g \in G_{\mathbf{k}}). \tag{2.153}$$

Thus, "small" representations $D_{\mathbf{k}}$ and $D'_{\mathbf{k}}$, on the basis of which representations $D^{(i)}$ and $D^{(i)\prime}$ of the space group G may be obtained, are connected each to other in a sufficiently complicated way, as it follows from the relation (2.152).

Case 3: $\overline{\mathbf{k}} \neq \mathbf{k}$, star $\{\overline{\mathbf{k}}\}$ does not coincide with the star $\{\mathbf{k}\}$. In this case, representations $D^{(i)}$ and $D^{(i)*}$ of the space group G belong to different stars, and they are not equivalent. The case (b) takes place, by Herring. Additional degeneracy consists in covering the energy bands due to time-reversal symmetry.

2.8 Corepresentations of a Symmetry Group

The time-reversal operator is an antiunitary operator. In spite of this it may be joined together with unitary operators g in a special group. Moreover, state vectors $|\psi\rangle$ and $\mathbf{T}\,|\psi\rangle$, if they are linearly independent, are also joined and form a representations of this group, called corepresentation, to remind about the sign of the complex conjugation in the basis relation, which determines the corepresentation (see below).

Corepresentations were first introduced by Wigner [222]. Properties of corepresentation differ from the properties of usual group representations. They were considered in details in [84, 222].

When the influence of the time-reversal symmetry on energy spectrum, rules of selections and other characteristics are considered, one may use at once these corepresentation, or begin with usual representations of space or point group and influence of the time-reversal symmetry is considered in addition. Certainly, both these approaches lead to the same results, however using the second method seems to be physically more visual [78]. In connection with this, below we consider very briefly bases of the method of the corepresentation theory without a detailed analysis.

The theory of representation of a group by linear transformations does not give a complete mathematical basis to consider a symmetry group, containing antiunitary operators. That is why it needs an extension. Before we pass to corepresentations, let consider briefly the method of introducing usual group representations, in order to see by comparison the distinction between corepresentations and representations. According to Wigner [222], the group, necessary to be considered in order to obtain consequences of the symmetry of some problem, is not the group of physical transformations, but the group of quantum mechanical operators, corresponding to these transformations.

By denoting elements of the symmetry group G of Schrödinger equation by $R, S, \ldots$, one may introduce linear operators $\mathbf{P}_{R,}\mathbf{P}_{S,}\ldots$ corresponding to these elements. In general case, when an eigenvalue E of the Hamiltonian with l linearly independent eigenfunctions $|\psi_1\rangle, |\psi_2\rangle, \ldots, |\psi_l\rangle$ are considered, these state vectors under the operator $\mathbf{P}$ are transformed under the representation D of the group G:

$$\mathbf{P}_R |\psi_\varkappa\rangle = \sum_{\varkappa=1}^{l} D(R)_{\varkappa\upsilon} |\psi_\varkappa\rangle . \tag{2.154}$$

If S belongs also to the symmetry group G, then we have

$$\mathbf{P}_S |\psi_\varkappa\rangle = \sum_{\lambda=1}^{l} D(S)_{\lambda\varkappa} |\psi_\lambda\rangle . \tag{2.155}$$

Further, one may show that

$$\mathbf{P}_S\mathbf{P}_R |\psi_\upsilon\rangle = \sum_{\lambda=1}^{l}\sum_{\varkappa=1}^{l} D(S)_{\lambda\varkappa} D(R)_{\varkappa\upsilon} |\psi_\lambda\rangle . \tag{2.156}$$

On the other hand,

$$\mathbf{P}_S\mathbf{P}_R |\psi_\upsilon\rangle = \mathbf{P}_{SR} |\psi_\upsilon\rangle = \sum_{\lambda=1}^{l} D(SR)_{\lambda\upsilon} |\psi_\lambda\rangle . \tag{2.157}$$

By comparing (2.156) and (2.157), we obtain

$$D(SR)_{\lambda\upsilon} = \sum_{\varkappa=1}^{l} D(S)_{\lambda\varkappa} D(R)_{\varkappa\upsilon}, \tag{2.158}$$

whence it follows that matrices $\mathbf{D}(S)$, $\mathbf{D}(R)$ and $\mathbf{D}(SR)$ satisfy the relation

$$\mathbf{D}(SR) = \mathbf{D}(S) \cdot \mathbf{D}(R). \tag{2.159}$$

These matrices form an irreducible representation of dimension l of the group G, relative to which the stationary Schrödinger equation $\mathbf{H}|\psi\rangle = E|\psi\rangle$ is invariant. In order to pass to corepresentations, we take into account that the time-reversal operator $\mathbf{T}$ commutes with operators $\mathbf{O}_R$ and $\mathbf{O}_U$, which correspond to rotations of the coordinate system:

$$\mathbf{O}_R\mathbf{T} = \mathbf{T}\mathbf{O}_R, \quad \mathbf{O}_U\mathbf{T} = \mathbf{T}\mathbf{O}_U, \tag{2.160}$$

where $\mathbf{U}$ is an element of the two-dimensional unimodular unitary transformation group.

We mention that if the number of electrons of the considered system is odd, then the group of quantum mechanical operators, which correspond to rotations, is isomorphic to the group of two-dimensional unitary unimodular transformations and is only homomorphic to rotation group. In this case, $\mathbf{U} = -\mathbf{1}$, not $\mathbf{U} = \mathbf{1}$ corresponds to operator $\mathbf{T}^2$. The complete group consists of unitary transformations $\mathbf{O}_U$ and antiunitary transformations $\mathbf{T}\mathbf{O}_U$ with multiplications rules [222]

$$\begin{aligned} \mathbf{O}_V \cdot \mathbf{O}_U &= \mathbf{O}_{VU}, \quad \mathbf{T}\mathbf{O}_V \cdot \mathbf{O}_U = \mathbf{T}\mathbf{O}_{VU}, \\ \mathbf{O}_V \cdot \mathbf{T}\mathbf{O}_U &= \mathbf{T}\mathbf{O}_{VU}, \quad \mathbf{T}\mathbf{O}_V \cdot \mathbf{T}\mathbf{O}_U = \mathbf{O}_{\pm VU}, \end{aligned} \tag{2.161}$$

where the upper sign in the last equality relates to even and the lower one to odd number of electrons. The multiplication rules (2.161) show that unitary operators form an invariant subgroup of index 2 and antiunitary operators form a coset of this subgroup.

Relations of type (2.154) and (2.155) remain valid also for antiunitary operators $\mathbf{T}\mathbf{O}_U$:

$$\mathbf{T}\mathbf{O}_U \left|\psi_\varkappa\right\rangle = \sum_\lambda D(\mathbf{T}\mathbf{O}_U)_{\lambda\varkappa} \left|\psi_\lambda\right\rangle . \tag{2.162}$$

It remains also in force the assertion about the fact that the matrix $\mathbf{D}(\mathbf{T}\mathbf{O}_U)$ is unitary, if state vectors $|\psi_\varkappa\rangle$ are orthogonal. This is a consequence of the fact that by virtue of (2.57) and (2.61), the relation

$$\langle \mathbf{T}\mathbf{O}_U\psi_\varkappa \mid \mathbf{T}\mathbf{O}_U\psi_\lambda\rangle = \langle \mathbf{O}_U\psi_\lambda \mid \mathbf{O}_U\psi_\varkappa\rangle = \langle \psi_\lambda \mid \psi_\varkappa\rangle = \delta_{\lambda\varkappa} \tag{2.163}$$

is fulfilled as well as for unitary operators.

We can see that the product of matrices $\mathbf{D}(\mathbf{T}\mathbf{O}_\mathbf{V})$ and $\mathbf{D}(\mathbf{O}_U)$ or $\mathbf{D}(\mathbf{T}\mathbf{O}_U)$ now will be not equal to $\mathbf{D}(\mathbf{T}\mathbf{O}_V\mathbf{O}_U) = \mathbf{D}(\mathbf{T}\mathbf{O}_{VU})$ or $\mathbf{D}(\mathbf{T}\mathbf{O}_V\mathbf{T}\mathbf{O}_U) = \mathbf{D}(\mathbf{O}_{\pm VU})$. In particular, if it is to apply $\mathbf{T}\mathbf{O}_V$ to (2.162), we obtain

$$\mathbf{T}\mathbf{O}_V\mathbf{T}\mathbf{O}_U\psi_\varkappa = \sum_\lambda D(\mathbf{T}\mathbf{O}_U)^*_{\lambda\varkappa}\mathbf{T}\mathbf{O}_V\psi_\lambda = \sum_{\lambda,\mu} D(\mathbf{T}\mathbf{O}_U)^*_{\lambda\varkappa} D(\mathbf{T}\mathbf{O}_V)_{\mu,\lambda}\psi_\mu, \tag{2.164}$$

whence it follows

$$\mathbf{D}(\mathbf{T}\mathbf{O}_V)\mathbf{D}(\mathbf{T}\mathbf{O}_U)^* = D(\mathbf{T}\mathbf{O}_V\mathbf{T}\mathbf{O}_U) = \mathbf{D}(\mathbf{O}_{\pm VU}). \tag{2.165}$$

In a similar way,

$$\mathbf{D}(\mathbf{T}\mathbf{O}_V)\mathbf{D}(\mathbf{O}_U)^* = D(\mathbf{T}\mathbf{O}_V\mathbf{O}_U) = \mathbf{D}(\mathbf{O}_{VU}). \tag{2.166}$$

Equations (2.165) and (2.166) show that matrices, which transform self-functions under group operations, do not form representations of this group, if the group contains antiunitary operators. The system of matrices **D**, satisfying equations (2.165) and (2.166), is not a representation of the group of unitary and antiunitary operators $\mathbf{O}_U$ and $\mathbf{TO}_U$ in a usual sense, but forms a corepresentation of this group. Wigner developed the method of reduction of corepresentations, that is a decomposition of reducible corepresentations to irreducible ones. However, due to the reasons mentioned at the beginning of this section, we will not consider this method, but confine ourselves to general concepts about corepresentations.

2.9 Time Reversal and Kramers Theorem Geometrical Interpretation

As it is known, the symmetry transformation connected with the time reversal is accepted to be called the time-reversal transformation. Under the time-reversal transformation the directions of motion of all particles, forming the system, are reversed. The system passes through the same set of states, in which it was successively the translational movement on time axis, but now just in the opposite direction.

In Chap. 1, we have noted and here we underline once more that a real reversal of the direction of movement of all particles of the system is not required for the existence of the time-reversal symmetry. For this, it is only required that such a reversal of the movement direction be possible in principle. In just the same way, it is not necessary to fulfill really the respective permutations of equivalent atoms in crystalline or molecular (claster) structures to make sure about the existence of one or other space or point symmetry group. It is sufficient only to make sure that such transformations are possible in principle. Sometimes the time-reversal transformation is called the time inversion by analogy with the space inversion, taking into account that at time inversion the point $t = 0$ on the time axis is fixed and is the center of inversion. For similar reasons this transformation of symmetry is called sometimes "the time reflection".

Using the group-theoretic approach, we shall show that the time-reversal transformation is an operation of antirotation in the space of basis state vectors. In a functional $(2S + 1)$-fold space, the state of the system with the total spin S corresponds to a point that, under the action of the time reversal operator, performs an antirotation around the axis passing through the coordinate system origin. In the simplest case of a particle with spin $S = \frac{1}{2}$ the antirotation axis of fourth order is orthogonal to the plane of a square with the vertices colored with two colors (red and blue, see Fig. 2.2). There exist a significant geometric difference between the systems of particles with half-integer total spin and a system of particles with integer spin. In the first case, the antirotation takes place by an angle of 90°, and the system returns to the initial state with the initial wave function after four consecutive antirotations by the same angle in the same direction. In the second case, the angle of antirotation

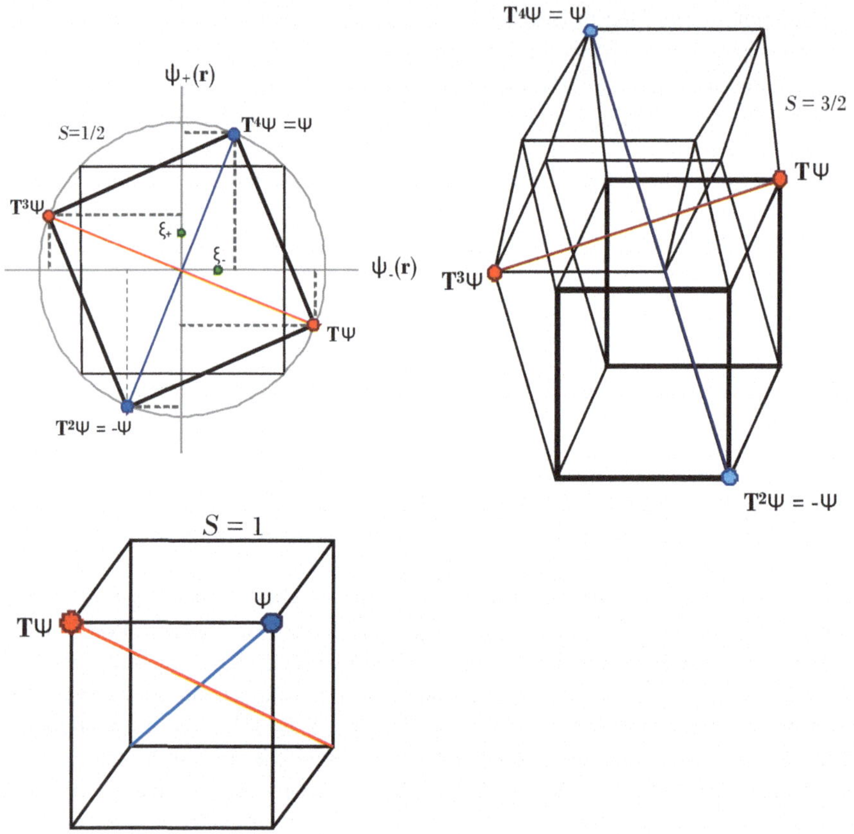

Fig. 2.2 The geometric interpretation of the wave function Ψ of systems with half-integer spins ($S = \frac{1}{2}, \frac{3}{2}$) and integer spin ($S = 1$) at time reversal in the absence of a magnetic field. The wave function is represented as an expansion in spinor basis vectors (in the case of a particle with spin $S = \frac{1}{2}$ these basis vectors are $\xi_+ = \binom{1}{0}$ and $\xi_- = \binom{0}{1}$). The quantum system state corresponds to a point at the vertex of the $(2S + 1)$-dimensional cube; blue bullet •—the state Ψ; red bullet ∘—the time-reversed state $\mathbf{T}\Psi$

is 180°, and the system is returned to the initial state after the second antirotation, that is opposite with respect to the first one.This is shown in Fig 2.2 for $S = \frac{1}{2}$, 1 and $\frac{3}{2}$. The operation of anti-identification corresponds to carrying out the complex conjugation of the wave function.

The different geometric behavior of the systems with integer and half-integer S at time-reversal is connected with the spin dependence of the shape of unitary part of the time-reversal operator [85]. Such a sharp difference in the behavior of spin systems with integer and half-integer spins under action of the time-reversal operator has a group-theoretic explanation. It consists in the fact that at an integer S (or an even number N of particles with spin $\frac{1}{2}$) there exists the Abelian group of the second order $2'$: $\{\mathbf{T}, \mathbf{T}^2 = \mathbf{1}\}$, while at a half-integer spin S (or an odd number N of particles with spin $\frac{1}{2}$) there exists the Abelian group of fourth order $4'$: $\{\mathbf{T}, \mathbf{T}^2= -\mathbf{1}, \mathbf{T}^3= -\mathbf{T}, \mathbf{T}^4= \mathbf{1}\}$.

Using a system of coordinate axes, we construct a $(2S + 1)$-dimensional cube, inscribed in a $(2S + 1)$-dimensional unit sphere centered at the origin, where the coordinate axes pass through the center of $2S$-dimensional faces. Then we rotate this cube so that the position of an arbitrary chosen vertex would correspond to a point on $(2S+1)$-dimensional sphere to correspond to representing the wave function $\Psi^{(s)}$ of the system. The position of this point is determined by the values of $2S+1$ functions $\psi_\sigma(\mathbf{r}_1, \mathbf{r}_2, \ldots, \mathbf{r}_N)$ playing the role of coefficients in the decomposition of N-particle wave function $\Psi^{(s)}$ in the basis spinors $\xi_\sigma^{(s)}$

$$\Psi^{(s)} = \sum_{\sigma=-S}^{S} \psi_\sigma(\mathbf{r}_1, \mathbf{r}_2, \ldots, \mathbf{r}_N)\xi_\sigma^{(s)} \tag{2.167}$$

that satisfies the normalization condition

$$\sum_{\sigma=-S}^{S} \int \left|\psi_\sigma(\mathbf{r}_1, \mathbf{r}_2, \ldots, \mathbf{r}_N)\right|^2 d\mathbf{r}_1 d\mathbf{r}_2 \ldots d\mathbf{r}_N = 1. \tag{2.168}$$

As a result of the time-reversal transformation, the vector of state, directed along one of the main diagonals of the cube, is antirotated with the transition to the other main diagonal. From the theorem of Kramers on additional degeneracy of energy levels, due to the symmetry of time reversal, the geometric theorem follows [85].

Theorem *In an n-dimensional Euclidean cube, there are no mutually orthogonal main diagonals if n is an odd number, and there exists a unique pair of such diagonals for even n, with an accuracy up to a group of motions of the n-dimensional cube.*

A point-behavior analysis, which describes the state of a system under consecutive time-reversal transformations, has allowed to clarify the geometric meaning of Kramers theorem on additional degeneracy energy levels due to time-reversal symmetry. It consists in the existence of a some pair of mutually orthogonal diagonals (with an accuracy up to a movement group specified in the Theorem) in the $(2S+1)$-dimensional Euclidean cube when S is half-integer. In this case, the wave functions Ψ and $\mathbf{T}\Psi$ are orthogonal (Fig. 2.2, $S = \frac{1}{2}, \frac{3}{2}$) that leads to an additional degeneracy of energy levels (Kramers degeneracy). If S is an integer number, then the state vectors Ψ and $\mathbf{T}\Psi$ are linear dependent (the absence of Kramers degeneracy) that was proved first by Wigner [68, 222]. The geometric interpretation of this result consists in the fact that at integer spin S none of main diagonals of the $(2S + 1)$-dimensional cube are orthogonal (Fig. 2.2, $S = 1$).

2.10 Non-conventional Time-Reversal Symmetry

The invariance of Schrödinger equation (2.74) and Hamiltonian **H** (2.77) under conventional time-reversal operator **T** take place if **H** does not contains the operator of interaction of the electron with the magnetic field. The Schrödinger equation (2.74), which corresponds to a non-relativistic connection between the energy E and the impulse **P** of a particle with the mass m

$$E = \frac{\mathbf{P}^2}{2m} + V(r), \tag{2.169}$$

may be obtained formally from (2.169) using transformations

$$E \to i\hbar\frac{\partial}{\partial t}, \quad \mathbf{P} \to -i\hbar\nabla. \tag{2.170}$$

If instead of (2.169), we start from the relativistic relation between the energy and the impulse (c is the light velocity in the vacuum)

$$\frac{E^2}{c^2} = \mathbf{P}^2 + m^2c^2 \tag{2.171}$$

and use the same transformations (2.170), then we come to the Klein–Gordon–Fock equation for a spinless particle [86–88]

$$i\hbar\frac{\partial\,|\Psi\rangle}{\partial t} = \mathbf{H}_f\,|\Psi\rangle\,, \tag{2.172}$$

which coincides by form with the Schrödinger equation with the Hamiltonian

$$\mathbf{H}_f = (\tau_3 + i\tau_2)\frac{\mathbf{P}^2}{2m} + mc^2\tau_3, \tag{2.173}$$

where

$$\tau_1 = \begin{pmatrix} 0 & 1 \\ 1 & 0 \end{pmatrix}, \quad \tau_2 = \begin{pmatrix} 0 & -i \\ i & 0 \end{pmatrix}, \quad \tau_3 = \begin{pmatrix} 1 & 0 \\ 0 & -1 \end{pmatrix}. \tag{2.174}$$

and the state vector $|\Psi\rangle$ is given by the one-column matrix

$$|\Psi\rangle = \begin{pmatrix} \varphi \\ \chi \end{pmatrix}, \tag{2.175}$$

of which component satisfies the following equation

$$\left.\begin{aligned} i\hbar\tfrac{\partial|\varphi\rangle}{\partial t} &= -\tfrac{\hbar^2}{2m}\nabla^2(|\varphi\rangle + |\chi\rangle) + mc^2\varphi \\ i\hbar\tfrac{\partial|\chi\rangle}{\partial t} &= -\tfrac{\hbar^2}{2m}\nabla^2(|\varphi\rangle + |\chi\rangle) - mc^2\varphi \end{aligned}\right\}. \tag{2.176}$$

In the absence of the magnetic field the Klein–Gordon–Fock equation (2.172) also is invariant under time-reversal operator. This equation may be reduced to the form [61]:

$$\frac{\hbar^2}{c^2}\frac{\partial^2 |\Psi\rangle}{\partial t^2} = (\hbar^2\nabla^2 - m^2c^2)\,|\Psi\rangle\,. \tag{2.177}$$

Since the equation (2.177) contains a second derivative of the state vector $|\psi\rangle$ with respect to time, then its invariance under time-reversal operator is obvious.

Another equation, of which behavior under time-reversal operator is considered here, is the Pauli equation obtained from Dirac equation in the approximation $v \ll c$ (v is the particle velocity) [76]:

$$\left\{mc^2 + \frac{\mathbf{P}^2}{2m} - \frac{\mathbf{P}^4}{8m^3c^2} + V(\mathbf{r}) + \frac{\hbar}{i}\frac{1}{4m^2c^2}(\nabla V, \mathbf{P}) + \frac{\hbar}{4m^2c^2}[\nabla V \times \mathbf{P}]\cdot\sigma - E\right\}|\Psi\rangle = 0, \tag{2.178}$$

where $\mathbf{P}$, $V = V(\mathbf{r})$ and E are the operator of linear momentum, the potential and the total energy of the electron; σ is the Pauli operator determined by Pauli matrices σ_x, σ_y and σ_z from (2.86). The state vector $|\psi\rangle$ is a two-component spinor, represented as a decomposition in basis spinors.

The first three terms in (2.178) are the energy of the free electron. The main term mc^2 is the proper energy of the electron, $p^2/2m$ and $p^4/8m^3c^2$ are the kinetic energy and mass-velocity correction at first approximation with respect to v^2/c^2. The fifth term describes the relativistic component of the potential and is called the Darvin correction. The seven term corresponds to the interaction of the electron spin (of which operator is $\boldsymbol{S} = \frac{1}{2}\boldsymbol{\sigma}$) with its orbital momentum.

Acting with the operator $\mathbf{T}$ from (2.90) on the equation (2.178), it is easy to see that the state vector $|\boldsymbol{T\Psi}\rangle$, as well as $|\boldsymbol{\Psi}\rangle$, satisfy Pauli equation and belong to the same eigenvalue E. The presence in (2.178) of terms containing operators $(\nabla V, \mathbf{P})$ and $[\nabla V \times \mathbf{P}]\cdot\boldsymbol{\sigma}$ does not violate the time-reversal invariance. In this case, as well as in the case of the Schrödinger equation for spin-$\frac{1}{2}$ particle, $\mathbf{T}^2 = -\mathbf{1}$ and state vectors $|\Psi\rangle$ and $|\mathbf{T}\Psi\rangle$ not only belong to the same level of energy, but are also mutually orthogonal (the Kramers degeneracy).[7]

The invariance of above indicated equations under time-reversal operator is due to transformations

$$\begin{aligned} \mathbf{TrT}^{-1} &= \mathbf{r}, \\ \mathbf{TPT}^{-1} &= -\mathbf{P}, \\ \mathbf{TST}^{-1} &= -\mathbf{S}. \end{aligned} \tag{2.179}$$

The time-reversal operator $\mathbf{T}$ satisfy also the relation

$$\langle \mathbf{T}\Psi_1 | \mathbf{T}\Psi_2\rangle = |\Psi_2 | \Psi_1\rangle, \tag{2.180}$$

[7]In the case of Klein–Gordon–Fock equation for a spinless particle $\mathbf{T}^2 = +\mathbf{1}$.

$$\mathbf{T}^2 = \pm \mathbf{1}. \tag{2.181}$$

These relations were introduced earlier ((2.67), (2.72), (2.84), (2.87) and (2.57), (2.63)).

The motivation for reminder of relations (2.179), (2.180) and (2.181) is that many Hamiltonians for practically important systems are invariant to conventional time reversal. Particularly,

$$[\mathbf{H}, \mathbf{T}] = 0, \quad [\mathbf{H}_f, \mathbf{T}] = 0, \quad [\mathbf{H}_p, \mathbf{T}] = 0, \tag{2.182}$$

where $\mathbf{H}$ and $\mathbf{H}_f$ are determined by (2.77) and (2.173), and $\mathbf{H}_p$ is the Hamiltonian of the Pauli equation (2.178).

In the presence of an external magnetic field the time-reversal symmetry takes place only if the direction of the magnetic induction vector **B** is changed on opposite ($\mathbf{B} \to -\mathbf{B}$). In this case, the Hamiltonian of the system containing moving charged particles, gains additional terms, which describe the Zeeman interaction of orbital and spin electron momenta and nuclear momenta with the magnetic field. But, besides, the modification of translation moving of particles and quasiparticles takes place because the classical generalized linear momentum **P** is changed by the operator $\mathbf{P} - \frac{e}{c}\mathbf{A}$, where **A** is the vector potential of the magnetic field.Thus, the conventional time-reversal is broken by an extremal magnetic field. In fact, the validity of (2.179) is not at all necessary for the classification of Hamiltonians according to their groups of canonical transformation and, in this context, the concept of a "nonconventional" time-reversal invariance was introduced. It turned out that many systems have Hamiltonians that commute with some unitary operators, obeying (2.180) and (2.181), but not (2.179) [70].

As a confirmation of the existence of a nonconventional time-reversal, we give below two examples from Haake book [70] related to hydrogen atom.

Let us consider an hydrogen atom in presence of a constant magnetic field. By choosing the field as $\mathbf{B} = (0, 0, B)$ and the vector potential as $\mathbf{A} = \frac{1}{2}[\mathbf{B} \times \mathbf{r}]$ and including the spin-orbit interaction, we obtain the Hamiltonian

$$\mathbf{H} = \frac{\mathbf{P}^2}{2m} - \frac{e^2}{r} - \frac{eB}{2mc}(L_z + gS_z) + \frac{e^2B^2}{8mc^2}(x^2 + y^2) + f(\mathbf{r})\mathbf{LS}. \tag{2.183}$$

Here **L** and **S** denote the orbital angular momentum and spin, respectively, while the operator of the total angular momentum is $\mathbf{J} = \mathbf{L} + \mathbf{S}$.

The Hamiltonian (2.183) is invariant not under the usual time-reversal operator, $\mathbf{T}_0 = \mathbf{UK} = e^{i\pi \mathbf{J}_y}\mathbf{K}$, but under the unconventional time-reversal operator

$$\mathbf{T} = e^{i\pi \mathbf{J}_x}\mathbf{T}_0, \tag{2.184}$$

where the units system with $\hbar = 1$ is used. If the spin is absent, $\mathbf{T}^2 = +\mathbf{1}$, whereas, $\mathbf{T}^2 = -\mathbf{1}$ with spin.

When in the (2.184), in addition to the magnetic field, there is a homogeneous electric field **E**, then the Hamiltonian ceases to be invariant under unconventional time-reversal operator **T** as a result of changing the sign of the electric dipole momentum $-e\mathbf{r}\mathbf{E}$.

In this case $\mathbf{T} = \mathbf{R}\mathbf{T}_0$ is an antiunitary symmetry, where the unitary operator **R** is a reflection in the plane of **B** and **E**. The component of the angular momentum lying in this plane changes the sign under the action of the reflection operator **R**, because the angular momentum is an axial vector. The Zeeman term in **H** changes its sign under both conventional and nonconventional time-reversal operators and under the reflection **R**. Despite this, it is left-invariant under the combined transformation $\mathbf{R}\mathbf{T}_0$. The electron–dipole term of the Hamiltonian, as well as all other terms in **H**, are symmetric with respect to both $\mathbf{T}_0$ and **R** operators, owing to which $[\mathbf{H}, \mathbf{R}\mathbf{T}_0] = 0$ [70].

2.11 Selection Rules Due to Time-Reversal Symmetry

Selection rules for passing between different quantum states are determined by of symmetry properties of initial and final states, as well as those of perturbation operator. In terms of group theory, the matrix element of the transition between states, which are transformed by representations Γ_i and Γ_f under the action of perturbation operator $\mathbf{H}'(t)$, which is transformed under the representation Γ', is not equal to the zero, if $\Gamma' \in \Gamma_i \otimes \Gamma_f$ or, in other words, if the unit representation Γ_1 is contained in the Kronecker product of representations Γ_i, $\Gamma'$and Γ_f of the symmetry group of Hamiltonian ($\Gamma_1 \in \Gamma_i \otimes \Gamma' \otimes \Gamma_f$). Apart from this, there are additional selection rules caused by the time-reversal symmetry. Following to monograph [78], we will consider, what additional conditions the time-reversal invariance imposes on matrix elements of perturbation operator $\mathbf{H}'(t)$, involving quantum transitions between states $|\Psi_1\rangle$ and $|\Psi_2\rangle$:

$$\langle\Psi_1 \left|\mathbf{H}'\right| \Psi_2\rangle = \int \Psi_1^* \mathbf{H}' \Psi_2 d\tau. \tag{2.185}$$

According to formulas (2.57) and (2.63), which are related to spin 1/2 and to the system with arbitrary spin S respectively, we have

$$\langle\Psi_1 \left|\mathbf{H}'\right| \Psi_2\rangle^* = \langle\mathbf{T}\Psi_1 \left|\mathbf{T}\mathbf{H}'\ \Psi_2\right\rangle = \langle\mathbf{T}\Psi_1 \left|\mathbf{T}\mathbf{H}'\mathbf{T}^{-1}\ \mathbf{T}\Psi_2\right\rangle. \tag{2.186}$$

On the other hand,

$$\langle\Psi_1 \left|\mathbf{H}'\right| \Psi_2\rangle^* = \langle\Psi_2 \left|\mathbf{H}'^{+}\right| \Psi_1\rangle. \tag{2.187}$$

Therefore,

$$\langle\Psi_1 \left|\mathbf{H}'\right| \Psi_2\rangle = \langle\mathbf{T}\Psi_1 \left|\mathbf{T}\mathbf{H}'\mathbf{T}^{-1}\ \mathbf{T}\Psi_2\right\rangle^* = \langle\mathbf{T}\Psi_2 \left|\mathbf{T}\mathbf{H}'^{+}\mathbf{T}^{-1}\ \mathbf{T}\Psi_1\right\rangle, \tag{2.188}$$

where it was taken into account that $(\mathbf{T}\mathbf{H}'\mathbf{T}^{-1})^{+} = \mathbf{T}\mathbf{H}'^{+}\mathbf{T}^{-1}$.

We introduce the operator $\theta\mathbf{H} = \mathbf{T}\mathbf{H}'^{+}\mathbf{T}^{-1}$ as a result of action of the time-reversal operator $\mathbf{T}$ on the operator $\mathbf{H}'^{+}$. Since $\theta^2\mathbf{H}' = \mathbf{H}'$, any operator $\mathbf{H}'$ may be decomposed in two operators:

$$\mathbf{H}' = \mathbf{H}'_{+} + \mathbf{H}'_{-}, \quad \mathbf{H}'_{\pm} = \frac{1}{2}(\mathbf{H}' \pm \theta\mathbf{H}'). \tag{2.189}$$

In this case

$$\theta\mathbf{H}'_{\pm} = \mathbf{T}\mathbf{H}'^{+}_{\pm}\mathbf{T}^{-1} = f\mathbf{H}'_{\pm}, \quad f = \begin{cases} +1 & \text{for } \mathbf{H}'_{+} \\ -1 & \text{for } \mathbf{H}'_{-}. \end{cases} \tag{2.190}$$

The operator $\mathbf{H}'_{+}$ is called an even operator and $\mathbf{H}'_{-}$ - odd one with respect to time reversal. These names are due to the fact that for even operators the mean value $\overline{H}'(t) = \langle\Psi\left|\mathbf{H}'\right|\Psi\rangle$ does not change at replacing t by $-t$ (or, in other words, at replacing $|\Psi\rangle$ in $|\mathbf{T}\Psi\rangle$), while for odd operators it changes the sign at such a replacement.

In fact, according to (2.188) and (2.190)

$$\mathbf{T}\overline{\mathbf{H}}' \equiv \langle\mathbf{T}\Psi\left|\mathbf{H}'\right|\mathbf{T}\Psi\rangle = \langle\mathbf{T}^2\Psi\left|\mathbf{T}\mathbf{H}'^{+}\mathbf{T}^{-1}\right|\mathbf{T}^2\Psi\rangle = f\langle\Psi\left|\mathbf{H}'\right|\Psi\rangle = f\overline{\mathbf{H}}'. \tag{2.191}$$

Therefore, the operators of observables that do not change the sign at time reversal (for example, the position and energy) are even operators, whereas odd operators correspond to observables changing the sign at time reversal (velocity, linear and angular momenta, currents). In this case the operator $\mathbf{H}'$ corresponding to real value $\overline{H}'$ is Hermitian operator ($\mathbf{H}'^{+} = \mathbf{H}'$). Note that if these operators do not act on spin variables, then $\mathbf{T}\mathbf{H}'^{+}\mathbf{T}^{-1} = \mathbf{T}\mathbf{H}'\mathbf{T}^{-1} = \mathbf{K}\mathbf{H}'\mathbf{K}^{-1} = \mathbf{H}'^{*}$ ($\mathbf{K}$ is the complex conjugation operator).

For even and odd operators, according to (2.188) and (2.190), the time-reversal invariance imposes additional restrictions on matrix elements:

$$\langle\Psi_i\left|\mathbf{H}'\right|\Psi_j\rangle = f\langle\mathbf{T}\Psi_j\left|\mathbf{H}'\right|\mathbf{T}\Psi_i\rangle, \tag{2.192}$$

where $f = \pm 1$.

On the basis of Kramers conjugated state vectors the following two useful relationships for matrix elements can be obtained. Thus, if $|\Psi_i\rangle = |\Psi_j\rangle$, then according to (2.191) we can write

$$\langle\mathbf{T}\Psi_i\left|\mathbf{H}'\right|\mathbf{T}\Psi_i\rangle = f\langle\Psi_i\left|\mathbf{H}'\right|\Psi_i\rangle. \tag{2.193}$$

If we assume now $|\Psi_j\rangle = |\mathbf{T}\Psi_i\rangle$, then from (2.191) it follows that

$$\langle\Psi_i\left|\mathbf{H}'\right|\mathbf{T}\Psi_i\rangle = f\langle\mathbf{T}^2\Psi_i\left|\mathbf{H}'\right|\mathbf{T}\Psi_i\rangle = f\mathbf{T}^2\langle\Psi_i\left|\mathbf{H}'\right|\mathbf{T}\Psi_i\rangle. \tag{2.194}$$

The relationship (2.194) shows that

$$\left\langle \Psi_i \left| \mathbf{H}' \right| \mathbf{T}\Psi_i \right\rangle = 0 \quad \text{at} \quad f\mathbf{T}^2 = -\mathbf{1}. \tag{2.195}$$

This means that matrix elements of an even operator ($f = +1$) on Kramers conjugated spinor functions are equal to zero, and for an old operator ($f = -1$) matrix elements are equals to zero on complex-conjugated coordinate functions.

2.12 Time Reversal and Detailed Balance Principle

In the case of collision processes, the time-reversal operation indicates the change of roles of colliding and scattered particles. The presence of time-reversal symmetry leads to relationships between probabilities of direct and inverse processes, known as "detailed balance principle" [51, 61]. A basis for understanding the detailed balance principle is the conception of scattering matrix, which is introduced in a following way. Let suppose that the time-dependent perturbation $\mathbf{H}'(t)$ acts on the system described by the time-independent Hamiltonian $\mathbf{H}_0$ during some time, so that the full Hamiltonian

$$\mathbf{H} = \mathbf{H}_0 + \mathbf{H}'(t) \tag{2.196}$$

is time-dependent and the Schrödinger equation

$$i\hbar \frac{\partial \left| \Psi \right\rangle}{\partial t} = \left(\mathbf{H}_0 + \mathbf{H}'(t) \right) \left| \Psi \right\rangle \tag{2.197}$$

has no stationary solutions.

Let the state vector $|\Psi_a\rangle$ be given at $t \to -\infty$ before the inclusion of interaction $\mathbf{H}'(t)$ (initial state). The problem of scattering theory consists in finding the state vector $|\Psi\rangle$ of the system a long interval of time after the inclusion of the interaction ($t \to \infty$). The state vector $|\Psi\rangle$ can be expressed through the initial state vector $|\Psi_a\rangle$ by means of the unitary operator $\mathbf{V}(t, t_0)$ describing the time evolution of the state vector $|\Psi\rangle$:

$$|\Psi\rangle = \mathbf{V}(t, t_0) |\Psi_a\rangle, \ \mathbf{V}(t, t_0) = e^{-\frac{i}{\hbar}\mathbf{H}(t-t_0)}, \ \mathbf{V}^+(t, t_0)\mathbf{V}(t, t_0) = \mathbf{1}. \tag{2.198}$$

The limit value of the operator $\mathbf{V}(t, t_0)$ is called the "scattering matrix" $\mathbf{S}$ (more exactly, scattering operator, corresponding to the scattering matrix) [51]

$$\mathbf{S} = \lim_{t_0 \to -\infty, t \to \infty} \mathbf{V}(t, t_0). \tag{2.199}$$

Thus, the scattering operator $\mathbf{S}$ transforms the initial state vector $|\Psi_a(-\infty)\rangle$ into final state vector $|\Psi(\infty)\rangle$:

$$|\Psi(\infty)\rangle = \mathbf{S} |\Psi_a(-\infty)\rangle. \tag{2.200}$$

Index a in (2.200) denotes the complete set of quantum numbers, determining the state of the system before the scattering. It is assumed that both in the initial and in the final state the particles are separated by a sufficiently large time distance from each other, so that one can neglect the interaction between them (adiabatic hypothesis).

Let decompose the state vector $|\Psi\rangle$ into a series by some complete orthonormed system state vectors $|\Psi_b\rangle$ including the state vector $|\Psi_a\rangle$ as a particular case at $b = a$, where b denotes the respective complete set of quantum numbers

$$|\Psi\rangle = \sum_b c_b |\Psi_b\rangle . \tag{2.201}$$

In (2.201) the symbol $\sum\limits_b$ means summation by quantum numbers, running a discrete set of values and integration by quantum numbers, which are changed continuously.

Taking into account (2.200), it is easy to see that coefficients of decomposition from (2.201) are expressed by matrix elements of the operator **S**:

$$c_b = \langle \Psi_b \mid \Psi\rangle = \langle \Psi_b \mid \mathbf{S}\Psi_a\rangle = \langle \Psi_b |\mathbf{S}| \Psi_a\rangle = \mathbf{S}_{ba}. \tag{2.202}$$

Since the operator $\mathbf{V}(t, t_0)$ is unitary, one can conclude on the basis of (2.198) and (2.199) that the operator **S** is unitary as well

$$\mathbf{S}^{+}\mathbf{S} = \mathbf{S}\mathbf{S}^{+} = \mathbf{1}, \tag{2.203}$$

where **1** denotes an unit operator.

If the complete Hamiltonian can be represented in the form (2.196), then in it is convenient to use the representation of the interaction in order to find explicit form of S-matrix. In this representation the state vector $|\varphi\rangle$ is determined by the equation

$$i\hbar \frac{\partial |\varphi\rangle}{\partial t} = \mathbf{H}'(t) |\varphi\rangle , \tag{2.204}$$

which represents the Schrödinger equation with the Hamiltonian $\mathbf{H}'(t)$ in the interaction representation

$$\mathbf{H}'(t) = e^{\frac{i}{\hbar}\mathbf{H}_0 t}\mathbf{H}' e^{-\frac{i}{\hbar}\mathbf{H}_0 t}. \tag{2.205}$$

The evolution operator $\mathbf{V}(t, t_0)$ determined by (2.198) by definition transfers the state vector given in the Schrödinger representation at the time moment t_0 into a state vector at the time moment t. It may be determined in interaction representation

$$|\varphi(t)\rangle = \mathbf{V}(t, t_0) |\varphi(t_0)\rangle . \tag{2.206}$$

By substituting (2.206) into (2.204), we obtain

$$i\hbar\frac{\partial \mathbf{V}(t,t_0)}{\partial t} = \mathbf{H}'(t)\,\mathbf{V}(t,t_0), \tag{2.207}$$

$$\mathbf{V}(t_0,t_0) = \mathbf{1}. \tag{2.208}$$

One can associate to the system of equations (2.207) and (2.208) the integral equation

$$\mathbf{V}(t,t_0) = \mathbf{1} - \frac{i}{\hbar}\int_{t_0}^{t} \mathrm{d}t'\mathbf{H}'(t')\,\mathbf{V}(t',t_0), \tag{2.209}$$

which may be solved by the method of successive approximations

$$\mathbf{V}(t,-\infty) = \mathbf{1} - \frac{i}{\hbar}\int_{-\infty}^{t} \mathrm{d}t_1\mathbf{H}'(t_1) + \left(-\frac{i}{\hbar}\right)^2 \int_{-\infty}^{t}\mathrm{d}t_1\int_{-\infty}^{t_1}\mathrm{d}t_2\mathbf{H}'(t_1)\,\mathbf{H}'(t_2) +$$
$$+\left(-\frac{i}{\hbar}\right)^3\int_{-\infty}^{t}\mathrm{d}t_1\int_{-\infty}^{t_1}\mathrm{d}t_2\int_{-\infty}^{t_2}\mathrm{d}t_3\mathbf{H}'(t_1)\,\mathbf{H}'(t_2)\,\mathbf{H}'(t_3) + \cdots \tag{2.210}$$

By using the chronological operator **P**, which puts in order the product of operators depending on time (by placing them from left to right in order of a chronological sequence of decreasing time $t_1 > t_2 > t_3 > \cdots > t_n$):

$$\mathbf{P}\mathbf{L}(t_1)\,\mathbf{M}(t_2) = \begin{cases} \mathbf{L}(t_1)\,\mathbf{M}(t_2) & \text{for } t_1 > t_2, \\ \mathbf{L}(t_2)\,\mathbf{M}(t_1) & \text{for } t_2 > t_1, \end{cases} \tag{2.211}$$

the operator $\mathbf{V}(t,-\infty)$ can be represented in the form [51]

$$\mathbf{V}(t,-\infty) = \mathbf{P}\exp\left\{-\frac{i}{\hbar}\int_{-\infty}^{t}\mathbf{H}'(t)\,\mathrm{d}t\right\}. \tag{2.212}$$

On the basic of formulas (2.199) and (2.212), it is easy to obtain

$$\mathbf{S} = \lim_{t\to\infty,\ t_0\to-\infty}\mathbf{V}(t,t_0) = \lim_{t\to\infty}\mathbf{V}(t,-\infty) = \mathbf{P}\exp\left\{-\frac{i}{\hbar}\int_{-\infty}^{\infty}\mathbf{H}'(t)\,\mathrm{d}t\right\}. \tag{2.213}$$

This formula for **S**-matrix is called the Dyson formula. It is an exact formula in the sense that summation is fulfilled over all series of the perturbation theory.

Under the action of time-reversal operator $\mathbf{T} = \mathbf{UK}$, the operator $\mathbf{V}(t,t_0)$ is transformed as follows:

$$\mathbf{T}\mathbf{V}(t,t_0)\,\mathbf{T}^{-1} = \mathbf{UK}e^{-\frac{i}{\hbar}\mathbf{H}(t-t_0)}\mathbf{KU}^{+} = e^{-\frac{i}{\hbar}\mathbf{H}(t_0-t)}. \tag{2.214}$$

The amplitude of transition from the initial state $|\Psi_a\rangle$ at the moment t_a into the final state $|\Psi_b\rangle$ at time moment t_b is equal to the corresponding amplitude of transition from the state $|\mathbf{T}\Psi_b\rangle$ at time moment t_a into the state $|\mathbf{T}\Psi_a\rangle$ at the time moment t_b. Indeed [89],

$$\begin{aligned}\left\langle\Psi_b\left|e^{-\frac{i}{\hbar}\mathbf{H}(t_b-t_a)}\right|\Psi_a\right\rangle &= \left\langle\mathbf{T}\Psi_b\left|\left(\mathbf{T}e^{-\frac{i}{\hbar}\mathbf{H}(t_b-t_a)}\mathbf{T}^{+}\right)\right|\mathbf{T}\Psi_a\right\rangle^* = \\ &= \left\langle\mathbf{T}\Psi_b\left|\left(e^{-\frac{i}{\hbar}\mathbf{H}(t_a-t_b)}\right)\right|\mathbf{T}\Psi_a\right\rangle^* = \\ &= \left\langle\mathbf{T}\Psi_a\left|\left(e^{-\frac{i}{\hbar}\mathbf{H}(t_b-t_a)}\right)\right|\mathbf{T}\Psi_b\right\rangle. \end{aligned} \tag{2.215}$$

At such a transformation, under the assumption of translational invariance in time, the coordinates remain unchanged and the signs of impulses, and that of orbital and spin moments are reversed.

On the basis of (2.199) and taking into account (2.214) and (2.215), we obtain

$$\langle\Psi_b\,|\mathbf{V}\,(t_b, t_a)|\,\Psi_a\rangle = \langle\mathbf{T}\Psi_a\,|\mathbf{V}\,(t_b, t_a)|\,\mathbf{T}\Psi_b\rangle\,. \tag{2.216}$$

Since between the scattering matrix and the operator $\mathbf{V}\,(t, t_0)$ the relation (2.199) is valid, then between the elements of the scattering matrix there exists also a relation similar to (2.216):

$$\langle\Psi_b\,|\mathbf{S}|\,\Psi_a\rangle = \langle\mathbf{T}\Psi_a\,|\mathbf{S}|\,\mathbf{T}\Psi_b\rangle\,. \tag{2.217}$$

The $\mathbf{S}$ matrix element between an initial state $|\Psi_a\rangle$ and a final state $|\Psi_b\rangle$ can be written as

$$\begin{aligned}\langle\Psi_b\,|\mathbf{S}|\,\Psi_a\rangle &= \left\langle\Psi_b\left|\mathbf{T}^{+}\mathbf{T}\mathbf{S}\mathbf{T}^{+}\mathbf{T}\right|\Psi_a\right\rangle = \left\langle\mathbf{T}\Psi_b\left|\mathbf{S}^{+}\right|\mathbf{T}\Psi_a\right\rangle^* = \\ &= \langle\mathbf{S}\mathbf{T}\Psi_b\mid\mathbf{T}\Psi_a\rangle^* = \langle\mathbf{T}\Psi_a\,|\mathbf{S}|\,\mathbf{T}\Psi_b\rangle\,.\end{aligned} \tag{2.218}$$

The relation (2.218) has been obtained using the definition of the adjoint (hermitian-conjugated) operator $\mathbf{A}^{+}$ for an antilinear operator $\mathbf{A}$:

$$\left\langle\Psi\mid\mathbf{A}^{+}\Phi\right\rangle = \langle\mathbf{A}\Psi\mid\Phi\rangle^* \text{ for any } |\Psi\rangle\,,|\Phi\rangle\,. \tag{2.219}$$

If $\langle\Psi| = \langle\Psi_b|\,,\ |\Phi\rangle = |\mathbf{T}\Psi_a\rangle$ and $\mathbf{A}^{+} = \mathbf{T}^{+}\mathbf{S}^{+}$, we have

$$\left\langle\Psi_b\left|\mathbf{T}^{+}\mathbf{S}^{+}\right|\mathbf{T}\Psi_a\right\rangle = \langle\mathbf{S}\mathbf{T}\Psi_b\mid\mathbf{T}\Psi_a\rangle^*\,. \tag{2.220}$$

Here, it was taken into account that the product of a linear operator $\mathbf{S}$ by the antilinear operator $\mathbf{T}$ is antilinear to operators $\mathbf{ST}$. In addition, it was taken into account the property of the inner product of state vectors $|\mathbf{T}\Psi_a\rangle$ and $|\mathbf{S}\mathbf{T}\Psi_b\rangle$:

$$\langle\mathbf{T}\Psi_a\mid\mathbf{S}\mathbf{T}\Psi_b\rangle = \langle\mathbf{S}\mathbf{T}\Psi_b\mid\mathbf{T}\Psi_a\rangle^*\,. \tag{2.221}$$

The operator $\mathbf{V}(t_0, t)$ satisfies (2.207) at the initial condition (2.208). We apply $\mathbf{T}$ to (2.207) and obtain

$$\mathbf{T} i\hbar \frac{d}{\mathrm{d}t} \mathbf{V}(t, t_0) \mathbf{T}^+ = \mathbf{T}\mathbf{H}'(t) \mathbf{T}^+ \mathbf{T}\mathbf{V}(t, t_0) \mathbf{T}^+. \tag{2.222}$$

Since $\mathrm{d}t$ is a real quantity, $\mathbf{T}$ and $\frac{d}{\mathrm{d}t}$ commute:

$$\begin{aligned} \mathbf{T}\frac{d}{\mathrm{d}t}\mathbf{V}(t, t_0) &= \mathbf{T}\frac{1}{\mathrm{d}t}\left[\mathbf{V}(t + \mathrm{d}t, t_0) - \mathbf{V}(t, t_0)\right] = \\ &= \frac{1}{\mathrm{d}t}\left[\mathbf{T}\mathbf{V}(t + \mathrm{d}t, t_0) - \mathbf{T}\mathbf{V}(t, t_0)\right] = \frac{d}{\mathrm{d}t}\left[\mathbf{T}\mathbf{V}(t, t_0)\right]. \end{aligned} \tag{2.223}$$

Taking account of (2.223) becomes

$$-i\hbar \frac{d}{\mathrm{d}t}\left[\mathbf{T}\mathbf{V}(t, t_0)\right] = \mathbf{H}'(-t)\left[\mathbf{T}\mathbf{V}(t, t_0)\mathbf{T}^+\right]. \tag{2.224}$$

On the other hand, by setting $t = -t'$ and $t_0 = -t_0'$ in (2.207), we obtain

$$-i\hbar \frac{d}{\mathrm{d}t}\mathbf{V}(-t, -t_0) = \mathbf{H}'(-t)\mathbf{V}(-t, -t_0), \tag{2.225}$$

where the primes on t and t_0 were removed. Since $\mathbf{T}\mathbf{V}(t_0, t_0)\mathbf{T}^+ = \mathbf{T}\mathbf{T}^+ = \mathbf{1}$ and $\mathbf{V}(-t_0, -t_0) = \mathbf{1}$, we can see that $\mathbf{T}\mathbf{V}(t, t_0)\mathbf{T}^+$ and $\mathbf{V}(-t, -t_0)$ satisfy the same initial condition and are described by the same first-order differential equation. Hence,

$$\mathbf{T}\mathbf{V}(t, t_0)\mathbf{T}^+ = \mathbf{V}(-t, -t_0). \tag{2.226}$$

Using $|\Psi(t)\rangle = \mathbf{V}(t, t_0)|\Psi(t_0)\rangle$, we obtain

$$|\Psi(t_0)\rangle = \mathbf{V}(t_0, t)|\Psi(t)\rangle = \mathbf{V}(t_0, t)\mathbf{V}(t, t_0)|\Psi(t_0)\rangle, \tag{2.227}$$

which should hold for any $|\Psi(t_0)\rangle$. Thus,

$$\mathbf{V}(t_0, t)\mathbf{V}(t, t_0) = \mathbf{1} \;\text{ or }\; \mathbf{V}(t_0, t) = \mathbf{V}^{-1}(t, t_0). \tag{2.228}$$

Using the unitary property of $\mathbf{V}(t, t_0)$,

$$\mathbf{V}(t_0, t) = \mathbf{V}^+(t, t_0), \tag{2.229}$$

the relationship (2.226) becomes

$$\mathbf{T}\mathbf{V}(t, t_0)\mathbf{T}^+ = \mathbf{V}^+(-t_0, -t). \tag{2.230}$$

Taking the limits $t \to \infty$ and $t_0 \to -\infty$, we get

$$\mathbf{T}\mathbf{V}(\infty, -\infty)\,\mathbf{T}^{+} = \mathbf{V}^{+}(\infty, -\infty) \tag{2.231}$$

and applying the definition $S \equiv \mathbf{V}(\infty, -\infty)$ to (2.231), we obtain that operator $\mathbf{S}$ satisfies relation $\mathbf{T}S\mathbf{T}^{+} = S^{+}$.

Thus, a common consequence of the T-invariance of (2.207) and (2.231) is the reciprocity theorem, expressed by the relationship (2.217). According to this theorem, the amplitude of the transition from the initial state $|\Psi_a\rangle$ to the final state $|\Psi_b\rangle$, which is determined by the matrix element $\langle\Psi_b|\mathbf{S}|\Psi_a\rangle$ of the scattering operator $\mathbf{S}$, is equal to the amplitude of transition of the "reversed" process, in which the sequence of initial and final states is inverted, and the states $|\Psi_a\rangle$ and $|\Psi_b\rangle$ are changed by $|\mathbf{T}\Psi_a\rangle$ and $|\mathbf{T}\Psi_b\rangle$.

The matrix elements of the $\mathbf{S}$-matrix may be written in the form [61]

$$\langle b|\mathbf{S}|a\rangle = -2\pi i\,\langle b|\mathbf{f}|a\rangle\,\delta(E_b - E_a), \tag{2.232}$$

where

$$\langle b|\mathbf{f}|a\rangle = \langle b|\mathbf{H}'|a\rangle + \sum_k \frac{\langle b|\mathbf{H}'|k\rangle\langle k|\mathbf{H}'|a\rangle}{E_a - E_k + i\eta} + \tag{2.233}$$
$$+ \sum_{k,k'} \frac{\langle b|\mathbf{H}'|k\rangle\langle k|\mathbf{H}'|k'\rangle\langle k'|\mathbf{H}'|a\rangle}{(E_a - E_k + i\eta)(E_a - E_{k'} + i\eta)} + \cdots$$

The matrix element $\langle b|\mathbf{f}|a\rangle$ is called the matrix element on the energetic surface. The equality (2.233) may be written in a operator form

$$\mathbf{f} = \mathbf{H}' + \mathbf{H}'(E_a - \mathbf{H}_0 + i\eta)^{-1}\mathbf{H}' + \tag{2.234}$$
$$+ \mathbf{H}'(E_a - \mathbf{H}_0 + i\eta)^{-1}\mathbf{H}'(E_a - \mathbf{H}_0 + i\eta)^{-1}\mathbf{H}' + \cdots,$$

which can be considered as a solution obtained of the operator equation

$$\mathbf{f} = \mathbf{H}' + \mathbf{H}'(E_a - \mathbf{H}_0 + i\eta)^{-1}\mathbf{f} \tag{2.235}$$

by the method of successive approximations [61]. Here $\mathbf{H}_0$ is the operator of infinitely distant parts of the system and $\mathbf{H}'$ is an operator of interaction in Schrödinger representation, determining quantum transition.

The probability of transition at infinitely large time is determined by the equality

$$W_{ba}(\infty) = |\langle b|\mathbf{S}|a\rangle|^2 = 4\pi^2\delta^2(E_b - E_a)\,|\langle b|\mathbf{f}|a\rangle|^2, \tag{2.236}$$

and the probability of transition (probability of scattering and reactions ($b \to a$) at time unit at $b \neq a$ may be written in form

$$P_{ba} = \frac{2\pi}{\hbar} |\mathbf{f}_{ba}|^2 \rho(E_b) = \frac{2\pi}{\hbar} |\langle b |\mathbf{f}| a\rangle|^2 \rho(E_b), \tag{2.237}$$

where $\rho(E_b)$ is the number of finite states, in the volume V, per an unit energy interval in neighborhood of level E_b.

Taking account of (2.232) it follows from (2.217) the relation between matrix elements of direct and reversed in time transitions:

$$\langle \Psi_b |\mathbf{f}| \Psi_a\rangle = \langle \mathbf{T}\Psi_a |\mathbf{f}| \mathbf{T}\Psi_b\rangle. \tag{2.238}$$

The same relation may also be obtained by the direct method. Indeed,

$$\langle \mathbf{T}\Psi_a |\mathbf{f}| \mathbf{T}\Psi_b\rangle = \langle \mathbf{UK}\Psi_a |\mathbf{f}| \mathbf{UK}\Psi_b\rangle = \langle \mathbf{U}\Psi_a^* |\mathbf{f}| \mathbf{U}\Psi_b^*\rangle = \langle \Psi_a^* |\mathbf{U}^+\mathbf{f}\mathbf{U}| \Psi_b^*\rangle =$$
$$= \langle \Psi_a^* |(\mathbf{f}^+)^*| \Psi_b^*\rangle = \left\langle \Psi_a^* \left|\tilde{\mathbf{f}}\right| \Psi_b^*\right\rangle = \langle \Psi_b |\mathbf{f}| \Psi_a\rangle. \tag{2.239}$$

If we denote the matrix element $\langle \Psi_b |\mathbf{f}| \Psi_a\rangle$ by f_{ba} and the matrix element $\langle \mathbf{T}\Psi_a |\mathbf{f}| \mathbf{T}\Psi_b\rangle$ by $f_{-a,-b}$, we have

$$f_{ba} = f_{-a,-b}. \tag{2.240}$$

The (2.240) is another form of (2.238) and (2.239), expressing the reciprocity theorem in terms of matrix elements of $\mathbf{f}$. According to (2.237), the probability of transition $a \to b$ over a time unit is expressed by the square of modulus of the matrix element $|f_{ba}|^2 = |\langle b |\mathbf{f}| a\rangle|^2$ and the density of final states $\rho(E_b)$. Therefore the relationship (2.240) expresses another formulation of the reciprocity theorem, connecting probabilities of direct and inverted in time transitions [61]:

$$\frac{P_{ba}}{\rho(E_b)} = \frac{P_{-a,-b}}{\rho(E_a)}, \tag{2.241}$$

where $\rho(E_a)$ is the number of final states, in the volume V, per an unit energy interval in the neighborhood of the level E_a. If the densities of states in both processes are equal to each other, ($\rho(E_b) = \rho(E_a)$), then probabilities of direct and reversed in time transitions are also equal.

If Hamiltonian is invariant with respect to space coordinates reversal, then at simultaneous performing of space and time reversal the velocities and impulses of particles are not changed and the components of the angular momentum change the sign. Therefore in systems not containing spin variables the states $|a\rangle$ and $|-a\rangle$ are equivalent, so that wave functions of these states may differ only by a phase factor. In this case, absolute values of matrix elements of direct $a \to b$ and inverse $b \to a$ transitions are equal ($|f_{ba}| = |f_{ab}|$) and matrix elements of the corresponding scattering matrix satisfy the equality $|S_{ba}| = |S_{ab}|$.

In such systems, a *detailed equilibrium* takes place at which the probabilities of direct and inverse transitions, related to one finite state ($\rho(E_b) = \rho(E_a)$), are equal:

$$\frac{P_{ba}}{\rho\left(E_b\right)} = \frac{P_{ab}}{\rho\left(E_b\right)}. \tag{2.242}$$

If particles forming the system or its subsystem posses spin, then projections of spins in states $|a\rangle$ and $|-a\rangle$ differ by sign and therefore the detailed equilibrium is realized only for probabilities averaged by spin projections of initial and final states.

In connection with this, we note that else Boltzmann pointed to the possibility of detailed equilibrium violation at classical description of collisions between non-spherical molecules.

If the interaction operator involving transitions is invariant with respect to space rotations, then a transition takes place between states $|a\rangle$ and $|b\rangle$ characterized by quantum numbers j and m, at conservation of the complete angular momentum and its projection on any direction. In this case, the matrix elements f_{ba} do not depend on magnetic quantum numbers. Since the states $|a\rangle$ and $|-a\rangle$ differ only by sign of magnetic quantum numbers m, we have

$$\left|f_{ba}\right| = \left|f_{-a,-b}\right| = \left|f_{ab}\right| \tag{2.243}$$

and the detailed balance takes place.

In this first Born approximation the detailed balance is fulfilled for all systems. This conclusion follows from the following transformations

$$f_{ba}^{(B)} = \left\langle\Psi_b\left|\mathbf{H}'\right|\Psi_a\right\rangle = \left\langle\Psi_a\left|\mathbf{H}'\right|\Psi_b\right\rangle^* = f_{ab}^{*\,(B)}, \tag{2.244}$$

hence

$$\left|f_{ab}^{(B)}\right|^2 = \left|f_{ba}^{(B)}\right|^2. \tag{2.245}$$

The reciprocity theorem (2.217) and unitarity (2.203) of the scattering matrix impose additional conditions to its matrix elements. For a reaction running on N possible channels, the complex scattering matrix contains $2N^2$ real parameters. In view of reciprocity theorem and unitarity of scattering matrix, only $\frac{1}{2}N\left(N+1\right)$ of these parameters are independent. The proof of this statement is contained in [61].

As an example, let consider nuclear reactions of type $a+A \rightleftarrows b+B$, where A and B are initial and final nuclei, a is the falling particle and b is the outgoing particle as a result of direct reaction. Similarly, B and A are initial and final nuclei, b is the falling particle and a is the outgoing particle as a result of converse reaction. We denote by $j_a, m_a, j_A, m_A, j_b, m_b, j_B, m_B$ total angular momenta and their projections for particles involved in the reaction. The differential cross-sections of the direct and inverse reactions $\frac{d\sigma_{ba}}{d\Omega_b}$ and $\frac{d\sigma_{ab}}{d\Omega_a}$ will be expressed by matrix elements of **S**-matrix with regard to elastic and nonelastic channels in the following way [51]:

$$\frac{d\sigma_{ba}}{d\Omega_b} = \frac{4\pi^2}{k_a^2}\left|\left\langle j_b, m_b, j_B, m_B; -\mathbf{n}_b\left|\mathbf{S}\right| j_a, m_a, j_A, m_A; \mathbf{n}_a\right\rangle\right|^2, \tag{2.246}$$

$$\frac{d\sigma_{ab}}{d\Omega_a} = \frac{4\pi^2}{k_a^2} \left|\langle j_a, m_a, j_A, m_A; -\mathbf{n}_a \left|\mathbf{S}\right| j_b, m_b, j_B, m_B; \mathbf{n}_b\rangle\right|^2, \tag{2.247}$$

where $k_a = \frac{1}{\hbar} p_a$ and $k_b = \frac{1}{\hbar} p_b$; $\mathbf{n}_a = \frac{\mathbf{p}_a}{p_a}$ and $\mathbf{n}_b = \frac{\mathbf{p}_b}{p_b}$ are unit vectors, oriented in directions of linear momenta of the relative motion $\mathbf{p}_a$ and $\mathbf{p}_b$; $d\Omega_b$ is an interval of the solid angle, in which vector $\mathbf{n}_b$ lies; $d\Omega_a$ is an interval of the solid angle, in which vector $\mathbf{n}_a$ lies.

In the initial states there are two particles. The process is considered in the system of a centroid. The wave function of the initial state $|\Psi_a\rangle$ describes a state with a given energy of the relative motion E_a and the direction of the linear momentum of relative motion $\mathbf{n}_a$. Since the linear momentum of relative motion of particles in finite state is direct from centroid of the system, therefore the sign minus was ascribed to it in formulas (2.246) and (2.247).

Due to the fact that the reciprocity theorem binds cross-sections of the direct process (2.246) and the reversed process, which differs from (2.247) by changed signs of projections of angular momenta m_a, m_A, m_b, m_B on opposite, it is impossible to write directly a relation between these cross-sections of reactions. However, one can write a relation between cross-sections summed on projections of angular momenta of final states and averaged on projections of angular momenta of initial states. Such sections do not depend on projections of angular momenta and for them the reciprocity theorem (2.217), or the equivalent form (2.240), leads to the relationship [51]

$$\frac{1}{k_b^2}(2j_a+1)(2j_A+1)\left\langle\frac{d\sigma_{ba}}{d\Omega_b}\right\rangle = \frac{1}{k_a^2}(2j_b+1)(2j_B+1)\left\langle\frac{d\sigma_{ab}}{d\Omega_a}\right\rangle, \tag{2.248}$$

where

$$\left\langle\frac{d\sigma_{ba}}{d\Omega_b}\right\rangle = \frac{1}{(2j_a+1)(2j_A+1)} \sum_{\substack{m_a, m_A \\ m_b, m_B}} \frac{d\sigma_{ba}}{d\Omega_b}, \tag{2.249}$$

and

$$\left\langle\frac{d\sigma_{ab}}{d\Omega_a}\right\rangle = \frac{1}{(2j_b+1)(2j_B+1)} \sum_{\substack{m_a, m_A \\ m_b, m_B}} \frac{d\sigma_{ab}}{d\Omega_a}. \tag{2.250}$$

Relations similar to (2.248) may be written also for complete cross-sections of direct and inverse reactions

$$k_a^2(2j_a+1)(2j_A+1)\langle\sigma_{ba}\rangle = k_b^2(2j_b+1)(2j_B+1)\langle\sigma_{ab}\rangle. \tag{2.251}$$

In view of applicability of the perturbation theory, one can also establish a connection between non-averaged cross-sections of direct and inverse reactions

$$\frac{1}{k_b^2}\frac{d\sigma_{ba}}{d\Omega_b} = \frac{1}{k_a^2}\frac{d\sigma_{ab}}{d\Omega_a}. \tag{2.252}$$

In this case, the effective cross-section determined by the square of modulus of matrix elements of perturbation Hamiltonian $\mathbf{H}'$, for which the relation $\left|H'_{ba}\right|^2 = \left|H'_{ab}\right|^2$ is fulfilled as a consequence of Hermiticity, whence the relation (2.252) follows.

The consequences of the T-invariance in collision processes including, in particular, the polarization in scattering of spin $\frac{1}{2}$ particles on unpolarized targets, have been carried out in [2].

2.13 Dynamic Matrix and Time-Reversal Operator

The time-reversal operator $\mathbf{T}$ with antilinear and antiunitary properties, that was applied spin containing quantum systems in this chapter, has been introduced in quantum mechanics on the basis of invariance of the Hamiltonian under time reversal [68]. In other words, the explicit form of $\mathbf{T}$ was obtained considering that the state vectors $|\Psi\rangle$ and $|\mathbf{T}\Psi\rangle$ satisfy the Schrödinger equation. In the Chap. 1 the consequences, following from the time-reversal symmetry in classical physics have been discussed without using the explicit form of the operator $\mathbf{T}$.

In this section, the quantum-mechanical operator $\mathbf{T}$ is applied to a dynamic matrix in classical theory of lattice vibrations on the basis of approach developed in [91]. Firstly we note that the classical theory of crystal lattice vibrations does not take into account spin. Therefore, in this case the time-reversal operator has the form $\mathbf{T} = \mathbf{K}$.

We denote by $\mathbf{u}\binom{l}{\varkappa}$ the vector of displacement of the atom from the equilibrium state in an elementary cell of the crystal, where l determines the cell and $\varkappa$ the number of atoms in lth cell. The equations of movement for the dynamic variables $u_\alpha\binom{l}{\varkappa}$ have the form

$$M_\varkappa \ddot{u}_\alpha\begin{pmatrix} l \\ \varkappa \end{pmatrix} + \sum_{l'\varkappa'\beta} \phi_{\alpha\beta}\begin{pmatrix} ll' \\ \varkappa\varkappa' \end{pmatrix} u_\beta\begin{pmatrix} l' \\ \varkappa' \end{pmatrix} = 0, \tag{2.253}$$

where $\alpha = x, y, z,\ M_\varkappa$ is the mass of the $\varkappa$th particle and

$$\phi_{\alpha\beta}\begin{pmatrix} ll' \\ \varkappa\varkappa' \end{pmatrix} = \left.\left(\frac{\partial^2\phi}{\partial u_\alpha\binom{l}{\varkappa}\partial u_\beta\binom{l'}{\varkappa'}}\right)\right|_{\substack{u_\alpha\binom{l}{\varkappa}=0 \\ u_\beta\binom{l'}{\varkappa'}=0}}. \tag{2.254}$$

The function ϕ depends on the instantaneous positions of atoms $\boldsymbol{\rho}\binom{l}{\varkappa}$. In its turn each of them depends on the position of the resting atom $\mathbf{r}\binom{l}{\varkappa}$ and on the instantaneous value of the displacement vector $\mathbf{u}\binom{l}{\varkappa}$:

$$\boldsymbol{\rho}\begin{pmatrix} l \\ \varkappa \end{pmatrix} = \mathbf{r}\begin{pmatrix} l \\ \varkappa \end{pmatrix} + \mathbf{u}\begin{pmatrix} l \\ \varkappa \end{pmatrix}. \tag{2.255}$$

The functions $\phi_{\alpha\beta}\begin{pmatrix} ll' \\ \varkappa\varkappa' \end{pmatrix}$ are called "force constants" and satisfy the relationship

$$\phi_{\alpha\beta}\begin{pmatrix} ll' \\ \varkappa\varkappa' \end{pmatrix} = \phi_{\beta\alpha}\begin{pmatrix} l'l \\ \varkappa'\varkappa \end{pmatrix}, \tag{2.256}$$

because the order of differentiation in (2.254) is not essential. Since all atoms are shifted at translation in equivalent states, and the tensor of force constants must remain invariant under symmetry operations of the lattice, for the force constants we have [91]:

$$\phi_{\alpha\beta}\begin{pmatrix} l-l' & 0 \\ \varkappa & \varkappa' \end{pmatrix} = \phi_{\alpha\beta}\begin{pmatrix} l & l' \\ \varkappa & \varkappa' \end{pmatrix}. \tag{2.257}$$

Let Fourier transformation are fulfilled for cartesian components of elementary displacements:

$$u_\alpha\begin{pmatrix} l \\ \varkappa \end{pmatrix} = \frac{1}{\sqrt{M_\varkappa}}\frac{1}{\sqrt{N}}\sum_{\mathbf{k}} W_\alpha(\varkappa|\,\mathbf{k})e^{i\mathbf{k}\mathbf{R}_L} \tag{2.258}$$

where N is the order of the translation group of the crystal at boundary cyclic conditions of Born–Karman, $\mathbf{k}$ is the wave vector in the Brillouin zone, $\mathbf{R}_L$ is the vector of the lattice, which determines the state of the elementary cell and $W_\alpha(\varkappa|\,\mathbf{k})$ are complex variables, which satisfy the relationship

$$W_\alpha^*(\varkappa|\,\mathbf{k}) = W_\alpha(\varkappa|-\mathbf{k}), \tag{2.259}$$

so that components $u_\alpha\binom{l}{\varkappa}$ of the displacement vector should be real.

In order to obtain equations for variables $W_\alpha(\varkappa|\,\mathbf{k})$, we express the potential and kinetic energies in these variables. Then, by substituting (2.258) in the expression of the potential energy

$$V = \frac{1}{2}\sum_{\substack{l\varkappa\alpha \\ l'\varkappa'\beta}} u_\alpha\begin{pmatrix} l \\ \varkappa \end{pmatrix}\phi_{\alpha\beta}\begin{pmatrix} ll' \\ \varkappa\varkappa' \end{pmatrix}u_\beta\begin{pmatrix} l' \\ \varkappa' \end{pmatrix}, \tag{2.260}$$

we obtain

$$V = \frac{1}{2}\sum_{\substack{l\varkappa\alpha \\ l'\varkappa'\beta}}\sum_{\mathbf{k}\mathbf{k}'}\phi_{\alpha\beta}\begin{pmatrix} ll' \\ \varkappa\varkappa' \end{pmatrix}\frac{1}{N\sqrt{M_\varkappa M_{\varkappa'}}}W_\alpha(\varkappa|\,\mathbf{k})W_\beta(\varkappa'|\,\mathbf{k}')\cdot\exp[i(\mathbf{k}\mathbf{R}_L+\mathbf{k}'\mathbf{R}'_L)]. \tag{2.261}$$

Now we use the relations

$$\sum_{\mathbf{R}_L}\left(\frac{\exp i\mathbf{k}\mathbf{R}_L}{\sqrt{N}}\right)\left(\frac{\exp i\mathbf{k}'\mathbf{R}_L}{\sqrt{N}}\right)^* = \delta_{\mathbf{k}\mathbf{k}'}, \tag{2.262}$$

$$\sum_{\mathbf{k}}\left(\frac{\exp i\mathbf{k}\mathbf{R}_L}{\sqrt{N}}\right)\left(\frac{\exp i\mathbf{k}\mathbf{R}_{L'}}{\sqrt{N}}\right)^* = \delta_{LL'}, \tag{2.263}$$

which may be obtained from the orthogonality and norming relations for rows and columns of irreducible representations of the translation group $\mathcal{T}$. These relations may be also considered as a set of equations of completeness for the totality of functions $\frac{1}{\sqrt{N}}\exp i\mathbf{k}\mathbf{R}_L$, which are defined for all $\mathbf{k}$ in the Brillouin zone (at fixed $\mathbf{R}_L$) and for all vectors of lattice (at fixed $\mathbf{k}$). Using (2.262) and (2.263), we obtain

$$V = \frac{1}{2}\sum_{\substack{\varkappa\alpha\\ \varkappa'\beta}}\sum_{\mathbf{k}} W_\alpha(\varkappa|-\mathbf{k}) D_{\alpha\beta}\begin{pmatrix}\mathbf{k}\\ \varkappa\varkappa'\end{pmatrix} W_\beta(\varkappa'|\mathbf{k}'), \tag{2.264}$$

where

$$D_{\alpha\beta}\begin{pmatrix}\mathbf{k}\\ \varkappa\varkappa'\end{pmatrix} \equiv \sum_{\lambda}\phi_{\alpha\beta}\begin{pmatrix}\lambda\\ \varkappa\varkappa'\end{pmatrix}\frac{\exp(-i\mathbf{k}\mathbf{R}_L)}{\sqrt{M_\varkappa M_{\varkappa'}}}. \tag{2.265}$$

The matrix $\mathbf{D}(\mathbf{k})$, of which the matrix elements are given in (2.265), is the *dynamic matrix* of the crystal. The matrix elements $\phi_{\alpha\beta}\begin{pmatrix} l & l'\\ \varkappa & \varkappa'\end{pmatrix}$ from (2.254) form the matrix of force constants ϕ. The dynamic matrix, which must be considered complex, is a Hermitian matrix

$$D_{\alpha\beta}\begin{pmatrix}\mathbf{k}\\ \varkappa\varkappa'\end{pmatrix} = D_{\beta\alpha}\begin{pmatrix}\mathbf{k}\\ \varkappa'\varkappa\end{pmatrix}^* \tag{2.266}$$

or

$$\mathbf{D}(\mathbf{k}) = (\mathbf{D}(\mathbf{k}))^+ = (\widetilde{\mathbf{D}(\mathbf{k})})^* \tag{2.267}$$

This follows from (2.256) or (2.257), which determines the symmetry of the matrix of force constants.

Similarly, by substituting (2.258) in

$$T = \frac{1}{2}\sum_{l\varkappa\alpha} M_\varkappa\left(\dot{u}_\alpha\begin{pmatrix} l\\ \varkappa\end{pmatrix}\right)^2, \tag{2.268}$$

for the kinetic energy we obtain (do not confuse the kinetic energy T with the time-reversal operator)

$$T = \frac{1}{2}\sum_{\varkappa\alpha}\sum_{\mathbf{k}} \dot{W}_\alpha(\varkappa|\mathbf{k})^*\dot{W}_\alpha(\varkappa|\mathbf{k}). \tag{2.269}$$

Thus, the Lagrangian will be written in the form

$$L = T - V = \frac{1}{2}\sum_{\varkappa\alpha} \dot{W}_\alpha\left(\varkappa|\,\mathbf{k}\right)^* \dot{W}_\alpha\left(\varkappa|\,\mathbf{k}\right) - \tag{2.270}$$

$$- \frac{1}{2}\sum_{\varkappa\varkappa'}\sum_{\alpha\beta}\sum_{\mathbf{k}} W_\alpha\left(\varkappa|\,\mathbf{k}\right)^* D_{\alpha\beta}\begin{pmatrix}\mathbf{k}\\ \varkappa\varkappa'\end{pmatrix} W_\beta\left(\varkappa'|\,\mathbf{k}\right).$$

If we take $W_\alpha\left(\varkappa|\,\mathbf{k}\right)^*$ as a coordinate, then the conjugate linear momentum corresponding to it will be

$$\pi_\alpha\left(\varkappa|\,\mathbf{k}\right) = \frac{\partial L}{\partial W_\alpha\left(\varkappa|\,\mathbf{k}\right)^*} = \dot{W}_\alpha\left(\varkappa|\,\mathbf{k}\right). \tag{2.271}$$

From the Hamilton equations it follows [91]

$$\dot{\pi}_\alpha\left(\varkappa|\,\mathbf{k}\right) = \ddot{W}_\alpha\left(\varkappa|\,\mathbf{k}\right) = -\sum_{\varkappa'\beta} D_{\alpha\beta}\begin{pmatrix}\mathbf{k}\\ \varkappa\varkappa'\end{pmatrix} W_\beta\left(\varkappa'|\,\mathbf{k}\right) \tag{2.272}$$

or

$$\ddot{W}_\alpha\left(\varkappa|\,\mathbf{k}\right) + \sum_{\varkappa'\beta} D_{\alpha\beta}\begin{pmatrix}\mathbf{k}\\ \varkappa\varkappa'\end{pmatrix} W_\beta\left(\varkappa'|\,\mathbf{k}\right) = 0. \tag{2.273}$$

Since the exponent in (2.258) is complex, a complication in calculations appears connected with the complex field of shifts $\mathbf{W}(|\mathbf{k})$. Since there are only $3r$ components $W_\alpha\left(\varkappa|\,\mathbf{k}\right)$ $(\alpha = 1, 2, 3;\ \varkappa = 1, 2, \ldots, r)$, $\mathbf{W}(|\mathbf{k})$ is a vector in the $3r$-space for fixed $\mathbf{k}$. Further we suppose that $W_\alpha\left(\varkappa|\,\mathbf{k}\right)$ depends harmonically on time:

$$W_\alpha\left(\varkappa|\,\mathbf{k}\right) = \xi_\alpha\left(\varkappa|\,\mathbf{k}\right) \exp i\omega\left(\mathbf{k}|\,j\right)t, \tag{2.274}$$

where $j = 1, 2, \ldots, 3r$.

By substituting (2.274) in (2.273), we obtain the equation

$$-\,\omega^2\left(\mathbf{k}|\,j\right)\xi_\alpha\left(\varkappa|\,\mathbf{k}\right) + \sum_{\varkappa'\beta} D_{\alpha\beta}\begin{pmatrix}\mathbf{k}\\ \varkappa\varkappa'\end{pmatrix} \xi_\beta\left(\varkappa'|\,\mathbf{k}\right) = 0 \tag{2.275}$$

for finding the eigenvalues and eigenvectors.

We denote $3r$ of eigenvectors of the dynamic matrix from (2.275) by

$$\mathbf{e}\left(\left|\begin{matrix}\mathbf{k}\\ j\end{matrix}\right.\right), \quad j = 1, 2, \ldots, 3r.$$

Every of such vectors has $3r$ components for fixed j:

$$e_\alpha\left(\varkappa\left|\begin{matrix}\mathbf{k}\\ j\end{matrix}\right.\right), \quad \alpha = 1, 2, 3; \quad \varkappa = 1, 2, \ldots, r.$$

The corresponding eigenvalues of the matrix $\mathbf{D}(\mathbf{k})$ are denoted by $\omega^2(\mathbf{k}|\,j)$, $j = 1, 2, \ldots, 3r$. For fixed $\mathbf{k}$ and j, the following equation take place

$$\sum_{\varkappa'\beta} D_{\alpha\beta}\left(\begin{matrix}\mathbf{k}\\ \varkappa\varkappa'\end{matrix}\right) e_\beta\left(\varkappa'\left|\begin{matrix}\mathbf{k}\\ j\end{matrix}\right.\right) = \omega^2(\mathbf{k}|\,j) e_\alpha\left(\varkappa\left|\begin{matrix}\mathbf{k}\\ j\end{matrix}\right.\right). \tag{2.276}$$

In absence of degeneracy the eigenvectors can be closed satisfying the Hermite conditions of orthogonality and normalization [92]

$$\sum_{\varkappa\alpha} e^*_\alpha\left(\varkappa\left|\begin{matrix}\mathbf{k}\\ j\end{matrix}\right.\right) e_\alpha\left(\varkappa\left|\begin{matrix}\mathbf{k}\\ j'\end{matrix}\right.\right) = \delta_{jj'}, \tag{2.277}$$

$$\sum_{j} e^*_\beta\left(\varkappa'\left|\begin{matrix}\mathbf{k}\\ j\end{matrix}\right.\right) e_\alpha\left(\varkappa\left|\begin{matrix}\mathbf{k}\\ j\end{matrix}\right.\right) = \delta_{\alpha\beta}\delta_{\varkappa\varkappa'}. \tag{2.278}$$

These relations are valid because the matrix $\mathbf{D}(\mathbf{k})$ is Hermitian. If degeneracy is present, it is necessary to ascribe to eigenvectors $\mathbf{D}(\mathbf{k})$ a supplementary index numbering the degenerated eigenvectors.Thus instead of (2.276), we obtain the equations for all l_m-multiple degenerated eigenvalues

$$\mathbf{D}(\mathbf{k})\mathbf{e}\left(\left|\begin{matrix}\mathbf{k}\\ j_\mu\end{matrix}\right.\right) = \omega^2(\mathbf{k}|\,j)\mathbf{e}\left(\left|\begin{matrix}\mathbf{k}\\ j_\mu\end{matrix}\right.\right); \quad \mu = 1, 2, \ldots, l_m. \tag{2.279}$$

In this case, it is necessary to rewrite equations (2.277) and (2.278) in the form

$$\sum_{\varkappa\alpha} e^*_\alpha\left(\varkappa\left|\begin{matrix}\mathbf{k}\\ j_\mu\end{matrix}\right.\right) e_\alpha\left(\varkappa\left|\begin{matrix}\mathbf{k}\\ j'_{\mu'}\end{matrix}\right.\right) = \delta_{jj'}\delta_{\mu\mu'}, \tag{2.280}$$

$$\sum_{j}\sum_{\mu} e^*_\beta\left(\varkappa'\left|\begin{matrix}\mathbf{k}\\ j_\mu\end{matrix}\right.\right) e_\alpha\left(\varkappa\left|\begin{matrix}\mathbf{k}\\ j'_{\mu'}\end{matrix}\right.\right) = \delta_{\alpha\beta}\delta_{\varkappa\varkappa'}. \tag{2.281}$$

Since the matrix $\mathbf{D}(\mathbf{k})$ is Hermitian, eigenvalues $\omega^2(\mathbf{k}|\,j)$ are real. Because the matrix of force constants ϕ is nonnegative, the dynamic matrix $\mathbf{D}(\mathbf{k})$ is also nonnegative and we have $\omega^2(\mathbf{k}|\,j) > 0$ (the condition of stability). Using a suitable set of the eigenvectors, it is possible to construct the following $(3r \times 3r)$-matrix

$$\mathbf{E}(\mathbf{k}) = \begin{pmatrix} e_1\left(1\left|\begin{matrix}\mathbf{k}\\ 1\end{matrix}\right.\right) \cdots e_1\left(1\left|\begin{matrix}\mathbf{k}\\ 3r\end{matrix}\right.\right) \\ \vdots \\ e_3\left(r\left|\begin{matrix}\mathbf{k}\\ 1\end{matrix}\right.\right) \cdots e_3\left(r\left|\begin{matrix}\mathbf{k}\\ 3r\end{matrix}\right.\right) \end{pmatrix}. \tag{2.282}$$

From (2.280) and (2.281), it follows that $\mathbf{E}(\mathbf{k})$ is an unitary matrix

$$(\mathbf{E}(\mathbf{k}))^{-1} = (\mathbf{E}(\mathbf{k}))^{+} = (\widetilde{\mathbf{E}(\mathbf{k})})^{*}. \tag{2.283}$$

Thus from (2.279) and (2.283), it follows that

$$(\mathbf{E}(\mathbf{k}))^{-1}\mathbf{D}(\mathbf{k})\mathbf{E}(\mathbf{k}) = \boldsymbol{\Delta}(\boldsymbol{k}), \tag{2.284}$$

where $\boldsymbol{\Delta}(\boldsymbol{k})$ is a diagonal matrix with $\mathbf{k}$-dependent real eigenvalues [91]

$$\boldsymbol{\Delta}(k) = \begin{pmatrix} \omega^2(\mathbf{k}|\,1) & 0 & \cdots & 0 \\ 0 & \omega^2(\mathbf{k}|\,2) & \cdots & 0 \\ \vdots & \vdots & \cdots & \vdots \\ 0 & 0 & \cdots & \omega^2(\mathbf{k}|\,j_m) \end{pmatrix} \tag{2.285}$$

In (2.285) each proper value is met so many times as is its degree of degeneracy. The complex eigenvector $\mathbf{e}\left(\left|\begin{matrix}\mathbf{k}\\ j\end{matrix}\right.\right)$, the components of which satisfy (2.276), and real eigenvectors $\mathbf{e}_j$ of which components $e_\alpha\left(\begin{matrix} l \\ \varkappa \end{matrix}\middle| j\right)$ satisfy the equations

$$-\omega_j^2 e_\alpha\left(\begin{matrix} l \\ \varkappa \end{matrix}\middle| j\right) + \sum_{l'\varkappa'\beta'} D_{\alpha\beta}\left(\begin{matrix} ll' \\ \varkappa\varkappa' \end{matrix}\right) e_\beta\left(\begin{matrix} l' \\ \varkappa' \end{matrix}\middle| j\right) = 0, \tag{2.286}$$

or, in operator form,

$$\mathbf{D}\mathbf{e}_j = \omega_j^2\mathbf{e}_j, \tag{2.287}$$

are connected with each other by the relation [91]

$$e_\alpha\left(\varkappa\middle|\begin{matrix}\mathbf{k}\\ j\end{matrix}\right) = \frac{1}{\sqrt{N}}\sum_{L} e^{-i\mathbf{k}\mathbf{R}_L} e_\alpha\left(\begin{matrix} l \\ \varkappa \end{matrix}\middle| j\right). \tag{2.288}$$

The relation converse to (2.288) has the form

$$e_\alpha\left(\begin{matrix} l \\ \varkappa \end{matrix}\middle| j\right) = \frac{1}{\sqrt{N}}\sum_{\mathbf{k}} e^{-i\mathbf{k}\mathbf{R}_L} e_\alpha\left(\varkappa\middle|\begin{matrix}\mathbf{k}\\ j\end{matrix}\right). \tag{2.289}$$

Now let us introduce complex normal coordinates $Q\begin{pmatrix}\mathbf{k}\\ j\end{pmatrix}$, writing amplitudes $W_\alpha(\varkappa|\,\mathbf{k})$ from (2.258) in the form

$$W_\alpha(\varkappa|\mathbf{k}) = \sum_{j=1}^{3r} e_\alpha\left(\varkappa\left|\begin{matrix}\mathbf{k}\\ j\end{matrix}\right.\right) Q\begin{pmatrix}\mathbf{k}\\ j\end{pmatrix}. \tag{2.290}$$

Using (2.277), we find the converse relation

$$Q\begin{pmatrix}\mathbf{k}\\ j\end{pmatrix} = \sum_{\varkappa\alpha} e_\alpha^*\left(\varkappa\left|\begin{matrix}\mathbf{k}\\ j\end{matrix}\right.\right) W_\alpha(\varkappa|\mathbf{k}). \tag{2.291}$$

By substituting (2.290) in (2.264) taking account of (2.259) and using (2.276), we find the potential energy of the oscillating crystalline lattice in the harmonic approximation

$$V = \frac{1}{2}\sum_{\mathbf{k}}\sum_{j} \omega^2(\mathbf{k}|\,j) Q\begin{pmatrix}\mathbf{k}\\ j\end{pmatrix}^* Q\begin{pmatrix}\mathbf{k}\\ j\end{pmatrix}. \tag{2.292}$$

The derivative on time from (2.290) is

$$\dot{W}_\alpha(\varkappa|\mathbf{k}) = \sum_{j=1}^{3r} e_\alpha\left(\varkappa\left|\begin{matrix}\mathbf{k}\\ j\end{matrix}\right.\right) \dot{Q}\begin{pmatrix}\mathbf{k}\\ j\end{pmatrix}. \tag{2.293}$$

By substituting (2.293) in (2.269), we find for the kinetic energy

$$T = \frac{1}{2}\sum_{\mathbf{k}}\sum_{j} \dot{Q}\begin{pmatrix}\mathbf{k}\\ j\end{pmatrix}^* \dot{Q}\begin{pmatrix}\mathbf{k}\\ j\end{pmatrix}. \tag{2.294}$$

Thus, the Hamiltonian of the system, expressed in normal coordinates, takes the form

$$H = \frac{1}{2}\sum_{\mathbf{k}}\sum_{j} \left\{ \dot{Q}\begin{pmatrix}\mathbf{k}\\ j\end{pmatrix}^* \dot{Q}\begin{pmatrix}\mathbf{k}\\ j\end{pmatrix} + \omega^2(\mathbf{k}|\,j) Q\begin{pmatrix}\mathbf{k}\\ j\end{pmatrix}^* Q\begin{pmatrix}\mathbf{k}\\ j\end{pmatrix} \right\}. \tag{2.295}$$

Starting from (2.295), one can obtain the following equation of motion for complex amplitudes:

$$\ddot{Q}\begin{pmatrix}\mathbf{k}\\ j\end{pmatrix} + \omega^2(\mathbf{k}|\,j) Q\begin{pmatrix}\mathbf{k}\\ j\end{pmatrix} = 0. \tag{2.296}$$

The solution to equation (2.296) for fixed j has the form

$$Q\begin{pmatrix}\mathbf{k}\\ j\end{pmatrix} = Q_0\begin{pmatrix}\mathbf{k}\\ j\end{pmatrix} \exp[\pm i\omega(\mathbf{k}|\,j)t], \tag{2.297}$$

where $Q_0\begin{pmatrix}\mathbf{k}\\ j\end{pmatrix}$ does not depend on time. In (2.292), (2.294) and (2.295), the summation is taken over all wave vectors $\mathbf{k}$ in Brillouin zone and over all j for each $\mathbf{k}$.

Physical shifts $U_\alpha\begin{pmatrix} l \\ \varkappa \end{pmatrix}$ can be expressed in the form

$$U_\alpha\begin{pmatrix} l \\ \varkappa \end{pmatrix} = \frac{1}{\sqrt{M_\varkappa N}}\sum_j\sum_{\mathbf{k}} e^{i\mathbf{k}\mathbf{R}_L} e_\alpha\left(\varkappa\middle|\begin{matrix}\mathbf{k}\\ j\end{matrix}\right) Q\begin{pmatrix}\mathbf{k}\\ j\end{pmatrix}. \tag{2.298}$$

The relation converse to (2.298) has the form

$$Q\begin{pmatrix}\mathbf{k}\\ j\end{pmatrix} = \frac{1}{\sqrt{N}}\sum_{\mathbf{L}}\sum_{\varkappa\alpha} e^{-i\mathbf{k}\mathbf{R}_L} e^*_\alpha\left(\varkappa\middle|\begin{matrix}\mathbf{k}\\ j\end{matrix}\right) U_\alpha\begin{pmatrix} l \\ \varkappa \end{pmatrix}\sqrt{M_\varkappa}. \tag{2.299}$$

Relations (2.298) and (2.299) show that normal coordinates $Q\begin{pmatrix}\mathbf{k}\\ j\end{pmatrix}$ are cartesian components of shifts multiplied by $\sqrt{M_\varkappa}$ and projected on the vector $e^{-i\mathbf{k}\mathbf{R}_L} e^*_\alpha \left(\varkappa\middle|\begin{matrix}\mathbf{k}\\ j\end{matrix}\right)$ [91].

Each space symmetry group G contains the subgroup of translations $\mathfrak{T}$ as a normal subgroup $G/\mathfrak{T}$. Since $\mathfrak{T}$ is Abelian group, its irreducible representations $D^{(\mathbf{k})}$ are one-dimensional and are characterized by the wave vector $\mathbf{k}$ and the Bloch state vector $|\psi_{\mathbf{k}}\rangle$ [93]. The set of admissible values $\mathbf{k}$ fills the first Brillouin zone of the crystal and characterizes all irreducible representations $D^{(\mathbf{k})}$ of the group $\mathfrak{T}$ (with regard to Born–Karman cyclic boundary conditions this number of irreducible representations is finite).

For each vector $\mathbf{k}$, the set of operators of the group G is determined, which transforms the state vector $|\psi_{\mathbf{k}}\rangle$ into a state vector with an equivalent value of the wave vector $\mathbf{k}$. As it was mentioned in Sect. 2.7, this totality of operators forms the group $G_{\mathbf{k}}$, named the group of the wave vector, which is a subgroup of the space group G.

The dynamic matrix $\mathbf{D}(\mathbf{k})$ commutes with each of matrices of representations $D^{(\mathbf{k})(e)}$ of the group $G_{\mathbf{k}}$ (e is the index of admissible representation). The form of the matrix $\mathbf{D}(\mathbf{k})$ is given by (2.265), from which it is seen that $\mathbf{D}(\mathbf{k})$ is not an identity matrix multiplied by a scalar. Since the matrix $\mathbf{D}(\mathbf{k})$ is not reducible to a constant one and commutes with all matrices of the representation $D^{(\mathbf{k})(e)}$ of the group $G_{\mathbf{k}}$, it follows from Schur lemma [68, 80, 94] that the representation $D^{(\mathbf{k})(e)}$ is reducible. Moreover, the dynamic matrix $\mathbf{D}(\mathbf{k})$ commutes with each of matrices of the reducible representation $D^{(\mathbf{k})(e)}$ of the group $G_{\mathbf{k}}$:

$$\mathbf{D}(\mathbf{k})\mathbf{D}^{(\mathbf{k})(e)} = \mathbf{D}^{(\mathbf{k})(e)}\mathbf{D}(\mathbf{k}). \tag{2.300}$$

Now if we fix the argument of the matrix $D^{(\mathbf{k})(e)}$, which is not indicated in (2.300) and which is an element of the wave vector group $G_{\mathbf{k}}$, and realize the unitary transformation using $\mathbf{E}(\mathbf{k})$ in both parts of (2.300) with regard to (2.284), we obtain [91]

$$(\mathbf{E}(\mathbf{k}))^{-1}\mathbf{D}^{(\mathbf{k})(e)}\mathbf{E}(\mathbf{k})\boldsymbol{\Delta}(\boldsymbol{k}) = \boldsymbol{\Delta}(\boldsymbol{k})(\mathbf{E}(\mathbf{k}))^{-1}\mathbf{D}^{(\mathbf{k})(e)}\mathbf{E}(\mathbf{k}). \tag{2.301}$$

Let us introduce the notation

$$\overline{\mathbf{D}^{(\mathbf{k})(e)}} = (\mathbf{E}(\mathbf{k}))^{-1}\,\mathbf{D}^{(\mathbf{k})(e)}\mathbf{E}(\mathbf{k}) = (\widetilde{\mathbf{E}(\mathbf{k})})^{*}\mathbf{D}^{(\mathbf{k})(e)}\mathbf{E}(\mathbf{k}). \tag{2.302}$$

Thus, taking account of (2.302), instead of (2.301) we obtain

$$\boldsymbol{\Delta}(\boldsymbol{k})\overline{\mathbf{D}^{(\mathbf{k})(e)}} = \overline{\mathbf{D}^{(\mathbf{k})(e)}}\boldsymbol{\Delta}(\boldsymbol{k}). \tag{2.303}$$

If we put together degenerated eigenvalues, the formula (2.285) can be rewritten in the form

$$\boldsymbol{\Delta}(\boldsymbol{k}) = \begin{pmatrix} \omega^2\,(\mathbf{k}|\,1)\cdot\mathbf{1}_1 & 0\cdots & 0\cdots & 0 \\ 0 & \omega^2\,(\mathbf{k}|\,2)\cdot\mathbf{1}_2 & & \vdots \\ \vdots & & \omega^2\,(\mathbf{k}|\,j)\cdot\mathbf{1}_j & \\ 0\cdots & & & \omega^2\,(\mathbf{k}|\,n)\cdot\mathbf{1}_{l_n} \end{pmatrix}, \tag{2.304}$$

where $\mathbf{1}_j$ is a l_j-dimensional identity matrix.

Since according to (2.302) and (2.303), every matrix $\overline{\mathbf{D}^{(\mathbf{k})(e)}}$ commutes with one and the same block of the diagonal matrix (2.304), any matrix $\overline{\mathbf{D}^{(\mathbf{k})(e)}}$ for each element of the group $G_{\mathbf{k}}$ is reduced to block form, in accordance with (2.304)

$$\overline{\mathbf{D}^{(\mathbf{k})(e)}} = \begin{pmatrix} \mathbf{D}^{(\mathbf{k})(1)} & 0 & 0 & \cdots \\ 0 & \mathbf{D}^{(\mathbf{k})(2)} & 0 & \\ 0 & 0 & \mathbf{D}^{(\mathbf{k})(3)} & \\ \vdots & & & \mathbf{D}^{(\mathbf{k})(n)} \end{pmatrix}. \tag{2.305}$$

Matrices $\mathbf{D}^{(\mathbf{k})(1)}, \ldots, \mathbf{D}^{(\mathbf{k})(n)}$ in (2.305) have dimensions $(l_1 \cdot l_1), \ldots, (l_n \cdot l_n)$. Each of these matrices is associated with its eigenvalue $\omega^2\,(\mathbf{k}|\,1), \omega^2\,(\mathbf{k}|\,2), \ldots, \omega^2\,(\mathbf{k}|\,n)$.

An important property of eigenvectors of the dynamic matrix $\mathbf{D}(\mathbf{k})$ is that they may be used as a basis for representations $D^{(\mathbf{k})(e)}$ of the wave vector group $G_{\mathbf{k}}$. This is due to the fact that $3r$ eigenvectors $\mathbf{e}\left(\left|\begin{matrix}\mathbf{k}\\ j\end{matrix}\right.\right)$ of dynamic matrix $(j = 1, 2, \ldots, 3r)$ are Bloch vectors, corresponding to the wave vector $\mathbf{k}$. It is easy to show that the dynamic matrix $\mathbf{D}(\mathbf{k})$ is invariant to transformations of the group $G_{\mathbf{k}}$, and this means that one can construct from its eigenvectors the bases of irreducible representations of the group $G_{\mathbf{k}}$.

Side by side with the representation $\mathbf{D}^{(\mathbf{k})(e)}$, one can also construct the representation $\mathbf{D}^{(\mathbf{k})(j)}$ of the group $G_{\mathbf{k}}$, which is interconnected with $\mathbf{D}^{(\mathbf{k})(e)}$ by an unitary transformation. This possibility is due to the fact that, besides $3r$ components $e_\alpha\left(\varkappa\left|\begin{matrix}\mathbf{k}\\ j\end{matrix}\right.\right)$ $(\alpha = 1, 2, 3;\quad \varkappa = 1, 2, \ldots, r)$ of the $3r$-dimensional vector $\mathbf{e}\left(\left|\begin{matrix}\mathbf{k}\\ j\end{matrix}\right.\right)$, there exist

$3r$ components $e_\alpha\left(\varkappa\middle|\mathbf{k}\right)$ $(\alpha = 1, 2, 3; \quad \varkappa = 1, 2, \ldots, r)$ of the $3r$-dimensional vector $\mathbf{e}_\alpha\left(\varkappa\middle|\mathbf{k}\right)$. The component $e_\alpha\left(\varkappa\middle|\begin{matrix}\mathbf{k}\\ j\end{matrix}\right)$ may be considered as an $(\alpha\varkappa)$-component of the vector $\mathbf{e}\left(\middle|\begin{matrix}\mathbf{k}\\ j\end{matrix}\right)$, or as a j-component of the vector $\mathbf{e}_\alpha\left(\varkappa\middle|\mathbf{k}\right)$. In addition, between the set of $3r$ vectors

$$\mathbf{e}\left(\middle|\begin{matrix}\mathbf{k}\\ j\end{matrix}\right), \quad j = 1, 2, \ldots, 3r \tag{2.306}$$

and the set of $3r$ vectors

$$\mathbf{e}_\alpha\left(\varkappa\middle|\mathbf{k}\right), \quad (\alpha = 1, 2, 3; \quad \varkappa = 1, 2, \ldots, r) \tag{2.307}$$

there is a linear dependence [91]

$$\mathbf{e}_\alpha\left(\varkappa\middle|\mathbf{k}\right) = \mathbf{U}\mathbf{e}\left(\middle|\begin{matrix}\mathbf{k}\\ j\end{matrix}\right), \tag{2.308}$$

where $\mathbf{U}$ is an unitary matrix:

$$\mathbf{U}^{-1} = \mathbf{U}^{+}. \tag{2.309}$$

Due to the fact that $\mathbf{U}$ is an unitary matrix, the relation converse to (2.308) has the form

$$\mathbf{e}\left(\middle|\begin{matrix}\mathbf{k}\\ j\end{matrix}\right) = \mathbf{U}^{+}\mathbf{e}_\alpha\left(\varkappa\middle|\mathbf{k}\right). \tag{2.310}$$

Since the bases of representations $D^{(\mathbf{k})(e)}$ and $D^{(\mathbf{k})(j)}$ are connected by a unitary transformation (see (2.308) and (2.310)), it follows that matrices of these representations $\mathbf{D}^{(\mathbf{k})(e)}$ and $\mathbf{D}^{(\mathbf{k})(j)}$ must be equivalent:

$$\mathbf{D}^{(\mathbf{k})(j)} = \mathbf{U}^{-1}\mathbf{D}^{(\mathbf{k})(e)}\mathbf{U}. \tag{2.311}$$

In order to find all consequences of essential degeneracy of oscillating modes, caused by operators of the complete space group G, it is necessary to obtain the set of representations $D^{(^*\mathbf{k})(m)}$ of G, using the representation $D^{(\mathbf{k})(e)}$ of the wave vector group $G_\mathbf{k}$, found by solving the dynamic problem.

The irreducible representation $D^{(^*\mathbf{k})(m)}$ is marked by the star $^*\mathbf{k}$ asterisk at the wave vector $\mathbf{k}$ (the set $^*\mathbf{k} \equiv \{\mathbf{k}, \mathbf{k}_2, \ldots, \mathbf{k}_s\}$ from s wave vectors, corresponding to the totality of s Bloch state vectors $\{|\psi_\mathbf{k}\rangle, |\psi_{\mathbf{k}_2}\rangle, \ldots, |\psi_{\mathbf{k}_s}\rangle\}$) and by the index of the admissible representation m. For values of s there are two possibilities: $s = p$ or $s < p$, where p is the index of the subgroup $\mathcal{T}$ of the group G, that is the order of the quotient group $G/\mathcal{T}$.

The presence of an even more complete significant degeneracy in the problem of dynamics of the crystalline lattice is connected with the complete space-time group $\mathcal{G}$ of the crystal, which consists of the sum of the crystalline space group G and the coset $\mathbf{T}G$

$$\mathcal{G} = G + \mathbf{T}G, \tag{2.312}$$

where $\mathbf{T}$ is the time-reversal operator.

Now, we can proceed to the consideration of the role of time-reversal symmetry in the classical problem of dynamics of crystalline lattice. As it was already noted, since spins of electrons and nuclei are not taken into account, when solving such problem, the time-reversal operator $\mathbf{T} = \mathbf{UK}$ is reduced to the operator of complex conjugation $\mathbf{K}$. The operator $\mathbf{K}$ transforms a function into a complex-conjugated one and plays a significant role as one of operators of symmetry of the dynamic matrix $\mathbf{D}(\mathbf{k})$, as it will be shown below.

The operator $\mathbf{K}$ does not change the real eigenvector $\mathbf{e}^{(R)}$

$$\mathbf{K}\mathbf{e}^{(R)} = \mathbf{e}^{(R)} \tag{2.313}$$

but changes the sign of the imaginary eigenvector $\mathbf{e}^{(I)}$:

$$\mathbf{K}\mathbf{e}^{(I)} = -\mathbf{e}^{(I)}. \tag{2.314}$$

If $\mathbf{e}$ is a complex eigenvector, in particular $\mathbf{e}\left(\left|\begin{matrix}\mathbf{k}\\ j'_\mu\end{matrix}\right.\right)$, satisfying the condition of a scalar product

$$\mathbf{e}\left(\left|\begin{matrix}\mathbf{k}\\ j_\mu\end{matrix}\right.\right)^* \cdot \mathbf{e}\left(\left|\begin{matrix}\mathbf{k}\\ j'_{\mu'}\end{matrix}\right.\right) = \delta_{jj'}\delta_{\mu\mu'}, \tag{2.315}$$

then

$$\left(\mathbf{Ke}\left(\left|\begin{matrix}\mathbf{k}\\ j_\mu\end{matrix}\right.\right)\right)^* \cdot \left(\mathbf{Ke}\left(\left|\begin{matrix}\mathbf{k}\\ j'_{\mu'}\end{matrix}\right.\right)\right) = \left(\mathbf{e}\left(\left|\begin{matrix}\mathbf{k}\\ j_\mu\end{matrix}\right.\right)^* \cdot \mathbf{e}\left(\left|\begin{matrix}\mathbf{k}\\ j'_{\mu'}\end{matrix}\right.\right)\right)^*. \tag{2.316}$$

The antilinear and antiunitary properties of the operator $\mathbf{K}$, expressed by relations (2.23), (2.57) and (2.316) make the operator $\mathbf{K}$ qualitatively different from operators of transformations of the space group G. Moreover, the operator $\mathbf{K}$ commutes with all operators of the group G. In this case, the complete space-time symmetry group of crystal is determined by the relation (2.312), in which $\mathbf{T} = \mathbf{K}$.

The completely depending on time solution to the dynamic equation (2.279) is obtained, if one takes into account that the depending on time eigenvector $\boldsymbol{\varepsilon}$ corresponding to the wave vector $\mathbf{K}$, is [91]

$$\boldsymbol{\varepsilon}\left(\left|\begin{matrix}\mathbf{k}\\ j_\mu\end{matrix}\right|t\right) \equiv \mathbf{e}\left(\left|\begin{matrix}\mathbf{k}\\ j_\mu\end{matrix}\right.\right)\exp\left\{-i\omega\left(\mathbf{k}|\, j_\mu\right)t\right\} \tag{2.317}$$

and satisfies the equation

$$\mathbf{D}(\mathbf{k}) \cdot \boldsymbol{\varepsilon}\left(\left|\begin{matrix}\mathbf{k}\\ j_\mu\end{matrix}\right| t\right) + \frac{\partial^2 \boldsymbol{\varepsilon}}{\partial t^2}\left(\left|\begin{matrix}\mathbf{k}\\ j_\mu\end{matrix}\right| t\right) = 0. \tag{2.318}$$

We act with the operator $\mathbf{K}$ on the left side of the (2.318):

$$\mathbf{K}\mathbf{D}(\mathbf{k})\mathbf{K}^{-1} \cdot \boldsymbol{K\varepsilon}\left(\left|\begin{matrix}\mathbf{k}\\ j_\mu\end{matrix}\right| t\right) + \mathbf{K}\frac{\partial^2}{\partial t^2}\mathbf{K}^{-1} \cdot \boldsymbol{K\varepsilon}\left(\left|\begin{matrix}\mathbf{k}\\ j_\mu\end{matrix}\right| t\right) = 0. \tag{2.319}$$

Since $\frac{\partial^2}{\partial t^2}$ is a real operator, we have

$$\mathbf{K}\left(\frac{\partial^2}{\partial t^2}\right)\mathbf{K}^{-1} = \frac{\partial^2}{\partial t^2}. \tag{2.320}$$

We act now with the operator $\mathbf{K}$ on $\mathbf{D}(\mathbf{k})$

$$\mathbf{K}\mathbf{D}(\mathbf{k})\mathbf{K}^{-1} = (\mathbf{D}(\mathbf{k}))^*, \tag{2.321}$$

using the definition (2.265) and the reality of the force constant. As a result we obtain

$$(\mathbf{D}(\mathbf{k}))^* = \mathbf{D}(-\mathbf{k}). \tag{2.322}$$

By comparing (2.322) with the condition the matrix $\mathbf{D}(\mathbf{k})$ to be Hermitian (2.267), it follows that the matrix $\mathbf{D}(\mathbf{k})$ at fixed $\mathbf{k}$ is a Hermitian matrix and (2.322) gives the relation between matrices $\mathbf{D}(\mathbf{k})$ and $\mathbf{D}(-\mathbf{k})$. From (2.320), (2.322) and (2.319), it follows

$$\mathbf{D}(-\mathbf{k}) \cdot \boldsymbol{K\varepsilon}\left(\left|\begin{matrix}\mathbf{k}\\ j_\mu\end{matrix}\right| t\right) = \frac{\partial^2}{\partial t^2} \cdot \boldsymbol{K\varepsilon}\left(\left|\begin{matrix}\mathbf{k}\\ j_\mu\end{matrix}\right| t\right). \tag{2.323}$$

Now consider the complete set of eigenvectors depending on time at a given value $-\mathbf{k}$:

$$\boldsymbol{\varepsilon}\left(\left|\begin{matrix}-\mathbf{k}\\ j_\lambda\end{matrix}\right| t\right) = \mathbf{e}\left(\left|\begin{matrix}-\mathbf{k}\\ j_\lambda\end{matrix}\right|\right) \exp\left\{-i\omega\left(-\mathbf{k}|\, j_\lambda\right)t\right\}. \tag{2.324}$$

Each eigenvector from this set satisfies the equation

$$\mathbf{D}(-\mathbf{k})\boldsymbol{\varepsilon}\left(\left|\begin{matrix}-\mathbf{k}\\ j_\lambda\end{matrix}\right| t\right) + \frac{\partial^2}{\partial t^2}\boldsymbol{\varepsilon}\left(\left|\begin{matrix}-\mathbf{k}\\ j_\lambda\end{matrix}\right| t\right) = 0. \tag{2.325}$$

By comparing (2.323) and (2.325), we see that the vector $\boldsymbol{K\varepsilon}\left(\left|\begin{matrix}\mathbf{k}\\ j_\mu\end{matrix}\right| t\right)$ is connected directly with the vector $\boldsymbol{\varepsilon}\left(\left|\begin{matrix}-\mathbf{k}\\ j_\lambda\end{matrix}\right| t\right)$. In order to establish this connection, we return to equations of motion non-depending of time for dynamic matrix. At a given value **k** these equations have the form (2.279) and equations for the wave vector $-\mathbf{k}$ have the form

$$\mathbf{D}(-\mathbf{k})\mathbf{e}\left(\left|\begin{matrix}-\mathbf{k}\\ \bar{j}_\lambda\end{matrix}\right.\right)=\omega^2\left(-\mathbf{k}|\,\bar{j}\right)\mathbf{e}\left(\left|\begin{matrix}-\mathbf{k}\\ \bar{j}_\lambda\end{matrix}\right.\right),\quad \lambda=1,2,\ldots,\bar{l}_j. \tag{2.326}$$

Acting with the operator **K** on both sides of the (2.279) with regard to (2.322), we obtain

$$\mathbf{D}(-\mathbf{k})\mathbf{K}\mathbf{e}\left(\left|\begin{matrix}\mathbf{k}\\ j_\mu\end{matrix}\right.\right)=\omega^2\left(\mathbf{k}|\,j\right)\mathbf{K}\mathbf{e}\left(\left|\begin{matrix}\mathbf{k}\\ j_\mu\end{matrix}\right.\right),\quad \mu=1,2,\ldots,l_j. \tag{2.327}$$

It follows that non-depending on time eigenvectors $\mathbf{K}\mathbf{e}\left(\left|\begin{matrix}\mathbf{k}\\ j_\mu\end{matrix}\right.\right)$ satisfy the same equations for eigenvalues, as $\mathbf{e}\left(\left|\begin{matrix}-\mathbf{k}\\ \bar{j}_\lambda\end{matrix}\right.\right)$ do, with the dynamic matrix $\mathbf{D}(-\mathbf{k})$.

In [91], it is proved that the relation

$$e^*_\alpha\left(\varkappa\left|\begin{matrix}\mathbf{k}\\ j_\lambda\end{matrix}\right.\right)\omega^2\left(\mathbf{k}|\,j\right)=\omega^2\left(-\mathbf{k}|\,\bar{j}\right)e_\alpha\left(\varkappa\left|\begin{matrix}-\mathbf{k}\\ \bar{j}_{\bar{\mu}}\end{matrix}\right.\right) \tag{2.328}$$

takes peace. Since real eigenvalues of matrices $\mathbf{D}(\mathbf{k})$ and $\mathbf{D}(-\mathbf{k})=\mathbf{D}(\mathbf{k})^*$ are the same, we can put

$$\omega^2\left(\mathbf{k}|\,j\right)=\omega^2\left(-\mathbf{k}|\,\bar{j}\right) \tag{2.329}$$

and then

$$e^*_\alpha\left(\varkappa\left|\begin{matrix}\mathbf{k}\\ j_\lambda\end{matrix}\right.\right)=e_\alpha\left(\varkappa\left|\begin{matrix}-\mathbf{k}\\ \bar{j}_{\bar{\mu}}\end{matrix}\right.\right), \tag{2.330}$$

or

$$\mathbf{K}\mathbf{e}\left(\left|\begin{matrix}\mathbf{k}\\ j_\lambda\end{matrix}\right.\right)=\mathbf{e}^*\left(\left|\begin{matrix}\mathbf{k}\\ j_\lambda\end{matrix}\right.\right)=\mathbf{e}\left(\left|\begin{matrix}-\mathbf{k}\\ \bar{j}_{\bar{\mu}}\end{matrix}\right.\right). \tag{2.331}$$

Formulas (2.331) were obtained in the [95] for the particular case of absence of degeneracy $\left(l_j=1 \text{ or } \bar{l}_j=1\right)$. It is necessary to note that, in general case, $j_{\bar{\mu}}$ and j_λ in (2.330), as well as in (2.331), are different.

It is easy to show that the relation

$$\boldsymbol{K}\boldsymbol{\varepsilon}\left(\left|\begin{matrix}\mathbf{k}\\ j_\mu\end{matrix}\right|t\right)=\boldsymbol{\varepsilon}\left(\left|\begin{matrix}-\mathbf{k}\\ \bar{j}_{\bar{\mu}}\end{matrix}\right|-t\right) \tag{2.332}$$

takes place. According to (2.332), the operator **K** transforms eigenvector corresponding to the shift with the wave vector **k** at the time moment t, into the eigenvector corresponding to the shift with the wave vector $-\mathbf{k}$ at time moment $-t$. That is way the eigenvector $\boldsymbol{K}\boldsymbol{\varepsilon}\left(\left|\begin{matrix}\mathbf{k}\\ j_\mu\end{matrix}\right|t\right)$ can be named time-reversed relative to the eigenvalue $\boldsymbol{\varepsilon}\left(\left|\begin{array}{c}\mathbf{k} \\ j_\mu\end{array}\right| t\right)$. It is necessary to note that the non-depending on time eigenvectors $\mathbf{e}\left(\left|\begin{array}{c}\mathbf{k} \\ j_\mu\end{array}\right.\right)$ and $\mathbf{e}^*\left(\left|\begin{array}{c}\mathbf{k} \\ j_{\bar{\mu}}\end{array}\right.\right)$ are concerned to the degenerated proper state. Therefore, even for non-depending on time eigenvectors one may consider that $\mathbf{e}^*\left(\left|\begin{array}{c}\mathbf{k} \\ j_{\bar{\mu}}\end{array}\right.\right)$ is time-reversed relative to $\mathbf{e}\left(\left|\begin{array}{c}\mathbf{k} \\ j_\mu\end{array}\right.\right)$.

Now we consider the complete set of eigenvectors, which determine the vector space and form a basis for the irreducible representation $D^{(^*\mathbf{k})(m)}$ of the space symmetry group G:

$$\sum^{(^*\mathbf{k})(j)} \equiv \left\{\mathbf{e}\left(\left|\begin{array}{c}\mathbf{k} \\ j_1\end{array}\right.\right), \ldots, \mathbf{e}\left(\left|\begin{array}{c}\mathbf{k} \\ j_2\end{array}\right.\right), \ldots, \mathbf{e}\left(\left|\begin{array}{c}\mathbf{k}_s \\ j_{l_j}\end{array}\right.\right)\right\} \tag{2.333}$$

The set $\left(s \cdot l_j\right)$ of complex normal coordinates

$$\sum^{(^*\mathbf{k})(j)} \equiv \left\{Q\left(\left|\begin{array}{c}\mathbf{k} \\ j_1\end{array}\right.\right), \ldots, Q\left(\left|\begin{array}{c}\mathbf{k} \\ j_2\end{array}\right.\right), \ldots, Q\left(\left|\begin{array}{c}\mathbf{k}_s \\ j_{l_j}\end{array}\right.\right)\right\} \tag{2.334}$$

may also be used as a basis of the irreducible representation $D^{(^*\mathbf{k})(j)}$.

As it was shown in [91], the complete set (2.333) of eigenvectors of the dynamic matrix $\mathbf{D}^{(\mathbf{k})}$ or the complex normal coordinates (2.334), forming a basis of the irreducible representation $D^{(^*\mathbf{k})(j)}$ of the space group G, may also serve as a basis for the irreducible representation $D^{(^*\mathbf{k})(j)^*}$ of the group G, if this basis is transformed using the time-reversal operator $\mathbf{K}$. If $D^{(^*\mathbf{k})(j)}$ is an irreducible representation of the group G, it follows that $D^{(^*\mathbf{k})(j)^*}$ will be also an irreducible representation of the same group.

Since the operator $\mathbf{K}$ is an operator of symmetry, the set of the eigenvalues of matrix $\mathbf{D}^{(\mathbf{k})}$

$$\left\{\omega^2\left(\mathbf{k} \mid j\right), \ldots, \omega^2\left(\mathbf{k} \mid j_{l_j}\right), \ldots, \omega^2\left(\mathbf{k} \mid j_1'\right), \ldots, \omega^2\left(\mathbf{k} \mid j'_{l_{j'}}\right)\right\} \tag{2.335}$$

is identically equal to the set of eigenvalues of the matrix $\mathbf{D}^{(\mathbf{k})^*} = \mathbf{D}(-\mathbf{k})$

$$\left\{\omega^2\left(-\mathbf{k} \mid j_1\right), \ldots, \omega^2\left(-\mathbf{k} \mid j_{l_j}\right), \ldots, \omega^2\left(-\mathbf{k} \mid j_1'\right), \ldots, \omega^2\left(-\mathbf{k} \mid j'_{l_{j'}}\right)\right\}. \tag{2.336}$$

Taking account of (2.335), it follows that the space generated by the set of all eigenvectors of the matrix $\mathbf{D}(\mathbf{k})$ or normal coordinates at a given $\mathbf{k}$:

$$\sum^{(\mathbf{k})(\{j\})} \equiv \left\{\mathbf{e}\left(\left|\begin{array}{c}\mathbf{k} \\ j_1\end{array}\right.\right), \ldots, \mathbf{e}\left(\left|\begin{array}{c}\mathbf{k} \\ j_{l_j}\end{array}\right.\right), \ldots, \mathbf{e}\left(\left|\begin{array}{c}\mathbf{k}_s \\ j'_{l_{j'}}\end{array}\right.\right)\right\} \tag{2.337}$$

or

$$\overset{(\mathbf{k})(\{j\})}{\sum} \equiv \left\{ Q\left(\left|\begin{matrix}\mathbf{k}\\ j_1\end{matrix}\right.\right), \ldots, Q\left(\left|\begin{matrix}\mathbf{k}\\ j_{l_j}\end{matrix}\right.\right), \ldots, Q\left(\left|\begin{matrix}\mathbf{k}\\ j'_{l'_j}\end{matrix}\right.\right) \right\}, \tag{2.338}$$

is degenerated with the space formed by corresponding sets at a given $-\mathbf{k}$:

$$\overset{(-\mathbf{k})(\{\mu\})}{\sum} = \overset{(\mathbf{k})(\{j\})^*}{\sum} \equiv \mathbf{K} \overset{(\mathbf{k})(\{j\})}{\sum}, \tag{2.339}$$

$$\overset{(\mathbf{k})(\{j\})^*}{\sum} \equiv \left\{ \mathbf{e}^*\left(\left|\begin{matrix}\mathbf{k}\\ j_1\end{matrix}\right.\right), \ldots, \mathbf{e}^*\left(\left|\begin{matrix}\mathbf{k}\\ j'_{l_{j'}}\end{matrix}\right.\right) \right\}, \tag{2.340}$$

or

$$\overset{(\mathbf{k})(\{j\})^*}{\sum} \equiv \left\{ Q^*\left(\left|\begin{matrix}\mathbf{k}\\ j_1\end{matrix}\right.\right), \ldots, Q^*\left(\left|\begin{matrix}\mathbf{k}\\ j'_{l_{j'}}\end{matrix}\right.\right) \right\}. \tag{2.341}$$

If the space $\sum^{(^*\mathbf{k})(j)}$ is real and satisfies the relationship

$$\mathbf{K} \overset{(^*\mathbf{k})(j)}{\sum} = \overset{(^*\mathbf{k})(j)}{\sum}, \tag{2.342}$$

then it may serve as a basis for the "physically irreducible representation" of the complete space-time group $\mathcal{G}$, which appears in the problem of dynamics of the crystal lattice with the space symmetry group G. Since the physical shifts $\mathbf{u}$ describing deviations of atoms of the crystal from equilibrium positions are real, there are possible only such physical irreducible representations, of which basis is the real space Σ [91].

Since every normal oscillation appearing in lattice dynamics is transformed as a basis for the irreducible representation of the group $\mathcal{G}$, it is possible to say that if

$$D^{(^*\mathbf{k})(j)} \equiv D^{(^*\mathbf{k})(j)^*}, \tag{2.343}$$

then the irreducible representation of the space group G is real and it can be used as a "physical" irreducible representation of the group $\mathcal{G}$. However if (2.343) does not take place, i.e.,

$$D^{(^*\mathbf{k})(j)} \not\equiv D^{(^*\mathbf{k})(j)^*}, \tag{2.344}$$

then the "physical" irreducible representation of the complete space-time group $\mathcal{G}$ is a direct sum

$$D^{(^*\mathbf{k})(j)} \oplus \mathbf{K}D^{(^*\mathbf{k})(j)} = D^{(^*\mathbf{k})(j)} \oplus D^{(^*\mathbf{k})(j)^*}, \tag{2.345}$$

which is, in this case, a real irreducible representation of the group G [91].

If the relation (2.343) is fulfilled, then accounting the time-reversal symmetry does not change the results of this analysis made by taking into account only the space symmetry group G. In the case when relations (2.344) and (2.345) are valid, taking account of the time-reversal operator leads to doubling the degeneracy of oscillation modes. Namely, it takes place a transition from the degeneracy due to the space group G that is characterized by the representation $D^{(^\star\mathbf{k})(m)}$ to a degeneracy connected with the complete space-time symmetry group $\mathcal{G}$, which is characterized by a direct sum of representations $D^{(^\star\mathbf{k})(m)} \oplus D^{(^\star\mathbf{k})(m)^*}$. In addition,

$$D^{(^\star\mathbf{k})(m)^*} = D^{(^\star-\mathbf{k})(\bar{m})}, \tag{2.346}$$

where $D^{(^\star-\mathbf{k})(\bar{m})}$ is the irreducible representation corresponding to the star $(^\star-\mathbf{k})$.

It is necessary to mention that the irreducible representation $D^{(^\star\mathbf{k})(m)}$ of the space group G is an induced representation constructed using the small representation of the wave vector group $G_{\mathbf{k}}$. Therefore, in order to study representation $D^{(^\star\mathbf{k})(m)}$, one may use Herring criteria mentioned in Sect. 2.7 (2.150), which allow to operate only with characters of elements of the group $G_{\mathbf{k}}$.

According to Frei [96], only three classes of wave vectors $\mathbf{k}$ and stars $^\star\mathbf{k}$ are distinguished

$$Class\ I: \mathbf{k} = -\mathbf{k}+2\pi\mathbf{B}_H, \quad {}^\star\mathbf{k} = {}^\star-\mathbf{k}. \tag{2.347}$$

$$Class\ II: \mathbf{k} \neq -\mathbf{k}+2\pi\mathbf{B}_H, \quad {}^\star\mathbf{k} = {}^\star-\mathbf{k}. \tag{2.348}$$

$$Class\ III: \mathbf{k} \neq -\mathbf{k}+2\pi\mathbf{B}_H, \quad {}^\star\mathbf{k} \neq {}^\star-\mathbf{k}. \tag{2.349}$$

In (2.347)–(2.349), $\mathbf{B}_H$ is the vector of the reciprocal lattice (to avoid ambiguity, here we do not use the notation $\mathbf{K}$ for the vector of the reciprocal lattice, introduced in Sect. 2.7, the formula (2.142), do not confuse this vector with the time-reversal operator $\mathbf{K}$ for spinless systems).

Each of three classes I, II, III requires a special consideration, which is done in [91] using a method different from Frei's method. Here we restrict ourselves only to the discussion of results.

Class I. If $D^{(\mathbf{k})(m)}$ is real, then $m = \bar{m}$ and a supplementary degeneracy of oscillating modes is absent. If $D^{(\mathbf{k})(m)}$ is complex, then $m \neq \bar{m}$ and representations $D^{(\mathbf{k})(m)}$ and $D^{(\mathbf{k})(\bar{m})}$ are joined. The admissible physical irreducible representations for phonons are representations $D^{(\mathbf{k})(m)} \oplus D^{(\mathbf{k})(\bar{m})}$, having a double dimension as compared with $D^{(\mathbf{k})(m)}$. This is an important case from the practical point of view since it corresponds to a local doubling of the degree of degeneracy relative to the degeneracy, caused by a poorly space symmetry group G.

Class II. If $m = \bar{m}$, then the supplementary degeneracy is absent. However, if $m \neq \bar{m}$, then the representations $D^{(\mathbf{k})(m)}$ and $D^{(\mathbf{k})(\bar{m})}$ are joined. In this case there is again a local doubling of the degeneracy comparing with degeneracy caused by the purely space symmetry group G.

Class III. For wave vectors of class III, the following relation between irreducible representations takes place

$$D^{(^{\star}\mathbf{k})(m)} \not\equiv D^{(^{\star}-\mathbf{k})(\bar{m})} \tag{2.350}$$

This relationship shows that the complete representations $D^{(^{\star}\mathbf{k})(m)}$ and $D^{(^{\star}-\mathbf{k})(\bar{m})}$ must be considered together as a single representation. Such a doubling the degeneracy multiplicity of oscillating modes is "global". It differs from a "local" join of the representations at a given wave vector **k**. In this case, the significant degeneracy is connected with the doubling of the number of representations of the space group G by means of the complete space-time symmetry group $\mathcal{G}$.

2.14 Time-Reversal Symmetry in Quantized Field Theory

Let the basis of the Hilbert space can be taken as [90]

$$|\{n_i, \mathbf{p}_i, \sigma_i\}\rangle \equiv a^+_{n_1 \mathbf{p}_1 \sigma_1} \ldots a^+_{n_k \mathbf{p}_k \sigma_k} |0\rangle , \tag{2.351}$$

which is a k-particle state, where the ith particle is specified by the particle type n_i, its momentum p_i, and its spin component along an axis σ_i. The n-type antiparticle will be specified by $\bar{n}$ included in n_i. In this case, the antiparticle will be considered just as another particle type. Bound states are included as separate particles. One can take different spin quantization axes for different particle types and momenta, but for simplicity we will consider the z axis for all particle types and momenta. In (2.351), $a^+_{n_i \mathbf{p}_i \sigma_i}$ is the creation operator of the ith particle and $|0\rangle$ is the vacuum state.

The invariance of the quantum field theory equations under time-reversal operator **T**, along with invariances of these equations under the parity operator **P** and charge conjugation operator **C**, have been analyzed by many authors [2, 70, 72, 89, 97–102].

In this section, a short presentation of the T-invariance in quantum field theory is given according to [70, 90].

In the definition of the basis state (2.351) it is implicitly assumed the following. If there are N particles with exactly the same type of momentum and spin, then there is a factor $1/\sqrt{N!}$ to ensure the normalization

$$\langle\{n_i \mathbf{p}_i \sigma_i\}| \; \{n_i \mathbf{p}_i \sigma_i\}\rangle = 1. \tag{2.352}$$

The complete set of basis states consists of all possible combinations of particle types, momenta, spins and particle numbers. Once the action of the antiunitary time-reversal operator **T** on each of the creation and destruction operators is given, its action on the basis state is defined and, consequently, the **T** operator is uniquely determined.

Under the time-reversal operator the 3-momentum **P** and the spin component σ change their signs, whereas the particle type n remains the same. Thus, the time-reversal operator **T** should transform a creation operator of the type n particle, momentum **P** and spin component σ to that for $(n, -p, -\sigma)$ up to the phase factor, which in general case depends on (n, p, σ):

$$\mathbf{T}a^{+}_{n\mathbf{p}\sigma}\mathbf{T}^{+} = \zeta_{n\mathbf{p}\sigma}a^{+}_{n-\mathbf{p}-\sigma}. \quad (|\zeta_{n\mathbf{p}\sigma}| = 1) \tag{2.353}$$

By analogy, for the annihilation operator $a_{n\mathbf{p}\sigma}$ we obtain

$$\mathbf{T}a_{n\mathbf{p}\sigma}\mathbf{T}^{+} = \zeta^{*}_{n\mathbf{p}\sigma}a_{n-\mathbf{p}-\sigma}. \tag{2.354}$$

In spite of the vacuum state $|0\rangle$ should transform to itself under time reversal up to a phase

$$\mathbf{T}\,|0\rangle = \zeta_{\mathrm{VAC}}\,|0\rangle\,, \tag{2.355}$$

the basis state $|\{n_i\mathbf{p}_i\sigma_i\}\rangle$ of (2.351) transforms as

$$\begin{aligned}\mathbf{T}\,|\{n_i\mathbf{p}_i\sigma_i\}\rangle &= \mathbf{T}a^{+}_{n_1\mathbf{p}_1\sigma_1}\mathbf{T}^{+}\mathbf{T}\ldots\ \ldots\mathbf{T}^{+}\mathbf{T}a^{+}_{n_k\mathbf{p}_k\sigma_k}\mathbf{T}^{+}\mathbf{T}\,|0\rangle = \\ &= \zeta_{n_1\mathbf{p}_1\sigma_1}\cdots\zeta_{n_k\mathbf{p}_k\sigma_k}\zeta_{\mathrm{VAC}}a^{+}_{n_1-\mathbf{p}_1-\sigma_1}\cdots a^{+}_{n_k-\mathbf{p}_k-\sigma}\,|0\rangle = \\ &= \zeta_{\{n_i\mathbf{p}_i\sigma_i\}}\zeta_{\mathrm{VAC}}\,|\{n_i-\mathbf{p}_i-\sigma\}\rangle\,,\end{aligned} \tag{2.356}$$

where we have defined

$$\zeta_{\{n_i\mathbf{p}_i\sigma_i\}} \equiv \zeta_{n_1\mathbf{p}_1\sigma_1}\cdots\zeta_{n_k\mathbf{p}_k\sigma_k}. \tag{2.357}$$

Here we have taken into account that ζ_{VAC} from (2.355) is equal to 1 ($\zeta_{\mathrm{VAC}} = 1$) from the following considerations: the phase associated with the vacuum state can be removed by redefining the overall phase of the operator $\mathbf{T}$. If we define $\mathbf{T} \equiv \zeta_{\mathrm{VAC}}\mathbf{T}'$, then

$$\mathbf{T}\,|0\rangle = \zeta_{\mathrm{VAC}}\,|0\rangle\,,\ \zeta_{\mathrm{VAC}}\mathbf{T}'\,|0\rangle = \zeta_{\mathrm{VAC}}\,|0\rangle \Rightarrow \mathbf{T}'\,|0\rangle = |0\rangle\,. \tag{2.358}$$

Because of the antilinearity of the operator $\mathbf{T}$, we can also remove the phase by redefining the vacuum state phase. Namely, if $\mathbf{T}$ transforms the state vector into itself up to a phase factor

$$\mathbf{T}\,|\psi\rangle = e^{i\phi}\,|\psi\rangle\,, \tag{2.359}$$

then by defining $|\psi\rangle = e^{-i\frac{\phi}{2}}\,|\psi'\rangle$, we have

$$\mathbf{T}e^{-i\frac{\phi}{2}}\,|\psi'\rangle = e^{i\phi}e^{-i\frac{\phi}{2}}\,|\psi'\rangle \Rightarrow \mathbf{T}\,|\psi'\rangle = |\psi'\rangle, \tag{2.360}$$

taking into account that $\mathbf{T}e^{-i\frac{\phi}{2}} = e^{i\frac{\phi}{2}}\mathbf{T}$. We see that the phase factor is an eigenvalue of the time-reversal operator $\mathbf{T}$. Since this factor has no physical meaning, it is assumed that the time-reversal phase of the vacuum state is $+1$ [90]:

$$\mathbf{T}\,|0\rangle = |0\rangle\,. \tag{2.361}$$

The matrix elements of **T** with respect to the basis state vectors are determined by

$$\left\langle \psi_{\{n_i \mathbf{p}_i \sigma_i\}} \right| \mathbf{T} \left| \psi_{\{n_j \mathbf{p}_j \sigma_j\}} \right\rangle = \zeta_{\{n_i \mathbf{p}_i \sigma_i\}} \delta_{\{n_i \mathbf{p}_i \sigma_i\}\{n_j - \mathbf{p}_j - \sigma_j\}}, \tag{2.362}$$

where the notation $\left|\psi_{\{n_i \mathbf{p}_i \sigma_i\}}\right\rangle = |\{n_i \mathbf{p}_i \sigma_i\}\rangle$ is used and $\delta_{\{n_i \mathbf{p}_i \sigma_i\}\{n_j - \mathbf{p}_j - \sigma_j\}}$ is taken to be $+1$, if there are only one type of particles and momenta and, in addition, spins are of opposite signs between the two groups, otherwise the value is zero. The huge matrix formed by matrix elements (2.362) is an unitary matrix containing only one nonzero element in each row or in each column. The antilinear time-reversal operator **T** which is represented by a product of unitary operator and operator of complex conjugation is an antiunitary operator.

We consider the Yukawa interaction Hamiltonian

$$h_Y(t) = \int d^3x \bar{\psi}\psi\phi \tag{2.363}$$

and the quantum electrodynamics (QED) interaction Hamiltonian

$$h_{\text{QED}} = \int d^3x A^{\mu}(x)\bar{\psi}(x)\gamma_{\mu}\psi(x), \tag{2.364}$$

where γ_μ is the Dirac 4-matrix, ψ and $\bar{\psi}$ are the 4-component spinor and the corresponding Hermitian conjugated spinor A^μ is the 4-vector potential of the electromagnetic field.

In accordance to [90], it will be shown below that it is possible to select the arbitrary time-reversal phases $\zeta_{n\mathbf{p}\sigma}$ from (2.353) in such a way that Hamiltonians $h_Y(t)$ and $h_{\text{QED}}(t)$ satisfy the condition of time-reversal invariance:

$$\mathbf{T}h(t)\mathbf{T}^{+} = h(-t). \tag{2.365}$$

For a spin-0 particle the spin index can be omitted and thus (2.353) and (2.354) are

$$\mathbf{T}a^{+}_{s\mathbf{p}}\mathbf{T}^{+} = \zeta_{s\mathbf{p}}a^{+}_{s-\mathbf{p}}, \; \mathbf{T}a_{s\mathbf{p}}\mathbf{T}^{+} = \zeta^{*}_{s\mathbf{p}}a_{s-\mathbf{p}}. \tag{2.366}$$

By applying **T** to the scalar spin-0 field

$$\phi(x) = \sum_{\mathbf{p}} \left(a_{s\mathbf{p}} e_{\mathbf{p}}(x) + a^{+}_{s\mathbf{p}} e^{*}_{\mathbf{p}}(x) \right) \tag{2.367}$$

with

$$e_{\mathbf{p}}(x) = \frac{e^{-ipx}}{2p^0 V}, \quad p^0 = \sqrt{\mathbf{p}^2 + m^2}, \tag{2.368}$$

where p and $\mathbf{p}$ are linear 4- and 3-momentum, m is the physical mass of the particle and V is the volume at quantization of the field in a limited volume, we obtain

$$\mathbf{T}\phi(x)\mathbf{T}^{+} = \sum_{\mathbf{p}} \left(\left(\mathbf{T}a_{s\mathbf{p}}\mathbf{T}^{+}\right) e^{*}_{\mathbf{p}}(x) + (\mathbf{T}a^{+}_{\bar{s}\mathbf{p}}\mathbf{T}^{+}) e_{\mathbf{p}}(x) \right) =$$

$$= \sum_{\mathbf{p}} \zeta^{*}_{s\mathbf{p}} a_{s-\mathbf{p}} e^{*}_{\mathbf{p}}(x) + \zeta_{\bar{s}\mathbf{p}} a^{+}_{\bar{s}-\mathbf{p}} e_{\mathbf{p}}(x) =$$

$$= \sum_{\mathbf{p}} \zeta^{*}_{s-\mathbf{p}} a_{s\mathbf{p}} e^{*}_{-\mathbf{p}}(x) + \zeta_{s-\mathbf{p}} a^{+}_{\bar{s}\mathbf{p}} e_{-\mathbf{p}}(x) =$$

$$= \sum_{\mathbf{p}} \zeta^{*}_{s-\mathbf{p}} a_{s\mathbf{p}} e_{\mathbf{p}}(x') + \zeta_{s-\mathbf{p}} a^{+}_{\bar{s}\mathbf{p}} e^{*}_{\mathbf{p}}(x'), \tag{2.369}$$

where we have used $x' \equiv Tx = (-x^0, \mathbf{x})$ and $e_{-\mathbf{p}}(x) = e^{*}_{\mathbf{p}}(x')$ in the last sum. In the third sum from (2.369) the substitution $\mathbf{p} \to -\mathbf{p}$ was fulfilled, which did not led to changing the size of this sum.

We require that the transformed field should be "proportional" to the original field [90]. Then, the phase factors should not depend on $\mathbf{p}$: $\zeta_{s\mathbf{p}} = \zeta_s$, $\zeta_{\bar{s}\mathbf{p}} = \zeta_{\bar{s}}$ and $\zeta_{\bar{s}} = \zeta^{*}_{s}$.

So we have

$$\mathbf{T}\phi(x)\mathbf{T}^{+} = \zeta^{*}_{s}\phi(x'), \quad (x' \equiv Tx). \tag{2.370}$$

In (2.370), the operator T which transforms x into x' belongs to the Lorentz group and it is explicitly written in the four dimension space-time as

$$T = \begin{pmatrix} -1 & & & \\ & 1 & & \\ & & 1 & \\ & & & 1 \end{pmatrix}. \tag{2.371}$$

For the fermion field, the transformation (2.353) gives

$$\mathbf{T}a^{+}_{f\mathbf{p}\sigma}\mathbf{T}^{+} = \zeta_{f\mathbf{p}\sigma} a^{+}_{f-\mathbf{p}-\sigma}. \tag{2.372}$$

An analogical relationship for $a_{f\mathbf{p}\sigma}$ is

$$\mathbf{T}a_{f\mathbf{p}\sigma}\mathbf{T}^{+} = \zeta^{*}_{f\mathbf{p}\sigma} a_{f-\mathbf{p}-\sigma}. \tag{2.373}$$

By applying $\mathbf{T}$ to the fermion field, $\psi(x)$,

$$\psi(x) = \sum_{\mathbf{p}\sigma} (a_{f\mathbf{p}\sigma} f_{\mathbf{p}\sigma}(x) + a^{+}_{\bar{f}\mathbf{p}\sigma} g_{\mathbf{p}\sigma}(x)), \tag{2.374}$$

where

$$f_{\mathbf{p}\sigma}(x) = u_{\mathbf{p}\sigma} e_{\mathbf{p}}(x), \; g_{\mathbf{p}\sigma}(x) = v_{\mathbf{p}\sigma} e^{*}_{\mathbf{p}}(x) \tag{2.375}$$

and $e_{\mathbf{p}}(x)$ is determined by (2.368), we obtain

$$\mathbf{T}\psi(x)\mathbf{T}^{+} = \sum_{\mathbf{p}\sigma}(\mathbf{T}a_{f\mathbf{p}\sigma}\mathbf{T}^{+}f^{*}_{\mathbf{p}\sigma}(x) + \mathbf{T}a^{+}_{\bar{f}\mathbf{p}\sigma}\mathbf{T}^{+}g^{*}_{\mathbf{p}\sigma}(x)) = \tag{2.376}$$
$$= \sum_{\mathbf{p}\sigma}(\zeta^{*}_{f\mathbf{p}\sigma}a_{f-\mathbf{p}-\sigma}f^{*}_{\mathbf{p}\sigma}(x) + \zeta_{\bar{f}\mathbf{p}\sigma}a^{+}_{\bar{f}-\mathbf{p}-\sigma}g^{*}_{\mathbf{p}\sigma}(x)) =$$
$$= \sum_{\mathbf{p}\sigma}(\zeta^{*}_{f-\mathbf{p}-\sigma}a_{f\mathbf{p}\sigma}f^{*}_{-\mathbf{p}-\sigma}(x) + \zeta_{\bar{f}-\mathbf{p}-\sigma}a^{+}_{\bar{f}\mathbf{p}\sigma}g^{*}_{-\mathbf{p}-\sigma}(x)),$$

where the following substitutions were used in the last sum: $\mathbf{p} \to -\mathbf{p}$, $\sigma \to -\sigma$.

Using $e^{*}_{-\mathbf{p}}(x) = e_{\mathbf{p}}(x')$ with $x' \equiv Tx = (-x^{0}, x)$ it was shown in [90] that

$$f^{*}_{-\mathbf{p}-\sigma}(x) = u^{*}_{-\mathbf{p}-\sigma}e^{*}_{-\mathbf{p}}(x) = u^{*}_{-\mathbf{p}-\sigma}e_{\mathbf{p}}(x'), \tag{2.377}$$

$$g^{*}_{-\mathbf{p}-\sigma}(x) = v^{*}_{-\mathbf{p}-\sigma}e_{-\mathbf{p}}(x) = v^{*}_{-\mathbf{p}-\sigma}e_{\mathbf{p}}(x'), \tag{2.378}$$

and correspondingly, $\mathbf{T}\psi(x)\mathbf{T}^{+}$ is

$$\mathbf{T}\psi(x)\mathbf{T}^{+} = \sum_{\mathbf{p}\sigma}(\zeta^{*}_{f-\mathbf{p}-\sigma}a_{f\mathbf{p}\sigma}u^{*}_{-\mathbf{p}-\sigma}e_{\mathbf{p}}(x') + \zeta_{\bar{f}-\mathbf{p}-\sigma}a^{+}_{\bar{f}\mathbf{p}\sigma}v^{*}_{-\mathbf{p}-\sigma}e^{*}_{\mathbf{p}}(x')). \tag{2.379}$$

The values of $u^{*}_{-\mathbf{p}-\sigma}$ and $v^{*}_{-\mathbf{p}-\sigma}$ in (2.379) are

$$u^{*}_{-\mathbf{p}-\sigma} = \text{sign}(\sigma)\delta\mathcal{T}u_{\mathbf{p}\sigma},\; v^{*}_{-\mathbf{p}-\sigma} = \text{sign}(\sigma)\bar{\delta}\mathcal{T}v_{\mathbf{p}\sigma}, \tag{2.380}$$

where δ and $\bar{\delta}$ are phase factors do not depended on σ. The (4×4)-matrix $\mathcal{T}$ unitary and antisymmetric

$$\mathcal{T}^{+}\mathcal{T} = 1, \quad \mathcal{T}^{T} = -\mathcal{T}, \tag{2.381}$$

and satisfies the relation

$$\mathcal{T}\gamma^{\mu}\mathcal{T}^{+} = \gamma^{*}_{\mu} \qquad (\mu = 0, 1, 2, 3). \tag{2.382}$$

We note that the (4×4)-matrix $\mathcal{T}$ is not antilinear and acts in the spinor space. It is necessary to distinguish it from the antilinear time-reversal operator $\mathbf{T}$, which operates in the Hilbert space, as well as from the Lorentz transformation matrix T, operating in the four-dimensional space-time.

In Dirac representation, the matrix $\mathcal{T}$ has the form

$$\mathcal{T} = \gamma_{3}\gamma_{1} = \begin{pmatrix} 0 & -1 & 0 & 0 \\ 1 & 0 & 0 & 0 \\ 0 & 0 & 0 & -1 \\ 0 & 0 & 1 & 0 \end{pmatrix} \tag{2.383}$$

We can see from (2.383) that the matrix $\mathcal{T}$ is antisymmetric. Using the relationship $\gamma^{+}_{\mu} = \gamma^{\mu}$, we obtain

$$\mathcal{T}^{+}\mathcal{T} = (\gamma_3\gamma_1)^{+}(\gamma_3\gamma_1) = \gamma_1^{+}\gamma_3^{+}\gamma_3\gamma_1 = \gamma^1\gamma^3\gamma_3\gamma_1 = \mathbf{1}. \tag{2.384}$$

Thus, $\mathcal{T}$ is an unitary and antisymmetric matrix.

On the basis of (2.380) the formula (2.379) can be transformed to the form [90]:

$$\mathbf{T}\psi(x)\mathbf{T}^{+} = \mathcal{T}\sum_{\mathbf{p}\sigma}(\zeta^{*}_{f-\mathbf{p}-\sigma}\operatorname{sign}(\sigma)\cdot\delta a_{f\mathbf{p}\sigma}f_{\mathbf{p}\sigma}(x') + \zeta_{f-\mathbf{p}-\sigma}\operatorname{sign}(\sigma)\cdot\bar{\delta}a^{+}_{f\mathbf{p}\sigma}g_{\mathbf{p}\sigma}(x')). \tag{2.385}$$

We require that the transformed fermion field should be "proportional" to the original one. Also the phase factors in front of $a_{f\mathbf{p}\sigma}$ and $a^{+}_{f\mathbf{p}\sigma}$ should not depend on $\mathbf{p}$, σ, and particle or antiparticle:

$$\zeta^{*}_{f-\mathbf{p}-\sigma}\operatorname{sign}(\sigma)\delta = \zeta_{\bar{f}-\mathbf{p}-\sigma}\bar{\delta} \equiv \zeta^{*}_{f}. \tag{2.386}$$

The relation (2.386) can be considered as a definition of ζ^{*}_{f}. So we have

$$\mathbf{T}\psi(x)\mathbf{T}^{+} = \zeta^{*}_{f}\mathcal{T}\psi(x')). \tag{2.387}$$

Under the time-reversal operator $\mathbf{T}$, the spin-1 vector field

$$A^{\mu}(x) = \sum_{\mathbf{p}\sigma}(a_{v\mathbf{p}\sigma}\varepsilon^{\mu}_{\mathbf{p}\sigma}e_{\mathbf{p}}(x) + a^{+}_{\bar{v}\mathbf{p}\sigma}\varepsilon^{\mu *}_{\mathbf{p}\sigma}e^{*}_{\mathbf{p}}(x)) \tag{2.388}$$

transforms as

$$\begin{aligned}\mathbf{T}A^{\mu}\mathbf{T}^{+} &= \sum_{\mathbf{p}\sigma}(\mathbf{T}a_{v\mathbf{p}\sigma}\mathbf{T}^{+}\varepsilon^{\mu}_{\mathbf{p}\sigma}e_{\mathbf{p}}(x) + \mathbf{T}a^{+}_{\bar{v}\mathbf{p}\sigma}\mathbf{T}^{+}\varepsilon^{\mu *}_{\mathbf{p}\sigma}e^{*}_{\mathbf{p}}(x)) = \\ &= \sum_{\mathbf{p}\sigma}(\zeta^{*}_{v\mathbf{p}\sigma}a_{v-\mathbf{p}-\sigma}\varepsilon^{\mu *}_{\mathbf{p}\sigma}e^{*}_{\mathbf{p}}(x) + \zeta_{\bar{v}\mathbf{p}\sigma}a^{+}_{\bar{v}-\mathbf{p}-\sigma}\varepsilon^{\mu}_{\mathbf{p}\sigma}e_{\mathbf{p}}(x)) = \\ &= \sum_{\mathbf{p}\sigma}(\zeta^{*}_{v-\mathbf{p}-\sigma}a_{v\mathbf{p}\sigma}\varepsilon^{\mu *}_{-\mathbf{p}-\sigma}e^{*}_{-\mathbf{p}}(x) + \zeta_{\bar{v}-\mathbf{p}-\sigma}a^{+}_{\bar{v}\mathbf{p}\sigma}\varepsilon^{\mu}_{-\mathbf{p}-\sigma}e_{-\mathbf{p}}(x)).\end{aligned} \tag{2.389}$$

When passing from the second double sum in (2.389) to the third one, substitutions $\mathbf{p} \to -\mathbf{p}$ and $\sigma \to -\sigma$ have been fulfilled, which do not change the sums.

Using the relationship [90]

$$\varepsilon^{\mu *}_{-\mathbf{p}-\sigma} = (-1)^{1+\sigma}\varepsilon_{\mathbf{p}\sigma\mu} \tag{2.390}$$

and taking into account that $e^{*}_{-\mathbf{p}}(x) = e_{\mathbf{p}}(x')$ $(x' \equiv Tx)$, we have

$$\mathbf{T}A^{\mu}\mathbf{T}^{+} = \sum_{\mathbf{p}\sigma}(\zeta^{*}_{v-\mathbf{p}-\sigma}(-1)^{1+\sigma}a_{v\mathbf{p}\sigma}\varepsilon_{\mathbf{p}\sigma\mu}e_{\mathbf{p}}(x') + \zeta_{\bar{v}-\mathbf{p}-\sigma}(-1)^{1+\sigma}a^{+}_{\bar{v}\mathbf{p}\sigma}\varepsilon^{*}_{\mathbf{p}\sigma\mu}e^{*}_{\mathbf{p}}(x'). \tag{2.391}$$

In order for this to be "proportional" to $A^{\mu}(x')$, the phase factors should be independent on particle type, momentum, and spin:

$$\zeta^{*}_{\bar{\nu}-\mathbf{p}-\sigma}(-1)^{1+\sigma} = \zeta_{\bar{\nu}-\mathbf{p}-\sigma}(-1)^{1+\sigma} \equiv \zeta^{*}_{\nu}. \tag{2.392}$$

Thus, the transformation of a spin-1 field becomes

$$\mathbf{T}A^{\mu}(x)\mathbf{T}^{+} = \zeta^{*}_{\nu}A_{\mu}(x') \qquad (x' = Tx) \tag{2.393}$$

To summarize, under time reversal the fields transforms as [90]

$$\begin{aligned} \mathbf{T}\phi(x)\mathbf{T}^{+} &= \zeta^{*}_{n}\phi(Tx) \quad (\text{spin} - 0) \\ \mathbf{T}\psi(x)\mathbf{T}^{+} &= \zeta^{*}_{n}\mathcal{T}\psi(Tx) \quad (\text{spin} - 1/2) \\ \mathbf{T}A^{\mu}(x)\mathbf{T}^{+} &= \zeta^{*}_{n}A_{\mu}(Tx) \quad (\text{spin} - 1), \end{aligned} \tag{2.394}$$

where the factors ζ_n in general case can differ for different particles, and $\mathcal{T}$ is ternary and antisymmetric (4×4)-matrix that satisfies the relationship $\mathcal{T}\gamma^{\mu}\mathcal{T}^{+} = \gamma^{*}_{\mu}$ $(\mu = 0, 1, 2, 3)$. We note that for the spin-1 field the Lorentz index μ changes from superscript to subscript under time-reversal operator $\mathbf{T}$.

It is shown in [90] that under time-reversal operator the current transforms as

$$\mathbf{T}j_{\mu}(x)\mathbf{T}^{+} = \psi^{+}(x')\gamma_0\gamma^{\mu}\psi(x') = j^{\mu}(x') \quad (x' = Tx). \tag{2.395}$$

Now let find how the operator $\mathbf{T}$ acts on Hamiltonian $h_{\text{QED}}(t)$ (2.364), beforehand presenting it in the form

$$h_{\text{QED}} = \int d^3x H_{\text{int}}(x) = \int d^3x A^{\mu}(x) j_{\mu}(x), \quad j_{\mu}(x) = \bar{\psi}(x)\gamma_{\mu}\psi(x). \tag{2.396}$$

Taking into account (2.396), (2.395) and (2.394), the transformed Hamiltonian $H_{\text{int}}(x)$ under time-reversal operator $\mathbf{T}$ can be represented as

$$\mathbf{T}H_{\text{int}}(x)\mathbf{T}^{+} = \mathbf{T}A^{\mu}(x)\mathbf{T}^{+}\mathbf{T}j_{\mu}(x)\mathbf{T}^{+} = \zeta^{*}_{\nu}A_{\mu}(x')j^{\mu}(x'), \tag{2.397}$$

where the index ν of the phase factor is related to the photon field $(n = \nu)$. By choosing the arbitrary phase factor ζ^{*}_{ν} to be unity, we obtain

$$\mathbf{T}H_{\text{int}}(x)\mathbf{T}^{+} = A_{\mu}(x')j^{\mu}(x') = H_{\text{int}}(x') \quad (x' = Tx) \tag{2.398}$$

The space-integrated Hamiltonian of QED interaction is transformed under time reversal as [90]

$$\mathbf{T}h_{\text{QED}}(t)\mathbf{T}^{+} = \int d^3x \mathbf{T}H_{\text{int}}(x)\mathbf{T}^{+} = \int d^3x H_{\text{int}}(x') = h_{\text{QED}}(-t) \quad (x' = Tx). \tag{2.399}$$

Similarly, the Yukawa coupling transforms under **T** as

$$\mathbf{T}\bar{\psi}(x)\psi(x)\phi(x)\mathbf{T}^{+} = \zeta_s^{*}\bar{\psi}(x')\psi(x')\phi(x') \quad (x' = Tx). \tag{2.400}$$

When the formula (2.400) was deduced, the first two relationships (2.394) were used. By choosing $\zeta_s^* = 1$, the interaction Hamiltonian $h_Y(t)$ satisfy the relationship [90]

$$\mathbf{T}h_Y(t)\mathbf{T}^{+} = h_Y(-t). \tag{2.401}$$

Thus, all processes caused by the Yakawa coupling are invariant under time-reversal.

At the end of this section, we note that applying **T** twice spin-0 and spin-1 fields recovers the original field

$$\mathbf{T}\mathbf{T}\phi(x)\mathbf{T}^{+}\mathbf{T}^{+} = \zeta_n\mathbf{T}\phi(Tx)\mathbf{T}^{+} = \zeta_n\zeta_n^{*}\phi(T^2x) = \phi(x). \tag{2.402}$$

$$\mathbf{T}\mathbf{T}A^{\mu}(x)\mathbf{T}^{+}\mathbf{T}^{+} = \zeta_n\mathbf{T}A_{\mu}(Tx)\mathbf{T}^{+} = \zeta_n\zeta_n^{*}A^{\mu}(T^2x) = A^{\mu}(x). \tag{2.403}$$

As for the spin-$1/2$ fields, applying **T** twice flips the sign:

$$\begin{aligned}\mathbf{T}\mathbf{T}\psi(x)\mathbf{T}^{+}\mathbf{T}^{+} &= \mathbf{T}\zeta_n^{*}\mathcal{T}\psi(Tx)\mathbf{T}^{+} = \zeta_n\mathcal{T}^{*}\mathbf{T}\psi(Tx)\mathbf{T}^{+} = \\ &= \zeta_n\zeta_n^{*}\mathcal{T}^{*}\mathcal{T}\psi(T^2x) = -\psi(x),\end{aligned} \tag{2.404}$$

because $\zeta_n\zeta_n^* = 1$, $\mathcal{T}^*\mathcal{T} = -1$, and $T^2x = x$.

The relation (2.404) represents the Kramers theorem, which takes place in the quantum field theory for spin-$1/2$ fields, as well as in quantum mechanics for a system formed by an odd number of particles with spin $1/2$.

2.15 The *CPT* Theorem

Relations (2.394) show how scalar field $\phi(x)$ ((2.367), spin equal to 0), fermion field $\psi(x)$ ((2.374), spin equal to $1/2$), and the charged vector field $A^{\mu}(x)$ ((2.388), spin equal to 1) are transformed under the time-reversal operator **T**. Similarly, (2.399) and (2.401) describe the transformations of space-integrated Hamiltonians of QED interaction and Yukawa coupling under **T**. In all these cases, we deal with the time-reversal symmetry or, in other words, with the invariance of interactions under time-reversal (T-invariance).

In a similar way, we can consider **P**- and **C**-invariances, connected with the operator of space-reversal **P** and with operator of charged conjugation **C**, respectively. In Table 2.1 it is indicated how some physical quantities are transformed at operations **P**, **T** and **C**.

Now, we consider the principle of invariance of physical laws relative to each of the symmetry transformations **P**, **T** and **C**. First of all, we note that the operation of

Table 2.1 Transformation of some physical quantities at operations **P**, **T** and **C** [103]

Initial quantity	Result of operation		
	P	**T**	**C**
Vector of position $\mathbf{r}(t)$	$-\mathbf{r}(t)$	$\mathbf{r}(-t)$	$\mathbf{r}(t)$
Impulse $\mathbf{p}(t)$	$-\mathbf{p}(t)$	$-\mathbf{p}(-t)$	$\mathbf{p}(t)$
Total angular momentum $\mathbf{J}(t)$	$\mathbf{J}(t)$	$-\mathbf{J}(-t)$	$\mathbf{J}(t)$
Electrical charge q	q	q	$-q$
Electric field strength $\mathbf{E}(\mathbf{r}, t)$	$-\mathbf{E}(-\mathbf{r}, t)$	$\mathbf{E}(\mathbf{r}, -t)$	$-\mathbf{E}(\mathbf{r}, t)$
Magnetic induction $\mathbf{B}(\mathbf{r}, t)$	$\mathbf{B}(-\mathbf{r}, t)$	$-\mathbf{B}(\mathbf{r}, -t)$	$-\mathbf{B}(\mathbf{r}, t)$
Electric dipole momentum $\mathbf{d}_E(t)$	$-\mathbf{d}_E(t)$	$\mathbf{d}_E(-t)$	$-\mathbf{d}_E(t)$
Magnetic dipole momentum $\boldsymbol{\mu}(t)$	$\boldsymbol{\mu}(t)$	$-\boldsymbol{\mu}(-t)$	$-\boldsymbol{\mu}(t)$

space reversal is a product of the mirror reflection in the plane xOy and the rotation by angle 180^0 around the axis Oz perpendicular to the given plane. Since there is no reason to doubt the invariance of laws relative to rotations, it follows that it is necessary to consider either space reversal or reflection. Until 1956, the principle of invariance of physical laws relative to the space reversal **P** remained immovable and nobody decided to doubt, for example, the fact the parity conservation law is realized in all interactions.[8] However, it was discovered soon that the weak interaction is not invariant relative to the space reversal. Wu et al. [105] discovered that parity at β-decay of ^{60}Co is not preserved, in the experiment suggested by Lee and Yang [106]. Another experimental proof of violation of the invariance principle relative to space reversal in weak interactions is the fact, that neutrino is met in nature with chirality $h_\upsilon = -1$, while photon may exist both in the state with chirality $+1$, and with chirality -1 (the chirality h_υ of particle or antiparticle is defined as $h_\upsilon = (\mathbf{s} \cdot \mathbf{p})/(|\mathbf{s}| \cdot |\mathbf{p}|)$, where $\mathbf{s}$ is the spin vector and $\mathbf{p}$ is the linear momentum vector). These two examples are sufficient to show that P-invariance is not universal (it is violated in weak interactions) and, at the same time, to doubt the universality of C- and T-invariances.

The operation of charge conjugation **C** consists in reversal of all electric, baryon and lepton charges (as well as strangeness) of particles without changing their position and velocity. In other words, other conditions being equal, these operations consists in a transformation of the particle system in a system of antiparticles.

Up to 1956, it was considered that all physical laws are invariant not only relative to operation **P**, but also to the operation of charge conjugation **C**. Now, this assertion

[8]Although in 1848 Paster called attention to the fact that some organic compounds are met in biological structures only in a left rotation form [104].

is confirmed for all interactions, excepting the weak interaction. The violation of **C**-invariance in the case of weak interactions is attested, in particular, by the existence, as it was already noted, of a unique state of chirality $h_\nu = -1$ at neutrino. If it is produced a "charge conjugation" of neutrino, one gets a fictitious anti-neutrino with chirality $h_{\bar{\nu}} = -1$, which does not exist in nature.

Thus, the weak interaction violates simultaneously P-invariance and C-invariance, but it may turn out that PC-invariance is not violated. In the existence of PC-invariance, one can be convinced by considering again the properties of the neutrino. Indeed, as a result of action on the neutrino by the operator of space reversal **P** a fictitious neutrino with chirality $h_\nu = +1$ is obtained (we remind that the real neutrino has the chirality $h_\nu = -1$). However, if on this imaginary object it is produced the operation of charge conjugation **C**, which in this case consists in changing the sign of the lepton charge, then we again obtain an object existing in nature, namely an anti-neutrino with chirality $h_{\overline{\nu}} = +1$.

This result is the basis of the so-called theory of two-component neutrino [103], where Dirac equation for neutrino with the mass taken equal to zero[9] has only two, not four linearly independent solutions. One of them corresponds to the positive energy and is connected with the neutrino in a state with chirality $h_\nu = -1$, and the other corresponds to the negative energy and describes the anti-neutrino with chirality $h_{\overline{\nu}} = +1$. The anti-neutrino, of which mass is zero in this theory, like the mass of neutrino, is spread, like a neutrino, with a light velocity.

However, the anti-neutrino differs significantly from the neutrino in that its spin is always directed along the motion, and not against, as in the case of neutrinos.

Thus, in a processes with the participation of a neutrino the C-invariance is violated.[10] However, the role of the neutrino is not restricted only to this fact, especially after the fundamental discovery of existence its rest mass. In particular, the data concerning the atmosphere neutrinos are consistent with $\nu_\mu \longleftrightarrow \nu_\tau$ oscillations (ν_μ is the muon neutrino and ν_τ is the tau neutrino). These oscillations show that neutrinos have some rest mass, although for a long time they were considered massless. The experimental data show a zenith angle v dependent deficit of muon neutrinos, which is inconsistent with the exceptions based on calculations of the atmosphere neutrinos flux. These data are consistent with $\nu_\mu \longleftrightarrow \nu_\tau$ oscillations with $\sin^2 2\theta > 0.82$ and $5 \cdot 10^{-4} < \Delta\mathrm{m}^2 < 6 \cdot 10^{-3}\,\mathrm{eV}^2/\mathrm{c}^2$ (c is the light speed in vacuum) at a 90% confidence level [108]. After this discovery the neutrino becomes an important participant in the Universe evolution scenario.

The observed now expansion of the Universe, along with the experimental discovery of the relic microwave background radiation confirm the Big Bang theory. Will the expansion of the Universe be continuous for ever, beginning with, or will it replaced by compression in some time moment, it depends on the amount of neutrino particles in the Universe. In spite of the very small rest mass of these particles, their

[9]Experimental researches, made often the suggestion of theory of two-component neutrino, showed that really the rest mass of neutrino is not equal to zero.

[10]Moreover, as noted in [107], in some processes with the participation of neutrinos even the *CPT* theorem is violated.

great amount in the Universe can be sufficient for the mean density of the matter in the Universe $\overline{\rho}$ to be greater than the critical one ($\overline{\rho} > \rho_{cr}$), at which the expansion of the Universe will be replaced by its compression. On the other hand, a recent publication [107] mentioned the role of neutrinos in the process of expansion of the Universe.

Numerous experimental data allow to conclude that the weak interaction is CP-invariant [103]. This means that the C-invariance is violated if P-invariance is violated (and vice versa). This is exactly what is observed experimentally in weak interactions. The evolution from P- to CP-invariance shows that in the case when a violation of CP-invariance is observed, it is necessary to require all physical laws to be invariant relative to the transformation CPT. The violation of the CP-invariance in weak interactions was observed as early as 1964 in the decay of K_L^0 meson into two π-mesons in one special process ($K_L^0 \to \pi^+ + \pi^-$) [109]. Though this process occurs very rarely (the relative probability $\sim 1.6 \cdot 10^{-3}$), it raises doubts about the existence of CP-invariance. In this situation, the requirement of CPT-invariance inevitably leads to violation of T-invariance. Different ways of violation of T-invariance (more exactly, the reduction of the T-symmetry from $T \equiv T_3$ to T_2 and T_1 as a consequence of the special structure of non-Abelian and Abelian groups, introduced in the Chap. 8), will be discussed in the Chap. 9.

Thus, all physical laws are invariant relative to the antiunitary involution $\boldsymbol{\theta} = \mathbf{CPT}$ (the order of succession of operators $\mathbf{C}, \mathbf{P}$ and $\mathbf{T}$ is not significant, because the operators are commuting). It is the famous Lüders–Pauli CPT theorem [110–112].

The CPT theorem is of a great importance, since it shows that the discrete symmetry appears not only in theories, which are symmetrically relative to operations $\mathbf{P}$, $\mathbf{C}$ and $\mathbf{T}$, but also in theories containing initially only an invariance relative to continuous symmetry groups.

According to the CPT theorem, even if the invariance relative to each of $\mathbf{C}, \mathbf{P}$ and $\mathbf{T}$ operators separately is not present, the invariance relative to their combination is fulfilled. Particularly, in the case of a charged field this means the existence of antiparticles with the same masses and spins as the respective particles. Moreover, in this case, the invariance of the theory relative to **CPT** transformations means also the equality of g-factors of particles and antiparticles. This makes it possible to experimentally test the CPT symmetry with high accuracy. The more exact value of the ratio of g-factors for the positron and electron is the following [89]:

$$\frac{g(e^+)}{g(e^-)} = 1 + (22 \pm 64) \cdot 10^{-22}.$$

In the case of neutral fields, the particles and antiparticles are identical.

The exceptionality of the Lüders–Pauli theorem is that its content during the time was deepened and methods of its proof were continuously improved, using a generalized functions and an axiomatic approach in the theory of quantum fields. In this connection, it should be noted especially the results, obtained by Jost, as well as by Streater and Wingtman in their books [98, 102]. The proof of the CPT theorem in an axiomatic approach was fulfilled by Jost [102], which has shown that every field theory satisfying all axioms possesses a supplementary discrete antilinear symmetry θ.

However, understanding the contents and proof of the *CPT* theorem in an axiomatic approach require a special mathematical training in the theory of generalized functions. Since our book is addressed to a wider range of readers, we shall not give here this proof, but will consider the transformation of fields under the antiunitary involution $\boldsymbol{\theta}$ [90]. For this, in addition to formulas (2.394), we will obtain similar formulas for the transformation of fields under the operators $\boldsymbol{P}$ and $\mathbf{C}$.

The parity operator $\mathbf{P}$, defined in the Hilbert space with the basis (2.351), should transform the creation operator of a particle of type n, momentum $\mathbf{p}$ and spin σ to one for $(n, -\mathbf{p}, \sigma)$ up to a phase factor that, in general case, depends on $(n, \mathbf{p}, \sigma)$:

$$\mathbf{P}a^+_{n\mathbf{p}\sigma}\mathbf{P}^{-1} = \eta_{n\mathbf{p}\sigma}a^+_{n-\mathbf{p}\sigma}, \quad (|\eta_{n\mathbf{p}\sigma}| = 1). \tag{2.405}$$

Taking into account that the vacuum state $|0\rangle$ should transform under a parity operator up to a phase factor

$$\mathbf{P}\,|0\rangle = \eta_{\text{VAC}}\,|0\rangle\,, \tag{2.406}$$

the basis state $|\{n_i, \mathbf{p}_i, \sigma_i\}\rangle$ of (2.351) transforms under the operator $\mathbf{P}$ as

$$\begin{aligned}\mathbf{P}\,|\{n_i, \mathbf{p}_i, \sigma_i\}\rangle &= \mathbf{P}a^+_{n_1\mathbf{p}_1\sigma_1}\mathbf{P}^{-1}\mathbf{P}\ldots\ldots\mathbf{P}^{-1}\mathbf{P}a^+_{n_k\mathbf{p}_k\sigma_k}\mathbf{P}^{-1}\mathbf{P}\,|0\rangle = \\ &= \eta_{n_1,\mathbf{p}_1,\sigma_1}\ldots\eta_{n_k\mathbf{p}_k\sigma_k}\eta_{\text{VAC}}a^+_{n_1-\mathbf{p}_1\sigma_1}\ldots a^+_{n_k-\mathbf{p}_k\sigma_k}\,|0\rangle = \\ &= \eta_{\{n_i,\mathbf{p}_i,\sigma_i\}}\eta_{\text{VAC}}\,|\{n_i, -\mathbf{p}_i, \sigma_i\}\rangle\,,\end{aligned} \tag{2.407}$$

where

$$\eta_{\{n_i,\mathbf{p}_i,\sigma_i\}} \equiv \eta_{n_1\mathbf{p}_1\sigma_1}\ldots\eta_{n_k\mathbf{p}_k\sigma_k}. \tag{2.408}$$

We suppose that the parity of the vacuum is taken to be $+1$ [90]. Then

$$\mathbf{P}\,|0\rangle = |0\rangle\,,$$

$$\mathbf{P}\,|\{n_i, \mathbf{p}_i, \sigma_i\}\rangle = \eta_{\{n_i,\mathbf{p}_i,\sigma_i\}}\,|\{n_i, -\mathbf{p}_i, \sigma_i\}\rangle\,. \tag{2.409}$$

We see that for the set of creation operators $\{a^+_{n_i\mathbf{p}_i\sigma_i}\}$ that correspond to all possible particle types, momenta and spins, one can choose an arbitrary set of phases for constructing a unitary operator $\mathbf{P}$ representing the space inversion in a Hilbert space.

Taking into account that $\mathbf{P}$ is an unitary operator, we can rewrite (2.405) in the form

$$\mathbf{P}a^+_{n\mathbf{p}\sigma}\mathbf{P}^+ = \eta_{n\mathbf{p}\sigma}a^+_{n-\mathbf{p}\sigma}. \tag{2.410}$$

Then we shall obtain an analogous expression for the transformation of an annihilation operator $a_{n\mathbf{p}\sigma}$:

$$\mathbf{P}a_{n\mathbf{p}\sigma}\mathbf{P}^+ = \eta^*_{n\mathbf{p}\sigma}a_{n-\mathbf{p}\sigma}. \tag{2.411}$$

The operators $a_{n\mathbf{p}\sigma}$ and $a^+_{n'\mathbf{p}'\sigma'}$ satisfy commutation relations

$$\left[a_{n\mathbf{p}\sigma}, a^+_{n'\mathbf{p}'\sigma'}\right] = \delta_{nn'}\delta_{\mathbf{p}\mathbf{p}'}\delta_{\sigma\sigma'} \tag{2.412}$$

for bosons and the anti-commutation relations

$$\left\{a_{n\mathbf{p}\sigma}, a^+_{n'\mathbf{p}'\sigma'}\right\} = \delta_{nn'}\delta_{\mathbf{p}\mathbf{p}'}\delta_{\sigma\sigma'} \tag{2.413}$$

for fermions.

Let firstly apply the parity operator to a spin 0 field from (2.367), where the index n is replaced by s to emphasize that this is a scalar field. For a neutral scalar field we have $\overline{s} = s$. Using (2.410), (2.411) and (2.367), we have [90]

$$\begin{aligned}\mathbf{P}\phi(x)\mathbf{P}^+ &= \sum_{\mathbf{p}}(\mathbf{P}a_{s\mathbf{p}}\mathbf{P}^+e_{\mathbf{p}}(x) + \mathbf{P}a_{\overline{s}\mathbf{p}}\mathbf{P}^+e^*_{\mathbf{p}}(x)) = \\ &= \sum_{\mathbf{p}}(\eta^*_{s\mathbf{p}}a_{s-\mathbf{p}}e_{\mathbf{p}}(x) + \eta_{\overline{s}\mathbf{p}}a^+_{\overline{s}-\mathbf{p}}e^*_{\mathbf{p}}(x)) = \\ &= \sum_{\mathbf{p}}\frac{1}{\sqrt{2p^0V}}(\eta^*_{s\mathbf{p}}a_{s-\mathbf{p}}e^{-i(p^0x^0-\mathbf{p}\mathbf{x})} + \eta_{\overline{s}\mathbf{p}}a^+_{\overline{s}-\mathbf{p}}e^{i(p^0x^0-\mathbf{p}\mathbf{x})}) = \\ &= \sum_{\mathbf{p}}\frac{1}{\sqrt{2p^0V}}(\eta^*_{s\mathbf{p}}a_{s-\mathbf{p}}e^{-i(p^0x'^0+\mathbf{p}\mathbf{x}')} + \eta_{\overline{s}\mathbf{p}}a^+_{\overline{s}-\mathbf{p}}e^{i(p^0x'^0+\mathbf{p}\mathbf{x}')}) = \\ &= \sum_{\mathbf{p}}\frac{1}{\sqrt{2p^0V}}(\eta^*_{s-\mathbf{p}}a_{s\mathbf{p}}e^{-ipx} + \eta_{\overline{s}-\mathbf{p}}a^+_{\overline{s}\mathbf{p}}e^{ipx}) = \\ &= \sum_{\mathbf{p}}(\eta^*_{s-\mathbf{p}}a_{s\mathbf{p}}e_{\mathbf{p}}(x') + \eta_{\overline{s}-\mathbf{p}}a^+_{\overline{s}\mathbf{p}}e^*_{\mathbf{p}}(x')),\end{aligned} \tag{2.414}$$

when passing from the third sum to the fourth one and from the fourth sum to the fifth one, the changes $\mathbf{p} \longrightarrow -\mathbf{p}$, $x' = (x^0, -\mathbf{x})$ and $px' = p^0x'^0 + \mathbf{p}\mathbf{x}'$ were used, respectively, and passing from the fifth sum to the sixth one the formula (2.368) was used.

Let's now apply the parity operator to a spin-1/2 fermion field from (2.374), where n is replaced by f to emphasize that this is a fermion field. Applying $\mathbf{P}$ to $\psi(x)$ from (2.374) and using (2.410) and (2.411), we get

$$\begin{aligned}\mathbf{P}\psi(x)\mathbf{P}^+ &= \sum_{\mathbf{p},\sigma}(\mathbf{P}a_{f\mathbf{p}\sigma}f_{\mathbf{p}\sigma}(x)\mathbf{P}^+ + \mathbf{P}a^+_{\overline{f}\mathbf{p}\sigma}g_{\mathbf{p}\sigma}(x)\mathbf{P}^+) = \\ &= \sum_{\mathbf{p},\sigma}\frac{1}{\sqrt{2p^0V}}(\eta^*_{f\mathbf{p}\sigma}a_{f-\mathbf{p}\sigma}u_{\mathbf{p}\sigma}e^{-ipx} + \eta_{\overline{f}\mathbf{p}\sigma}a^+_{\overline{f}-\mathbf{p}\sigma}v_{\mathbf{p}\sigma}e^{ipx}) = \\ &= \sum_{\mathbf{p},\sigma}\frac{1}{\sqrt{2p^0V}}(\eta^*_{f-\mathbf{p}\sigma}a_{f\mathbf{p}\sigma}u_{-\mathbf{p}\sigma}e^{-ipx'} + \eta_{\overline{f}-\mathbf{p}\sigma}a^+_{\overline{f}\mathbf{p}\sigma}v_{-\mathbf{p}\sigma}e^{ipx'}),\end{aligned} \tag{2.415}$$

when passing from the second sum to the third one, the changes $x' = (x^0, -\mathbf{x})$ and $\mathbf{p} \to -\mathbf{p}$ were used, taking into account definitions of functions $f_{\mathbf{p}\sigma}(x)$ and $g_{\mathbf{p}\sigma}(x)$ by formulas (2.375) and that of the function $e_{\mathbf{p}}(x)$ by the formula (2.368).

Using relationships [90]

$$\begin{aligned} \gamma^0 u_{\mathbf{p}\sigma} &= u_{-\mathbf{p}\sigma}, \\ \gamma^0 v_{\mathbf{p}\sigma} &= -v_{-\mathbf{p}\sigma}, \end{aligned} \tag{2.416}$$

where γ^0 is the Dirac matrix

$$\gamma^0 = \begin{pmatrix} \mathbf{I} & \mathbf{0} \\ \mathbf{0} & -\mathbf{I} \end{pmatrix}, \quad \mathbf{I} = \begin{pmatrix} 1 & 0 \\ 0 & 1 \end{pmatrix}, \quad \mathbf{0} = \begin{pmatrix} 0 & 0 \\ 0 & 0 \end{pmatrix}, \tag{2.417}$$

the formulas (2.415) may be written in the form

$$\begin{aligned} \mathbf{P}\Psi(x)\mathbf{P}^+ &= \sum_{\mathbf{p},\sigma} \frac{1}{\sqrt{2P^0V}} (\eta^*_{f-\mathbf{p}\sigma} a_{f\mathbf{p}\sigma} \gamma^0 u_{\mathbf{p}\sigma} e^{-ipx'} - \eta_{\overline{f}-\mathbf{p}\sigma} a^+_{\overline{f}\mathbf{p}\sigma} \gamma^0 v_{\mathbf{p}\sigma} e^{ipx'}) = \\ &= \gamma^0 \sum_{\mathbf{p},\sigma} (\eta^*_{f-\mathbf{p}\sigma} a_{f\mathbf{p}\sigma} f_{\mathbf{p}\sigma}(x') - \eta_{\overline{f}-\mathbf{p}\sigma} a^+_{\overline{f}\mathbf{p}\sigma} g_{\mathbf{p}\sigma}(x')). \end{aligned} \tag{2.418}$$

Taking into consideration the relations $\eta_{\overline{s}} = \eta^*_s$ and $\eta_{\overline{f}} = -\eta^*_f$ in (2.414) and (2.415), the transformations of spin-0 and spin-1/2 fields under the parity operator $\mathbf{P}$ are written as [90]

$$\begin{aligned} \mathbf{P}\phi(x)\mathbf{P}^+ &= \eta^*_s \phi(x'), \\ \mathbf{P}\psi(x)\mathbf{P}^+ &= \eta^*_f \gamma^0 \psi(x'), \quad \left(x' \equiv Px\right) \end{aligned} \tag{2.419}$$

where the operator of the space inversion P belongs to the Lorentz group. This operator can be explicitly written in the space-time four dimension as

$$P = \begin{pmatrix} 1 & & & \\ & -1 & & \\ & & -1 & \\ & & & -1 \end{pmatrix}. \tag{2.420}$$

We will proceed in an analogical way for the charged vector field (2.388), where we set $n = v$ indicating that it is a vector field and that $\overline{v} = v$ for a neutral spin-1 field. Applying $\mathbf{P}$ to (2.388) and fulfilling the change $\mathbf{p} \longrightarrow -\mathbf{p}$, we obtain

$$\begin{aligned} \mathbf{P}A^\mu(x)\mathbf{P}^+ &= \sum_{\mathbf{p},\sigma} (\eta^*_{v\mathbf{p}\sigma} a_{v-\mathbf{p}\sigma} \varepsilon^\mu_{\mathbf{p}\sigma} e_{\mathbf{p}}(x) + \eta_{\overline{v}\mathbf{p}\sigma} a^+_{\overline{v}-\mathbf{p}\sigma} \varepsilon^{\mu *}_{\mathbf{p}\sigma} e^*_{\mathbf{p}}(x)) = \\ &= \sum_{\mathbf{p}\sigma} (\eta^*_{v-\mathbf{p}\sigma} a_{v\mathbf{p}\sigma} \varepsilon^\mu_{-\mathbf{p}\sigma} e_{\mathbf{p}}(x') + \eta_{\overline{v}-\mathbf{p}\sigma} a^+_{\overline{v}\mathbf{p}\sigma} \varepsilon^{\mu *}_{-\mathbf{p}\sigma} e^*_{\mathbf{p}}(x')) = \end{aligned} \tag{2.421}$$

$$= -\sum_{\mathbf{p}\sigma}(\eta^*_{v-\mathbf{p}\sigma} a_{v\mathbf{p}\sigma} \varepsilon_{\mathbf{p}\sigma\mu} e_{\mathbf{p}}(x') + \eta_{\bar{v}-\mathbf{p}\sigma} a^+_{\bar{v}\mathbf{p}\sigma} \varepsilon^*_{\mathbf{p}\sigma\mu} e^*_{\mathbf{p}}(x')),$$

where x' is determined by (2.419) and (2.420). Here it was taken into account the relations

$$\varepsilon^{\mu}_{-\mathbf{p}\sigma} = -\varepsilon_{\mathbf{p}\sigma\mu}, \quad \varepsilon^{\mu *}_{-\mathbf{p}\sigma} = -\varepsilon^*_{\mathbf{p}\sigma\mu}. \tag{2.422}$$

Using $\eta_{\bar{v}} = \eta^*_v$ in (2.421), the transformation of a vector field under the parity operator is now written as

$$\mathbf{P}A^{\mu}(x)\mathbf{P}^{+} = -\eta^*_v A_{\mu}(x'), \tag{2.423}$$

where x' is determined by (2.419) and (2.420), and the Lorentz index μ changed its position.

Thus, the transformation properties of spin-0, spin-1/2 and spin-1 fields under the parity operator $\mathbf{P}$ are [90]

$$\begin{aligned}
\mathbf{P}\phi(x)\mathbf{P}^{+} &= \eta^*_n \phi(Px) \quad (\text{spin } S = 0) \\
\mathbf{P}\psi(x)\mathbf{P}^{+} &= \eta^*_n \gamma^0 \psi(Px) \quad (\text{spin } S = 1/2) \\
\mathbf{P}A^{\mu}(x)\mathbf{P}^{+} &= -\eta^*_n A_{\mu}(Px) \quad (\text{spin } S = 1) \\
\eta_{\overline{n}} &= \eta^*_n \quad (\text{spin } S = 0, 1) \\
\eta_{\overline{n}} &= -\eta^*_n \quad (\text{spin } S = 1/2).
\end{aligned} \tag{2.424}$$

Equation (2.424) satisfy the relationships $\eta^*_n = \eta_{\overline{n}}$ for spins $S = 0, 1$ and $\eta^*_n = -\eta_{\overline{n}}$ for $S = 1/2$. For self-conjugate (Hermitian) spin-0 and spin-1 fields, we have $\overline{n} = n$. Comparing this equality with $\eta_{\overline{n}} = \eta^*_n$ from (2.424), we obtain $\eta_n = \eta^*_n$, from which for the phase factor η_n we find $\eta_n = \pm 1$. The transformations of spin-0 and spin-1 fields under the parity operator are represented in the Table 2.2, where the phase factor η_n is noted by η.

For charged or non-Hermitian fields, η is not restricted by itself to ± 1, excepting the case when charged particles are related to neutral particles characterized by the isospin.

Table 2.2 The transformations of self-conjugate (Hermitian) spin-0 and spin-1 fields under the parity operator at $\eta = \pm 1$

S	$\eta = +1$	$\eta = -1$
0	$\mathbf{P}\phi(x)\mathbf{P}^{+} = \phi(Px)$	$\mathbf{P}\phi(x)\mathbf{P}^{+} = -\phi(Px)$
	scalar	pseudo-scalar
1	$\mathbf{P}A^{\mu}(x)\mathbf{P}^{+} = -A_{\mu}(Px)$	$\mathbf{P}A^{\mu}(x)\mathbf{P}^{+} = A_{\mu}(Px)$
	axial vector	vector

Finally, we consider transformations of the spin-0 filed $\phi(x)$, spin-$1/2$ field $\psi(x)$ and spin-1 field $A^{\mu}(x)$ under the operator of the charge conjugation **C**. The operator **C** changes the particle and the antiparticle without changing the momentum and the spin:

$$\mathbf{C}a^{+}_{n\mathbf{p}\sigma}\mathbf{C}^{+} = \xi_{n\mathbf{p}\sigma}a^{+}_{\overline{n}\mathbf{p}\sigma}, \tag{2.425}$$

where the phase factor $\xi_{n\mathbf{p}\sigma}$, in general, depends on the particle type, momentum, and spin. With such a definition, the particle and antiparticle are treated as separate particles. Namely, the above definition includes (with $\overline{\overline{n}} = n$)

$$\mathbf{C}a^{+}_{\overline{n}\mathbf{p}\sigma}\mathbf{C}^{+} = \xi_{\overline{n}\mathbf{p}\sigma}a^{+}_{n\mathbf{p}\sigma}, \tag{2.426}$$

In the analogy of **P** and **T** operators, the set of phases $\{\xi_{n\mathbf{p}\sigma}\}$ completely determines the operator **C** that it must be unitary:

$$\mathbf{C}^{+}\mathbf{C} = \mathbf{1}. \tag{2.427}$$

Similarly to **P** and **T**, the overall phase of **C** is defined by requiring the vacuum to be invariant under **C** :

$$\mathbf{C}\left|0\right\rangle = \left|0\right\rangle, \tag{2.428}$$

$$\mathbf{C}^{+}\left|0\right\rangle = \left|0\right\rangle, \tag{2.429}$$

where (2.429) is obtained by applying $\mathbf{C}^{+}$ to both sides of the (2.428). Then, applying (2.425) to the vacuum state, we obtain

$$\mathbf{C}\left|n\mathbf{p}\sigma\right\rangle = \xi_{n\mathbf{p}\sigma}\left|\overline{n}\mathbf{p}\sigma\right\rangle. \tag{2.430}$$

Under charge conjugation operator **C** the multi-particle states transform with a multiplicative phase factor

$$\begin{aligned}\mathbf{C}\left|\{n_i,\mathbf{p}_i,\sigma_i\}\right\rangle &= \mathbf{C}a^{+}_{n_1\mathbf{p}_1\sigma_1}\mathbf{C}^{+}\mathbf{C}\ldots\ldots\mathbf{C}^{+}\mathbf{C}a^{+}_{n_k\mathbf{p}_k\sigma_k}\mathbf{C}^{+}\mathbf{C}\left|0\right\rangle = \\ &= \xi_{n_1\mathbf{p}_1\sigma_1}\ldots\xi_{n_k\mathbf{p}_k\sigma_k}a^{+}_{\overline{n}_1\mathbf{p}_1\sigma_1}\ldots a^{+}_{\overline{n}_k\mathbf{p}_k\sigma_k}\left|0\right\rangle,\end{aligned} \tag{2.431}$$

where the implicit normalization factor is different from unity when there are more particles with identical quantum numbers. In this case,

$$\mathbf{C}\left|\{n_i,\mathbf{p}_i,\sigma_i\}\right\rangle = \xi_{\{n_i,\mathbf{p}_i,\sigma_i\}}\left|\{\overline{n}_i,\mathbf{p}_i,\sigma_i\}\right\rangle, \tag{2.432}$$

where

$$\xi_{\{n_i,\mathbf{p}_i,\sigma_i\}} \equiv \xi_{n_1\mathbf{p}_1\sigma_1}\ldots\xi_{n_k\mathbf{p}_k\sigma_k}. \tag{2.433}$$

For spin-0 field, we drop the spin index σ, and write (2.426) and the Hermitian conjugate of (2.425) as

$$\mathbf{C}a^+_{\bar{s}\mathbf{p}}\mathbf{C}^+ = \xi_{\bar{s}\mathbf{p}}a^+_{s\mathbf{p}}, \quad \mathbf{C}a_{s\mathbf{p}}\mathbf{C}^+ = \xi^*_{s\mathbf{p}}a_{\bar{s}\mathbf{p}}, \tag{2.434}$$

where $n = s$ indicates that it takes place for a spin-0 particle. Applying the operator **C** to (2.367) and using the above equation, we obtain

$$\begin{aligned}\mathbf{C}\phi(x)\mathbf{C}^+ &= \sum_{\mathbf{p}}(\mathbf{C}a_{s\mathbf{p}}\mathbf{C}^+e_{\mathbf{p}}(x) + \mathbf{C}a^+_{\bar{s}\mathbf{p}}\mathbf{C}^+e^*_{\mathbf{p}}(x)) = \\ &= \sum_{\mathbf{p}}(\xi^*_{s\mathbf{p}}a_{\bar{s}\mathbf{p}}e_{\mathbf{p}}(x) + \xi_{\bar{s}\mathbf{p}}a^+_{s\mathbf{p}}e^*_{\mathbf{p}}(x)).\end{aligned} \tag{2.435}$$

The right side of (2.435) contains $a_{\bar{s}\mathbf{p}}$ and $a^+_{s\mathbf{p}}$, while $\phi(x)$ contains $a_{s\mathbf{p}}$ and $a^+_{\bar{s}\mathbf{p}}$ (see (2.367)). This leads to the conclusion that the phases $\xi_{s\mathbf{p}}$ and $\xi_{\bar{s}\mathbf{p}}$ should not depend on the momentum:

$$\xi_{s\mathbf{p}} = \xi_s, \quad \xi_{\bar{s}\mathbf{p}} = \xi_{\bar{s}} \tag{2.436}$$

and the charge-conjugation phase of the antiparticle is the complex conjugate of the particle phase:

$$\xi_{\bar{s}} = \xi^*_s. \tag{2.437}$$

The Hermitian conjugate of (2.367) is

$$\phi^+(x) = \sum_{\mathbf{p}}(a^+_{s\mathbf{p}}e^*_{\mathbf{p}}(x) + a_{\bar{s}\mathbf{p}}e_{\mathbf{p}}(x)). \tag{2.438}$$

Multiplying both sides of (2.438) by ξ^*_s, one obtains

$$\xi^*_s\phi^+(x) = \sum_{\mathbf{p}}(\xi^*_s a^+_{s\mathbf{p}}e^*_{\mathbf{p}}(x) + \xi^*_s a_{\bar{s}\mathbf{p}}e_{\mathbf{p}}(x)). \tag{2.439}$$

By comparing (2.439) and (2.435), we see that the right sides of these equations coincide, if one takes into account (2.436) and (2.437). The transformation of a spin-0 field is

$$\mathbf{C}\phi(x)\mathbf{C}^+ = \xi^*_s\phi^+(x). \tag{2.440}$$

Thus, the choice of charge-conjugation phases reduces to assigning a phase to a spin-0 particle. For a self-conjugate particle, the relation $\xi_{\bar{s}} = \xi^*_s$ take place. Therefore ξ_s is real. In this case, $\xi_s = \pm 1$. By setting $\bar{s} = s$ in (2.430), one can be argued that this is the eigenvalue of the charge conjugation operator **C** for a single-particle state with any momentum

$$\mathbf{C}\,|s\mathbf{p}\rangle = \xi_s\,|s\mathbf{p}\rangle \qquad (s = \bar{s}). \tag{2.441}$$

Such ξ_s for a self-conjugate field is called the intrinsic parity conjugation of the particle [90].

We consider now the transformation of a fermion field under the operator of charge conjugation **C**. First, we write (2.426) and the Hermitian conjugate of (2.425) as

$$\mathbf{C}a^+_{\bar{f}\mathbf{p}\sigma}\mathbf{C}^+ = \xi_{\bar{f}\mathbf{p}\sigma}a^+_{f\mathbf{p}\sigma}, \quad \mathbf{C}a_{f\mathbf{p}\sigma}\mathbf{C}^+ = \xi^*_{f\mathbf{p}\sigma}a_{\bar{f}\mathbf{p}\sigma}, \tag{2.442}$$

where we put $n = f$ for fermions. Applying **C** to the (2.374), in which n and $\overline{n}$ are changed by f and $\bar{f}$, we obtain

$$\begin{aligned}\mathbf{C}\psi(x)\mathbf{C}^+ &= \sum_{\mathbf{p},\sigma}(\mathbf{C}a_{f\mathbf{p}\sigma}\mathbf{C}^+ f_{\mathbf{p}\sigma}(x) + \mathbf{C}a^+_{\bar{f}\mathbf{p}\sigma}\mathbf{C}^+ g_{\mathbf{p}\sigma}(x)) = \\ &= \sum_{\mathbf{p},\sigma}(\xi^*_{f\mathbf{p}\sigma}a_{\bar{f}\mathbf{p}\sigma}f_{\mathbf{p}\sigma}(x) + \xi_{\bar{f}\mathbf{p}\sigma}a^+_{f\mathbf{p}\sigma}g_{\mathbf{p}\sigma}(x)),\end{aligned} \tag{2.443}$$

where $f_{\mathbf{p}\sigma}(x)$ and $g_{\mathbf{p}\sigma}(x)$ are determined by (2.375).

According to [90], there exists a (4×4)-matrix Γ, which is unitary and symmetric (the upper index T in (2.444) is a transposition superscript)

$$\Gamma^+\Gamma = 1, \qquad \Gamma^T = \Gamma, \tag{2.444}$$

and satisfies

$$\Gamma\gamma^*_\mu\Gamma^+ = -\gamma_\mu \qquad (\mu = 0, 1, 2, 3), \tag{2.445}$$

where γ_μ are Dirac matrices. Using such matrix Γ, $f_{\mathbf{p}\sigma}(x)$ and $g_{\mathbf{p}\sigma}(x)$ from (2.443) can be represented as [90]

$$\begin{aligned}f_{\mathbf{p}\sigma}(x) &= u_{\mathbf{p}\sigma}e_{\mathbf{p}}(x) = \Gamma v^*_{\mathbf{p}\sigma}e_{\mathbf{p}}(x) = \Gamma g^*_{\mathbf{p}\sigma}(x), \\ g_{\mathbf{p}\sigma}(x) &= v_{\mathbf{p}\sigma}e^*_{\mathbf{p}}(x) = \Gamma u^*_{\mathbf{p}\sigma}e_{\mathbf{p}}(x) = \Gamma f^*_{\mathbf{p}\sigma}(x).\end{aligned} \tag{2.446}$$

Using (2.446), the transformation of $\psi(x)$ (2.443) becomes

$$\begin{aligned}\mathbf{C}\psi(x)\mathbf{C}^+ &= \sum_{\mathbf{p},\sigma}(\xi^*_{f\mathbf{p}\sigma}a_{\bar{f}\mathbf{p}\sigma}\Gamma g^*_{\mathbf{p}\sigma}(x) + \xi_{\bar{f}\mathbf{p}\sigma}a^+_{f\mathbf{p}\sigma}\Gamma f^*_{\mathbf{p}\sigma}(x)) = \\ &= \Gamma\sum_{\mathbf{p},\sigma}(\xi_{\bar{f}\mathbf{p}\sigma}a^+_{f\mathbf{p}\sigma}f^*_{\mathbf{p}\sigma}(x) + \xi^*_{f\mathbf{p}\sigma}a_{\bar{f}\mathbf{p}\sigma}g^*_{\mathbf{p}\sigma}(x)).\end{aligned} \tag{2.447}$$

By reasons analogous to those that were discussed in the case of transformation of the spin-0 field $\phi(x)$ under the **C** operator (formulas (2.436) and (2.437)), the phases $\xi_{f\mathbf{p}\sigma}$ and $\xi_{\overline{f}\mathbf{p}\sigma}$ should not depend on the momentum and spin, and there are the following relationships:

$$\xi_{f\mathbf{p}\sigma} = \xi_f, \qquad \xi_{\bar{f}\mathbf{p}\sigma} = \xi_{\bar{f}} \tag{2.448}$$

and

$$\xi_{\bar{f}} = \xi_f^*. \tag{2.449}$$

Now, the transformation (2.447) can be represented as

$$\mathbf{C}\psi(x)\mathbf{C}^+ = \xi_f^* \Gamma \psi^*(x). \tag{2.450}$$

The action of **C** on a charged spin-1 field is analogous to the action of **C** on a spin-0 field. By analogy with (2.425) and (2.426), the transformations of creation and annihilation operators of the charged spin-1 field are

$$\mathbf{C}a^+_{\bar{\nu}\mathbf{p}\sigma}\mathbf{C}^+ = \xi_{\bar{\nu}\mathbf{p}\sigma} a^+_{\nu\mathbf{p}\sigma}, \; \mathbf{C}a_{\nu\mathbf{p}\sigma}\mathbf{C}^+ = \xi^*_{\nu\mathbf{p}\sigma} a_{\bar{\nu}\mathbf{p}\sigma}, \tag{2.451}$$

where we put $n = \nu$ to specify that they refer to a particle with spin 1. In this case, the μ-component of spin-1 field from (2.388) transforms as

$$\begin{aligned} \mathbf{C}A^\mu(x)\mathbf{C}^+ &= \sum_{\mathbf{p},\sigma}(\mathbf{C}a_{\nu\mathbf{p}\sigma}\mathbf{C}^+ \varepsilon^\mu_{\mathbf{p}\sigma} e_{\mathbf{p}}(x) + \mathbf{C}a^+_{\bar{\nu}\mathbf{p}\sigma}\mathbf{C}^+ \varepsilon^{\mu *}_{\mathbf{p}\sigma} e^*_{\mathbf{p}}(x)) = \\ &= \sum_{\mathbf{p},\sigma}(\xi^*_{\nu\mathbf{p}\sigma} a_{\bar{\nu}\mathbf{p}\sigma} \varepsilon^\mu_{\mathbf{p}\sigma} e_{\mathbf{p}}(x) + \xi_{\bar{\nu}\mathbf{p}\sigma} a^+_{\nu\mathbf{p}\sigma} \varepsilon^{\mu *}_{\mathbf{p}\sigma} e^*_{\mathbf{p}}(x)) = \\ &= \sum_{\mathbf{p},\sigma}(\xi^*_{\bar{\nu}\mathbf{p}\sigma} a_{\nu\mathbf{p}\sigma} \varepsilon^\mu_{\mathbf{p}\sigma} e_{\mathbf{p}}(x) + \xi_{\nu\mathbf{p}\sigma} a^+_{\bar{\nu}\mathbf{p}\sigma} \varepsilon^{\mu *}_{\mathbf{p}\sigma} e_{\mathbf{p}}(x))^+. \end{aligned} \tag{2.452}$$

This can be proportional to $A^{\mu +}(x)$, if the phases do not depend on **p** and σ [90]:

$$\xi_{\nu\mathbf{p}\sigma} = \xi_\nu, \qquad \xi_{\bar{\nu}\mathbf{p}\sigma} = \xi_{\bar{\nu}}, \tag{2.453}$$

and

$$\xi_{\bar{\nu}} = \xi_\nu^*. \tag{2.454}$$

The transformation of $A^{\mu +}(x)$ under the operator of the charge conjugation **C** can be written as

$$\mathbf{C}A^\mu(x)\mathbf{C}^+ = \xi_\nu^* A^{\mu +}(x). \tag{2.455}$$

By analogy with the spin-0 field, if A^μ is Hermitian or the particle is self-conjugate, that is equivalent, then the relation $\xi_{\bar{\nu}} = \xi_\nu^*$ leads to $\xi_\nu = \pm 1$, which becomes an eigenvalue of the operator **C** for a single-particle state $|\nu\mathbf{p}\sigma\rangle$. In this case, the ξ_ν is the intrinsic charge parity of the self-conjugate particle with spin 1.

On the grounds of (2.440), (2.450) and (2.455), the transformation properties of the fields under **C** are [90]

$$\begin{aligned} \mathbf{C}\phi(x)\mathbf{C}^+ &= \xi_n^* \phi^+(x) \quad &&(\text{spin } S = 0) \\ \mathbf{C}\psi(x)\mathbf{C}^+ &= \xi_n^* \Gamma \psi^+(x) \quad &&(\text{spin } S = 1/2) \\ \mathbf{C}A^\mu(x)\mathbf{C}^+ &= \xi_n^* A^{\mu +}(x) \quad &&(\text{spin } S = 1) \end{aligned} \tag{2.456}$$

$$\xi_{\overline{n}} = \xi_n^*. \qquad \text{(spin } S = 0,\ 1/2,\ 1)$$

Now, when we know how fields are transformed under the action of every of operators **C**, **P** and **T** (see (2.456), (2.424) and (2.394), correspondingly), it is easy to find how fields are transformed under the action of the product of these operators. For this, we introduce the operator

$$\boldsymbol{\theta} = \mathbf{CPT} = \mathbf{PTC} = \mathbf{TCP} = \mathbf{CTP} = \mathbf{TPC} = \mathbf{PCT}. \tag{2.457}$$

The operator equalities (2.457) are due to the commutation of operators **C** and **P**, **P** and **T**, **T** and **C**.

Let us find out how fields $\phi(x)$, $\psi(x)$ and $A^\mu(x)$ are transformed under **CPT**. Using the transformations under **T** (2.394), **P** (2.424) and **C** (2.456), we obtain

$$\boldsymbol{\theta}\phi(x)\boldsymbol{\theta}^+ = \mathbf{CPT}\phi(x)\mathbf{T}^+\mathbf{P}^+\mathbf{C}^+ = \xi_n^*\eta_n^*\zeta_n^*\phi^+(-x), \tag{2.458}$$

$$\begin{aligned}\boldsymbol{\theta}\psi(x)\boldsymbol{\theta}^+ &= \mathbf{CPT}\phi(x)\mathbf{T}^+\mathbf{P}^+\mathbf{C}^+ = \mathbf{CP}\zeta_n^*\mathfrak{T}\psi(Tx)\mathbf{P}^+\mathbf{C}^+ = \\ &= \zeta_n^*\mathfrak{T}\mathbf{CP}\psi(Tx)\mathbf{P}^+\mathbf{C}^+ = \zeta_n^*\mathfrak{T}\mathbf{C}\eta_n^*\gamma^0\psi(-x)\mathbf{C}^+ = \\ &= \zeta_n^*\mathfrak{T}\eta_n^*\gamma^0\mathbf{C}\psi(-x)\mathbf{C}^+ = \zeta_n^*\mathfrak{T}\eta_n^*\gamma^0\Gamma\psi^+(-x) = \\ &= \zeta_n^*\eta_n^*\zeta_n^*(\mathfrak{T}\gamma^0\Gamma)\psi^+(-x).\end{aligned} \tag{2.459}$$

The spin-1 vector field transforms under the antiunitary involution $\boldsymbol{\theta}$ as

$$\boldsymbol{\theta}A^\mu\boldsymbol{\theta}^+ = \mathbf{CPT}A^\mu\mathbf{T}^+\mathbf{P}^+\mathbf{C}^+ = -\zeta_n^*\eta_n^*\xi_n^*A^{\mu+}(-x). \tag{2.460}$$

Taking into account that $\Gamma\gamma_\mu^*\Gamma^+ = -\gamma_\mu$ and $\mathfrak{T}\gamma^\mu\mathfrak{T}^+ = \gamma^*$, one can show that the matrix, which anti-commutes with all γ_μ must be γ_5 up to a phase factor. Therefore, on the basis of (2.458), (2.459) and (2.460), the transformations of spin-0, spin-1/2 and spin-1 fields under $\boldsymbol{\theta}$ are [90]:

$$\begin{aligned}\boldsymbol{\theta}\phi(x)\boldsymbol{\theta}^+ &= \theta_n^*\phi^+(-x), && \text{(spin } S = 0) \\ \boldsymbol{\theta}\psi(x)\boldsymbol{\theta}^+ &= \theta_n^*\gamma_5^*\psi^*(-x), && \text{(spin } S = 1/2) \\ \boldsymbol{\theta}A^\mu(x)\boldsymbol{\theta}^+ &= -\theta_n^*A^{\mu+}(-x), && \text{(spin } S = 1)\end{aligned} \tag{2.461}$$

where $\theta_n^* = \xi_n^*\eta_n^*\zeta_n^*$.

The *CPT* theorem is one of the most fundamental theorems in the quantum field theory, which states that the interaction Hamiltonian density satisfies the relation [90]

$$\boldsymbol{\theta}\boldsymbol{H}_{\text{int}}(x)\boldsymbol{\theta}^+ = \boldsymbol{H}_{\text{int}}(-x), \tag{2.462}$$

where the space-time argument x is sign-flipped for each of the four components: $PTx = -x$. Here the space inversion P belongs to the Lorentz group. It is explicitly written in the four dimension space-time as (2.420). The time inversion T also belongs to the Lorentz group and in the same space-time four dimension is explicitly written as (2.371).

The universality of *CPT* theorem consists in that it is based on the microscopic causality, Lorentz invariance and continuity of field operators.

Since $\boldsymbol{\theta}$ is an antilinear operator, it does not have physical eigenvalues. Thus, it does not lead to quantum numbers conservation, unlike the cases of **P** and **C** operators, for which the quantum numbers are conserved.

When creating the quantum field theory it was assumed that it is symmetric with respect to **P**, **C** and **T** transformation separately. As it was noted above, in 1957 it was found that the parity is violated in the β-decay of the nucleus ^{60}Co. In 1964, it was found that combined CP-symmetry is also violated in neutral kaon systems. The CPT-symmetry is still considered sacred because there is no reliable experimental evidence that it is violated.

Chapter 3
Magnetic Symmetry Point Groups

In this chapter, we consider the magnetic symmetry point groups, which are obtained by extending the classical point symmetry groups using the time-reversal operator **T**. In this case, the operator **T** is not an element of the extended symmetry group, but it is contained in the group as a product of elements of the original point group.

By extending the 32 classic point symmetry groups using the second-order cyclic group $\{\mathbf{T}, \mathbf{T}^2 = e\}$, where e is a unit group element, one obtains 58 nontrivial point groups of magnetic symmetry. These groups along with the original 32 point groups (which are trivial in terms of magnetic symmetry) form 90 point groups of magnetic symmetry. However, this conclusion is valid only for systems containing an even number of electrons. As it will be shown in this chapter, for systems with Kramers degeneracy of energy levels, the expansion of classic point groups should be carried out using a fourth-order cyclic group (or a non-cyclic one, in a more general case) that contains the operator **T**. In this case, the magnetic symmetry point groups are not two-color (black and white), but four-color ones. The number of nontrivial four-color magnetic symmetry groups is four, instead of 58 in the case of two-color groups.

The time-reversal operation also allows expanding the Fyodorov's space groups to magnetic space groups of symmetry, as it is the case for point groups. In magnetic crystals, there are magnetic space groups in which the time-reversal operator enters in combination with ordinary rotations, mirror rotations and translations, but it is not itself an element of symmetry.

3.1 Magnetic Two-Color Point Symmetry Groups for Non-Kramers Systems

The first publication related to point groups of magnetic symmetry date back to 1929 and belongs to Heesh [113], who gave a complete list of magnetic point groups. The next publication in this field, which appeared 27 years later, is the paper of Tavger

I. I. Geru, *Time-Reversal Symmetry*, Springer Tracts in Modern Physics 281,
https://doi.org/10.1007/978-3-030-01210-6_3

and Zaitsev [114]. Around the same time the theory of magnetic space groups began to be developed by different methods. The fundamentals of this theory will not be discussed in this book.

As for the magnetic point groups of symmetry, in deriving these groups by extension of 32 point groups using the Abelian group $\{\mathbf{T}, \mathbf{T}^2 = e\}$, it was assumed that they belong to both non-Kramers systems ($\mathbf{T}^2 = e$) and systems with Kramers degeneracy of energy levels ($\mathbf{T}^2 = -e$). In any case, nowhere in the early and in the modem scientific literature is it not emphasized that 58 nontrivial point groups of magnetic symmetry are only related to non-Kramers systems. To draw attention to this fact, the title of this Section emphasizes that considered magnetic symmetry groups are related to non-Kramers systems. The four-color point groups of magnetic symmetry for Kramers systems will be derived in the next Section.

In order to understand how the point groups of magnetic symmetry are introduced in physics, it should be noted that point groups describe a possible point symmetry of the charge density in the equilibrium state of the crystal, averaged over time. In this equilibrium state, one can also define the time-averaged current density, $\mathbf{j}$. The vector of the current density satisfies the equation

$$\operatorname{div} \mathbf{j} = 0 \tag{3.1}$$

since the charge sources and drains are absent in the equilibrium.

For antiferro- and ferromagnetic materials, $\mathbf{j} \neq 0$. If the direction of the current is reversed at any point, the equilibrium state is not changed. Now, if we act on the operator $\mathbf{j}$ by the time-reversal operator $\mathbf{T}$, it will cause a change in the sign of $\mathbf{j}$ at each point of the space without changing the spatial coordinates. The element $\mathbf{T}$ has the order 2 ($\mathbf{T}^2 = e$) and commutes with all spatial rotations and reflections. If among the elements of the point group G there is an element $g \in G$, under the action of which the vector $\mathbf{j}$ changes its direction to the opposite, then the vector $\mathbf{j}$ remains invariant with respect to the operator $\mathbf{T}g = g\mathbf{T}$. In other words, $\mathbf{T}g$ is an element of the extended group G' ($\mathbf{T}g \in G'$), which is one of point groups of magnetic symmetry. The name "point group of magnetic symmetry" is a reminder that an electric current of a density $\mathbf{j}$ creates a magnetic field in the surrounding space. In this case, the magnetic moments of paramagnetic ions localized in some of nodes of the crystal lattice have such mutual orientations, which cannot be described by the initial point group G. Instead, they are adequately described by the extended point group G' containing elements of $\mathbf{T}g$ type.

It is easy to see that the magnetic symmetry point group G' cannot be obtained by a simple addition of the time-reversal operator $\mathbf{T}$ to one of the usual 32 point groups. Indeed, if a group contains the element $\mathbf{T}$, it means that $\mathbf{j} = -\mathbf{j}$, which implies that $\mathbf{j} = 0$, is the contradiction with the fact that $\mathbf{j} \neq 0$ for point groups of magnetic symmetry. In other words, according to standard assumptions of the electromagnetic theory, the magnetic moment changes its sign under time reversal. Hence, the point groups of magnetic symmetry cannot have the time-reversal operator as an element. Note that all 32 classical point groups are admissible symmetry groups at $\mathbf{j} \neq 0$ together with the extended groups of the type G'.

The above consideration is based on the distribution of the current density **j**, but one could consider equally well the distribution of the magnetization **M**:

$$\mathbf{M} = \frac{\sum_k \boldsymbol{\mu}_k}{\Delta V}, \tag{3.2}$$

where $\boldsymbol{\mu}_k$ is the vector of the magnetic moment of kth paramagnetic ion, ΔV is a physically small volume, i.e., ΔV is much smaller than the volume V of the system, but it is large enough in order the macroscopic quantities (in this case, the magnetization) to be determined within the volume ΔV. The only feature that should be kept in mind is that **M** is an axial vector, where as **j** is a polar vector.

Before presenting the method of constructing the point groups of magnetic symmetry for systems consisting of an even number of particles with the spin $1/2$, we will make a few general comments and consider one simple example.

The extended point group of symmetry G' is a direct product of an usual point group G by identity transformation and a time-reversal operation that commutes with all operations of the point group. However, this statement may be true only for non-magnetic crystals, since the operator **T** reverses the direction of currents and spins. As a consequence, the magnetization direction in the magnetically ordered crystals is also reversed. In this case, the magnetic crystals would have only the symmetry of usual point groups. This is not always the case, because some of the magnetic crystals can be invariant with respect to the product of the operation **T** on a rotation, although they are not invariant with respect to the operation **T**. For example, in a ferromagnetic crystal with a magnetization along the z axis, the $\mathbf{T}R$ operation (where R is a rotation about the x axis by π) is a symmetry operation, since R reverses the direction of the magnetization, whereas **T** restores its original direction. As it was already mentioned, the groups of symmetry of this type comprising a time reversal operation only in conjunction with a rotation or reflection are called magnetic symmetry point groups.

Let us consider as an example a ferromagnetic crystal, which is described by the symmetry group D_3 in the Shenfliss notation (or 32 in the international notation). In a magnetically disordered state, above the transition temperature T_c, at which the magnetic order is established, the addition of the time-reversal operation leads to a larger group of symmetry $D_3 \otimes \{e, \mathbf{T}\}$ due to the absence of the magnetic moment. Below the temperature T_c the crystal becomes ferromagnetic, and its magnetic moment is directed along the third-order axis. Since the three C_2 rotations by angle π around the axes in the xy plane lead to the reversal of the magnetization direction, they are no more symmetry operations. The same applies to the time-reversal operator **T**. However, the products $\mathbf{T}C_2$ retain the value of symmetry operations. In this case, the new symmetry group includes the e, C_3, and C_3^2 operators and three operators $\mathbf{T}C_2$ containing the time-reversal transformation [115]. Note that the point group of magnetic symmetry contains the subgroup C_3 of the original point group D_3, and the C_3 subgroup is a normal index 2 subgroup of the D_3 group that is designated as

$D_3(C_3)$. It should also be noted that in the direct product of the groups $\mathcal{H}$ and $\mathcal{K}$, $\mathcal{G} = \mathcal{H} \otimes \mathcal{K}$ the elements of the group $\mathcal{G}$ commute with the elements of the group $\mathcal{K}$.

Now we introduce the so-called semi-direct product $\mathcal{G} = \mathcal{H} \wedge \mathcal{K}$. In this case, the products $G_{ia} = H_i K_a$ are elements of the group $\mathcal{G}$, where H_i is the element of the subgroup $\mathcal{H}$ and K_a is the element of the subgroup $\mathcal{K}$. The elements of $\mathcal{H}$ do not necessarily have to commute with the elements of $\mathcal{K}$, however the following relation should be satisfied

$$K_a H_i K_a^{-1} = H_f, \tag{3.3}$$

where H_f is the other element of the subgroup $\mathcal{H}$. This definition means that all the elements conjugated with any of H_i relative to any element of the group $\mathcal{G}$, belong to the subgroup $\mathcal{H}$.

The definition (3.3) is not symmetrical with respect to subgroups $\mathcal{H}$ and $\mathcal{K}$. The subgroup $\mathcal{H}$ is called a normal subgroup of the group $\mathcal{G}$. Along with the term "normal subgroup" it is also used the terms "invariant subgroup" and "self-conjugated subgroup". The subgroup $\mathcal{K}$ is a factor-group of the group $\mathcal{G}$ with respect to the subgroup $\mathcal{K}$. The relationship between $\mathcal{K}$, $\mathcal{G}$, and $\mathcal{H}$ is given by the formula $\mathcal{K} = \mathcal{G}/\mathcal{H}$.

If we denote by g the order of the group G (not to be confused with the earlier designation by the same symbol the element of the point group), and by h and k the number of elements of the subgroups $\mathcal{H}$ and $\mathcal{K}$, it is easy to see that $g = hk$, where k is called the index of the normal subgroup $\mathcal{H}$ in the group $\mathcal{G}$ [116].

In this context, $\mathcal{G}$, $\mathcal{H}$, and $\mathcal{K}$ represent some abstract groups, $\mathcal{H}$ and $\mathcal{K}$ being subgroups of the group $\mathcal{G}$. Further in this section, the space group of symmetry will be designated by $\mathcal{G}$.

We recall here some information from the abstract theory of groups that will be needed for the construction of magnetic point groups.

In particular, it can be shown that each point group of magnetic symmetry of non-Kramers systems possesses a normal (invariant, self-conjugated) subgroup of index 2. Finding of these invariant subgroups allows us to identify all the nontrivial point groups of magnetic symmetry, the number of which is equal to 58, as already mentioned.

First of all, note that if the two-color magnetic point group G' contains the element $\mathbf{T}g$, where g is an element of the point group $G (g \in G)$, the order of the element g must not be odd, because otherwise among all degrees of $\mathbf{T}$, which contains G', would be an element $\mathbf{T}$. For example, $(\mathbf{T}C_3)^3 = \mathbf{T}^3 e = \mathbf{T}$, taking into account that e is the unit element of the group G and that for non-Kramers systems with integer spin $\mathbf{T}^2 = +\mathbf{1}$, where $\mathbf{1}$ is a unit operator defined in the spin space. For Kramers system $\mathbf{T}^2 = -\mathbf{1}$ and $(\mathbf{T}C_3)^2 = -\mathbf{T}$. Consequently, among the elements of the group G' the $\mathbf{T}C_3$ or $\mathbf{T}S_3$ elements cannot exist (S_3 is the mirror rotation by $2\pi/3$).

The group G' cannot simultaneously contain the elements $g \in G$ and $\mathbf{T}g$, since otherwise G would also contain the element $\mathbf{T}g \cdot g^{-1} = \mathbf{T}$. In accordance with this result, we denote the elements of the group G' by g_k $(k = 1, 2, \ldots, m)$ where all the geometrical operations are different. It is easy to see that if $\mathbf{T}$ is replaced by a unit operator, then n elements of g_k $(k = 1, 2, \ldots, m)$ and $g_i (i = m + 1, m + 2, \ldots, n)$ form one of 32 point groups.

Now, let us take any of the point symmetry groups G and find the subgroup $\mathcal{H}$ of this group with elements g_k $(k = 1, 2, \ldots, m)$. Multiplying by $\mathbf{T}$ all the elements g_i $(i = m + 1, m + 2, \ldots, n)$ of the set $G \setminus \mathcal{H}$ (in other words, all the elements of the group G, which do not belong to the subgroup $\mathcal{H}$), we obtain $M_i = Tg_i$. If M_i and g_k form a group, then it is one of the point groups of magnetic symmetry that we are looking for. Such a method of constructing new point groups of magnetic symmetry is a very tedious one. However, the problem of finding the point groups of magnetic symmetry becomes easier, if the following theorem holds [117].

Theorem *It is necessary and sufficiently the subgroup $\mathcal{H}$ in the group G to have the index 2, in order that $\mathbf{T}g_i$ and g_k to form a group (g_i and g_k are the elements of any point group G).*

Proof If the subgroup $\mathcal{H}$ has the index 2 in the group G, then

$$G = \mathcal{H} + g_i\mathcal{H}, \tag{3.4}$$

where g_i is one of the elements of the set $G \setminus \mathcal{H}$. Now, let us choose a new set G', which satisfies the relationship:

$$G' = \mathcal{H} + \mathbf{T}g_i\mathcal{H} \tag{3.5}$$

The set G' is a group, because

$$\mathcal{H} \cdot \mathcal{H} = \mathcal{H},\ \mathbf{T}g_i \cdot \mathcal{H} \cdot \mathcal{H} = \mathbf{T}g_i\mathcal{H},\ \mathbf{T}g_i\mathcal{H}\cdot\mathbf{T}g_i\mathcal{H} = \mathcal{H}, \tag{3.6}$$

This is just one of the point groups of magnetic symmetry we are looking for.

Conversely, if the elements g_k and $\mathbf{T}g_i$ form the group G', then as a result of multiplying the m elements of the subgroup $\mathcal{H}$ by any of the elements $\mathbf{T}g_i$, we get m different elements of $\mathbf{T}g_i$ type. However, if we multiply the $\mathbf{T}g_i$ elements by any of them, we get n different elements of g_k type [117]. Therefore the subgroup $\mathcal{H}$ has the index 2 in the group G' and as a result $\mathcal{H}$ has also the index 2 in the group G. □

Thus, the above method of finding nontrivial point groups of magnetic symmetry is applied as follows. First of all, we need to select an arbitrary point group G. In the group G one selects any subgroup $\mathcal{H}$ of index 2. The elements of the set $G \setminus H$ are to be multiplied by the time-reversal operator $\mathbf{T}$. In this case we have the following relation

$$G' = \mathcal{H} + \mathbf{T}(G \setminus \mathcal{H}), \tag{3.7}$$

where G' is one of the sought-for two-color point groups of magnetic symmetry.

The above prescription of constructing a point group of magnetic symmetry requires finding firstly the subgroups of index 2 in a point group. This prescription is an adaptation of the well-known way of finding all subgroups of the group of proper and improper rotations, which does not contain the space inversion as an element. The group of all proper and improper rotations is a direct product group

$\mathcal{R} \otimes \mathcal{I}$, where $\mathcal{R}$ is the group of all proper rotations and $\mathcal{I}$ is a group consisting of the identity and the space inversion. Therefore, one has only to make G to correspond to $\mathcal{R}$ and $\{e, \mathbf{T}\}$ to $\mathcal{I}$, in order to obtain the correspondence between the two prescriptions [118].

The problem of finding all subgroups $\mathcal{H}^G$ of index 2 of an arbitrary group G is equivalent to finding all alternating representations of the group G. Actually, the factor group $G/\mathcal{H}^G$ has only two representations: the identical representation and the alternating one. Therefore, each subgroup $\mathcal{H}^G$ of G will generate one alternating representation of G. Conversely, if all alternating representations of G are known, one can find all subgroups $\mathcal{H}^G$. Namely, one can find the subgroup $\mathcal{H}^G$, which generates each alternating representation, by simply picking out from G all those elements, to which $+1$ corresponds to the alternating representation in question [6]. The equivalence of the two problems has been noticed by Indenbom [119] and Niggli [120] in the case when G is a point group or a space one, respectively.

Returning to the problem of finding all possible subgroups of index 2 for each of the 32 point groups, which are necessary for constructing the point groups of magnetic symmetry, we note that a full set of these subgroups is known and can be found in International Tables for X-Ray Crystallography [121]. Therefore the procedure of finding the point groups of magnetic symmetry is greatly simplified and reduced, in essence, to the application of (3.7), where the right-hand side contains the known elements of the invariant subgroup $\mathcal{H}$ and that of the set $G \setminus \mathcal{H}$, as well as the known time-reversal operator $\mathbf{T}$.

Regarding the operator $\mathbf{T}$, it is necessary to note the following. The elements of a magnetic group G', containing $\mathbf{T}$ included in the product $\mathbf{T}g$, act on the space-time variables, while the elements of the corresponding magnetic group of quantum-mechanical operators act on state vectors. In the last case, the time-reversal operator is defined in the space of vector states of the total angular momentum for systems consisting of particles with spin and in the space of vector states of the total orbital angular momentum for systems consisting of spinless particles. Since $\mathbf{T}$ is an antiunitary operator, the transformation of state vectors under a nontrivial magnetic group of quantum-mechanical operators generates corepresentations of the group, not representations of the group (see Sect. 2.8 and [122–124]).

The application of (3.7), where $\mathcal{H}$ is an invariant subgroup of index 2, leads to 58 nontrivial black-and-white point groups of magnetic symmetry [113, 114], which together with 32 point groups of symmetry form a system of 90 point groups of magnetic symmetry [113, 114]. These magnetic point groups are referred to as Heesh's groups, after Heesh published a complete list of these groups [113]. The Heesh groups have been considered by many authors [113–115, 117–120, 122–128] and their complete list can be found, for example, in [117, 118, 127].

The list of black-and-white magnetic point groups arranged in families [118] is given in the Table 3.1. The denotations used in the list have been introduced by Belov et al. [129] as a generalization of international symbols used for the ordinary point groups. The international symbol of a point group denotes the generating elements of the group. The symbols of generating elements in a magnetic point group have a prime as an upper right superscript. The bar denotes mirror rotations and the prime

– antirotations, if it is referred to a rotation axis, or antreflections if it is referred to a reflexion plan. By omitting the primes in the symbol of a magnetic point group, one obtains the symbol of the ordinary point group, to whose the family of magnetic point group belongs [118].

Excepting the point groups 1 (identity), 3 (trigonal), and 23 (tetrahedral), all point groups contain subgroups of index 2. In the list on Table 1, each line gives all magnetic point groups belonging to one family, if it is used the term family of magnetic point groups. The family of magnetic point group includes a point group G and the magnetic point groups constructed from it by means of the above prescription from all subgroups $\mathcal{H}^G$ of G, which in the case of non-Kramers systems are the subgroups of index 2.

The magnetic groups belonging to the same family are isomorphic. This implies that all of them have the same irreducible representations as the trivial magnetic groups of the family (Table 3.1).

In Table 3.1, 31 magnetic point groups are marked with asterisk. These are the so-called admissible magnetic point groups. The concept of admissible magnetic point groups is connected with the construction of an invariant spin arrangement and it will be discussed in Sect. 3.2.

Now, let us give a nontrivial example of constructing the family of a point group [118]. Let this group belongs to the class of point groups denoted by 422 and let it consists of the elements 1, 4_z, 2_z, $4_z^{\,3}$, 2_x, 2_y, 2_{xy}, $2_{\bar{x}y}$. The symbol 1 stands for the identity of the group. The symbols 4_z, 2_z, and $4_z^{\,2}$ stand for rotations by 90°, 180°, and 270° around the z-axis. The remaining four symbols stand for rotation by 180° around the x- and y-axis, and around the two bisectors of angles between the x- and y-axis.

It can be seen that the point group 422 consists of three subgroups of index 2 (see International Tables of X-ray Crystallography [121]). These are the groups consisting of elements 1, 4_z, 2_z, $4_z^{\,3}$; 1, 2_z, 2_x, 2_y, and 1, 2_z, 2_{xy}, $2_{\bar{x}y}$, respectively. The first group belongs to the class of point groups denoted by 4, and the last two groups are equivalent and belong to the class of point groups 222. Therefore the family of this point group contains three nontrivial magnetic point groups. The first of them contains the unprimed elements 1, 4_z, 2_z, $4_z^{\,3}$ and the primed elements $2'_x$, $2'_y$, $2'_{xy}$, $2'_{\bar{x}y}$. The second magnetic point group consists of unprimed elements 1, 2_z, 2_x, 2_y, and the primed elements $4'_z$, $4_z^{3'}$, $2'_{xy}$, $2'_{\bar{x}y}$. At last, the third magnetic point group consists of the unprimed elements 1, 2_z, 2_{xy}, $2_{\bar{x}y}$, and the primed elements $4'_z$, $4_z^{3'}$, $2'_x$, $2'_y$. The first of these groups belongs to the class of magnetic point groups denoted by $42'2'$. The last two groups are equivalent and belong to the class of magnetic point groups, which are denoted by $4'22'$ or by $4'2'2$.

Thus, we can conclude that the family of the point group 422 consists of magnetic point groups 422, $42'2'$, and $4'2'2$.

In conclusion, let us consider another aspect of magnetic point groups. The formulas (3.5) and (3.7) are two equivalent ways of defining the magnetic point group G'. In these formulas the time-reversal operator $\mathbf{T}$ cannot be an element of the group G', but it can enter the group as a product $\mathbf{T}g$, where g is an element of the initial

Table 3.1 List of magnetic point groups arranged by families

^{*1}					
$^*\bar{1}$	$\bar{1}'$				
^{*2}	$^*2'$				
$^*m$	$^*m'$				
$^*2/m$	$2'/m$	$2/m'$	$^*2'm'$		
222	$^*2'2'2$				
$mm2$	$^*m'm2'$	$^*m'm'2$			
mmm	$m'mm$	$^*m'm'm$	$m'm'm'$		
^{*4}	$4'$				
$^*\bar{4}$	$\bar{4}'$				
$^*4/m$	$4'/m$	$4/m'$	$4'/m'$		
422	$4'22'$	$^*42'2'$			
$4mm$	$4'm'm$	$^*4m'm'$			
$\bar{4}2m$	$\bar{4}'2'm$	$\bar{4}'2m'$	$^*\bar{4}2'm'$		
$4/mmm$	$4/m'mm$	$4'/mm'm$	$4'/m'm'm$	$^*4/mm'm'$	$4/m'm'm'$
^{*3}					
$^*\bar{3}$	$\bar{3}'$				
32	$^*32'$				
$3m$	$^*3m'$				
$\bar{3}m$	$\bar{3}'m$	$\bar{3}'m'$	$^*\bar{3}m'$		
^{*6}	$6'$				
$^*\bar{6}$	$\bar{6}'$				
$^*6/m$	$6'/m$	$6/m'$	$6'/m'$		
622	$6'2'2$	$^*62'2'$			
$6mm$	$6'm'm$	$^*6m'm'$			
$\bar{6}2m$	$\bar{6}'m'2$	$\bar{6}'m2'$	$^*\bar{6}m'2'$		
$6/mmm$	$6/m'mm$	$6'/mm'm$	$6'/m'm'm$	$^*6/mm'm'$	$6/m'm'm'$
23					
$m3$	$m'3$				
432	$4'32'$				
$\bar{4}3m$	$\bar{4}'3m'$				
$m3m$	$m'3m$	$m3m'$	$m'3m'$		

point group. Thus, one can expand normal point groups to magnetic point groups using the time-reversal operator **T**. However, this can also be done in a different way.

Let us assume that the crystal faces are colored in two colors—black (B) and white (W). We introduce the operation R that changes colors - W by B and B by W. In other words, we introduce the cyclic group $\{e, R\}$, where e is a unit element of this group and $R^2 = e$, instead of the second-order cyclic group $\{e, T\}$ for non-Kramers systems, when $\mathbf{T}^2 = +e$. Now, let us consider the elements Rg in addition to g operations of geometrical symmetry that lead to a shift of faces, but do not change their colors. For example, if g is a rotation transforming the face F into the face F', then Rg will transform F into F' and make the color of the face F' opposite to the color of F. We also require that no one face was painted in two colors, so that the element R should not belong to the group. Such groups are called colored. In the case of two colors (black and white), when the operation R changes one color to another, the groups are called black-white. The black-and-white symmetry groups (also called antisymmetry groups) were introduced by Shubnikov [130, 131].

The expansion of classical groups to the Shubnikov black and white symmetry groups is performed using the second-order cyclic group $\{e, R\}$, in complete analogy with the way it is done using the group $\{e, \mathbf{T}\}$. In both cases, the process of constructing new groups begins with finding the invariant subgroups of the original group of index 2. If one of the point or space symmetry groups is an original group, the resulting new groups are called respectively the Shubnikov black-and-white point or spatial symmetry groups.

Although the use of cyclic groups $\{e, \mathbf{T}\}$ or $\{e, R\}$ in the construction of magnetic symmetry groups leads to the same results, it seems more physically obvious the use of the cyclic group $\{e, \mathbf{T}\}$, as magnetically ordered crystals contain actual localized atoms (ions) with nonzero magnetic moments, whose directions are reversed by the time-reversal operator. These changes are compensated by geometrical transformations of the magnetic symmetry group, thus preserving its symmetry.

It is necessary to make one more remark that a point group of symmetry is a group of (proper or proper and improper) rotations, which leaves the lattice invariant. Such a group is called a crystallographic point group. Similarly, the magnetic point groups in non-Kramers systems, as defined in this Subsection, could also be called crystallographic magnetic point groups of symmetry.

3.2 Invariant Spin Arrangement and Admissible Magnetic Point Groups for Non-Kramers Systems

A spin arrangement in a group of magnetic atoms (ions) is defined, if the spin of each atom is specified. In this case, a spin arrangement is a vector function $S(\mathbf{r}_i)$ defined on a finite set $\mathbf{r}_i$ $(i = 1, 2, \ldots)$ of position vectors of all magnetic atoms that correspond to the symmetry space group of the crystal $\mathcal{G}$ [118]. If the magnetic atom, along with a spin, also has an orbital momentum, we will conventionally mean by spin the total angular momentum.

In order to determine the invariant spin arrangements, it is necessary to know how a vector function $S(\mathbf{r}_i)$ is transformed under each element g_m of the magnetic space group $\mathcal{G}'$. Each element g_m of the group $\mathcal{G}'$ is presented as $g_m = gA$, where $g \in \mathcal{G}'$, $A = e$ or $A = \mathbf{T}$ and e is a unit 2×2-matrix defined in the spin space with the basis vectors $|1/2, 1/2\rangle$ and $|1/2, -1/2\rangle$ (e and $\mathbf{T}$ form a second-order cyclic group). There are three types of transformations *(i)* the transformation of $S(\mathbf{r}_i)$ from $\mathbf{r}_i$ to the position $g_m\mathbf{r}_i = g\mathbf{r}_i$ $(A = e)$;

(ii) the rotation of $S(\mathbf{r}_i)$ in the same way as $\mathbf{r}_i$ is rotated, taking into account that $S(\mathbf{r}_i)$ is an axial vector, unlike $\mathbf{r}_i$, which is a polar vector;

(iii) the transformation

$$\mathbf{TS}(\mathbf{r}_i)\,\mathbf{T}^{+} = -\mathbf{S}(\mathbf{r}_i)\,, \tag{3.8}$$

where the time-reversal operator $\mathbf{T}$ is given by (2.109).

Thus,

$$g_m\mathbf{S}(g\mathbf{r}_i) = \varepsilon \det R \cdot R\mathbf{S}(\mathbf{r}_i)\,,\ i = 1, 2, \ldots \tag{3.9}$$

where R is a point transformation of the space group $\mathcal{G}$; $\det R$ is the determinant of the matrix corresponding to this element; $\varepsilon = 1$, if $A = e$ and $\varepsilon = -1$, if $A = \mathbf{T}$.

We should consider that the spin arrangement $\mathbf{S}(\mathbf{r}_i)\,,\ i = 1, 2, \ldots$ is invariant under $g_m \in \mathcal{G}'$, if for all i

$$g_m\mathbf{S}(\mathbf{r}_i) = \mathbf{S}(\mathbf{r}_i)\,. \tag{3.10}$$

A necessary and sufficient condition for an existence of a spin arrangement invariant under $\mathcal{G}'$ belonging to the family of $\mathcal{G}$ generated by $\mathcal{G}$ from an atom at $\mathbf{r}$ is the existence of a spin vector $\mathbf{S}(\mathbf{r})$ invariant under "the magnetic site point group of $\mathbf{r}$" $G'(\mathbf{r})$ [118]. It is therefore important to know, which of the magnetic point group leaves a spin vector invariant. Those groups that have this property are called admissible magnetic point groups and their invariant spin vectors are called admissible ones.

The number n of linearly independent spin vectors, which are invariant under all elements of the group $G'(\mathbf{r})$, is equal to the number of times the identical representation occurs in the standard representation of the group $G'(\mathbf{r})$ [118]:

$$n = \frac{1}{h} \sum_{R_m \in G'(\mathbf{r})} \chi(R_m)\,, \tag{3.11}$$

where $\chi(R_m)$ is the trace of the matrix $\varepsilon(R_m)\det R_m \cdot R_m$, h is the order of the magnetic site point group $G'(\mathbf{r})$, and the sum is extended over all its elements. The magnetic point group is admissible, if $n = 1, 2$ or 3 [118].

It was shown that 31 of the 90 magnetic point groups related to non-Kramers systems are admissible. The list of groups (firstly published by Tavger [132]) and their invariant spin vectors with comments of authors of the [118] are given in the Table 3.2.

No magnetic point groups of the cubic system are admissible. Each of the 31 admissible magnetic point groups is a subgroup of the infinite non-crystallographic magnetic group $(\infty/m)\left(2'/m'\right)$ [118]. To find out the special role of this group for the invariant spin arrangement, it should be noted that along with the continuous group $\mathcal{R}$ of all proper rotations, there exist five infinite continuous proper subgroups of the group $\mathcal{R} \otimes \mathcal{I}$ of all proper and improper rotations ($\mathcal{I}$ is the group consisting of identity and space inversion).

Table 3.2 List of admissible magnetic point groups

Magnetic point groups					Admissible spin directions
					$n = 3$
1	$1'$				Any direction
					$n = 2$
$2'$	$2'/m'$	$m'm2'$			Perpendicular to the axis
m'					Any direction in the plan
					$n = 1$
m					Perpendicular to the plan
$m'm'm$					Perpendicular to the unprimed plan
$2'2'2$					Along the unprimed axis
2	$2/m$	$m'm'2$			Along the axis
4	$\bar{4}$	$4/m$	$42'2'$		Along the axis of higher order
$4m'm'$		$\bar{4}2'm'$	$4/mm'm'$		Along the axis of higher order
3	$\bar{3}$	$32'$	$3m'$	$\bar{3}m'$	Along the axis of higher order
6	$\bar{6}$	$6/m$	$62'2'$		Along the axis of higher order
$6m'm'$		$\bar{6}m'2'$	$6/mm'm'$		Along the axis of higher order

In its turn, four of these subgroups have subgroups of index 2, which give rise to six nontrivial infinite non-crystallographic magnetic groups of rotations [133]. One of these groups, namely $(\infty/m)\left(2'/m'\right)$, is particularly important, because it leaves invariant the spin momentum vector and the magnetic field vector. The unprimed elements of this group consist of all proper and improper rotations around a fixed axis, while its primed elements are obtained from unprimed ones by combining the latter with the rotation by 180° around an axis perpendicular to the fixed axis. Spin and magnetic moments are invariant under $(\infty/m)\left(2'/m'\right)$, if only they are directed along the fixed axis.

Thus, if the magnetic point group G' is not admissible, then no spin arrangement invariant under all elements of the group $G'(\mathbf{r})$ can exist. If G' is an admissible magnetic point group, all invariant spin arrangements can be obtained using the standard procedure. Each invariant spin arrangement is uniquely characterized by G' and $\mathbf{S}(\mathbf{r})$.

3.3 Magnetic Four-Color Point Groups of Kramers Systems

In the academic and scientific literature, the Kramers and non-Kramers systems are usually not separated when studying their symmetry properties. In particular, it is accepted to consider that all 58 nontrivial magnetic point groups are, in general, related to localized spin systems without indicating more precisely whether the system has or has not a Kramers degeneracy of energy levels. However, these groups must actually be related only to non-Kramers systems, which is a consequence of the way they have been deduced. In fact, the magnetic point groups have been obtained by extending the 32 classical point groups of symmetry using the second-order cyclic group $G_2 : \{\mathbf{T}, \mathbf{T}^2 = e\}$, where $\mathbf{T}$ is the time-reversal operator and e is the identity element of the group [117]. The same result can be obtained, if the cyclic group of second order $2' : \{2'_z, 2'^2_z = e\}$ is used instead of the group G_2, where $2'_z$ is the antirotation by 180° around the antirotation axis of the second order.

However, such an extension is not correct in the case of Kramers systems and one should use fourth-order cyclic groups, instead of G_2 or $2'$, which are isomorphic with G_2. The difference in the number and types of magnetic point groups for Kramers and non-Kramers systems is due to the fact that the behavior of systems with half-integer spin under time-reversal is different from that of systems with integer spin. There are two ways to deduce the magnetic point groups of Kramers systems:

1. a purely geometric approach based on the symmetry and antisymmetry properties of a square with colored vertices;
2. an algebraic approach followed by a geometric image of the result of action of operators $\mathbf{T}$, $\mathbf{T}^2$, $\mathbf{T}^3$, and $\mathbf{T}^4$ on the total wave function, which depends on the space and spin coordinates.

At the beginning let us consider the first approach, consisting in selection of an arbitrary square with two opposite vertices colored in a color, and the adjacent vertices colored in differently colors. The opposite vertices of the square are equivalent, while any adjacent vertices, which are colored differently, are nonequivalent (Fig. 3.1). Let's draw through the center of the square an axis perpendicular to its plane, which is simultaneously a second-order rotation axis and a fourth-order antirotation axis.

The operation of antirotation by 90° consists in a rotation of the vertice of the square by 90° around an axis passing through the center of the square perpendicular to the plane, in which it is given, followed by changing of color into the "opposite" one. Four successive antirotations of the square with colored vertices by 90° around the axis of antirotation return the system to its initial state. The four antirotations form a fourth-order cyclic group $4' : \{4'_z, 4'^2_z, 4'^3_z, 4'^4_z = e\}$, which can be used to perform an extension of the 32 classical point groups and obtain magnetic point groups in the case of Kramers systems [85].

As it was shown in [85, 134], the group $4'$ can be used to extend the classical point symmetry groups to the following four-colors symmetry groups: $4^{(4'_z)}$, $\overline{4}^{(4'_z)}$, $4^{(4'_z)}/m^{(1)}$ and $4^{(4'_z)}/m^{(2)}$, where the overscribed bar indicates a mirror rotation and the prime denotes the antirotation. The first two groups are generated using the groups 4 and $\overline{4}$, respectively, while the other two groups are generated with the use of the group $4/m$. The symbols $m^{(1)}$ and $m^{(2)}$ denote the generating elements of the groups $4^{(4'_z)}/m^{(1)}$ and $4^{(4'_z)}/m^{(2)}$, which generate the symmetry transformations $4'^4_z m = em$, $4'^2_z m = 2_z m$, $4'_z m = 4'^3_z 2_z m$, and $4'^3_z m = 4'_z 2_z m$, respectively.

If the other four generalized symmetry elements of the square with nonequivalent adjacent vertices are taken into account (two mirror reflections $m^{(i)}_{xy}$ from the diagonals and two antireflections $m'^{(i)}_x$ from the straight lines passing through the centers of the opposite sides; $i = 1, 2$), in addition to the antirotations forming the cyclic

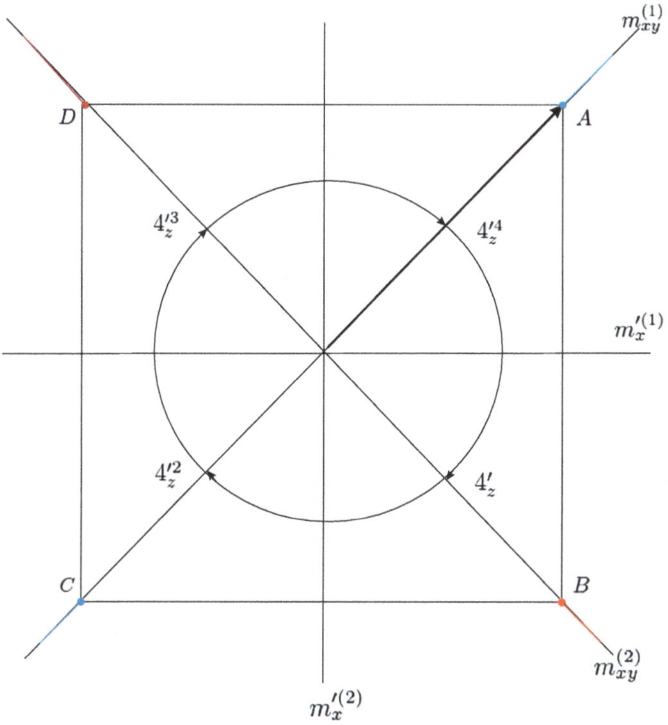

Fig. 3.1 Two symmetry and six antisymmetry elements of a square with colored vertices. Vertices A and B, as well as C and D are colored differently and are nonequivalent, while vertices A and C, and B and D are equivalent

group $4'$, then the $4'm'm$ group of eight order should be used for extending the classical point groups. Among all the 32 point groups of symmetry, only the groups 422, $4mm$ and $\overline{4}2m$ can be extended using the $4'm'm$ group. In this more general case, there also exist four point generalized-symmetry groups (four-color symmetry point groups [85, 134–136]): $4^{(4_z')}2^{(m_x')}2^{(m_{xy})}$, $4^{(4_z')}m^{(m_x')}m^{(m_{xy})}$, $\overline{4}^{(4_z')}2^{(m_x')}m^{(m_{xy})}$, and $\overline{4}^{(4_z')}2^{(m_{xy})}m^{(m_x')}$. The first two four-color symmetry groups are generalized using the groups 422 and $4mm$, respectively, while other two groups have been generalized by using the $\overline{4}2m$ group.

It is seen that the magnetic point groups of Kramers systems obtained by extension of 32 classical point groups by means of the cyclic group $4'$ or, in a more general case, by using the non-Abelian group $4'm'm$, are very different from the magnetic point groups of non-Kramers systems. This difference is related to both types of corresponding groups and their number (four instead of 58).

Let us now consider a second way to deduce the four-color point groups consisting in the analysis of behavior of the time-dependent total wave function in the form of a spinor under time-reversal operator **T**. Let's expand the full wave function of the system with total spin S in terms of spinor basis vectors [see (2.167)]:

$$\Psi = \sum_{\sigma=-S}^{S} \psi_\sigma \xi_\sigma^{(S)}, \tag{3.12}$$

where

$$\xi_S^{(S)} = |S, S\rangle, \xi_{S-1}^{(S)} = |S, S-1\rangle, \dots, \xi_{-S+1}^{(S)} = |S, -S+1\rangle, \xi_{-S}^{(S)} = |S, -S\rangle. \tag{3.13}$$

Under the action of the operator **T**, the wave function in the (3.12) is transformed into:

$$\mathbf{T}\Psi = \sum_{\sigma=-S}^{S} \psi_\sigma^* \mathbf{U} \xi_\sigma^{(S)}, \tag{3.14}$$

where the matrix of the unitary operator **U** is given in (2.120).

In the case of half-integer spin S, the action of operators $\mathbf{T}$, $\mathbf{T}^2 = -e$, $\mathbf{T}^3 = -\mathbf{T}$, and $\mathbf{T}^4 = e$ on the wave function Ψ corresponds to four successive antirotations of the point representing the state Ψ by 90° in the function space spanning the basis vectors $\xi_\sigma^{(S)}$. In Fig. 2.2, it is represented the action of these operators on the wave function Ψ in the case of half-integer spins $S = 1/2$ and $3/2$ (as well as in the case of the integer spin $S = 1$).

In the case of spin $S = 1/2$ the axis of antirotation passes through the origin of coordinates and is perpendicular to the plane of the square, whose corners correspond to the states $\mathbf{T}\Psi$, $\mathbf{T}^2\Psi = -\Psi$, $\mathbf{T}^3\Psi = -\mathbf{T}\Psi$, and $\mathbf{T}^4\Psi = \Psi$. The opposite corners of the square $\left(\Psi, -\Psi, \mathbf{T}^3\Psi, -\mathbf{T}\Psi\right)$ are equivalent, while any adjacent corners are nonequivalent and colored differently. Such coloration of the square corners, as it was shown above in this section, is a way to deduce the magnetic point groups of Kramers systems. In contrast to this, in the case of integer spin, the coloration only of a two corners of a $(2S + 1)$-dimensional cube, corresponding to the states Ψ and $\mathbf{T}\Psi$, is need to deduce the magnetic point groups of non-Kramers systems.

It is easy to see that there is an univocal correspondence between the elements $4'_z$, $4'^2_z$, $4'^3_z$, $4'^4_z$ of the group $4'$ and the elements $\mathbf{T}$, $\mathbf{T}^2$, $\mathbf{T}^3$, $\mathbf{T}^4$, which form a fourth-order cyclic group G_4. Therefore, the group $4'$ and G_4 are isomorphic. The isomorphic correspondence between the elements of groups $4'$ and G_4 is shown in Fig. 2.2 for $S = 1/2$. Such a correspondence cannot be represented graphically in the case of half-integer spin $S > 1/2$, that is, in a space of dimensionality higher than three. Nevertheless, using the properties of the unitary part of the time-reversal operator **T**, it can be shown that the geometric transformations corresponding to the elements of the group $4'$ are also four antirotations $4'_z$, $4'^2_z$, $4'^3_z$ and $4'^4_z$ [135, 136].

By comparing the Fig. 2.2 (the part related to the case $S = 1/2$) and Fig. 3.1, one can conclude that along with the isomorphism of the groups $4'$ and G_4, an isomorphic correspondence should also exist between the elements $m_x^{(1)}$, $m_x^{(2)}$, $m_{xy}^{(1)}$, $m_{xy}^{(2)}$ of the group $4'm'm$ and the four elements of the eight-order non-Abelian group G_8. These elements do not coincide with the elements of the group G_4, which is a subgroup of the G_8 group. Indeed, reflections $m_{xy}^{(i)}$ and antireflections $m_x^{\prime(i)}$ $(i = 1, 2)$ in the symmetry group $4'm'm$ are in an isomorphic correspondence with the operators σ_x, $-\sigma_x$, and $\sigma_z\mathbf{K}$, $-\sigma_z\mathbf{K}$, in the spinor basis for the spin $S = 1/2$. Here, σ_x and σ_z are real Pauli operators and $\mathbf{K}$ is a complex-conjugated operator, as earlier. Therefore, the group $4'm'm$ is isomorphic to the eighth-order non-Abelian group G_8, which is composed of operators $i\sigma_y\mathbf{K}$, $-i\sigma_y\mathbf{K}$, e, $-e$, σ_x, $-\sigma_x$, $\sigma_z\mathbf{K}$ and $-\sigma_z\mathbf{K}$, where σ_y is an imaginary Pauli operator.

The magnetic point groups of Kramers systems can be also obtained by the method based on the (3.7), which was used in the Sect. 3.1 to deduce the magnetic point groups of non-Kramers systems. However, unlike the non-Kramers systems, where $\mathcal{H}$ is a subgroup of index 2 of the classical point group G, in Kramers systems $\mathcal{H}$ must be a subgroup of index 4 of the group G.

Thus, the magnetic point groups of Kramers systems, obtained by extension of 32 classical point groups by means of the cyclic group $4'$ (isomorphic to the group G_4) or, in a more general case, by means of the non-Abelian group $4'm'm$ (isomorphic to the group G_8), are very different from the magnetic point groups of non-Kramers systems. This difference refers to both the type of corresponding groups (four colors instead of two colors) and the number of nontrivial magnetic groups (four instead of 58).

The isomorphism of groups $2'$, $4'$ and $4'm'm$ with groups G_2, G_4 and G_8 indicates the important role of the time-reversal operator $\mathbf{T}$ for any types of magnetic symmetry groups, including magnetic space groups, which are not considered in this book. For obtaining all classes of magnetic space groups, the Shubnikov's theory of black–white symmetry [137–142], the Belov's theory of color symmetry [143–147], and Zamorzaev's method for constructing the magnetic space groups [148–151] have been used.

Extension of 230 space groups to magnetic space groups can be carried out also on the basis of general formula (3.7), which is used for deducting of the magnetic point groups. However, in this case G is the classical space group, G' is the magnetic space group and H is the subgroup of the space group G. The group G contains a point group and a translation group as subgroups. The index of the subgroup H is 2 or 4 depending on the structure of the corresponding point subgroup (related to non-Kramers or Kramers systems). In other respects, the procedure of deducting the magnetic space groups is the same as in the case of magnetic point groups. Certainly, the number of nontrivial magnetic space groups is much bigger as compared with the number of magnetic point groups. It is caused not only by the fact that the number of the space groups is much bigger as compared with the number of point groups (230 instead of 32), but also by more complex structure of the space groups.

The mentioned isomorphism of geometric symmetry groups and corresponding groups of quantum-mechanical operators, including the time-reversal operator $\mathbf{T} = i\sigma_y\mathbf{K}$, indicates the fundamental role of the time-reversal symmetry in all magnetic phenomena, that has been long ago understood by Landau and Lifshits [3].

The magnetic point groups discussed in Sects. 3.1 and 3.3 do not include as trivial magnetic groups the icosahedral groups $\mathcal{J}$ and $\mathcal{J}_h = \mathcal{J} \otimes \{e, I\}$, where I is the space inversion. The group $\mathcal{J}$ consists of 60 rotations around the axes of symmetry of a right

20-polyhedron with triangular faces (of the icosahedron) and regular 12-polyhedron with pentagonal faces (of the pentagonal dodecahedron). These rotations are made around 6 axes of fifth order, 10 axes of third order and 15 axes of second order. The group $\mathfrak{I}_h$, which contains a spatial inversion, is a complete transformation of the icosahedron [3].

In the list of magnetic point groups there are not included the groups, which can be obtained by expansion of point groups $\mathfrak{I}$ and $\mathfrak{I}_h$ using the prescription for constructing magnetic point groups.

Since opening of the C_{60} fullerene molecule [152], the irreducible representations of the groups $\mathfrak{I}$ and $\mathfrak{I}_h$ have been widely used to identify the allowed transitions in the optical absorption and Raman spectra of fullerenes in the liquid and solid state [153].

For the study of the endohedral $C_{60} : Re^{3+}$ fullerene molecule, containing a rare earth $4f$-ion in the center of the molecule, there is no need to expand the group $\mathfrak{I}_h$ to magnetic point groups. Such an extension would be required to study the ordering of the magnetic moments of the nuclei ^{13}C at very low temperatures, if one synthesizes a molecule C_{60} consisting entirely or mostly of atoms of the isotope ^{13}C, the natural abundance of which is only 1,108%. It hardly makes sense to perform such an expensive experiment.

However, if for any practical goals one needs magnetic point groups constructed by expanding the group $\mathfrak{I}_h$, then such groups can be easily found. One can say in advance that in such magnetic point groups there will be forbidden rotations around axes of the third and fifth orders in combination with a time-reversal transformation that leads to structural distortions in the molecule $^{13}C_{60}$. The reason for this structural distortion, as well as for other magnetic point groups, is that the time-reversal operator **T** enters in the magnetic symmetry group only in combination with other elements of the original point group. In other words, the structural distortions are a consequence of the time-reversal symmetry, because if we do not take into account the operator **T**, the magnetic symmetry group becomes trivial and structural distortions are absent.

Chapter 4
Kramers Trimer Clusters and Time-Reversal Symmetry

In this chapter is shown that structural distortions of trihomonuclear Kramers clusters occur as a consequence of the time-reversal symmetry. These distortions are caused by the fact that the time-reversal operator **T** is not a separate element of the magnetic point group G', but forms a combined $\mathbf{T}g$ element with other g elements of the original point group G. The order of the element g cannot be an odd number, otherwise it would lead to $\mathbf{T} \in G'$, which contradicts the definition of the group G'. Forbidden rotations by $2\pi/3$ and $2\pi/5$ in the group G' lead to the fact that metal ions of homonuclear magnetic clusters in a magnetic ordered state cannot occupy the vertices of an equilateral triangle or those of a regular pentagon. The distortion of the regular triangle or pentagon occurs only if the time-reversal operator **T** is taken into account. Hence these structural distortions are due to the time-reversal symmetry. There are presented and discussed experimental data concerning the anomalous behavior of chromium(II), iron(III), copper(II), vanadium(II), and cobalt(II) trihomonuclear clusters determined by their four-color symmetry.

The chapter offers an overview of experimental data related to structural and magnetic properties [156–219], excepting the Sect. 4.1, which contains some theoretical results obtained by the author concerning time-reversal symmetry [85, 134, 155], on the basis of the analysis reproduced from [155], with the permission of AIP Publishing.

4.1 The Structural Asymmetry of Trihomonuclear Kramers Clusters as a Consequence of Time-Reversal Symmetry

The spatial arrangement of isolated groups of atoms, forming complex molecules or clusters is determined by the corresponding point group symmetry. If the atoms that form an isolated system, are "introduced" in a crystal, then the initial symmetry can

I. I. Geru, *Time-Reversal Symmetry*, Springer Tracts in Modern Physics 281,
https://doi.org/10.1007/978-3-030-01210-6_4

change, and the symmetry of the corresponding crystalline class may be lower the original point group symmetry. This can occur due to the interaction of "introduced" atoms with the atoms surroundings when the latter occupy a less symmetrical position in the crystal lattice.

Another reason for lowering the symmetry of the initial point group is the Jahn–Teller effect [154], which is observed in isolated and nonlinear structure of polyatomic molecules, clusters and crystals, if the electronic ground state is orbitally degenerated.

As is known, a decrease of symmetry also occurs at phase transitions of the second kind, regardless the physical nature of the order parameter. It is confirmed by many experimental data (first, by X-ray diffraction data), but it is not found why there is a decrease of symmetry. However, this does not apply to phase transitions in magnetically ordered states. In this case, the lowering of the symmetry can be understood proceeding from the structure of magnetic point groups [85, 118, 134, 155] discussed in the Sects. 3.1 and 3.3.

As it was mentioned, the $\mathbf{T}g$ element of the magnetic point group G' cannot contain an element g of odd order g belongs to the classical point group G. Otherwise, among the $\mathbf{T}g$ that contains G' an element $\mathbf{T}$ would be found, which is prohibited in G'. Therefore, among the elements G' cannot be the elements $\mathbf{T}C_3$ or $\mathbf{T}S_3$ (S_3 is the mirror rotation by the angle $2\pi/3$), as well as $\mathbf{T}C_5$ or $\mathbf{T}S_5$ (S_5 is the mirror rotation by the angle $2\pi/5$), if the magnetic point groups obtained by expanding the $\mathfrak{I}$ and $\mathfrak{I}_h$ groups are taken into account. In fact,

$$(C_3\mathbf{T})^3 = C_3^3\mathbf{T}^3 = e\mathbf{T}^2 \cdot \mathbf{T} = \begin{cases} \mathbf{T} & \text{for non-Kramers systems} \quad (4.1) \\ -\mathbf{T} & \text{for Kramers systems} \quad (4.2) \end{cases}$$

and

$$(C_5\mathbf{T})^5 = C_5^5\mathbf{T}^5 = e\mathbf{T}^2 \cdot \mathbf{T}^2 \cdot \mathbf{T} = \mathbf{T} \tag{4.3}$$

for both Kramers and non-Kramers systems.

As (4.1)–(4.3) show, if one admits that rotations C_3 or C_5 are allowed for the magnetic point group G', then the operator $\mathbf{T}$ will also be allowed as an element of G', in contradiction with the definition of G'. Thus, in both cases, structural distortions would arise as a consequence of the time-reversal symmetry. This means, in particular, that magnetic atoms (ions) with an integer spin of a trihomonuclear or pentahomonuclear cluster cannot occupy the vertices of an equilateral triangle or regular pentagon. Such interdictions and associated structural distortions of the equilateral triangle and the regular pentagon with magnetic atoms in the vertices are due to the structure of the point group of magnetic symmetry. Since for non-Kramers systems the magnetic symmetry group G' is obtained by expanding of classic point group G by a cyclic group $\{\mathbf{T}, \mathbf{T}^2 = e\}$, it can be concluded that resulting structural distortions are due to the existence of time-reversal symmetry.

Analogically, no one from the point groups of magnetic symmetry for Kramers systems (Sect. 3.3) contains simple or mirror rotations of the third-order- or fifth-order (if magnetic point groups due to expansion of $\mathfrak{I}$ and $\mathfrak{I}_h$ groups are taken into consideration; in this book such expansion is not examined). This applies to groups $4^{(4'_z)}$, $\overline{4}^{(4'_z)}$, $4^{(4'_z)}/m^{(1)}$ and $4^{4'_z}/m^{(2)}$, which are obtained by extending the classical point groups by using the cyclic group $4'$, and to groups $4^{(4'_z)}2^{(m'_x)}2^{(m_{xy})}$, $4^{(4'_z)}m^{(m'_x)}m^{(m_{xy})}$, $\overline{4}^{(4'_z)}2^{(m'_x)}m^{(m_{xy})}$ and $\overline{4}^{(4'_z)}2^{(m_{xy})}m^{(m'_x)}$, which are obtained by a general expansion using the non-cyclic group $4'm'm$. Because the group $4'$ is isomorphic to the cyclic group G_4 and the group $4'm'm$ is isomorphic to the non-Abelian group G_8, both containing the time-reversal operator, it means that considered structural distortions for trihomonuclear and pentagonal spin clusters are due to the time-reversal symmetry.

There are many experimental data related to the anomalous behavior of Cr(II), Fe(III), Cu(II), V(IV) and Co(II) trihomonuclear clusters due to their four-color symmetry that will be discussed in the Sects. 4.2–4.4 on the basis of the analysis reproduced from [155], with permission of AIP Publishing.

4.2 Trinuclear Chromium(III) and Iron(III) Carboxylate Clusters

The trimer clusters with an anomalous temperature behavior of their magnetic properties were discovered in 1928, when for the first time Welo [156] measured the magnetic susceptibility of polynuclear salts containing three atoms of iron and chromium in the temperature range from 200 to 400 K, and found that these materials follow the Curie–Weiss law with a large Curie–Weiss constant (100−600 K). Later, Kambe [157] tried to explain the large Curie–Weiss constant by introducing isotropic exchange interactions in the spin-trimer model, where three spin operators $\mathbf{S}_1$, $\mathbf{S}_2$, and $\mathbf{S}_3$ of spin 3/2 form an equilateral triangle in the crystal. However, the insufficiency of the model, was soon recognized. It was shown that the thermomagnetic data [158, 159], as well as data on specific heat of triads in Cr-acetate $\left[Cr_3O\,(CH_3COO)_6\,(OH)_2\right]Cl\cdot 8H_2O$ [160] need a model in the supposition of a significant lowering of the triad symmetry from D_{3h} with the transformation of the equilateral triangle into an isosceles one.

Further studies demonstrated that this supposition is confirmed by the X-ray structure data of metal–acetate complexes obtained at room temperature [161, 162]. In this case, a X-ray structure analysis of chromium acetate compounds shows that the chromium complex contains metal atoms as vertices of the equilateral triangle with an oxygen atom in the center. The X-ray single diffraction pattern for a monochloroacetate complex is similar to patterns for the chromium and iron complexes.

Further studies have shown that the trinuclear chromium acetate has a phase transition at 215 K with possible change of the molecular structure [163]. Actually, the earlier X-ray investigations were carried out at temperatures above the point of phase transition, while the magnetic susceptibility and specific heat data showing the lowering of symmetry were obtained at low temperatures.

The trinuclear iron(III) carboxylate compounds of the type $\left[Fe_3O\,(RCO_2)_6\,(OH)_2\right]\cdot X\cdot nH_2O$ have been studied by magnetic susceptibility and Mössbauer

Table 4.1 The best-fit values of J and J' for five iron(III) carboxylate complexes

Complex	$J'(cm^{-1})$	$J(cm^{-1})$	ΔE_Q(mm/s)
Formiate	-14.0 ± 1.0	61.0 ± 2.0	0.50 ± 0.10
Acetate	-12.0 ± 1.0	64.0 ± 2.0	0.43 ± 0.10
Propionate	-14.0 ± 1.0	62.0 ± 2.0	0.72 ± 0.10
Monochloroacetate	-10.0 ± 1.0	54.0 ± 2.0	0.69 ± 0.10
Triochloroacetate	-10.0 ± 1.0	58.5 ± 2.0	0.73 ± 0.10

spectroscopy methods [164]. The results of a magnetic study of similar trinuclear Fe(III) carboxylate complexes are also given in [165]. All the compounds are high-spin and trinuclear in solid state with the iron atoms in almost equivalent sites.

The best fit of experimental data was obtained using the spin Hamiltonian $H = J\,(\mathbf{S}_1\mathbf{S}_2{+}\mathbf{S}_2\mathbf{S}_3{+}\mathbf{S}_3\mathbf{S}_1) + J'\mathbf{S}_3\mathbf{S}_1$. The best-fit values of exchange integrals J and J' are given in Table 4.1 [164].

The Mössbauer spectrum of each of the complexes given in Table 4.1 has only two relatively broad absorption lines with approximately the same intensity. The probability is small that these two lines are caused by two different iron sites, since in this case the intensity ratio of the lines would be 2 : 1. The two lines are most likely caused by interactions of identical iron nuclei with the gradient of the electric field resulting in quadruple splitting ΔE_Q. The values of ΔE_Q from Table 1 are characteristic for high-spin iron(III) compounds. However, the width of the absorption lines in Mössbauer spectra is broader than one might expect. It can be caused by the iron atoms that are situated in a slightly different sites. This assumption is confirmed by magnetic measurements data. Since small structural distortions at highly symmetric environment are detected with high accuracy in the quadrupole splitting ΔE_Q [166], two broad lines are to be expected. The results obtained for carboxylate complexes investigated by means of magnetic susceptibility measurements and Mössbauer spectra are consistent with the trihomonuclear complex with one of the sites in a slightly different environment than the other two.

As it follows on the basis of a few experimental data, the data from Table 4.1 are typical for all trihomonuclear magnetic clusters. They are as follows: an equilateral triangle with equivalent magnetic ions in corners at high temperatures, which transforms into an isosceles one at low temperatures. The above-mentioned structural phase transition in chromium acetate at 215 K [163] was found at 211 K by using more accurate structural experiments on X-ray diffraction at 190 K [167]. The high temperature (HT) and low temperature (LT) space groups are $P2_12_12$ and $P2_12_12_1$. The intracluster Cr–Cr distance is 3.27 Å, while the intercluster one is equal to 5.78 Å. The LT magnetic and spectroscopic properties are due to structural distortions of trimer clusters.

In [168], the experimental results on the temperature behavior of trihomonuclear iron complexes have been evaluated in two different ways, using the spin Hamiltonian

$$H = -2J\,(\mathbf{S}_1\mathbf{S}_2 + \mathbf{S}_2\mathbf{S}_3) - 2J'\mathbf{S}_3\mathbf{S}_1. \tag{4.4}$$

In the first method (it is assumed $J = J'$), the value of J is varied to optimize the agreement between theoretical and experimental magnetic moments. In the second method, J and J' are varied independently to optimize the agreement. It was shown that for formiate, acetate, and monochloroacetate compounds a reasonable agreement can be obtained between calculated and experimental moments, if only an equivalent coupling between the three iron atoms is supposed.

For trichloroacetate, phenylacetate, and benzoate a reasonable agreement between experimental and theoretical data can be obtained, if J and J' vary independently. It was also shown that the line width for the two quadrupole split Mössbauer lines is considerably larger (from 0.31 and 0.32 mm/s for acetate and trichloroacetate to 0.37 mm/s for phenylacetate at room temperature), than expected for all equivalent iron atoms. These broadened line widths would be expected, if the three iron atoms were nonequivalent as the magnetic results indicate.

Using the spin Hamiltonian, (4.4), the magnetic properties and Mössbauer spectra of trinuclear iron(III) complexes of malonic, succinic, fumarie, and phthalic acids have been investigated [169]. In order to describe the magnetic behavior of these complexes, seven different models were discussed. There were investigated the following models: "the equilateral triangle (EqT), equilateral triangle plus a monomeric component ($EqT + Z$), isosceles triangle ($ISOS$), biquadratic exchange (BqX), biquadratic exchange plus a monomeric component ($BqX + Z$), equilateral triangle with intertrimer exchange ($EqTX$), and equilateral triangle with intertrimer exchange plus a monomeric component ($EqTX + Z$)". The best-fit parameters for molar static magnetic susceptibility vs. temperature for acetate and suscinate complexes were obtained for the $\{ISOS;\ BqX;\ EqTX + Z\}$ and $EqTX + Z$ models, respectively.

In [170] there have been determined the crystal structures of the isomorphic prototypic oxo-centered trihomonuclear coordination compounds $[Cr_3O(OOCCH_3)_6(H_2O)_3]Cl \cdot 6H_2O$ and $[Fe_3(OOCCH_3)_6(H_2O)_3]Cl \cdot 6H_2O$, CRAC and FEAC (**1,2**). These compounds are prototypes of the general class of trihomonuclear basic carboxylates. Authors argue that metal ions in the oxo-centered units are antiferromagnetically coupled, but the results of all investigations presented in [171–175] point to the fact that the three exchange coupling constants are different. It is not clear whether the inequalities are due to differences in metal–metal bond distances, metal–oxygen distances, or bond angles. More accurate structural data are necessary to solve this problem. Such data were reported in [170] as a complete structure determination of both CRAC and FEAC. In Table 4.2, the selected bond distances for both **1**(CRAC) and **2**(FEAC) are listed.

The results of Table 4.2 are in agreement with the magnetic symmetry of trinuclear magnetic clusters with the Kramers degeneracy of energy levels (see Sects. 3.3 and 4.1).

The study of the crystalline structure of polynuclear iron(III) pivalates [176] by X-ray diffraction analysis demonstrate that the metal core of the hexanuclear trimethylacetate $[Fe_6(O)_2(OH)_2(OOCCMe_3)_{12} \cdot (HOOCCMe_3)(THF)] \cdot 1.5C_6H_6$

Table 4.2 Selected bond distances (Å) **1** (CRAC) and **2** (FEAC)

	1 (CRAC)	**2** (FEAC)
M(1)–M(2)	3.281(1)	3.285(1)
M(1)–M(3)	3.288(1)	3.295(1)
M(2)–M(3)	3.279(1)	3.291(1)

complex consists of two triangular Fe_3O fragments. Fe–Fe distances are 3.423(4), 3.291(4), and 3.260(4) Å in one triangle and 3.258(4), 3.439(4), and 3.264(4) Å in another triangle, i.e., neither triangle is equilateral.

4.3 Trinuclear Copper(II) Clusters

The magnetic properties of the intramolecularly coupled trinuclear $(Cu^{II})_3$ compound have been summarized by Kahn [177]. He analyzed the different arrangement of copper(II) ions. Spin-frustration effects in trimeric units were also discussed.

A number of studies [178–194] have reported symmetric and asymmetric trinuclear copper complexes. In particular, two types of trinuclear copper(II) clusters with similar magnetic behavior were reported [193, 195]. The first type is a trinuclear Schiff base $[Cu_3L(OH)]^{+3}$ complex with Cu–Cu separations of 3.62, 4.92, and 5.89 Å [193]. The second trihomonuclear copper(II) cluster in the compound $\{[Cu_3L(\mu\text{-}OH)(\mu_3\text{-}HPO_4)(H_2O)][PF_6]_3\,3H_2O\}_n$ consists of trinuclear metal sites in which three copper(II) centers in a distorted square-pirammidal geometry are linked by an HPO_4^{2-} phosphate bridge. The Cu–Cu separations in the triangle of copper(II) ions are 4.56, 5.47, 3.56 Å [195].

In the case of the equilateral triangular arrangement of antiferromagnetically coupled metal ions, the description of the magnetic properties is hindered by the spin-frustration effects [177]. For the spin $S = 1/2$, of each component of a trimer cluster, the ground state for this arrangement is a generated doublet of Kramers doublets. In absence of spin-frustration, this degeneracy is not removed by Coulomb interactions, and it can arise in the second-order perturbation theory due to Dzyaloshinsky-Moria antisymmetric exchange [196, 197]. The effect of the isotropic antisymmetrical exchange is to slope the neighboring spins that leads to a weak magnetization in the direction orthogonal to the antiferromagnetic alignment axis. The estimated value of the effective antisymmetric exchange constant for a $S = 1/2$ trimeric system is $G_{\mathrm{eff}} = 5.5\,\mathrm{cm}^{-1}$ [198]. This antisymmetric exchange theoretical model confirms the assumption that in an antiferromagnetically coupled cyclic trimer, the spin-frustration leads to a ground state doublets splitting. The removal of the triads ground states fourfold degeneracy due to antisymmetric exchange has also been taken into account to interpret magnetic susceptibility data and *ESR* spectra for trinuclear chromium(III) and iron(III) carboxylates [199, 200].

In the [201] there have been reported two types of cyclic trinuclear copper (II) complexes $[Cu_3(\mu_3\text{-}OH)(aaat)_3(H_2O)_3]\,[NO_3]_2 \cdot H_2O$ (**1**) (where Haaat = 3-acetylamino-5-amino-1,2,4-triazole) and $[Cu_3\text{-}(\mu_3\text{-}OH)(aat)_3(\mu_3\text{-}SO_4)] \cdot 6H_2O$

Table 4.3 Selected bond distances (Å) for **1** and **2** complexes

	1	**2**
Cu(1)–Cu(2)	3.347(2)	3.337(4)
Cu(1)–Cu(3)	3.374(2)	3.364(4)
Cu(2)–Cu(3)	3.393(2)	3.337(5)

(**2**) (where Haat = 3-acetylamino-1,2,4-triazole). In Table 4.3, the selected bond lengths for **1** and **2** that were determined by using single-crystal X-ray diffraction data at room temperature, are given.

The temperature dependence of $\chi_m T$ product (where χ_m is the molar magnetic susceptibility per [Cu_3] entity) for complexes **1** and **2** exhibits that the $\chi_m T$ values (0.72 and 0.75 $cm^3\,mol^{-1}$ K for **1** and **2**, respectively, at room temperature) are lower than expected for three uncoupled $S = 1/2$ spins (ca. 1.2 $cm^3 mol^{-1} K$). These values are decreasing steadily at increasing temperature. Based on the temperature dependence curves, it was concluded that it does exist an intratrimer antiferromagnetic coupling. The magnetic behavior of **1** and **2** in the 90–300 K temperature range is explained within the framework of the isotropic Heisenberg-Dirac-van Vleck (HDVV) Hamiltonian formalism with a single exchange constant ($J_{12} = J_{13} = J_{23} = J$, $H_{HDVV} = -J(\mathbf{S}_1\mathbf{S}_2 + \mathbf{S}_2\mathbf{S}_3 + \mathbf{S}_3\mathbf{S}_1)$).

The best-fit parameters [201] are $J = -194.6\,cm^{-1}$ and $g = 2.08$ with $R = 2.1 \times 10^{-6}$ for **1**; $J = -185.1\,cm^{-1}$ and $g = 2.10$ with $R = 2.1 \times 10^{-6}$ for **2** (here R is the agreement factor).

To rationalize the magnetic behavior below 90 K the distortion of the equilateral triangle (C_{3v} point group symmetry) to give an isosceles one (C_{2v} point group symmetry) as well as the antisymmetric exchange interaction were taken into account

$$\hat{H} = -J\,(\mathbf{S}_1\mathbf{S}_2 + \mathbf{S}_1\mathbf{S}_3) - j\mathbf{S}_2\mathbf{S}_3 + \hat{H}_{as}, \tag{4.5}$$

$$\hat{H}_{as} = \sum_{ij} \mathbf{G}_{ij}\left[\mathbf{S}_i \times \mathbf{S}_j\right]. \tag{4.6}$$

To avoid over parameterization in the fitting process, there were supposed a C_{2v} symmetry and an identical g-factor for all three Cu(II) ions ($g_{1u} = g_{2u} = g_{3u}$, where $u = x, y, z$) and a common antisymmetric factor, G_{ij}, for all three interesting ion pairs ($\mathbf{G}_{12} = \mathbf{G}_{13} = \mathbf{G}_{23} = \mathbf{G}$). In this case, the relevant parameters are J , j, $g_\perp$, $g_\parallel$ and G and the resulting best-fit parameters are [201] $J = -191.0\,cm^{-1}$, $j = -156.4\,cm^{-1}$, $G = 27.8\,cm^{-1}$, $g_\parallel = 2.09$, $g_\perp = 2.00$ (fixed in the fitting process), and the agreement factor $R = 2.7 \times 10^{-6}$ for **1**; and $J = -175.4\,cm^{-1}$, $j = -153.2\,cm^{-1}$, $G = 31.0\,cm^{-1}$, $g_\parallel = 2.10$, $g_\perp = 2.10$ (fixed in the fitting process), and $R = 3.9 \times 10^{-6}$ for **2**.

These experimental results clearly confirm that the existence of a disagreement between the high symmetry of triads in X-ray data at room temperature (equilateral triangle) and lowered symmetry (isosceles triangle) at low temperatures is not accompanied by a structural phase transition. It is a general characteristic for all trimer magnetic clusters with the Kramers degeneracy of energy levels.

The study of the structure and magnetism of $Na_9[Cu_3(H_2O)_9(\alpha\text{-}AsW_9O_{32})_2]\cdot 26H_2O$ (**Na**$_9$-**2**) over $2-300$ K [202] has shown that **Na**$_9$-**2** has antiferromagnetically coupled Cu^{2+} ions, with $J=-1.36\pm 0.01\,\text{cm}^{-1}$ (using isotropic exchange Hamiltonian). The ground state total spin in **Na**$_9$-**2** is $S_T=1/2$. In fact, the polyanion **2** has three Cu^{2+} ions arranged in a triangular fashion with Cu–Cu distances [203] $Cu_1-Cu_2=Cu_1-Cu_3=4.696$ Å and $Cu_2-Cu_3=4.689$ Å. The possible super exchange path between any two Cu^{2+} ions of the polyanion **2** are the same (Cu–O–W–O–W–O–Cu), therefore the Cu–Cu distances are equal and the Cu_3 system is an equilateral triangle.

The hyperfine interactions of the exchange-coupled $3d^9$-electrons of copper(II) dimer clusters with the $^{63,65}Cu$ nuclei are neglected in the susceptibility analysis, because these interactions are not resolved in the *EPR* spectra. Probably, this is due to the fact that the rate of the spin exchange process is higher than the hyperfine splitting (in frequency units). There was also neglected the zero-field splitting constants D and E, since these constants are much smaller than the constant of exchange interaction J, as it follows in *ESR* experiments. The temperature behavior of the molar susceptibility χ_m for **Na**$_9$-**2** at $H=1000$ Oe has shown that in the range of 60–200 K the $\chi_m T$ value is nearly constant ($\sim$1.34 emu $\cdot$ K $\cdot$ mol^{-1}). This indicates an admixture of both $S_T=1/2$ and $S_T=3/2$ spin states. In the case of $S_1=S_2=S_3=1/2$ and $J'=J$, the total spin takes the values $1/2\,(\uparrow\uparrow\downarrow)$, $1/2\,(\uparrow\downarrow\uparrow)$, or $3/2\,(\uparrow\uparrow\uparrow)$, with energies $3J/2$, $3J/2$ or $-3J/2$, respectively. The single-crystal *EPR* experiments on **Na**$_9$-**2** at Q-band ($\nu=34$ GHz) give the following values of spin Hamiltonian constants: $g_{\parallel}=2.117\pm 0.005$ and $g_{\perp}=2.254\pm 0.005$ for $S_T=1/2$ (ground state) and $g_{\parallel}=2.060\pm 0.005$, $g_{\perp}=2.243\pm 0.005$ and $|D|=0.023\pm 0.005\,\text{cm}^{-1}$ for $S_T=3/2$ (excited state). The W-band ($\nu=93.165$ GHz) powder *EPR* spectrum of **Na**$_9$-**2** at 4 K, as well as the temperature dependence of the W-band ($\nu=93.165$ GHz) *EPR* spectrum, confirm the Q-band single-crystal *EPR* data. This is a rare case *EPR* observed from both the ground and excited states [202]. We note that although the assumption $J'=J$ was correctly argued and the best J parameter was obtained [202], the experimental values of $g_{\parallel}$ and $g_{\perp}$ $(g_{\parallel}\neq g_{\perp})$ agree with distortions of the trinuclear copper(II) cluster.

The equilateral triangle with Cu^{2+} ions in the corners contained in the $Cu_3(O_2C_{16}H_{23})_6\cdot 1.2C_6H_{12}$ compound [204] shows an unexpected structure and an example of spin frustrations. The structure of the crystal consists of discrete molecules $Cu_3(O_2C_{16}H_{23})_6$ with a triangular arrangement of copper(II) ions with a 3-fold axis of symmetry passing through the center of the equilateral triangle. The Cu–Cu separations are 3.131(3) Å, precluding from any direct bonding. Magnetic susceptibility measurements on a polycrystalline sample at 1000 G in the temperature range from 1.8 to 380 K are consistent with the presence of a doublet ground state with a g value equal to 2.071(5). The magnetic moment μ_{eff} is constant between 10 and 50 K at $1.80\mu_B$. Above 50 K, μ_{eff} gradually increases without saturation from $1.80\mu_B$ to $2.69\mu_B$ at 380 K. At low temperatures (below 10 K) a small decrease is probable due to some very small antiferromagnetic interaction ($|\theta|<0.1$ K) between

Cu(II) trimers. Since Cu_3 is an equilateral triangle, the magnetic susceptibility measurements data were explained on the basis of the spin Hamiltonian describing the isotropic Zeeman and exchange interactions. The best-fit parameters [204] are $J = -108$ K and $g = 2.07$.

Detailed Q-band ($\nu = 34$ GHz) single-crystal *EPR* experiments were carried [205] out to clarify the peculiarities of the magnetic behavior of this high-symmetric trimer cluster. The main result obtained lead to the conclusion that at high temperatures (above 50 K) *EPR* data are in agreement with an equilateral triangle model of trinuclear cluster (confirmed by X-ray diffraction data), while at low temperatures (below 50 K) the structural distortions of trinuclear copper (II) clusters are detected and, as a consequence, the model of equilateral triangle is not acceptable.

4.4 Trinuclear Vanadium(IV) and Cobalt(II) Clusters

The trinuclear vanadium(IV) magnetic clusters with Kramers degeneracy of energy levels are contained in two species of polyoxovanadate-based magnetic molecules, labeled **V-1** and **V-2**, with the chemical formulas [206] $(CN_3H_6)_4\,Na_2\left[H_4V_6^{IV}O_8\,(PO_4)_4\,\{(OCH_2)_3\,CCH_2OH\}_2\right]\cdot 14H_2O$ and $Na_6\left[H_4V_6^{IV}O_8\,(PO_4)_4\,\{(OCH_2)_3\,CCH_2OH\}_2\right]\cdot 18H_2O$. A single-crystal X-ray analysis has established [207] that in **V-1** the distances between V^{IV} ions are $3.218(V1-V2)$, $3.222(V1-V3)$ and $3.364(V2-V3)$ Å. In **V-2**, the corresponding distances are 3.212, 3.253 and 3.322 Å. These structural data allow us to suppose that only two exchange constants are necessary and sufficient for defining the Heisenberg model Hamiltonian $\left(J_{12} = J_{13} = J,\, J_{23} = J'\right)$. The best-fit parameters for interpreting the magnetization versus temperature data are [206] $J = -32.6 \pm 0.5$ K and $J' = -3.5 \pm 1$ K for **V-1**, and $J = -31.9 \pm 0.6$ K and $J' = -2.9 \pm 1.1$ K for **V-2**.

The proton spin-lattice relaxation rate as a function of temperature, as well as the ^{23}Na spin-lattice relaxation rate as a function of temperature confirm the magnetic susceptibility data for both **V-1** and **V-2** compounds. The temperature dependence of spin-lattice relaxation rate, $1/T_1$, is described by a phenomenological formula [207] giving $1/T_1$ proportional to $T \cdot \chi\,(T)$, if the condition $2a/b = 1$ holds. Here a and b are coefficients in the (18) from [206]. These constants can be expressed in terms of magnetic dipole–dipole interaction tensor elements. In the case of protons, $a = 4.94$ and $b = 13.45$, and $a = 1.45$ and $b = 3.97$ for **V-1** and **V-2**, respectively, that is $2a/b \approx 0.73$ for both **V-1** and **V-2**. For ^{23}Na the results are $a = 0.525$ and $b = 1.252$, and $a = 0.6$ and $b = 1.09$ for **V-1** and **V-2**, respectively. We see that $2a/b$ is equal to 0.81 for **V-1** and to 1.11 for **V-2**, in a satisfactory agreement with the condition $2a/b = 1$.

Note that $V-V$ distances 3.212 and 3.253 Å in **V-2** are not so close as distances 3.218 and 3.222 Å in **V-1**. Therefore, more accurately, **V-1** must not be considered an isosceles triangle, but a general one $(J_{12} \neq J_{23} \neq J_{31})$. The magnetic contribution to the specific heat, as well as the inelastic neutron structure factor confirm this conclusion for these and other complexes with trimer magnetic clusters having the spin ground state $S_T = 1/2$ [207], including $K_6\,[V_{15}As_6O_{42}\,(H_2O)] \cdot 8H_2O$ [208–212] (symmetric trimer + capping hexamers), **Na₉-2** (isosceles trimer) [213], $\left[Cu_3\,(cpse)_3\,(H_2O)_3\right] \cdot 8.5H_2O$ (symmetric trimer) [213] and $Cu_3(O_2C_{16}H_{23})_6 \cdot 1.2C_6H_{12}$ (symmetric trimer) [204]. The magnetic contribution to the specific heat is usually masked by much larger phonon contributions [207] and it may be separated experimentally by subtracting the specific heats in zero and non-zero magnetic fields. As for inelastic neutron cross-sections for the symmetric trimer with two degenerate $S_T = 1/2$ ground states and a single $S_T = 3/2$ ex citate state, there are two different inelastic neutron structure factors, but only one single transition energy. The same structure factors follow also for the isosceles trimers. However, in that case the two $S_T = 1/2$ states are nondegenerated (the two ground state Kramers doublets are separated). The 15-cobalt-substituted polyoxotungstate $[Co_6\,(H_2O)_{30}\,|Co_9Cl_2\,(OH)_3\,(H_2O)_9\,(\beta\text{-}SiW_8O_{31})_3 \sim]^{5-}$ (**Co-1**) and $Na_5\,[Co_6\,(H_2O)_{30}\,|Co_9Cl_2\,(OH)_3\,(H_2O)_9\,(\beta\text{-}SiW_8O_{31})_3 \sim] \cdot 37H_2O$ (**Co- 2**) have been studied [214]. The trimeric polyanion **Co-1** has a core of nine Co^{II} ions encapsulated by three unexampled (β-SiW_8O_{31}) fragments and two Cl^- ligands. The 9 exchange-coupled Co^{II} ions in the core of **Co-1** can be considered as three triangular units belonging to three Keggin fragments that are connected by two chlorobridges. The $\left(Co^{II}\right)_9$ core is the same for both **Co-1** and **Co-2**. The magnetic properties of **Co-2** in the low temperature range ($T < 30$ K) can be understood if we neglect the sufficiently small difference between $Co_1 - Co_2$(3.037 Å) and $Co_1 - Co_3$(3.141 Å) distances, e.g., taking into account that each triangle $\left(Co^{II}\right)_3$ is an equilateral one. The total static magnetic susceptibility is presented as a sum of contributions of the $\left(Co^{II}\right)_9$ core and the six uncoupled peripheral Co^{II} ions with effective spins $S = 1/2$. Taking into consideration that the bond angles and the bond lengths of the $\left(Co^{II}\right)_9$ core are very similar to that in $\left[Co_9\,(OH)_3\,(H_2O)_6\,(HPO_4)_2\,(PW_9O_{34})_3\right]^{16-}$, the full anisotropic Ising model with J_{1z} and J_{2z} exchange constants used by Coronado and co-workers [215] has been employed. The best fit parameters are $J_{1z} = 17.0 \pm 1.5\,\text{cm}^{-1}$ and $J_{2z} = -13.1 \pm 1.5\,\text{cm}^{-1}$, which are in good agreement with the values presented in literature for $\left[Co_9\,(OH)_3\,(H_2O)_6\,(HPO_4)_2\,(PW_9O_{34})_3\right]^{16-}$ ([215]). The powder *EPR* spectra of **Co-2** in Q-band (34 GHz) and W-band (93 GHz) at 4 K exhibit an asymmetric transition at low magnetic fields, that can be associated with an anisotropic ground state Kramers doublet [216]. The effective g-values associated with this doublet are $g_{\text{eff}}\,(xx) = 2.63$, $g_{\text{eff}}\,(yy) = 3.89$ and $g_{\text{eff}}\,(zz) = 5.72$, as expected for Co^{II} coordination compounds [216–219]. This effective $S = 1/2$ ground state is in a good agreement with the static magnetic susceptibility data.

We see that low temperature magnetic susceptibility and *EPR* data for **Co-1** and **Co-2** can be satisfactory explained in the framework of the symmetric trimer model, although the Co_3 triangle is in fact distorted.

The experimental results discussed in this subsection show that the trinuclear vanadium(IV) clusters have a temperature behavior similar to that for trihomonuclear clusters formed by Cr(III), Fe(III), Cu(II) or Co(II) atoms. For all symmetric trimer magnetic clusters, the model of an equilateral triangle is in a good agreement with the magnetic susceptibility, *ESR*, specific heat and inelastic neutron scattering data at high temperatures (above 50 K), but it is in contradiction with the experiment at low temperatures (below 50 K).

4.5 Concluding Remarks

A relative large number of trinuclear species have been described in this chapter by neglecting local anisotropies and anisotropic interactions between spins. However, only a very limited number of trimer clusters with the arrangement of magnetic atoms in the vortexes of an equilateral triangle are known at present. Excepting the above-mentioned copper(II), full-symmetric trimers with the ground state total spin $S_T = 1/2$, $Cu_3(O_2C_{16}H_{23})_6 \cdot 1.2C_6H_{12}$, $\left[Cu_3(cpse)_3(H_2O)_3\right] \cdot 8.5H_2O$, and **Na$_9$-2**, there is also one equilateral triangle related to the triangular cation $\left[Cu_3(pao)_3OH\right]^{2+}$ with H_{pao} = pyridine-2-aldehyde-oxime [177]. It should be remembered that the **Na$_9$-2** complex (see the notation above) was considered [207] as containing equilateral triangles Cu_3 while the crystalline structure data demonstrated that there are nonequilateral triangles [202, 203]. Thus, only three triangles are considered as equilateral and they are all formed by copper atoms [177, 207, 213].

It is curious to note that the first discovered trinuclear clusters Fe_3 and Cr_3 were considered for long time as clusters with a symmetric arrangement of the iron and chromium atoms, respectively, in the vortexes of an equilateral triangle [156–163, 177, 207]. This consideration was based on Figgis, Robertson [161] and Chang, Geffrey [162] structural data published in 1964 and 1970, respectively. Later, it was established that conclusions about the equilateral triangles have been based a rather low accuracy analysis of the single-crystal X-ray diffraction. A more exact structural data determined by Schent and Güdel in 1982 by low temperature (190 K) X-ray diffraction brought to distortions of triangular cluster Cr_3 contained in the $\left[Cr_3O(CH_3COO)_6(H_2O)_3\right]Cl \cdot 6H_2O$ complex [167]. In 1997, it was done more precise determination of the crystalline structure of this complex on the basis of single-crystal X-ray diffraction at room temperature, as well as of the isomorpheous iron containing trinuclear complex (denoted in Sect. 3.2 as CRAC **1** and FEAC **2**, respectively). It was shown [170] that there are isosceles triangles for both CRAC **1** and FEAC **2**.

According to the results given in Sects. 3.1, 3.3, all trihomonuclear magnetic clusters must be characterized by generalized point groups (black-and-white groups for non-Kramers systems and four-color magnetic point groups for Kramers systems) with a low-symmetry atom arrangement. This follows from specific properties of the black-and-white groups and four-color magnetic point groups, taking into account the possible correlation between the spin configurations and the corresponding arrangements of atoms (ions) with spins. The structure of black-and-white and four-color point groups of magnetic symmetry depends on the type of simultaneous transformations of coordinate and spin parts of the full wave function (depending on the space, spin, and time variables) under action of the time-reversal operator. We certify that no any structural distortions appear, if the time-reversal operator is not taken into account, i.e., structural distortions under consideration are due to the time-reversal symmetry.

The three types of Cu_3 triangles, considered now as equilateral triangles [177, 204, 213] on the basis of single-crystal X-ray diffraction data, are actually not equilateral, in magnetic ordered phase at low temperature, because there are no four-color groups of magnetic symmetry with threefold rotation axes. The single-crystal X-ray diffraction experiments lead to the conclusion about symmetric copper trimers in these cases as a consequence of small geometrical distortions of equilateral triangles. In the case when experimentally measured sides of the triangle (or exchange interaction constants between magnetic ions) appear to be equal, the magnetic ions are nonequivalent, within the experimental errors. This can be revealed by more precise measurements. There are no doubts, that in the future such measurements will be made and the nonequivalence of copper atoms in trinuclear clusters Cu_3 will also be discovered, in analogy with the early-discovered nonequivalence of iron and chromium atoms.

Chapter 5
Time-Reversal Symmetry of Quantum Systems with Quasi-energy Spectrum

In this section, it will be shown how the wave function of a non-stationary state of a quantum system is transformed under the action of the time-reversal operator **T**. Despite the fact that in the case of non-stationary states the total energy of the system is uncertain, it does not mean violating the law of energy conservation, since the average energy is conserved. It is found that the wave functions $\Psi(\xi, t)$ and $\mathbf{T}\Psi(\xi, t)$, on average, belong to the same energy level and are linearly dependent, if $\mathbf{T}^2 = \mathbf{1}$. However, if $\mathbf{T}^2 = -\mathbf{1}$, then the wave functions $\Psi(\xi, t)$ and $\mathbf{T}\Psi(\xi, t)$ belonging in average to the same level, are orthogonal (quasi-degeneracy of energy levels due to time-reversal symmetry).

The Schrödinger equation for the Green function is invariant relative to time reversal, and the function itself behaves under time reversal like wave functions of non-stationary states.

It is demonstrated that Kramers degeneracy of quasi-energy states is present (absent) at a half-integer (integer) value of the angular momentum. An additional (compared with Kramers one) degeneracy of quasi-energy states is found for cases when the time-reversal operator doesn't commutates with time-translation operators.

The main results presented in this Chapter, have been published in [135, 220, 221] and are related, in particular, to the reduced Brillouin zone in the space of quasi-energies and to supplementary degeneracy of quasi-energy levels due to non-commutation of time-reversal and time-translation operators.

I. I. Geru, *Time-Reversal Symmetry*, Springer Tracts in Modern Physics 281,
https://doi.org/10.1007/978-3-030-01210-6_5

5.1 Non-stationary States of Quantum System Under Time-Reversal Operator

The Wigner time-reversal operator **T** has been introduced in the non-relativistic quantum mechanics on the basis of invariance of the time-dependent Schrödinger equation under the reversal of time [222]. From the requirement of this invariance it follows inevitably that the time-independent Hamiltonian **H** in the coordination representation is also invariant with respect to **T**. Another consequence of this invariance is that the operator **T** should be represented as the product of an unitary operator $\mathbf{U}$ $\left(\mathbf{U}^{+}\mathbf{U} = \mathbf{1}\right)$, having the property

$$\mathbf{U}\mathbf{H}^{*} = \mathbf{H}\mathbf{U}, \tag{5.1}$$

and the operator of complex conjugation **K**. The last operator is nonlinear by definition that leads to the nonlinearity of the operator **T**. On the other hand, the operator of time reversal is an anti-unitary operator, since the unitary operator **U**, satisfying the relation (5.1), is also part of **T**.

For systems consisting of spinless particles, $\mathbf{U} = 1$ and $\mathbf{T} = \mathbf{K}$, if each of particles does not have orbital momentum. If spinless particles forming the system have nonzero orbital moments, the matrix of the operator **U** is a symmetric one with alternating elements $+1$ and -1, located only on the secondary diagonal. The remaining matrix elements are zero. The number of nonzero matrix elements is equal to $2l + 1$ or $2L + 1$, where l is the orbital momentum of a single particle, and L is the total orbital angular momentum of the system of particles.

Since the operator **T** is defined up to a phase constant, of which module is equal to 1, it does not matter what element ($+1$ or -1) is located on the end of the secondary diagonal; it is important that the selected phase factor at the operator **T** remains unchanged. Thus, in this case, there is no additional degeneracy of energy levels due to time-reversal symmetry. The matrix of the operator **U** has the same structure, if the orbital momentum is zero, and the spin of particles (paramagnetic ions, atomic nuclei) and, correspondingly the total spin of the system of particles is an integer (see Sect. 2.4).

In the case of systems formed of an odd number of half-integer spin particles and, accordingly, having a half-integer total angular momentum as a whole, the matrix of the operator **U** has a similar structure, with the only difference that the secondary diagonal now contains an even number of blocks $(+1, -1)$, arranged symmetrically relative to its center, excepting the case of the spin $S = 1/2$, to which corresponds only one such a block.

In the first case, there is no degeneracy of energy levels due to time-reversal symmetry. On the contrary, in the second case, such a degeneracy is present in systems containing an odd number of particles with half-integer spin. In the presence of an electric field of any strength it does not occur a complete splitting of energy levels, at least their double degeneracy remains (Kramers theorem [69]).

In a system consisting of an odd number of particles with spin-orbit interaction that have both an orbital and a spin angular momentum, there exist also an additional degeneration of energy levels caused by the time-reversal symmetry.

It is appropriate to make a remark concerning the change in the form of time-reversal operator at the transition from the coordinate representation to the momentum one. In the coordinate representation, the Hamiltonian **H** is a real operator, so that $\mathbf{H}^* = \mathbf{H}$. Therefore, according to (5.1) we have the relation

$$\mathbf{UH} = \mathbf{HU}, \tag{5.2}$$

that is, the operators **U** and **H** commute. The commutativity of operators **U** and **H** is broken in the momentum representation, in which $\mathbf{H}^* \neq \mathbf{H}$. Since $\widehat{r} = i\hbar\nabla_{\mathbf{p}}$ and $\widehat{\mathbf{p}}= \mathbf{p}$, in this representation the time-reversal operator for a system of N particles with spin 1/2, but without the orbital angular momentum, is of the form

$$\mathbf{T} = i^N \mathbf{O}_{1p}\mathbf{O}_{2p}\ldots\mathbf{O}_{Np}\sigma_{1y}\sigma_{2y}\ldots\sigma_{Ny}\mathbf{K}, \tag{5.3}$$

where $\mathbf{O}_{ip}$ is an unitary operator, which changes the direction of the momentum of ith particle to opposite one, and σ_{iy} it is the imaginary Pauli operator of ith particle.

The problems related to the time-reversal symmetry of stationary states (eigenfunctions of the operator **H**) have been discussed in Sects. 2.3–2.7 and 2.9–2.13.

Let us now consider how the wave functions of non-stationary states of a quantum system are transformed under the time-reversal operator **T**. We assume that the system at time $t = 0$ is in a state, described by a square-integrable complex wave function (see (5.6) below) $\Psi(\xi, 0) = \Phi(\xi)$, which depends on $3N$ dynamic variables designated by $\xi : \xi \equiv \{\xi_1, \xi_2, \ldots, \xi_N\}$. In this case, the wave function of the system consisting of N particles, which depends both on the spatial coordinates and the spin variables, is given by

$$\Psi(\xi, t) = \Psi(\mathbf{r}_1, \mathbf{r}_2, \ldots, \mathbf{r}_N; \sigma_1, \sigma_2, \ldots, \sigma_N; t). \tag{5.4}$$

The set of all square-integrable complex functions of N real variables forms an infinite linear Hilbert space.

If the function $\Phi(\xi)$ coincides with one of the eigenfunctions $\Psi_E(\xi)$ of the Hamiltonian **H** of the system, it means that the system is in a stationary state with the total energy E. In this case, the change with time of the wave function is determined by the well-known harmonic law

$$\Psi_E(\xi, t) = \Psi_E(\xi)\exp\left(-\frac{i}{\hbar}Et\right). \tag{5.5}$$

Now we suppose that $\Phi(\xi)$ is not an eigenfunction of the Hamiltonian. In this case, in order to find the wave function of the non-stationary state of the system, $\Psi(\xi, t)$ at $t > 0$, we shall use the wave Schrödinger equation

$$i\hbar\frac{\partial\Psi(\xi,t)}{\partial t}=\mathbf{H}\Psi(\xi,t) \tag{5.6}$$

at the initial condition

$$\Psi(\xi,0)=\Phi(\xi)\,. \tag{5.7}$$

We take into account that all functions of stationary states of the system along with all "generalized" eigenfunctions of the Hamiltonian (including factors depending on the time) form a complete set. Any solution of the Schrödinger equation can be represented as an expansion over this set [223]:

$$\Psi(\xi,t)=\sum_n c_n\Psi_{\varepsilon_n}(\xi)\exp\left(-\frac{i}{\hbar}\varepsilon_n t\right)+\int C(\xi)\,\Psi_\varepsilon(\xi)\exp\left(-\frac{i}{\hbar}\varepsilon t\right)\mathrm{d}\varepsilon, \tag{5.8}$$

where c_n and $C(\xi)$, in general case, are complex numbers. The wave functions of the stationary states $\Psi_{\varepsilon_n}(\xi)$ and $\Psi_\varepsilon(\xi)$ from (5.8) satisfy the stationary Schrödinger equation:

$$\mathbf{H}\Psi_{\varepsilon_n}(\xi)=\varepsilon_n\Psi_{\varepsilon_n}(\xi)\,, \tag{5.9}$$

$$H\Psi_\varepsilon(\xi)=\varepsilon\Psi_\varepsilon(\xi)\,. \tag{5.10}$$

Further, we require the wave function of the non-stationary state $\Psi(\xi,t)$ from (5.8) to satisfy the initial condition (5.7)

$$\sum_n c_n\Psi_{\varepsilon_n}(\xi)+\int C(\xi)\,\Psi_\varepsilon(\xi)\,\mathrm{d}\varepsilon=\Phi(\xi)\,, \tag{5.11}$$

where the summation is over all points of the discrete spectrum and the integration – over all points of the continuous spectrum. Due to orthonormality conditions

$$\left\langle\Psi_{\varepsilon_n}|\Psi_{\varepsilon_{n'}}\right\rangle=\delta_{nn'} \tag{5.12}$$

and

$$\int\Psi_\varepsilon^*(\xi)\,\Psi_{\varepsilon'}(\xi)\,\mathrm{d}\xi=\delta\left(\varepsilon-\varepsilon'\right), \tag{5.13}$$

the coefficients c_n and $C(\xi)$ of (5.11) are defined as follows:

$$c_n=\left\langle\Psi_{\varepsilon_n}|\Phi\right\rangle=\int\Psi_{\varepsilon_n}^*(\xi)\,\Phi(\xi)\,\mathrm{d}\xi, \tag{5.14}$$

$$C(\xi)=\langle\Psi_\varepsilon|\Phi\rangle=\int\Psi_\varepsilon^*(\xi)\,\Phi(\xi)\,\mathrm{d}\xi. \tag{5.15}$$

After substituting (5.14) and (5.15) into (5.8), it can be considered that the wave function of the non-stationary state is completely defined.

For a system consisting of N particles with the spin $1/2$ and without an orbital angular momentum, the time-reversal operator in the coordinate representation has the form:

$$\mathbf{T} = i^N \sigma_{1y}, \sigma_{2y}, \ldots, \sigma_{Ny}\mathbf{K}, \tag{5.16}$$

where σ_{iy} is determined in (5.3).

The time-reversal operator $\mathbf{T}$ in the coordinate representation for a system consisting of N particles (5.16) differs from the operator $\mathbf{T}$ in the momentum representation (5.3) by the absence of the product of operators $\mathbf{U}_{ip}(i = 1, 2, \ldots, N)$.

It is easy to show that acting on the wave function of the non-stationary state $\Psi(\xi, t)$ from (5.8) by the operator $\mathbf{T}^2$ (where $\mathbf{T}$ is defined by (5.16)), we obtain

$$\mathbf{T}^2\Psi(\xi, t) = \mathbf{1} \cdot \Psi(\xi, t), \mathbf{T}^2 = +\mathbf{1} \tag{5.17}$$

for systems consisting of an even number of particles with spin $1/2$, and

$$\mathbf{T}^2\Psi(\xi, t) = -\mathbf{1} \cdot \Psi(\xi, t), \mathbf{T}^2 = -\mathbf{1} \tag{5.18}$$

for systems consisting of odd number of particles with the spin $1/2$. In (5.17) and (5.18), $\mathbf{1}$ is a unitary operator, which is a product of operators σ_{iy}^2 $(i = 1, 2, \ldots, N)$.

Equations (5.17) and (5.18) are analogical to corresponding equations for stationary states $\Psi_{\varepsilon_n}(\xi)$ and $\Psi_{\varepsilon}(\xi)$ from equation (5.8):

$$\mathbf{T}^2\Psi_{\varepsilon_n}(\xi) = \mathbf{1} \cdot \Psi_{\varepsilon_n}(\xi), \mathbf{T}^2 = \mathbf{1} \tag{5.19}$$

$$\mathbf{T}^2\Psi_{\varepsilon}(\xi) = \mathbf{1} \cdot \Psi_{\varepsilon}(\xi), \mathbf{T}^2 = \mathbf{1} \tag{5.20}$$

for non-Kramers systems and

$$\mathbf{T}^2\Psi_{\varepsilon_n}(\xi) = -\mathbf{1} \cdot \Psi_{\varepsilon_n}(\xi), \mathbf{T}^2 = -\mathbf{1} \tag{5.21}$$

$$\mathbf{T}^2\Psi_{\varepsilon}(\xi) = -\mathbf{1} \cdot \Psi_{\varepsilon}(\xi), \mathbf{T}^2 = -\mathbf{1} \tag{5.22}$$

for Kramers systems, where $\mathbf{T}$ is the Wigner time-reversal operator.

It can be seen from the comparison of the (5.17) with (5.19) and (5.20), as well as (5.18) with (5.21) and (5.22) that wave functions of non-stationary states $\Psi(\xi, t)$ and $\mathbf{T}\Psi(\xi, t)$ are the solutions of the same time-dependent Schrödinger equation.

Similarly, the wave functions of stationary states $\Psi_{\varepsilon_n}(\xi)$ and $\mathbf{T}\Psi_{\varepsilon_n}(\xi)$, as well as $\Psi_{\varepsilon}(\xi)$ and $\mathbf{T}\Psi_{\varepsilon}(\xi)$ are solutions of the Schrödinger equation independent of time for systems with discrete and, respectively, continuous energy spectrum (5.9) and (5.10).

However, unlike, for example, the functions $\Psi_{\varepsilon_n}(\xi)$ and $\mathbf{T}\Psi_{\varepsilon_n}(\xi)$ belonging to the same eigenvalue of the operator $\mathbf{H}$, in the case of non-stationary states the total energy of the system is indefinite. The uncertainty of the energy in a non-stationary state does not mean that the energy of this state is not conserved. It is conserved in the average.

5.2 Time-Reversal Invariance of Schrödinger Equation for Green Function

Let us express the wave function of the non-stationary state $\Psi(\xi, t)$ in (5.8) through the Green function $G(\xi, \xi', t)$. To do this, we introduce the expressions for the coefficients c_n and $C(\xi)$ from (5.14) and (5.15) into the (5.8), which defines the function $\Psi(\xi, t)$, and change the order of integration with respect to ξ and summation (integration) over the energy spectrum:

$$\Psi(\xi, t) = \int d\xi' \left[\sum_n \Psi_{\varepsilon_n}(\xi) \Psi^*_{\varepsilon_n}(\xi') \exp\left(-\frac{i}{\hbar}\varepsilon_n t\right) + \right. \tag{5.23}$$
$$\left. + \int \Psi_\varepsilon(\xi) \Psi^*_\varepsilon(\xi') \exp\left(-\frac{i}{\hbar}\varepsilon t\right) \mathrm{d}\varepsilon \right] \Phi(\xi').$$

If we now introduce the Green function $G(\xi, \xi', t)$ for the Schrödinger equation (5.6)

$$G(\xi, \xi', t) = \sum_n \Psi_{\varepsilon_n}(\xi) \Psi^*_{\varepsilon_n}(\xi') \exp\left(-\frac{i}{\hbar}\varepsilon_n t\right) + \int \Psi_\varepsilon(\xi) \Psi^*_\varepsilon(\xi') \exp\left(-\frac{i}{\hbar}\varepsilon t\right) \mathrm{d}\varepsilon, \tag{5.24}$$

the function $\Psi(\xi, t)$ will be is expressed in terms of the function $G(\xi, \xi', t)$ as follows:

$$\Psi(\xi, t) = \int G(\xi, \xi', t) \Phi(\xi') \mathrm{d}\xi', \tag{5.25}$$

where the function $\Phi(\xi)$ is defined by (5.11).

The Green's function $G(\xi, \xi', t)$ satisfies the Schrödinger equation [223]

$$i\hbar \frac{\partial G(\xi, \xi', t)}{\partial t} = \mathbf{H} G(\xi, \xi', t) \tag{5.26}$$

with the initial condition

$$G(\xi, \xi', 0) = \delta(\xi - \xi'). \tag{5.27}$$

If we denote the function $G(\xi, \xi', t)$ that satisfies the Schrödinger equation (5.26) by G_α, one can introduce a "time reversed" function $G_{-\alpha}$, which is obtained from the function G_α under the action of the time-reversal operator.

Since the Hamiltonian $\mathbf{H}$ is invariant under the time-reversal operator, the function $G_{-\alpha}$ satisfies the Schrödinger equation (5.26), in which t should be replaced by $-t$:

$$-i\hbar\frac{\partial G_{-\alpha}}{\partial t}=\mathbf{H}G_{-\alpha}. \tag{5.28}$$

Now we consider an equation, complex conjugate to the (5.26):

$$-i\hbar\frac{\partial G_{\alpha}^{*}}{\partial t}=\mathbf{H}^{*}G_{\alpha}^{*} \tag{5.29}$$

If there exists a unitary operator $\mathbf{U}$ $\left(\mathbf{U}^{+}\mathbf{U}=\mathbf{U}^{+}\mathbf{U}=\mathbf{1}\right)$ that satisfies the condition

$$\mathbf{U}\mathbf{H}^{*}=\mathbf{H}\mathbf{U}, \tag{5.30}$$

it can be seen that if this operator acts on the both sides of equation (5.29), we obtain the equation

$$-i\hbar\frac{\partial\left(\mathbf{U}G_{\alpha}^{*}\right)}{\partial t}=\mathbf{H}\left(\mathbf{U}G_{\alpha}^{*}\right). \tag{5.31}$$

By comparing this equation with the (5.28), one can see that

$$G_{-\alpha}=\mathbf{U}G_{\alpha}^{*}=\mathbf{U}\mathbf{K}G_{\alpha}=\mathbf{T}G_{\alpha}, \tag{5.32}$$

where $\mathbf{K}$ is a operator of complex conjugation and $\mathbf{T}$ is the time-reversal operator.

It is easy to see that under the action of the operator $\mathbf{T}=\mathbf{U}\mathbf{K}$, the Green function $G\left(\xi,\xi',t\right)$ from the (5.29) is transformed as follows:

$$\mathbf{T}G\left(\xi,\xi',t\right)=\sum_{n}\mathbf{U}\Psi_{\varepsilon_n}^{*}\left(\xi\right)\Psi_{\varepsilon_n}\left(\xi'\right)\exp\left(\frac{i}{\hbar}\varepsilon_n t\right)+\mathbf{U}\int\Psi_{\varepsilon}^{*}\left(\xi\right)\Psi\left(\xi'\right)\exp\left(\frac{i}{\hbar}\varepsilon t\right)d\varepsilon. \tag{5.33}$$

Acting by $\mathbf{T}$ operator on both sides of equation (5.33), we obtain

$$\mathbf{T}^{2}G\left(\xi,\xi',t\right)=\mathbf{U}^{2}G\left(\xi,\xi',t\right),\mathbf{T}^{2}=\mathbf{U}^{2}. \tag{5.34}$$

If the system, of which function[1] $G\left(\xi,\xi',t\right)$ is determined, is characterized by an integer angular momentum, then $\mathbf{U}^{2}=+\mathbf{1}$, since the matrix of the operator $\mathbf{U}$ is symmetrical (see Sect. 2.4). On the contrary, for systems with half-integer angular momentum it is true that $\mathbf{U}^{2}=-\mathbf{1}$. As can be seen from (5.24), although the Green's function itself $G\left(\xi,\xi',t\right)$ is not a wave function of a non-stationary state $\Psi\left(\xi,t\right)$, it is transformed analogously to the non-stationary state function under the action of the operator $\mathbf{T}^{2}$ (see (5.17) and (5.18)):

[1] ξ and ξ' include both space coordinates and spin variables.

$$\mathbf{T}^2 G\left(\xi, \xi', t\right) = \mathbf{1} \cdot G\left(\xi, \xi', t\right), \mathbf{T}^2 = +\mathbf{1} \tag{5.35}$$

in the case of systems with an integer angular momentum and

$$\mathbf{T}^2 G\left(\xi, \xi', t\right) = -\mathbf{1} \cdot G\left(\xi, \xi', t\right), \mathbf{T}^2 = -\mathbf{1} \tag{5.36}$$

in the case of systems with an half-integer angular momentum.

The comparison of (5.19), (5.20), (5.17) and (5.35), as well as (5.21), (5.22), (5.18) and (5.36) shows that wave functions of stationary states $\Psi_{\varepsilon_n}(\xi)$ and $\Psi_{\varepsilon}(\xi)$, the wave function of the non-stationary state $\Psi(\xi, t)$ and the Green's function $G\left(\xi, \xi', t\right)$ are transformed in a similar way under the action of $\mathbf{T}^2$ operator. In this case the states $\Psi_{\varepsilon_n}(\xi)$, $\mathbf{T}\Psi_{\varepsilon_n}(\xi)$ and $\Psi_{\varepsilon}(\xi)$, $\mathbf{T}\Psi_{\varepsilon_n}(\xi)$ belong in pairs to the same energy levels. For systems with a half-integer angular momentum, this leads to the Kramers degeneration of energy levels. Non-stationary states $\Psi(\xi, t)$ and $\mathbf{T}\Psi(\xi, t)$ belong in average to the same energy level, since the energy is conserved only on average. In this case for systems with half-integer angular momentum, instead of Kramers degeneration, a quasi-degeneracy of energy levels occurs, of which position is determined on average.

As for the Green's function $G\left(\xi, \xi', t\right)$ and the time reversed function $\mathbf{T}G\left(\xi, \xi', t\right)$, there is no such simple interpretation for them. Nevertheless, it is curious that the behavior of functions $\Psi_{\varepsilon_n}(\xi)$, $\Psi_{\varepsilon}(\xi)$, $\Psi(\xi, t)$ and $G\left(\xi, \xi', t\right)$ under the action of $\mathbf{T}^2$ is similar.

5.3 Quasi-energy Spectrum and Brillouin Zone in Quasi-energy Space

So far we have considered the transformation of wave functions of stationary and non-stationary states of a quantum system under the action of the time-reversal operator. According to (5.8), the wave function of a non-stationary state is defined as a linear superposition of wave functions of stationary states of the discrete and continuous spectrum of the operator **H**.

In both cases, according to the Heitler quantum theory of radiation [224], the quantum system and the radiation field have been considered as independent subsystems in the zero approximation. In this case, the radiation field was considered a perturbation and the quantum transition was described as the result of the perturbation operator action on the wave functions.

Of course, there are of particular interest, one-, two-, three-, and multi-quantum transitions between stationary energy levels of the system allowed by the selection rules. It is possible, in principle, to talk about quantum transitions between non-stationary states, although such transitions are not of great practical interest.

Let us now consider the situation when the allowed resonant transitions between two or more levels of energy (or energy bands, in the case of elementary excitations such as excitons, biexcitons, magnetic excitons, magnons et al. in crystals) occur at deep saturation. Under the concept of deep saturation of quantum transitions one implies so high values of the alternating field strength, at which the quantum system and the radiation field should be treated as a whole.

After the discovery of the laser in 1961, this task has become actual in the experiment, that stimulated the implementation of the first theoretical studies in this field [225–227].

Let us consider the behavior of a quantum system in the periodic field of large strength, using the concept of quasi-energy introduced by Zeldovich [226].

In the presence of an external periodic field of large strength with a frequency ω, the Hamiltonian of the system has the property of time-periodicity:

$$\mathbf{H}(\mathbf{x}, t + \tau) = \mathbf{H}(\mathbf{x}, t), \tau = 2\pi/\omega. \tag{5.37}$$

Let $\Psi(\mathbf{x}, t)$ be a solution of the Schrödinger equation dependent on time ($\mathbf{x}$ includes both spatial coordinates and spin variables) of the form

$$\Psi(\mathbf{x}, t) = \varphi(\mathbf{x}, t) e^{-i\varepsilon t/\hbar}, \tag{5.38}$$

where the function $\varphi(\mathbf{x}, t)$ is square integrable (i.e., the integral $\int |\varphi(\mathbf{x}, t)|^2 \mathbf{dx}$ has a finite value), and has the property of periodicity in time

$$\varphi(\mathbf{x}, t + \tau) = \varphi(\mathbf{x}, t). \tag{5.39}$$

The function $\varphi(\mathbf{x}, t + \tau)$ satisfies the Schrödinger equation

$$\mathcal{H}(t)\varphi(\mathbf{x}, t) = \mathcal{E}\varphi(\mathbf{x}, t), \tag{5.40}$$

where

$$\mathcal{H}(t) = \mathbf{H}(t) - i\hbar\frac{\partial}{\partial t} \tag{5.41}$$

and $\mathcal{E}$ is the eigenvalue of the operator $\mathcal{H}(t)$ (quasi-energy).

The self-functions of the operator $\mathcal{H}(t)$ (5.41) are referred to as Floquet states, named after the author, who as early as in 1883 found the solution of the first degree differential equation containing an operator depending periodically on time, long before the discovery of quantum mechanics [228].

The time-translation operator transforms the total wave function $\Psi(\mathbf{x}, t)$ (5.38) into the function $\Psi(\mathbf{x}, t + m\tau)$

$$\mathbf{T}_{m\tau}\Psi(\mathbf{x}, t) = \Psi(\mathbf{x}, t + m\tau), \quad m = 0, \pm 1, \pm 2, \ldots \tag{5.42}$$

In this section, the time-translation operator is designated by $\mathbf{T}_{m\tau}$ (not to be confused with the time-reversal operator, denoted in this book by $\mathbf{T}$).

Taking into account (5.42) and (5.38), we obtain:

$$\mathbf{T}_{m\tau}\Psi(\mathbf{x},t) = \varphi(\mathbf{x},t+m\tau)\exp\left[-\frac{i}{\hbar}\mathcal{E}(t+m\tau)\right], \quad m = 0, \pm 1, \pm 2, \ldots \tag{5.43}$$

It is easy to see that this equation reduces to the form

$$\mathbf{T}_{m\tau}\Psi(\mathbf{x},t) = \exp\left(-\frac{i}{\hbar}\mathcal{E}m\tau\right)\Psi(\mathbf{x},t), \quad m = 0, \pm 1, \pm 2, \ldots \tag{5.44}$$

if we introduce (5.38) into (5.43) and take into account that an equality of wave functions $\varphi(\mathbf{x},t)$ and $\varphi(\mathbf{x},t+\tau)$ takes place along with the equality of wave functions $\varphi(\mathbf{x},t)$ and $\varphi(\mathbf{x},t+m\tau)$ (where $m = 0, \pm 1, \pm 2, \ldots$).

This result can be interpreted on the basis of the theory of groups. The time-translation operators $\mathbf{T}_{m\tau}(m = 0, \pm 1, \pm 2, \ldots)$ commute with the operator $\mathcal{H}(t)$ from (5.41) and form an Abelian symmetry group, of which irreducible representations must be numbered by quasi-energy values $\mathcal{E}$. Here τ is fixed and given by the resonance frequency ω of high amplitude vibrations of the external generator whose action on the quantum system leads to the formation of the quasi-energy spectrum.

In general, a set of wave functions $\varphi_{lm}(\mathbf{x},t)$ corresponds to the quasi-energy value $\mathcal{E}$

$$\varphi_{lm}(\mathbf{x},t) = f_l(\mathbf{x})\exp(im\omega t), \quad (l = 1, 2, \ldots; m = 0, \pm 1, \pm 2, \ldots). \tag{5.45}$$

The functions $\varphi_{lm}(\mathbf{x},t)$ can be considered as vectors of states in the generalized Hilbert space $\mathcal{R}+\mathcal{T}$ [229]. In this case, the functions $f_l(\mathbf{x})$ are defined in the space $\mathcal{R}$ and the functions $\exp(im\omega t)$ – in the space $\mathcal{T}$. In the space $\mathcal{R}+\mathcal{T}$, the operator $-i\hbar\frac{\partial}{\partial t}$ is Hermitian and the Hamiltonian $\mathbf{H}(t)$ is Hermitian as well. Therefore, their sum $\mathbf{H}(t)-i\hbar\frac{\partial}{\partial t}$, representing the operator $\mathcal{H}(t)$ (see (5.41)), is also a Hermitian operator in this space. Consequently, each value of the quasi-energy is a real number and the functions $\varphi_{lm}(\mathbf{x},t)$ belonging to different values of quasi-energy are orthogonal.

It is useful to make an analogy with the Brillouin zones in the one-dimensional crystals. If $\mathcal{E}$ is the eigenvalue of the operator $\mathcal{H}(t)$(the quasi-energy) and $\varphi(\mathbf{x},t)$ is its eigenfunction, then

$$\mathcal{E}' = \mathcal{E}+m\hbar\omega, \varphi'(\mathbf{x},t) = \varphi(\mathbf{x},t)\exp(im\omega t) \tag{5.46}$$

also, respectively, the eigenvalue and the eigenfunction of the operator $\mathcal{H}(t)$ for any integer m. At the same time, the total wave function of the system is not changed

$$\varphi(\mathbf{x},t)\exp\left(-\frac{i}{\hbar}\mathcal{E}t\right) = \varphi'(\mathbf{x},t)\exp\left(-\frac{i}{\hbar}\mathcal{E}'t\right). \tag{5.47}$$

Thus, all the solutions (5.46) of the (5.40) for any integer m are equivalent.

It is important to note that any values of the quasi-energy can be reduced to the points in the first Brillouin zone in the space of quasi-energies (similar to the first Brillouin zone in the case of one-dimensional crystal lattice) by defining the real

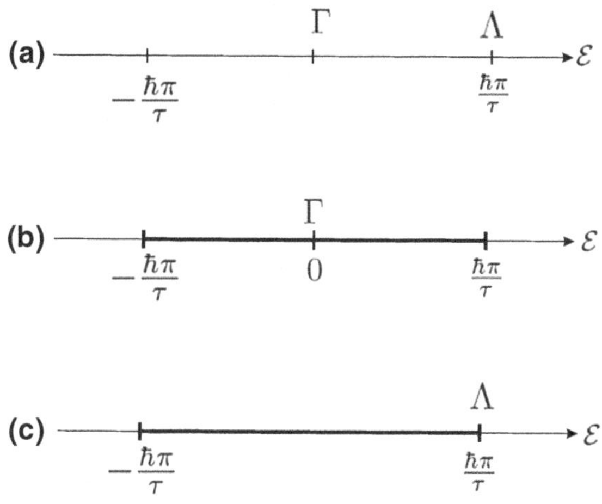

Fig. 5.1 The Brillouin zone with singular points Γ and Λ in the space of quasi-energies

value of the energy E

$$\left(E - \frac{1}{2}\hbar\omega\right) \leq \mathcal{E} \leq \left(E + \frac{1}{2}\hbar\omega\right). \tag{5.48}$$

The choice of E and hence the choice of the Brillouin zone in the space of quasi-energies is arbitrary.

Figure 5.1a shows the first Brillouin zone with the singular points Γ and Λ in the space of quasi-energies. The energy value E from (5.48) is selected here as the origin of quasi-energies.

We note that all points lying in the first Brillouin zone, with the exception of the point Γ, are in a general position and therefore none of them is a singular point. This is due to the fact that neither in the Brillouin zone, nor on its borders there is not a singular point, which would be equivalent to a point occupying a general position in the zone.

In the next section, we will analyze the influence of symmetry properties of point groups of singular points Γ and Λ of the reduced Brillouin zone on the presence or absence of supplementary degeneration of the quasi-energy levels due to the time-reversal symmetry.

5.4 Time-Reversal Symmetry at Commuting Time-Reversal and Quasi-energy Operators

We consider the transformation of the function $\varphi(\mathbf{x}, t)$, which is a solution of the equation (5.40) with the "Hamiltonian" (5.41), under the time reversal. Equation

(5.40) is not invariant under the Wigner time-reversal operator $\mathbf{T} = \mathbf{UK}$, as operators $\mathbf{T}$ and $\mathcal{H}(t)$ are non-commuting ones

$$[\mathbf{T}, \mathcal{H}(t)] \neq 0. \tag{5.49}$$

Indeed,

$$\left[\mathbf{T}, \mathbf{H}(t) - i\hbar\frac{\partial}{\partial t}\right] = [\mathbf{T}, \mathbf{H}(t)] + i\hbar\left(\mathbf{T}\frac{\partial}{\partial t} + \frac{\partial}{\partial t}\mathbf{T}\right). \tag{5.50}$$

In order that the operator $\mathcal{H}(t)$ was invariant under time reversal, it is necessary the operator $\mathbf{T} = \mathbf{UK}$ to be supplemented by a factor in the form of the operator $\mathbf{I}_t$, which under acting on any operator $\mathbf{A}(t)$ or any function $F(t)$ depending explicitly on time, changes t by $-t$

$$\mathbf{I}_t\mathbf{A}(t)\mathbf{I}_t^{-1} = \mathbf{A}(-t), \mathbf{I}_t F(t) = F(\mathbf{I}_t^{-1}t) = F(-t), \mathbf{I}_t^2 = 1. \tag{5.51}$$

The operator $\mathbf{I}_t$ is similar to the operator of space inversion ($\mathbf{I_r} f(\mathbf{r}) = f(-\mathbf{r})$, $\mathbf{I_r}^2 = 1$).

Thus, if the operator $\mathcal{H}(t)$ of quasi-energy and its eigenfunction $\varphi(\mathbf{x}, t)$ are depending periodically on time, the time-reversal operator has the form [135]

$$\mathbf{T}' = \mathbf{T}\mathbf{I}_t = \mathbf{UK}\mathbf{I}_t. \tag{5.52}$$

In this case

$$\mathbf{T}'\left(\mathbf{H}(t) - i\hbar\frac{\partial}{\partial t}\right)\mathbf{T}'^{-1} = \mathbf{TH}(-t)\,\mathbf{T}^{-1} - i\hbar\frac{\partial}{\partial t}, \tag{5.53}$$

i.e. the "Hamiltonian" $\mathcal{H}(t)$ is invariant under the time-reversal operator $\mathbf{T}'$

$$\mathbf{T}'\mathcal{H}(t)\,\mathbf{T}'^{-1} = \mathcal{H}(t), \tag{5.54}$$

if the following relation takes place:

$$\mathbf{TH}(-t)\,\mathbf{T}^{-1} = \mathbf{H}(t). \tag{5.55}$$

The relation (5.55) is actually valid for the Hamiltonian of interaction of the quantum system with a time-periodic field of frequency ω and large amplitude (macrofilled mode).

As an example, we consider the problem of resonant interaction of hyper- or terasound wave of frequency $\omega_{\mathbf{q}}$ with excitons having the wave vectors $\mathbf{k}$ and $\mathbf{k}+\mathbf{q}$, which are in the exciton bands λ and λ', respectively. For the macrofilled hyper- or terasound mode the temporal dependence of the operator $b_{\mathbf{q}}$ of phonon destruction may be represented in the form

$$b_{\mathbf{q}} = B_{\mathbf{q}} \exp\left(-i\omega_{\mathbf{q}} t\right), \quad B_{\mathbf{q}} = \sqrt{N_{\mathbf{q}}}, \tag{5.56}$$

where $N_{\mathbf{q}}$ is the macroscopic number of hyper- or terasound phonons with the wave vector $\mathbf{q}$ in the macrofilled mode. In a similar way, the temporal dependence of the operator $b_{\mathbf{q}}^{+}$ of phonon creation may be represented in the form

$$b_{\mathbf{q}}^{+} = B_{\mathbf{q}}^{+} \exp\left(i\omega_{\mathbf{q}} t\right), \quad B_{\mathbf{q}}^{+} = \sqrt{N_{\mathbf{q}}}. \tag{5.57}$$

In (5.56) and (5.57) $B_{\mathbf{q}}$ and $B_{\mathbf{q}}^{+}$ are c -numbers.

The Hamiltonian of the interaction of the macrofilled phonon mode with excitons in the λ and λ' zones has the form [135]

$$\mathbf{H}_{ex-ph}(t) = \sqrt{N_{\mathbf{q}}} \sum_{\mathbf{k}} \left[g_{\lambda\lambda'}(\mathbf{q})\, C_{\mathbf{k}+\mathbf{q},\lambda'}^{+} C_{\mathbf{k},\lambda} \exp\left(-i\omega_{\mathbf{q}} t\right) + H.c. \right], \tag{5.58}$$

where $C_{\mathbf{k}+\mathbf{q},\lambda'}^{+}$ and $C_{\mathbf{k},\lambda}$ are the exciton creation and destruction operators in exciton bands λ' and λ, correspondingly, and $g_{\lambda\lambda'}(\mathbf{q})$ is the constant of exciton–phonon coupling. We see that the Hamiltonian $\mathbf{H}_{ex-ph}(t)$ is periodically dependent on time

$$\mathbf{H}_{ex-ph}(t) = \mathbf{H}_{ex-ph}(t + m\tau), \quad m = 0, \pm 1, \pm 2, \ldots \tag{5.59}$$

as

$$\exp\left[\mp i\omega_{\mathbf{q}}(t + m\tau)\right] = \exp\left(\mp i\omega_{\mathbf{q}} t\right)\exp\left(\mp i \cdot 2\pi m\right) = \exp\left(\mp i\omega_{\mathbf{q}} t\right), \quad m = 0, \pm 1, \pm 2, \ldots \tag{5.60}$$

Thus, the operator $\mathbf{H}_{ex-ph}(t)$ satisfies the relationship of the type (5.55)

$$\mathbf{T}\mathbf{H}_{ex-ph}(-t)\,\mathbf{T}^{-1} = \mathbf{H}_{ex-ph}(t). \tag{5.61}$$

Acting by the time-reversal operator $\mathbf{T}'$ from (5.52) directly on wave functions (5.45) one can show that the functions $\varphi_{lm}(\mathbf{x}, t)$ and $\mathbf{T}'\varphi_{lm}(\mathbf{x}, t)$ belong to the same value of the quasi-energy $\mathcal{E}$. If $\mathbf{x}$ does not include the spin coordinate, then $\varphi_{lm}(\mathbf{x}, t)$ and $\mathbf{T}'\varphi_{lm}(\mathbf{x}, t)$ are linearly dependent and the supplementary degeneracy of quasi-energy levels due to time-reversal symmetry is absent. Taking into account the spin, it is possible to have an additional degeneracy of quasi-energy levels due to the time-reversal symmetry. In spin systems, the matrix of the operator $\mathbf{U}$ has matrix elements on the secondary diagonal equal, in turn, to $+1$ and -1, and other matrix elements equal to zero [230]. If the total angular momentum is an integer, then the matrix of the operator $\mathbf{U}$ is symmetric and no additional degeneracy exists. The additional degeneracy of quasi-energy levels due to time-reversal symmetry exists only in the case of the half-integer total angular momentum, when $\mathbf{U}^2 = -\mathbf{1}$ and correspondingly $\mathbf{T}'^2 = -\mathbf{1}$. In the presence of a crystal field, it is necessary to use the operator $\mathbf{U}$ in the space of basis functions of ordinary irreducible representations of crystallographic point groups of symmetry. In this case the evident form of the matrix of the operator **U** is given in the Appendix A.

We now consider the singular points Γ and Λ in the Brillouin zone in the space of quasi-energies (Fig. 5.1) and consequences relating to the additional degeneration of quasi-energy levels due to the time-reversal symmetry by neglecting the spin. These consequences are determined by symmetry properties of the point groups related to the points Γ and Λ.

We apply a criteria, which are similar to the Herring criteria for the wave vector group (Sect. 2.7), to the point group, of which elements Q_0 transfer the point $\mathcal{E}$ in $-\mathcal{E}$ in the first Brillouin zone in the space of quasi-energies, while Q_0^2 does not change $\mathcal{E}$. We obtain

$$\sum_{Q_0} \chi\left(Q_0^2\right) = \begin{cases} n - \text{the case (a)} \\ 0 - \text{the case (b)} \\ -n - \text{the case (c)} \end{cases}, \tag{5.62}$$

where $\chi\left(Q_0^2\right)$ is the character of the element Q_0^2 and n is the order of the group.

(1). Point Γ (Fig. 5.1b). The star Γ consists of one point $\mathcal{E} = 0$ and is irreducible. It is connected with the point Γ of the point symmetry group $\overline{1}$ in international notation or $S_2(C_i)$ in notation of Schoenflies. This group contains the unitary element e and the inversion operator $I_{\mathcal{E}}$ in the space of quasi-energies. It is characterized by one-dimension symmetric and antisymmetric representations A_g and A_u. The application of (5.62) to the point group $\overline{1}$ gives

$$\sum_{Q_0} \chi\left(Q_0^2\right) = 2, \tag{5.63}$$

which corresponds to the case (a), i.e., to the absence of an additional degeneracy of quasi-energy levels caused by the time-reversal symmetry.

(2). Point Λ (Fig. 5.1c). The point Λ, located on the border of the one-dimensional Brillouin zone in the space of quasi-energies has a corresponding equivalent point located on the opposite boundary of the zone. The quasi-energy values $\mathcal{E}_1 = +\hbar\pi/\tau$ and $\mathcal{E}_2 = -\hbar\pi/\tau$ differ from each other by the "reciprocal lattice vector" $2\pi\hbar/\tau$ in the space of quasi-energies

$$\mathcal{E}_1 - \mathcal{E}_2 = 2\pi\hbar/\tau. \tag{5.64}$$

The star of the point Λ is irreducible and consists of "vectors" $\mathcal{E}_1$ and $\mathcal{E}_2$. It should be emphasized once again that the corresponding points on opposite ends of the Brillouin zone must be considered as points belonging to the same quasi-energy $\mathcal{E}$. For example, the symmetry group of the point Λ contains the transformation $I_{\mathcal{E}}$, despite the fact that it transfers the point $\hbar\pi/\tau$ into the point $-\hbar\pi/\tau$. As a result, the symmetry group that owns the singular point Λ is the group $\overline{1}$. Therefore, we have once again the relation (5.63), which corresponds to the absence of degeneracy of the quasi-energy levels due to time-reversal symmetry (the case (a)).

The fact that the same point symmetry group corresponds to singular points Γ and Λ of the Brillouin zone in the space of quasi-energies, is not an exception. A similar situation occurs, for example, for space singular points Γ and M of a square lattice in the quasi-momentum space (wave vectors), to whom the same point symmetry group corresponds [9].

The analysis carried out in this section corresponds to the case when the spin of particles is not taken into account. Moreover, the translation operators $\mathbf{T}_{m\tau}$ $(m = 0, \pm 1, \pm 2, \ldots)$ and the time-reversal operator $\mathbf{T}'$ are considered separately, not jointly. In the next section, a joint consideration of these operators will be presented.

5.5 Quasi-energy Doublets Due to Non-commuting Time-Reversal and Time-Translation Operators

The "Hamiltonian" $\mathcal{H}(t)$ (5.41) is simultaneously invariant under both the time-reversal and time-translation operators

$$\left[\mathcal{H}(t), \mathbf{T}'\right] = \left[\mathcal{H}(t), \mathbf{T}_{m\tau}\right] = 0, \ (m = 0, \pm 1, \pm 2, \ldots). \qquad (5.65)$$

Let us consider a new group of transformations created by operators $\mathbf{T}'\mathbf{T}_{m\tau}$. This new expanded group of symmetry is not an Abelian one, like the translations group, because $\mathbf{T}'\mathbf{T}_{m\tau}$ and $\mathbf{T}_{m\tau}\mathbf{T}'$ are not coinciding operators.

Taking into account (5.38), (5.43), (5.44), and (5.52), it is easy to show that the following operator equality takes place

$$\mathbf{T}'\mathbf{T}_{-m\tau} = \mathbf{T}_{m\tau}\mathbf{T}', \ (m = 0, \pm 1, \pm 2, \ldots). \qquad (5.66)$$

Indeed,

$$\mathbf{T}'\mathbf{T}_{-m\tau}\Psi(\mathbf{x}, t) = \mathbf{U}\varphi^*(\mathbf{x}, -t - m\tau)\exp\left[-\frac{i}{\hbar}\mathcal{E}(t + m\tau)\right], \ (m = 0, \pm 1, \pm 2, \ldots), \qquad (5.67)$$

and

$$\mathbf{T}_{m\tau}\mathbf{T}'\Psi(\mathbf{x}, t) = \mathbf{U}\varphi^*(\mathbf{x}, -t + m\tau)\exp\left[-\frac{i}{\hbar}\mathcal{E}(t + m\tau)\right], \ (m = 0, \pm 1, \pm 2, \ldots). \qquad (5.68)$$

Due to the periodicity in time of the function $\varphi(\mathbf{x}, t)$, we have the equality

$$\varphi^*(\mathbf{x}, -t - m\tau) = \varphi^*(\mathbf{x}, -t + m\tau), \ (m = 0, \pm 1, \pm 2, \ldots), \qquad (5.69)$$

which leads to the coincidence of right sides of equations (5.67) and (5.68), and correspondingly, to the relationship

$$\left(\mathbf{T}'\mathbf{T}_{-m\tau} - \mathbf{T}_{m\tau}\mathbf{T}'\right)\Psi(\mathbf{x}, t) = 0, \ (m = 0, \pm 1, \pm 2, \ldots), \qquad (5.70)$$

that proves the validity of equality (5.66).

Now we find the commutation relation for operators $\mathbf{T}'$ and $\mathbf{T}_{m\tau}$:

$$\mathbf{T}'\mathbf{T}_{m\tau}\Psi(\mathbf{x},t) = \mathbf{U}\varphi^*(\mathbf{x},-t+m\tau)\exp\left(-\frac{i}{\hbar}\mathcal{E}t\right)\exp\left(\frac{i}{\hbar}m\mathcal{E}\tau\right), \tag{5.71}$$

$$\mathbf{T}_{m\tau}\mathbf{T}'\Psi(\mathbf{x},t) = \mathbf{U}\varphi^*(\mathbf{x},-t+m\tau)\exp\left(-\frac{i}{\hbar}\mathcal{E}t\right)\exp\left(-\frac{i}{\hbar}m\mathcal{E}\tau\right). \tag{5.72}$$

Subtracting (5.72) from (5.71) and taking into account that, according to (5.38) and (5.52)

$$\mathbf{T}'\Psi(\mathbf{x},t) = \mathbf{U}\varphi^*(\mathbf{x},-t)\exp\left(-\frac{i}{\hbar}\mathcal{E}t\right), \tag{5.73}$$

and that the functions $\varphi^*(\mathbf{x},-t)$ and $\varphi^*(\mathbf{x},-t+m\tau)$(where $m = 0,\pm1,\pm2,\ldots$) coincide due to their periodicity in time, we obtain

$$\left[\mathbf{T}',\mathbf{T}_{m\tau}\right] = 2i\sin\left(m\mathcal{E}\tau/\hbar\right)\mathbf{T}'\Psi(\mathbf{x},t), \quad (m = 0,\pm1,\pm2,\ldots) \tag{5.74}$$

In the absence of additional degeneration of quasi-energy levels due to the time-reversal symmetry, wave functions $\Psi(\mathbf{x},t)$ and $\mathbf{T}'\Psi(\mathbf{x},t)$ are linearly dependent

$$\mathbf{T}'\Psi(\mathbf{x},t) = c\Psi(\mathbf{x},t), \tag{5.75}$$

where c is a constant. In this case, according to (5.74) and (5.75), the commutation relation for operators $\mathbf{T}'$ and $\mathbf{T}_{m\tau}$ is

$$\left[\mathbf{T}',\mathbf{T}_{m\tau}\right] = 2ic\sin\left(2\pi m\frac{\mathcal{E}}{\hbar\omega}\right), \quad (m = 0,\pm1,\pm2,\ldots), \tag{5.76}$$

where it was considered that $\tau = 2\pi/\omega$.

The commutation relation (5.76) shows that if the ratio of the quasi-energy $\mathcal{E}$ to the energy $\hbar\omega$ of the quantum of monochromatic periodic field of large amplitude, $\mathcal{E}/\hbar\omega$, is equal to a integer (the case $m = 0$ is not relevant), the operators $\mathbf{T}'$ and $\mathbf{T}_{m\tau}$ commute. In this case, there is no additional degeneracy of quasi-energy levels.

On the contrary, if the ratio $\mathcal{E}/\hbar\omega$ is not a integer or half-integer, then the commutator of the operators $\mathbf{T}'$ and $\mathbf{T}_{m\tau}$ from (5.76), which are elements of the extended non-Abelian symmetry group, is different from zero. However, it is easy to see that each of the non-commuting operators $\mathbf{T}'$ and $\mathbf{T}_{m\tau}$ commute with the quasi-energy operator $\mathcal{H}(t)$. In other words, there is an invariance of the operator $\mathcal{H}(t)$ with respect to the operator $\mathbf{T}'$ (see (5.54)) and similarly with respect to the operator $\mathbf{T}_{m\tau}$ ($\mathbf{T}_{m\tau}\mathcal{H}\mathbf{T}_{m\tau}^{-1} = \mathcal{H}(t)$). The situation when two operators do not commute, but each separately commutes with the Hamiltonian of the system, corresponds in quantum mechanics to the presence of a twofold degeneracy of energy levels [231]. Since in our case the operators $\mathbf{T}'$ and $\mathbf{T}_{m\tau}$ commute with the quasi-energy operator $\mathcal{H}(t)$, it must be concluded that there is an additional twofold degeneracy of quasi-energy levels, of which nature is caused by the simultaneous existence of both a time-reversal and a time-translation symmetry. In contrast to the Kramers degeneracy of energy levels, which exists only in systems containing an odd number of particles with the spin $1/2$, the new additional degeneracy of quasi-energy levels we predicted is partly due to the time-reversal symmetry and partly to the time-translation symmetry. Such a specific quasi-energy levels degeneracy takes place not only when the spin is taken into account, but also in systems formed from spinless particles. This new type of quasi-energy levels degeneration has been discussed for the first time in [221, 232].

Chapter 6
Transformation of Antiferromagnetic Type of Exchange Interaction into Ferromagnetic One in Dimer Clusters

We consider coordination compounds of ions with unfilled up $3d$- and $4f$-shell in a weak or intermediary crystalline field, containing dimer clusters in the case when the distance between clusters is more than the distance between the ions forming the cluster. In this case it is possible to neglect the exchange interaction between electrons of different clusters and the structure of the ground state of separate dimer is determined by the intercluster exchange, spin orbital, magnetic dipole–dipole and superhyperfine interactions.

Supplementary simplification consists in neglecting effects of covalency, conditioned by overlapping of orbits of cation electrons and electrons of neighbor diamagnetic ions or atoms. By this we exclude from consideration complexes of $4d$- and $5d$-groups, as well a cyanides of $3d$-group and complexes containing ions with partially filled up $5f$-shell (actinides). However, even after these restrictions the number of complex compounds satisfying above-mentioned demands is sufficiently large.

It is also required that for the considered $3d$- and $4f$-shells effects of intraconfigurational interaction of electrons are slightingly weak. In Sects. 6.1 and 6.2 we bring up experimental data confirming this supposition. In these conditions the wave function of the hole in unfilled up shell of paramagnetic ion may be obtained from the wave function of electron of the same shell with the help of time-reversal operator. Within the framework of this approach it will be shown that it is possible to realize transformation of antiferromagnetic type of exchange interaction in dimer cluster into ferromagnetic one, for which the ground state of dimer cluster is spin degenerate.

I. I. Geru, *Time-Reversal Symmetry*, Springer Tracts in Modern Physics 281,
https://doi.org/10.1007/978-3-030-01210-6_6

6.1 Magnetic Dimer Clusters in Coordination Compounds

The first experimental manifestations of exchange interaction between paramagnetic ions in dimer clusters of coordination compounds were lowered values (approximately as many as twice) of the magnetic moment of the ion Cu^{2+} in salts of greasy acids [233] and the temperature dependence of the magnetic susceptibility of copper monohydrate acetate $Cu_2\,(CH_3COO)_4 \cdot 2H_2O$, which has a maximum near the room temperature and decreases rapidly with decreasing temperature, becoming zero at 50 K [234]. However, authors of [233, 234] limited themselves only to the establishing the anomaly without explaining it. That is why it is necessary to consider that the first publications, that started the research of dimer clusters in coordination compounds, are papers of Bleany and Bowers [235, 236]. On the basis of studied EPR spectra of the copper monohidrat acetate single crystal, they suggested that the abnormal paramagnetism is due to the presence of isolated pairs, in which copper ions are coupled by exchange forces. A similar result was obtained in [237, 238] on the basis of EPR spectra of the same single crystal. In this case the interaction of two spins leads to arising an upper triplet state with a summary spin $S = 1$ and a lower singlet state with $S = 0$ [235–241]. The roentgenostructural study of above mentioned crystals [239] confirmed that really there are dimer clusters in crystalline lattice (Fig. 6.1).

The six nearest neighbors of the copper atom are four atoms of oxygen belonging to four distinct acetate groups, copper atom and oxygen atom of water molecule. These six atoms are located in vertices of a distorted octahedron with the copper atom in center All the distances between the copper atom and oxygen of carboxyl groups are approximately equal to 1.97 Å. Every of four acetate groups is planar. The distance between two neighbor atoms of copper is very small and equals 2.64 Å,

Fig. 6.1 A structure fragment of the binuclear copper(II) acetate monohydrate

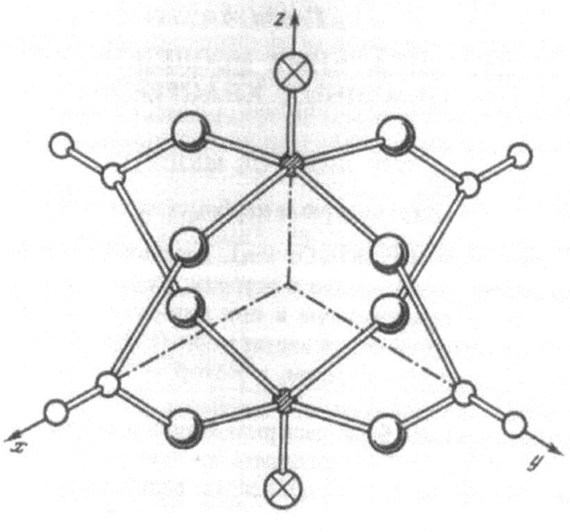

which is only a little more than distances in metallic copper (2.56 Å). The distance Cu–O(H_2) is 2.20 Å.

The study of EPR spectra of crystals of copper(II) propionate monohydrate $Cu_2(CH_3CH_2COO)_4 \cdot 2H_2O$ [242], copper(II) *n*-butyrate monohydrate $Cu_2(C_4H_7O_2)_4 \cdot 2H_2O$ and copper(II) monochloroacetate monohydrate $Cu_2(CH_2Cl\,COO)_4 \cdot 2H_2O$ [243] has shown that ions of copper in these compounds also form dimer clusters.

A review of research results on dimer magnetic clusters in coordinating compounds obtained over 50 years since their discovery is contained in the book [241]. Quadratic on spin operators interaction between uncoupled electrons of two ions of magnetic clusters with orbital non-degenerate ground states has the form [241]

$$\mathcal{H} = \mathcal{H}_{\text{ex}}^{\text{iso}} + \mathcal{H}_{\text{ex,anis}}^{s} + \mathcal{H}_{\text{ex,anis}}^{\text{as}}, \tag{6.1}$$

where

$$\mathcal{H}_{\text{ex}}^{\text{iso}} = J\mathbf{S}_1\mathbf{S}_2 \tag{6.2}$$

is the Heisenberg–Dirac–van Vleck operator of the isotropic exchange interaction (J is the exchange integral; $J > 0$ and $J < 0$ correspond to the exchange interaction of antiferromagnetic and ferromagnetic types, respectively), and

$$\mathcal{H}_{\text{ex,anis}}^{s} = \sum_{j,k} D_{jk}^{\text{ex}}\left(S_{1j}S_{2k} + S_{1k}S_{2j}\right) \tag{6.3}$$

describes the symmetric part of the anisotropic exchange interaction,

$$\mathcal{H}_{\text{ex,anis}}^{\text{as}} = \mathbf{G}_{\text{ex}}[\mathbf{S}_1 \times \mathbf{S}_2] \tag{6.4}$$

is the Dzyaloshinsky–Moryia antisymmetric exchange [244], leading to the appearance of the vector of magnetization perpendicular to the plane, in which spins of two interacting ions are disposed. D_{jk}^{ex} from (6.3) and $\mathbf{G}$ from (6.4) are determined by

$$D_{jk}^{\text{ex}} = \frac{1}{2}\left(J_{jk} - J_{kj}\right) - \delta_{jk}J, \tag{6.5}$$

$$J = \frac{1}{3}\sum_i J_{ii}; \quad j, k = x, y, z,$$

$$G_x^{\text{ex}} = \frac{1}{4}\left(J_{yz} - J_{zy}\right), \quad G_y^{\text{ex}} = \frac{1}{4}\left(J_{zx} - J_{xz}\right), \quad G_z^{\text{ex}} = \frac{1}{4}\left(J_{xy} - J_{yx}\right). \tag{6.6}$$

As Moryia [245] has shown, the spin Hamiltonian (6.1) without the last term can be obtained in the framework of the perturbation theory if as a perturbation one takes into account the spin–orbital coupling inside one ion and the isotropic exchange interaction between both ions of the cluster. Under this consideration the operator $\mathbf{G}_{ex}[\mathbf{S}_1 \times \mathbf{S}_2]$ appears in the second order of perturbation theory and for the constant G of antisymmetric exchange, which gives contribution to the splitting of spin multiplets, the relation $G \sim (\lambda/\Delta)\,J'$ may be obtained, where λ is the constant of spin–orbital coupling, Δ is the difference of energies between the levels of the ground and excited states of exchange-coupled pair, J' is the constant of isotropic exchange interaction in the excited state. It is to be taken into account that $G \neq 0$ only for the clusters without inversion centers [241]. The theory of antisymmetrical exchange is considered in [241, 246, 247].

The intramolecular exchange changes the structure of energy levels of interacting ions, leading to splitting the spin levels corresponding to different values of the summary spin. In different exchange pairs of iron group the intervals between levels due to the isotropic part of exchange interactions J change their values in wide limits: from fractions of cm^{-1} to $1000\,cm^{-1}$ and more. Anisotropic exchange interactions cause the splitting of spin multiplets in intervals $\leq 1\,cm^{-1}$.

The magnetic dipole–dipole interaction between paramagnetic ions of dimer clusters leads to the splitting of spin multiplets ($\sim 0.2 \pm 0.1\,cm^{-1}$ for the distance between ions of metal $R_{Me-Me} \sim 2.5 - 3.0\,Å$), which is much smaller than splittings caused by anisotropic exchange interactions and the splittings of spin levels of separate centers with $S \geq 1$ in the crystalline electric field [241]. That is why the operator of dipole–dipole interaction between magnetic moments of ions of dimer clusters is not included in the spin Hamiltonian (6.1). It is also not necessary to take into account in (6.1) the operators of isotropic and anisotropic hyperfine interactions when solving the problem of changing the degree of spin degeneration of the energy level of the ground state of a magnetic dimer cluster discussed in this chapter.

In the case of high spin interacting ions ($S_1 > 1/2,\ \ S_2 > 1/2$) it is necessary to take into account in the spin Hamiltonian not only quadratic in spin operator summands, but also scalar products of spin operators of higher degrees. For example, that is the operator of biquadratic exchange $\mathfrak{J}\,(\mathbf{S}_1\mathbf{S}_2)^2$, which describes double jumpings of electrons in exchange processes. However, experimental data show that for exchange pairs Mn (II)–Mn(II) and Cr (III)–Cr(III) [248] we have $\mathfrak{J}/J \sim 0.01 - 0.05$, therefore processes of biquadratic exchange interaction may be neglected, as compared with other processes of exchange interaction. This refers to the case when cluster ions interacting with one another exist separately in the singlet ground state. In the presence of orbital degeneration of the cluster ground state, the exchange anisotropy sharply increases, and the role of biquadratic effects is also extended [249].

The spin Hamiltonian (6.1) is the input in order to describe interactions between electrons of different ions of the magnetic dimer cluster. It is necessary to analyze three types of exchange interactions: direct (under immediate overlapping of magnetic orbits), indirect (by orbits of unmagnetic ligands) and hybrid (as a result of the first two). By analyzing values of exchange interaction in clusters, it is necessary to consider all three types of exchange. However, since ions of metals are usually located at a sufficiently large distance from each other, the main contribution is due to the exchange interaction of indirect and hybrid types, in which bridge ligands are involved.

The mechanism of indirect exchange interaction with allowance for the wave functions of intermediary anions in the third order of the perturbation theory was first suggested by Kramers [250]. This interaction leads to a weak effect of ferromagnetic ordering. In order to explain the antiferromagnetic ordering, it was suggested a mechanism, according to which two electrons with opposite spins are simultaneously separated from the diamagnetic ion and transferred to a half-filled state of magnetic ions [251, 252]. Later on it was shown that a more essential antiferromagnetic contribution is due to a mechanism that was called "Anderson superexchange" [253, 254]. In this model, it is necessary firstly to find wave functions of individual centers and thereupon take into account their interaction. According to Anderson theory, the indirect exchange interaction consists of two contributions: from electrostatic interaction (potential exchange) and from virtual jumpings (kinetic exchange). The kinetic exchange leads to a antiferromagnetic contribution to the parameter of exchange interaction, which is proportional to the square of modulus of the transfer integral of the electron $|b_{12}|^2$ from one center to another:

$$J_{\text{kin}} = 2|b_{12}|^2 U, \tag{6.7}$$

where U is the excited state energy, when there are two electrons with opposite spins on one center. According to Anderson's estimations, the potential exchange (of ferromagnetic type) is an order of magnitude less than the kinetic one. Therefore, the expression (6.7) describes the main contribution in J, i.e., one may conclude that between electrons of distinct ions of a magnetic dimer cluster there is an exchange interaction of antiferromagnetic type ($J > 0$). The theory of indirect exchange evoked a great interest [255–267]. It is a particular importance the question of the mechanism of indirect exchange and its role in determining the exchange parameter. Anderson's ideas were widely adopted and formed the basis of most essential methods for calculating the parameter J [268–274]. Some of these mechanisms, as well as ab initio direct calculations of the singlet–triplet separation in dimer clusters were discussed in [241, 261, 275–277].

According to the microscopic theory of superexchange [278], actually there are more probable schemes of electron jumpings, than Anderson has shown. These supplementary schemes are realized with the participation not only of half-filled orbits of magnetic ions, but also of empty and filled orbits. In Eremin approach [278–282], the Hamiltonian of indirect exchange interaction is constructed simultaneously with finding the ground states of paramagnetic ions. This is an significant difference from the Anderson's theory. The Eremin's theory of indirect exchange interaction of paramagnetic ions in dielectrics at arbitrary relations between energies of electron transfer from cation to cation and from cation to anion contains all results of Anderson's theory as a particular case of high energy electron transfer from a ligand to a paramagnetic ion. Moreover, the theoretical analysis is performed on one-electron wave functions, orthogonalized by Bogolyubov method [257]. From all possible perturbations there are taken into account: (1) transfer of the electron from a ligand to the neighbor paramagnetic ion; (2) carry of the electron from a metal ion to another one inside the cluster; (3) simultaneous transfer of two electrons from a ligand, one by one, to every of neighbor paramagnetic ions of the cluster.

In [278–282] calculations are carried out in the representation of the secondary quantization using the perturbations theory up to fifth order, inclusively, in energies of the electron carry from an anion to a cation. Taking into account covalency effects, this approach permitted to get the most complete Hamiltonian of exchange interaction with allowance for all main mechanisms, including direct, indirect and hybrid interactions. Schematically this Hamiltonian can be represented in the following form (superscripts of operator show the order of perturbations theory):

$$\mathcal{H}_{\text{ex}} = \mathcal{H}^{(1,2)}_{\text{dir.}} + \mathcal{H}^{(3)}_{\text{dir.}} + \mathcal{H}^{(3)}_{\text{hib.}} + \mathcal{H}^{(4)}_{\text{hib.}} + \mathcal{H}^{(3)}_{\text{indir.}} + \mathcal{H}^{(4)}_{\text{indir.}} + \mathcal{H}^{(5)}, \tag{6.8}$$

where

$$\mathcal{H}^{(5)} = \mathcal{H}^{(5)}_{c} + \mathcal{H}^{(5)}_{a} + \mathcal{H}^{(5)}_{b} \tag{6.9}$$

describes the exchange interaction of ferromagnetic type between paramagnetic ions of the dimer cluster.

In (6.8), $\mathcal{H}^{(1,2)}_{\text{dir.}}$ and $\mathcal{H}^{(3)}_{\text{dir.}}$ describe the direct exchange interaction between the ions of cluster, which appears in the first, second, and third orders of the perturbation theory [241]. It is obtained in the supposition that the energy of electron transfer is much more than intervals between the terms of individual electronic configurations of ions. This approximation is well fulfilled for the ground states, but it turned out to be incorrect for strongly excited states of paramagnetic ions. $\mathcal{H}^{(3)}_{\text{hib.}}$ and $\mathcal{H}^{(4)}_{\text{hib.}}$ describe different hybrid interactions. In particular, $\mathcal{H}^{(4)}_{\text{hib.}}$ is the operator of the hybrid exchange interaction of ferromagnetic type when the direct jumpings of electrons are realized in empty orbits and from filled to half-filled ones.

The operators $\mathcal{H}^{(3)}_{\text{indir.}}$ and $\mathcal{H}^{(4)}_{\text{indir.}}$ describe the indirect exchange interaction in the third and fourth orders of the perturbations theory. The first summand in (6.9) is the potential exchange in Anderson's theory. Two other summands in (6.9) characterize processes of virtual interactions of electrons at the supplementary consideration of interactions of electrons with ions a and b, respectively (in (6.9), c is the anion with a filled orbit).

The process of electron transfer, corresponding to the fifth order of perturbation theory (the operator $\mathcal{H}^{(5)}$ from (6.8) and (6.9)) becomes important when corrections of third and fourth orders vanish. Expressions for all summands of the Hamiltonian (6.8) are given in [279]. As similar attempt to develop the theory of superexchange in higher orders of perturbation theory was earlier undertaken by Fukuchi [283]. Using the method developed in [257, 284], he derived a series of cumbersome expressions in one s-electron model. However, these formulas have not been discussed and no physical conclusions have been made on their basis.

Experimental data belonging to iron group compounds, which are mostly related to binuclear copper(II) compounds, confirm the existence of magnetic dimer clusters in coordinating compounds. Some of these results are given below (Tables 6.1, 6.2, 6.3, 6.4, 6.5, 6.6, 6.7, 6.8 and 6.9).

Table 6.1 Singlet–triplet splitting J and parameters D and E of crystalline field for copper(II) acetate monohydrate $Cu_2(CH_3COO)_4 \cdot 2H_2O$ [236, 238, 240]

Effective spin S	Temperature, K	J, cm^{-1} (EPR data)	J, cm^{-1} (χ data)	$\|D\|$, cm^{-1}	E, cm^{-1}
1	90	260 ± 50	310 ± 15	0.34(3)	0.10(5)
1	320			0.345(5)	0.005(3)

Table 6.2 Singlet–triplet splitting J, anisotropic symmetric exchange parameter D_{ex}, constants D and E of electric crystalline field of axial and rhombic symmetries and distance R_{Cu-Cu} between copper ions in dimers for adducts $(RCOO)_4Cu_2Cl \cdot 2L$ [241]

L	R	J, cm^{-1}	D_{ex}, cm^{-1}	D, cm^{-1}	E, cm^{-1}	R_{Cu-Cu}, Å
NC_9H_7 (chin)	H	600		0.429		2.652
	CH_3	342	+0.521 −0.161	0.341	0.003	2.652
	$C_6H_5CH_2$	328		0.351	0.002	2.724
	$ClCH_2$	371	+0.538 −0.204	0.371	0.002	
	FCH_2	364		0.395	0.002	
	Cl_2CH	294		0.389	0.003	
	F_3C	310		0.425		2.886
2-Cl--C_5H_5N (2-Cl--py)	$C_6H_5CH_2$	309		0.348	0.001	
	$ClCH_2$	331		0.362	0.001	
	FCH_2	400		0.380	0.002	
	Cl_2CH	311		0.379	0.002	
	Cl_3C	183		0.366	0.003	
$OC(NH_2)_2$ (urea)	CH_3	270	+0.512 −0.144	0.328	0.002	
	$C_6H_5CH_2$	–	+0.526 −0.158	0.342	0.002	2.625
	$ClCH_2$	311	+0.533 −0.171	0.353	0.003	2.643
	FCH_2	355	+0.546 −0.190	0.368	0.002	2.665

py = C_5H_5N (pyridine)

Table 6.3 Singlet–triplet splitting J and parameters D and E of crystalline field for some copper arilcarboxilates [241]

Compound	J, cm^{-1}	D, cm^{-1}	E, cm^{-1}
$(C_6H_5COO)_2Cu$	312	0.344	0.014
$(C_6H_5COO)_2Cu \cdot C_6H_5COOH$	315	0.336	0.006
$(C_6H_5COO)_2Cu \cdot$ chin	306	0.356	0.002
(p-Br-$C_6H_4COO)_2Cu$	302	0.339	0.012
(p-Br-$C_6H_4COO)_2Cu \cdot CH_3OH$	293	0.343	0.002
(m-I-$C_6H_4COO)_2Cu \cdot C_4H_9OH$	321	0.352	0.001
(m-NO_2-$C_6H_4COO)_2Cu$	–	0.345	0.005
(m-$NO_2C_6H_4COO)_2Cu \cdot CH_3OH$	296	0.348	0.002
(o-NO_2-$C_6H_4COO)_2Cu$	300	0.352	0.011
(3, 5-$(NO_2)_2$-$C_6H_3COO)_2Cu$	–	0.355	0.001
(3, 5-$(NO_2)_2$-$C_6H_3COO)_2Cu \cdot$ chin	–	0.368	0.001
(2, 4-$(OCH_3)_2$-$C_6H_3COO)_2Cu$	–	0.327	0.005
$(C_6H_5CH{=}CHCOOH)_2Cu$	300	0.317	0.005

Table 6.4 Singlet–triplet splitting J and parameters D and E of crystalline field for copper dimers with bicarbonic acids [241]

Compound	J, cm^{-1}	D, cm^{-1}	E, cm^{-1}
$[O_2C(CH_2)_2CO_2]Cu$	312	0.335	0.002
$[O_2C(CH_2)_3CO_2]Cu$	320	0.337	0.002
$[O_2C(CH_2)_3CO_2]Cu \cdot$ py	337	0.358	0.001
$[O_2C(CH_2)_4CO_2]Cu$	348	0.338	0.005
$[O_2C(CH_2)_4CO_2]Cu \cdot$ py	342	0.357	0.002

Table 6.5 Distances R_{Cu-Cu} in the dimeric complexes $[Cu_2L_2L'_2]$ and $[Cu_2L_2L'_2] \cdot 2Y$ [$L = 1, 10$ phenathroline (phen), or 2, 2′ bipyridyl (bipy), or di-2-pyridylamine (dpa)]; [L' = phenylpropionato (PPr), or salicylaldehydato (salal), or diisopropylsalicylato (DIPS), or salycilato (sal)]; [$Y = H_2O$, or H_2 sal, or H_2 DIPS]

Complex	R_{Cu-Cu}, Å	References
$[Cu_2(phen)_2(sal)_2] \cdot 2H_2O$	3.242(2)	[285]
$[Cu_2(bipy)_2(sal)_2] \cdot 2H_2O$ sal $\cdot 4H_2O$	3.188(3)	[286]
$[Cu_2(bipy)_2(DIPS)_2] \cdot 2H_2O$ DIPS	3.168(1)	[287]
$[Cu_2(dpa)_2(salal)_2(ClO_4)_2]$	3.343(1)	[288]
$[Cu_2(phen)_2(salal)_2(NO_3)_2]$	3.486(1)	[289]
$[Cu_2(phen)_2(salal)_2(ClO_4)_2]$	3.369(1)	[290]
$[Cu_2(bipy)_2(salal)_2(ClO_4)_2]$	3.454(1)	[291]
$[Cu_2(phen)_2(PPr)_2(H_2O)_2]^{2+} \cdot 2NO_3$	3.054(1)	[292]

Table 6.6 Singlet–triplet splitting J for some binuclear copper(II) complexes containing 6-(benzylamino)purines as bridging ligands: HL_1 = 6-[(2-methoxybenzyl)amino]purine; HL_2 = 6-[(4-methoxybenzyl)amino]purine; HL_3 = 6-[(2, 3-dimethoxybenzyl)amino]purine; HL_4 = 6-[(3, 4-dimethoxy benzyl)amino]purine; AdeH = Adeline; AzabH = 4-azabenzimidazole; 2-ClBapH = 6-[(2-chlorobenzyl)amino]purine; 3-ClBapH = 6-[(3 -chlorobenzyl)amino]purine; 4-ClBapH = 6-[(4-chlorobenzyl)amino]purine; Nphtd = 1, 8-naphthyridine; H_2 oda = diglycolic acid

Compound	J, cm^{-1}	References
$[Cu_2(\mu\text{-AzabH})_4Cl_2]Cl_2 \cdot 3CH_3OH$	329	[294]
$[Cu_2(\mu\text{-AdeH})_4Cl_2]Cl_2 \cdot 6H_2O$	285	[295, 296]
$[Cu_2(\mu\text{-AdeH})_4(H_2O)_2](ClO_4)_4 \cdot 2H_2O$	312	[296, 297]
$\{[Cu_2(\mu\text{-AdeH})_4(H_2O)_2]\cdot$ $[Cu\,(oda)(H_2O)]_4\} \cdot 6H_2O$	274	[298]
$[Cu_2(\mu\text{-HL}_1)_4Cl_2]Cl_2 \cdot 2H_2O$	266	[293]
$[Cu_2(\mu\text{-HL}_2)_4Cl_2]Cl_2 \cdot 2H_2O$	256	[293]
$[Cu_2(\mu\text{-HL}_3)_4Cl_2]Cl_2 \cdot 2H_2O$	261	[293]
$[Cu_2(\mu\text{-HL}_4)_4Cl_2]Cl_2 \cdot 2H_2O$	269	[293]
$[Cu_2(\mu\text{-Nphtd})_2(\mu\text{-Cl}_2)Cl_2]$	139	[299, 300]
$[Cu_2(\mu\text{-HL}_1)_2(\mu\text{-Cl}_2)Cl_2]$	78, 9	[293]
$[Cu_2(\mu\text{-HL}_2)_2(\mu\text{-Cl}_2)Cl_2]$	114	[293]
$[Cu_2(\mu\text{-HL}_3)(\mu\text{-Cl}_2)Cl_2]$	119	[293]
$[Cu_2(\mu\text{-3-ClBapH})_2(\mu\text{-Cl}_2)Cl_2]$	82	[301]
$[Cu_2(\mu\text{-4-ClBapH})_2(\mu\text{-Cl}_2)Cl_2]$	114	[302]
$[Cu_2(\mu\text{-2-ClBapH})_2(\mu\text{-Cl}_2)\cdot$ $(2\text{-ClBapH})_2Cl_2] \cdot 2H_2O$	111	[303]
$[Cu_2(\mu\text{-3-ClBapH})_2(\mu\text{-Cl}_2)\cdot$ $(3\text{-ClBapH})_2Cl_2] \cdot 2H_2O$	123	[301]
$[Cu_2(\mu\text{-2-ClBapH})_2(\mu\text{-Cl}_2)(H_2O)_2]$	77	[303]
$[Cu_2(\mu\text{-3-ClBapH})_2(\mu\text{-Cl}_2)(H_2O)_2]$	100	[303]

Table 6.7 Singlet–triplet splitting J and μ_{eff}/μ_B at different temperatures for binuclear copper(II) complexes **1a**-**3b** [303]

Compound	J, cm^{-1}	μ_{eff}/μ_B
1a	110.6(6)	1.87 (294 K)–1.22 (82 K)
1b	110.0(3)	1.98 (296 K)–1.28 (80 K)
2a	76.8(6)	1.94 (295 K)–1.36 (80 K)
2b	99.4(4)	1.64 (286 K)–1.20 (82 K)
3a	15.2(2)	1.94 (291 K)–1.82 (83 K)
3b	41.6(5)	1.92 (294 K)–1.63 (80 K)

Table 6.8 Experimental and theoretical values of magnetic moments of ions of iron group [318]

Ion	Electron configuration	Term	$g\sqrt{J(J+1)}$, (μ_B)	$2\sqrt{s(s+1)}$, (μ_B)	$\mu_{eff}^{exp}(\mu_B)$
Ti^{3+}, V^{4+}	$3d^1$	$^2D_{3/2}$	1.55	1.73	1.8
V^{3+}	$3d^2$	$^3F_2$	1.63	2.83	2.8
Cr^{3+}, V^{2+}	$3d^3$	$^4F_{3/2}$	0.77	3.87	3.8
Mn^{3+}, Cr^{2+}	$3d^4$	$^5D_0$	0	4.90	4.9
Fe^{3+}, Mn^{2+}	$3d^5$	$^6S_{5/2}$	5.92	5.92	5.9
Fe^{2+}	$3d^6$	$^5D_4$	6.70	4.90	5.4
Co^{2+}	$3d^7$	$^4F_{9/2}$	6.63	3.87	4.8
Ni^{2+}	$3d^8$	$^3F_4$	5.59	2.83	3.2
Cu^{2+}	$3d^9$	$^2D_{5/2}$	3.55	1.73	1.9

Table 6.9 Singlet–triplet splittings for VO(II)–VO(II), Cr(III)–Cr(III) and Fe(III)–Fe(III) dimers [241]

Compound	$S_1 = S_2$	S	J, cm^{-1}	References
[tmen $(H_2O)VO(C_2O_4)VO\cdot$ (H_2O) tmen]$(ClO_4)_2 \cdot 1.25\, H_2O$	1/2	0; 1	5.7 (χ data)	[319]
$[LCr(OH)_3CrL](ClO_4)_3 \cdot 3\, H_2O$ $L = (\mu$-N, N′-bis(6-ethil-3,6-diazooctil)oxamidato (2)-$N^1, N^3, N^6, O : N^1, N^3, N^{6'}, O')$	3/2	0; 1 2; 3	128 (χ data)	[320]
DPACr (chrome di-μ-diphenilphosh phinatoceto-nato)	3/2	0; 1 2; 3	2.4 ± 0.4 (EPR data) 2.4 ± 0.05 (χ data)	[321]
$[(FeHEDTA)_2O] \cdot 6\, H_2O$	5/2	5; 4 3; 2 1; 0	190 (EPR data) 185 (χ data)	[322]
$(acac)_2Fe(C_2O_4)Fe$ $(acac)_2\, 0.5H_2O$	5/2	5; 4 3; 2 1; 2	7.2 (χ data)	[323]

tmen = N, N, N′, N′-tetramethilethilenediamine,
HEDTA = nitrilotriacetic acid,
acac = acetylacetone

6.1.1 *Copper(II) Dimers*

As it was earlier mentioned, dimer clusters have been detected for the first time in coordination compounds, in copper(II) monohydrate acetate [235]. In Table 6.1, there is given the constant J of isotropic exchange interaction of $3d$-electrons belonging to different ions of Cu^{2+} dimer clusters. It was found on the basis of EPR spectra measurements and static magnetic susceptibility χ. Also, constants of axial and rhombic crystalline field (D, E) for this compound are presented.

Other experimental data, related to magnetic and structure characteristics of the copper(II) dimer clusters, are presented in the Tables 6.2, 6.3, 6.4, 6.5, 6.6, 6.7, 6.8 and 6.9.

There are many other copper(II) coordination compounds containing Cu(II)–Cu(II) dimer clusters. Particularly, a set of binuclear 1, 10-phenantroline and salicylate ternary copper(II) complexes contain Cu(II)–Cu(II) dimers with the distance $R_{\text{Cu-Cu}}$ between Cu(II) ions given in the Table 6.5.

As it is seen from Table 6.5, the distance $R_{\text{Cu-Cu}}$ is a little changed when passing from one of the coordination compounds, given in this table, to another. In accordance with this, the constant of the isotropic exchange interaction J is also a little changed, as it takes place for other classes of binuclear copper(II) complexes too (Tables 6.1, 6.2, 6.3 and 6.4). Such a relatively weak dependence of the constant J on the type of coordination compound is connected with the fact that in these cases the main contribution to the exchange interaction between ions Cu(II) in dimers is due to indirect interactions with the participation of one diamagnetic atom. But in the cases when an indirect exchange interaction is realized, for example, by means of 6-(benzylamino)purines as bridging ligands [293], the variations of the constant J value are more significant (Table 6.6).

Note: The exchange constant J from the spin Hamiltonian used in [293] has an opposite sign relative to J in comparison with J from (6.2).

From the physical point of view, there are of interest the magnetic properties of the dimer copper(II) complexes with 6-(2-chlorobenzylamino)purine (HL_1) and 6-(3-chlorobenzylamino)purine (HL_2)[1]:

$[Cu_2(\mu\text{-}HL_1)_2(\mu\text{-}Cl_2)_2(HL_1)_2Cl_2]$ (**1a, b**),

$[Cu_2(\mu\text{-}Cl)_2(\mu\text{-}L_1)_2(H_2O)_2]$ (**2a**), $[Cu_2(\mu\text{-}Cl)_2(\mu\text{-}L_2)_2(H_2O)_2]$ (**2b**),

$[Cu_2(\mu\text{-}Cl)_2(HL_1)_2Cl_2]$ (**3a**), $[Cu_2(\mu\text{-}Cl)_2(HL_2)_2Cl_2]$ (**3b**).

The components **1a** and **1b** have the same chemical formula $C_{48}H_{40}N_{20}Cl_8Cu_2 \cdot H_2O$, but they were synthesized in two different ways. The method of preparation **1a** consists in adding the organic ligand HL_1 (0.5 mmol, 130 mg) to a solution of $CuCl_2 \cdot 2H_2O$ (1 mmol, 170 mg) in 2M HCl. For preparing **1b** the above procedure was used, but with an inverse molar ratio of the reactants, i.e., $CuCl_2 \cdot 2H_2O$ (0.5 mmol) and HL_1 (1 mmol) [303]. Surprisingly, **1a** and **1b** having the same chemical formula exhibit slightly different magnetic properties (Table 6.7). The singlet–triplet splitting J and the ratio of the measured effective magnetic moment per one Cu^{2+} ion to Bohr magneton, μ_{eff}/μ_B, at different temperatures are presented in the Table 6.7.[2]

The magnetic moments per one copper(II) ion for **1a** and **1b** are 1.87 μ_{eff}/μ_B at 294 K, and 1.98 μ_{eff}/μ_B at 296 K, respectively, and gradually decrease to 1.22 μ_{eff}/μ_B (for **1a**) and 1.28 μ_{eff}/μ_B (for **1b**) with decreasing temperature near 80 K.

---

[1] Do not confuse these notations of HL_1 and HL_2 with those given in the title of the Table 6.6.

[2] The exchange constant J from the spin Hamiltonian used in [303] is twice as many and has an opposite sign in comparison with J from (6.2).

In contrast to this, the molar magnetic susceptibility values increase up to a maximum near 100 K (Neel temperature) and consequently decrease [303], thus indicating the antiferromagnetic interaction between the Cu(II) ions.

The room temperature magnetic moments of **2a** and **2b** per one copper(II) ion are 1.94 and 1.64 μ_{eff}/μ_B. These magnetic moments gradually decrease with decreasing temperature, suggesting the presence of an antiferromagnetic interaction between the Cu(II) ions.

Compounds **3a** and **3b** have the room temperature magnetic moments of 1.94 and 1.92 μ_{eff}/μ_B, respectively. However, the μ_{eff}/μ_B values decrease as the temperature is reduced to the temperature of liquid nitrogen, indicating the presence of antiferromagnetic exchange.

The binuclear complex of copper(II) with coumarin-3-carboxylic acid (CcaH) is an exotic system containing Cu(II)–Cu(I) dimers. This complex supports the stoichiometry $[Cu_2)(Cca)_4(H_2O)_2] \cdot 2H_2O$ [304].

The $UV - Vis$ reflectance diffuse spectrum exhibits two bands closely analogous to spectra of other copper(II) dimers with carboxyl bridges [305]. The band at 625 nm may be assigned to the spin allowed transitions $(d_{xy}, d_{yz}) \rightarrow (d_{x^2-y^2})$ [304, 306, 307], whereas the band at 386 nm is referred as a characteristic of the brinding systems [308].

^1H NMR data for this complex show that the ligand is coordinated to copper, resulting in slight up-field chemical shifts of the aromatic protons relative to the free ligand. The free ligand show a peak at 8.74 ppm due to carboxylic protons, which is absent in the NMR spectrum of the complex, due to the deprotonation of the carboxylic group upon coordination to the copper atom. The paramagnetism of the copper centers results in broadening of the NMR lines due to the aromatic proton in the complex [304]. However, on the basis of NMR data it is impossible to determine why the narrowing of NMR lines happens: due to the interaction of protons with single ions Cu^{2+} or as a consequence of action on protons of the magnetic field, created by Cu(I)–Cu(II) dimers in the excited triplet state with $S = 1$. For that independent experiments are required, among them EPR investigations have an important role. The X-band EPR powder spectra of the complex exhibit features typical for those dimeric complexes with axial symmetry: $g_{\|} \approx 2.3$ and $g_{\perp} = 2.09$ [304] that are similar to those reported for other binuclear Cu(II) carboxylate adducts ($g_{\|} \approx 2.3$ and $g_{\perp} \approx 2.08$ [309–311]), indicating that this complex has an axially elongated octahedral geometry with Jahn-Teller distortions. The parameters of axial and rhombic crystalline fields calculated by the method of Watson et al. [312] are $|D| = 0.368\,\text{cm}^{-1}$ and $E = 0.0043\,\text{cm}^{-1}$ [304]. These values are in agreement with axial and rhombic zero field splittings for a large number of binuclear copper(II) complexes (Tables 6.2, 6.3 and 6.4). The room temperature magnetic moment of the complex ($\mu_{\text{eff}} = 1.3\,\mu_B$) is substantially smaller than the spin-only value for Cu^{2+} ion, suggesting that copper(II) ions are coupled by exchange interaction into dimers with $S = 0$ in ground state and $S = 1$ in an excited one.

The results of investigations of other dimer copper(II) clusters with bridging ligands of different types (including hidroxyl and metroxyl bridges, Schiff's bases and other) can be found, for example, in [241, 313–316]. It is to be noted that the authors [316] have studied the structure of coordination polyhedron of complex of the same copper(II) acetate with 2-[2-hydroxiphenil]-4, 4-diphenil-1, 2-dihydro-4H-3, 1 -benzoxazin in solid state and in solutions of ethanol and chloroform almost 60 years later after the discovery by Bleaney and Bowers [235] of the existence of dimer clusters in coordination compounds on the example of copper acetate. They have detected an exchange interaction of antiferromagnetic type between two ions of copper(II) in a dimer cluster ($J = 57.8\,\mathrm{cm}^{-1}$).[3]

6.1.2 Dimer Clusters of Other 3d-Elements

It is known [317, 318] that ions of transit metals of iron group are characterized so much the paramagnetic $3d$-shell of such ions lies closer to the surface of the ion, where it is screened more efficiently from the nucleus. As a consequence, $3d$-electrons are not screened from the electric crystalline field and are exposed to its direct action. In its turn, the interaction of electrons with the crystalline field is stronger than the spin-orbital interaction (the case of intermediary crystalline field). On the other hand, contributions of above-discussed mechanisms of indirect and direct exchanges to the total exchange interaction between ions in a dimer depend on the form of wave functions of $3d$-electrons in the crystalline field of different symmetries (with regard to covalency effects). These contributions are determined by overlaps of wave function tails of interacting electrons.

Another common property of $3d$-ions is that in such ions the Russell–Saunders coupling of angular momenta is realized (a separate addition of orbital and spin-angular moments of electrons: $\mathbf{L} = \sum_i \mathbf{l}_i$, $\mathbf{S} = \sum_i \mathbf{s}_i$).

Finally, for $3d$-ions in crystalline field the effect of freezing of the orbital angular momentum is also characteristic [317, 318]. This effect is a consequence of the following theorem [317].

Theorem *The matrix element of the orbital angular momentum, taken between two non-degenerated states, has an arbitrary phase. In particular, it may be real or pure imaginary.*

The proof of this theorem is carried out using the time-reversal operator $\mathbf{T}$, acting on non-generated state Ψ neglecting the spin ($\mathbf{T}\Psi = \Psi^*$, $\Psi^* = e^{i\varphi}\varphi$, where φ is a real quantity) and taking into account that $\mathbf{L}$ is a pure imaginary operator and $\mathbf{T}$ is an antiunitary operator.

This theorem has an important corollary.

Corollary 1 *The mean value of the operator* $\mathbf{L}$ *for non-generated states is equal to zero.*

[3]The exchange constant J from the spin Hamiltonian used in [316] is twice as many and has an opposite sign in comparison with the sign from (6.2).

According to the theorem, the diagonal matrix element of the operator of orbital angular momentum $\langle n|\mathbf{L}|n\rangle$ must be pure imaginary. However, since all physical observable values (including the orbital angular momentum) must be real, it follows that $\langle n|\mathbf{L}|n\rangle = 0$.

Thus, if the crystalline field has a sufficiently low symmetry to cut the orbital degeneration, then in the first approach of the perturbations theory the orbital momentum is equal to zero. This is just the circumstance, which explains the fact that $g\sqrt{J(J+1)}$ (where J is the complete angular momentum of the ion and g is Lande-factor) is not consistent with experimental values of magnetic momentums of $3d$-ions, while pure spin values $2\sqrt{s(s+1)}$ lead to close coincidence with experimental data (Table 6.8).

Up to now, binuclear clusters of other ions of iron group are by far less studied than the above discussed copper dimers. Some information on magnetic properties of such clusters with bridging ligands of different types (including hidroxyl and methoxyl bridges, Shiff's bases and others), and also information about the dimers, in which one of ions have the orbital angular momentum, are given in [241]. Below some of these properties according to mentioned work will be discussed.

In Table 6.9, the values of exchange interaction constants in homonuclear dimers, constructed from iron group ions, are given. These data are obtained on the basis of EPR spectra and static magnetic susceptibility measurements of respective compounds.

In the Table 6.9 S denotes the value of the total spin of the magnetic dimer. The dimer fragments for five compounds indicated in the Table 6.9 are presented below in the order they appear on a vertical in this table as *a–e* in the following figures:

(a)

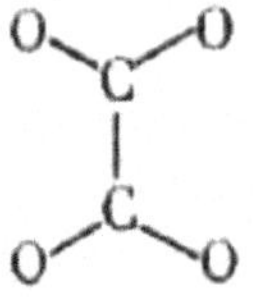

(b)

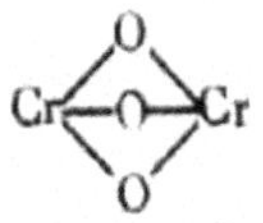

(c)

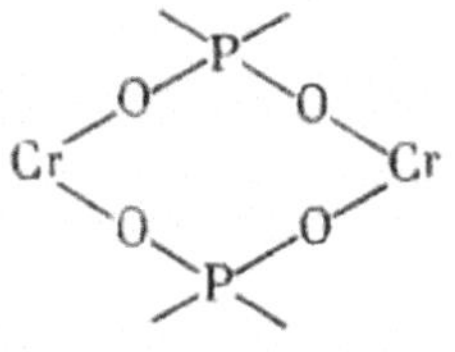

(d)

(e)

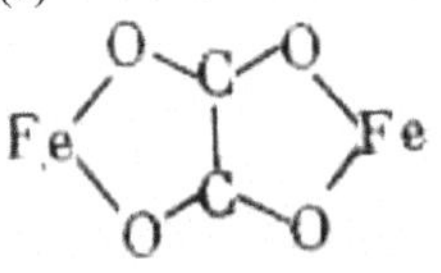

VO(II)–VO(II) ($3d^1$; the principal term of single center is ^2D, $S_i = 1/2$). The cluster structure appeared in EPR spectra of vanadyl dimer with tridental Schiff bases [324, 325] in the form of a forbidden transition with $\Delta M_S = 2$ [319]. Using the spectral position of EPR line corresponding to this transition, it was estimated the constant of axial crystalline field D, which turned out to be significantly less than for analogous copper dimers, though the isotropic exchange interaction constants are comparable for both cases.

Cr(III)–Cr(III) ($3d^3$; the main term of single center is ^4F, $S_i = 3/2$). Such dimers there exist in monocrystals of the compound $[LCr(OH)_3CrL](ClO_4)_3 \cdot 3\,H_2O$ [320] (see Table 6.9 and dimer fragment *c*) as three hydrobridged dimers with distance $R = 2.642\,Å$ (the least known distance for dimers Cr(III)–Cr(III) [241]) and, respectively, with a large constant of the isotropic exchange interaction ($J = 128\,cm^{-1}$).

Co(II)–Co(II) ($3d^7$; ^4F, $S_i = 3/2$). In the crystalline field of the trigonal bipyramid the high spin slate of the ion Co(II) with a frozen orbital momentum is realized if distorted trigonal bipyramid exists around the ion Co(II) in the compound $[Co_2(tren)_2X_2](BPh_4)_2$ [326], which is a dimer with weak exchange interaction ($J \lesssim 1\,cm^{-1}$) in accordance with measurements of the magnetic susceptibility. Here *tren* = 2, 2′, 2″-triaminotriethylamine, BPh_4 = tetraphenylborate.

Mn(II)–Mn(II) ($3d^5$; ^6S, $S_i = 5/2$). In [327] EPR spectra of dimer compounds Mn(II) $[Mn_2(tren)_2X_2](BPh_4)_2$ and $[Mn_2(tren)_2(NCO)_2](BPh_4)_2$, where $X = NCS^-$, OCN^-, were studied. It was found that these *EPR* spectra are due to exchange pairs Mn(II)–Mn(II) and that all splittings in spectra (thin structure of the EPR spectra of single centers, Zeeman and exchange splittings) are comparable in magnitude.

Fe(III)–Fe(III) ($3d^5$; ^6S, $S_i = 5/2$). The last two compounds in the Table 6.9 are examples of iron(III) dimer with one oxygen bridge and dimer formed by oxalate bridge. Iron dimer with oxalate bridge was studied on polycrystalline patterns and in frozen solution [323]. It was shown that in the studied iron dimer, where $R_{Fe-Fe} \gtrsim 5\,Å$, the parameter J found from temperature dependence of static magnetic susceptibility is comparable in magnitude with J for dihydroxodimers of iron(III), where $R_{Fe-Fe} \sim 3\,Å$.

Thus, data presented in Tables 6.2, 6.3, 6.4, 6.6, 6.7, and 6.9 show that the exchange interaction between ions of iron group, which form homonuclear dimer clusters, has an antiferromagnetic character. The exchange interaction constant depends on many factors, including the geometry encirclement of paramagnetic ions forming dimer clusters, and decreases with the increase of extension of polyatomic bridged fragments (Table 6.9 and corresponding dimer fragments).

The main conclusion that can be drawn from the analysis of data given in Sects. 6.1.1 and 6.1.2, is that there is an exchange interaction of antiferromagnetic type in homonuclear dimer magnetic clusters formed by ions of iron group with electron configuration $3d^n$ ($n \neq 5$). In principle, an exception is possible, when it is necessary to take into account the contribution of excited states leading to weak ferromagnetism in homonuclear clusters [241]. However, this happens very rarely. As to dimers of iron group ions with the electron configuration $3d^5$ (Mn(II)–Mn(II), Fe(III)–Fe(III), Mn(II)–Fe(III)), the probability of formation of such dimers with the exchange interaction of antiferromagnetic or ferromagnetic type is apparently approximately the same (Sect. 6.4).

It is to pay attention to unusual magnetic behavior of the heteronuclear exchange pair Cu(II)–VO(II). While according to above-mentioned empiric rule, the homonuclear dimers Cu(II)–Cu(II) and VO(II)–VO(II) shows an exchange interaction of antiferromagnetic type, the exchange interaction between ions Cu(II) and VO(II) has a ferromagnetic character. The theoretical interpretation of the nature of exchange interaction of ferromagnetic type in such a dimer will be discussed in Sect. 6.3. In Sect. 6.2 it will be introduced the notion of "partial time-reversal". Within this concept the transition from the antiferromagnetic ordering to the ferromagnetic one (and vice versa) is a natural consequence of the combined transformation of symmetry "partial time-reversal" + inversion of sign of exchange integral.

6.1.3 Dimer Clusters of 4*f*-Elements

In the rare-earth group 4*f*-shell is built, in which the maximal number of electrons attains fourteen. As a result of chemical interactions, atoms of rare-earth elements lose three electrons and trivalent ions are formed with the external electron configuration $5s^2 5p^6$ [318].

Unlike the unpaired *d*-electrons of transition metal ions and *s*, *p* electrons of organic radicals, the unpaired *f*-electrons of lanthanide ions are highly contracted and energetically shielded deeply by outer 5*s*, 5*p* shells. Therefore, the magnetic interaction between 4*f*-electrons of the rare-earth ions coupled into exchange pairs is very weak, the value of coupling constant J being normally smaller than 1 cm^{-1} [328–365].

The *f* orbitals of the lanthanide ions have a strong unquenched orbital angular momentum, which can induce effective spin-orbit coupling. The trivalent ions are characterized by f^n configurations, which give rise to $^{2S+1}L_J$ multiplets, further split by spin–orbit interaction to give J states

$$E\left(^{2S+1}L_J\right) = (\lambda/2)[J(J+1) - L(L+1) - S(S+1)], \tag{6.10}$$

where J is defined by the angular momentum summation rules $|L - S| \leq J \leq |L + S|$, and $\lambda = \pm\xi/2S$. ξ is the spin-orbit coupling constant, which ranges from 600 to 3000 cm^{-1} the highest values in lanthanide series corresponding to heaviest lanthanide ions [329]. The (+) sign applies for $n < 7$ and the (−) sign for $n > 7$. The sign of the spin–orbit coupling constant λ implies that in the ground state $J = L - S$ for $n < 7$ and $J = L + S$ for $n > 7$. For f^7 ions such as Gd(III) and Eu(II), $L = 0$ and $S = 7/2$, the orbital momentum is completely absent in the ground state. In the Table 6.10 the trivalent rare-earth ions in the order of building their electron shells are presented [318].

In the third column of Table 6.10 the spectral terms of trivalent rare-earth ions determined accordingly to Hund rules are presented. In the fourth and fifth columns

Table 6.10 Electron configurations, spectroscopic terms, calculated and measured magnetic moments of rare-earth ions

Ion	Electron configuration	Spectroscopic term	μ_{eff}^{theor}, Bohr magnetons	μ_{eff}^{exp}, Bohr magnetons
Ce^{2+}	$4f^1 5s^2 5p^6$	$^2F_{5/2}$	2.54	2.4
Pr^{3+}	$4f^2$	$^3H_4$	3.58	3.5
Nd^{3+}	$4f^3$	$^4I_{9/2}$	3.62	3.5
Pm^{3+}	$4f^4$	$^5I_4$	2.68	–
Sm^{3+}	$4f^5$	$^6H_{5/2}$	0.84	1.5
Eu^{3+}	$4f^6$	$^7F_0$	0	3.4
Gd^{3+}	$4f^7$	$^8S_{7/2}$	7.94	8.0
Tb^{3+}	$4f^8$	$^7F_6$	9.72	9.5
Dy^{3+}	$4f^9$	$^6H_{15/2}$	10.63	–
Ho^{3+}	$4f^{10}$	$^5I_8$	10.60	10.4
Er^{3+}	$4f^{11}$	$^4I_{15/2}$	9.59	9.5
Tu^{3+}	$4f^{12}$	$^3H_6$	7.57	7.3
Yb^{3+}	$4f^{13}$	$^2F_{7/2}$	4.54	4.5

the values of magnetic moments calculated using formulas $\mu_{eff}^{theor} = g[J(J+1)]^{1/2}$ and $\mu_{eff}^{exp} = (1/\mu_B)(3\chi_0 k_B T/n)^{1/2}$ are given. Here μ_B is the Bohr magneton, J is the total angular momentum of the ion, χ_0 is the static magnetic susceptibility, $k_B T$ is the thermal energy, n is the concentration of paramagnetic centers and g is the Landé-factor

$$g = \frac{3}{2} + \frac{S(S+1) - L(L+1)}{2J(J+1)}. \tag{6.11}$$

As we can see in Table 6.10, the model of **LS**-interaction leads to an adequate description of static paramagnetic properties of trivalent rare-earth ions, excepting one or two cases, the most characteristic one being the ion Eu^{3+}.

The Ln(III) ions have both a large spin-angular momentum part and a large orbital angular momentum part associated with the internal nature of the valence f orbitals. Compared with the d-electrons of transition-metal ions, the orbital component of the magnetic moment is much more important for the rare-earth ions, because the crystal-field effects are smaller and spin–orbit coupling larger for f-electrons.

In contrast to the ions of transition metals, there are no simple rules to predict the occurrence of ferro- or antiferromagnetic coupling, unlike those of Goodenough and Kanamori for compounds of d-elements [330].

Let us consider magnetic properties of binuclear coordination compounds, beginning with the most studied Gd(III)–Gd(III) dimer in different compounds. The magnetic behavior of the single crystals of $Gd(HF_2CCOO)_3(H_2O)_2 \cdot H_2O$ were investigated in the temperature range 1.7–300 K. The magnetic data were interpreted considering exchange interaction effects in the binuclear Gd(III)–Gd(III) units. The exchange coupling constant $J = -0.024\,cm^{-1}$ ($g =$

Table 6.11 Comparison of exchange coupling constants J $[cm^{-1}]$ and Landé-factors g for dimeric gadolinium compounds [331]

Compound	J	g	References
$Gd(HF_2CCOO)_3(H_2O)_2 \cdot H_2O$	−0.024	1.99	[331]
$Gd(H_3CCOO)_3(H_2O)_2 \cdot 2H_2O$	+0.05	1.99	[331]
$Gd(H_3CCOO)_3(H_2O)_2 \cdot 2H_2O$	+0.06	2	[332]
$Gd(L)_2 \cdot 2CHCl_3$	−0.09	2	[333]
$Gd_2(L^1)_2(NO_3)_2 \cdot 2MeOH$	−0.198	1.975	[334]
$Gd_2(L^2)(NO_3)_2 \cdot 4H_2O$	−0.211	1.975	[335]
$Gd_2L^3(NO_3)_2 \cdot 3H_2O$	−0.104	1.999	[336]
$[Gd_2(L^4\text{-}3H)(NO_3)_2](NO_3) \cdot 1.5H_2O$	−0.194	2.023	[337]
$[Gd(L^5)_3(H_2O)]_2$	+0.05	1.98	[338]
$[Gd_2(\mu_2\text{-}OOCFc)_2(OOCFc)_4 \cdot$ $(MeOH)_2(H_2O)_2 \cdot 2MeOH \cdot 2H_2O]$	+0.006	2	[339]
$Gd_2salen_3 \cdot 2H_2O$	−0.14	2.025	[340]
$Gd_2dsp_3 \cdot 2H_2O$	−0.1	1.976	[340]
$[Gd_2(CH_3CO_2)_6(phen)_2]$	−0.053	2	[341]

1.99) shows an antiferromagnetic interaction, whereas $J = +0.05\,cm^{-1}$ ($g = 1.99$) for $Gd(H_3CCOO)_3(H_2O)_2 \cdot 2H_2O$ a ferromagnetic one [331]. These data were obtained using the spin Hamiltonian $\mathbf{H} = -J\mathbf{S}_{Gd1}\mathbf{S}_{Gd2}$, where $S_{Gd1} = S_{Gd2} = 7/2$, g is Landé-factor.

Some binuclear complexes of gadolinium are available for comparison [332–341]. The constant J of these compounds ranges between +0.06 and $-0.211\,cm^{-1}$ (Table 6.11).

In Table 6.11 the following notations are introduced:

H_3L = tris(((2-hydroxybenzol)amino)ethyl)amine;

H_3L^1 = N-salicylidene-2-(bis-(2-hydroxyethyl)amino)ethylamine;

H_4L^2 = Schiff bases obtained by condensation of 3-formylsalicilic acid and 4-N-dodecycldiethylenetriamine;

H_3L^3 = tris-[4-(2-hydroxy-3-methoxyphenyl)-3-aza-3 -butenyl]amine;

L^4-3H = cryptante, given in [105]; HL^5 = salicylic acid;

Fc = ferrocene; H_2salen = N, N'-diaminoethane-salicylaldimine;

H_2dps = disalicylidene-1, 2-phenylenediamine; phen = o-phenanthroline.

The different types of interaction in these acetates might be due to the nature of bridging in the Gd(III)–Gd(III) dimers. Whereas $Gd(H_3CCOO)_3(H_2O)_2 \cdot 2H_2O$ shows the bridging mode μ_2O'; k^2O, O' (Fig. 6.2b), the compound $Gd(HF_2CCO)_3(H_2O)_2 \cdot H_2O$ shows the bridging mode μ_2-carboxylato-k^1O: k^1O' (Fig. 6.2a) [331].

In these two dimeric compounds the constants of isotropic exchange interaction J have opposite signs, i.e. in one dimer there is the exchange interaction of antiferromagnetic type between Gd(III) ions, while in the other - the exchange interaction of ferromagnetic type. It is due to the substitution of H_3 from the first dimer by HF_2

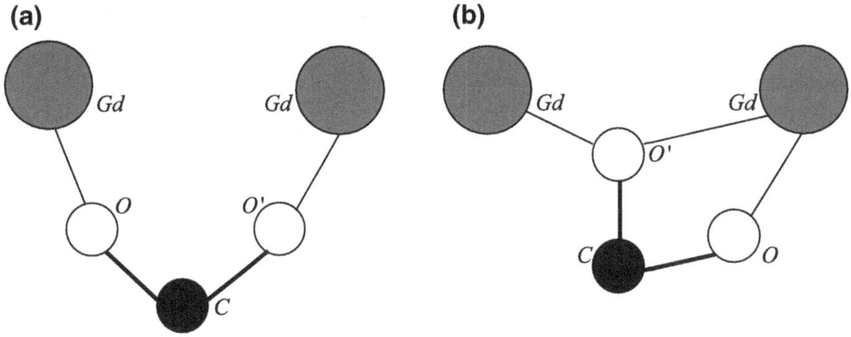

Fig. 6.2 Bounding and bridging types of the carboxylate groups: **a** - bidental (μ_2-carboxylato-k^1O: k^1O′); **b** - tridental (μ_2O′; k^2O, O′)

Table 6.12 Comparison of J $[cm^{-1}]$, bridging mode and $R_{\text{Gd-Gd}}$ distance [Å] for $Gd_2(ClH_2CCOO)_6(bipy)_2$ (**1**), where bipy = 2, 2′-bipyridyl, $Gd_2(CF_2HCOO)_3(phen)$ (**2**), $[NH_3C_2H_5][Gd(Cl_2HCCOO)_4]$ (**3**), $[NH_3CH_3][Gd(Cl_2HCCOO)_4]$ (**4**), $Gd(H_3CCOO)_3(H_2O)_2 \cdot 2H_2O$ (**5**) and $Gd(F_2HCCOO)_3(H_2O)_2 \cdot H_2O$ (**6**) [342]

Compound	J	Bridging mode	$R_{\text{Gd-Gd}}$, Å	References
1	−0.020	μ_2O′; k^2O, O′ and μ_2-carboxylato-k^1O: k^1O′	3.990	[342]
2	−0.016	μ_2O′; k^2O, O′ and μ_2-carboxylato-k^1O: k^1O′	4.034	[343]
3	+0.029	μ_2O′; k^2O, O′	4.181	[344]
4	+0.023	μ_2O′; k^2O, O′	4.184	[345]
5	+0.025	μ_2O′; k^2O, O′	4.206	[332, 346]
6	−0.012	μ_2-carboxylato-k^1O: k^1O′	4.456	[331]
4	−0.007	μ_2-carboxylato-k^1O: k^1O′	4.516	[345]

in the second one, as well as to different numbers of H_2O molecules in the outer sphere (two in the first case and only one in the second one). Further investigation of gadolinium compounds, where the Gd(III) ions are differently bridged to dimers, confirmed the assumption about a correlation between the types of bridging modes and types of exchange interactions (Table 6.12) [342].

The values for the constant of exchange interaction J from Table 6.12 were obtained considering that the intramolecular interaction within the dimeric unit is described by the Heisenberg model with the spin Hamiltonian $\mathbf{H} = -2J\mathbf{S}_{Gd1}\mathbf{S}_{Gd2}$. Therefore, to compare the values J given in Tables 6.12 and 6.11, it is necessary to double the data from Table 6.12.

The crystal structure of **1** is built up by discrete dimers Gd(III)–Gd(III) ($R_{\text{Gd-Gd}} =$ 3.990 Å) with two kinds of bridging carboxylate groups (μ_2-carboxylato-k^1O: k^1O′ and μ_2O′; k^2O, O′) and besides a chelating monochloroacetate ion [331]. The Gd^{3+} ion is ninefold coordinated by seven oxygen atoms originating from the carboxylate groups and two nitrogen atoms stemming from a 2, 2′-bipyridyl molecule.

As it can be seen from Table 6.12, the $\mu_2O'; k^2O, O'$-bridging mode is responsible for a ferromagnetic interaction, whereas the μ_2-carboxylato-k^1O: k^1O' -bridging mode leads to an antiferromagnetic interaction (see also [331, 332, 343–347]).

Such a magnetic behavior of dimer clusters formed by rare-earth ions with the electron configuration $3d^7$ is not surprising. As noted earlier, in this case the electron shell of each ion is half-filled both with electrons and holes. In such conditions, unlike the homonuclear dimer clusters, where equivalent ions have different numbers of electrons and holes, the exchange interaction of both antiferromagnetic and ferromagnetic type can be realized.

According to data from Table 6.12, the Gd(III)–Gd(III) dimers are characterized by bidentate (μ_2-carboxylato-k^1O: k^1O')-bridging mode (A) and tridentate (μ_2O'; k^2O, O')-bridging mode (B). Besides the pure bridging A and B, where only bidentate or tridentate modes are involved, there is one another, in which both modes are involved (bridging C) (Table 6.13).

The Hamiltonian used for finding J is $\mathbf{H} = -2J\mathbf{S}_{Gd1}\mathbf{S}_{Gd2}$.

The magnetic data given in Table 6.13 show that antiferromagnetic interactions are found for A and C, and a ferromagnetic one for B. These data, as well as the data from Table 6.11 show very small values for the exchange interaction constant J for binuclear units Gd(III)–Gd(III) in different coordination compounds. This is due to the fact that the electrons of $4f$-shells of gadolinium ions are relatively strong localized that weakens the exchange interaction between the $4f$ -electrons in the cluster Gd(III)–Gd(III).

The binuclear lanthanide(III) complexes described by the general formula $[Ln_2(CTPHA)(Mephen)_4(ClO_4)_2](ClO_4)_2$ (Ln = La, Ce, Nd, Sm, Eu, Gd, Tb, Dy, Ho, Er), where *Mephen* stands for 5-methyl-1, 10-phenanthroline, and CTPHA are the 2-chloroterephthalate dianions, have been synthetized and characterized in [354]. It was shown that these complexes are CTPHA-bridged structures consisting of two lanthanide(III) ions. The complex with Ln = Gd is an ideal model for studying the magnetic interaction between Ln(III) and Ln(III) ions among the binuclear complexes. The reason for the choice of gadolinium is that Gd(III) ion and its complex are quite simple for magnetic study. The Gd^{3+} ion has a $^8S_{7/2}$ ground state, which is located at some $3 \times 10^4\,cm^{-1}$ below the first excited state and is not perturbed by crystal field effects [355–357].

The effective magnetic moment of the binuclear complex $[Gd_2(CTPHA)(Mephen)_4(ClO_4)_2](ClO_4)_2$ at room temperature is 11.15 μ_B, which is slightly less than the spin-only value, 11.23 μ_B, calculated from the equation $\mu_{\mathrm{eff}} = (\mu^2_{\mathrm{Gd(III)}} + \mu^2_{\mathrm{Gd(III)}})^{1/2}$, in the absence of the exchange interaction for a binuclear Gd(III) system, and is nearly constant over the 30–300 K region, but it sharply decreases below 30 K [355]. This behavior is characteristic for the weak intramolecular antiferromagnetic spin–exchange interaction between the paramagnetic ions in the dimeric cluster. Two spin states of Gd(III) ions are coupled in the dimeric cluster to yield the total spin states, $S = 7, 6, 5, 4, 3, 2, 1, 0$ ($D_{7/2} \otimes D_{7/2} = \sum_{S=0}^{7} D_S$, D_S

Table 6.13 Exchange parameter J and bridging motifs for Gd^{3+} carboxylates [348]

Compound	J, cm^{-1}	Bridging	References
$Gd_2(CF_2ClCOO)_6 \cdot (H_2O)_6$	−0.006	A	[347]
$Gd_2(CCl_3COO)_6(bipy)_2 \cdot (H_2O)_2 \cdot 4\,bipy$	−0.004	A	[348]
$Gd_2(CH_3COO)_6(H_2O)_4 \cdot 4\,H_2O$	+0.03	B	[332]
$Gd_2(OOCFc)_6(CH_3OH)_2 \cdot (H_2O)_2 \cdot 2CH_3OH \cdot 2\,H_2O$	+0.003	B	[339]
$Gd_2(bz)_6(phen)_2$	−0.215	C	[349]
$Gd_2(oet)_6(DMF)_2(H_2O)_2$	−0.034	C	[350]
$Gd_2(CClH_2COO)_6(bipy)_2$	−0.02	C	[342]
$Gd_2(CF_2HCOO)_6(phen)_2$	−0.016	C	[343]
$(CH_3NH_3)_2[Gd_2(CCl_3COO)_6 \cdot (H_2O)_6](CCl_3COO)_2 \cdot 2CCl_3COOH$	−0.0106	C	[348]
$Gd_2(crot)_6(phen)_2 \cdot 2\,H_2O$	AFM [a]	C	[351]
$Gd_2(crot)_6(bipy)_2$	AFM [a]	C	[352]
$Gd_2(crot)_6(H_2O)_4 \cdot 4bpa$	AFM [a]	C	[352]
$Gd_2(CH_3COO)_6(phen)_2$	+0.0265	–	[353]
$Gd_2(sal)_6(H_2O)_2$	+0.025	–	[338]

[a]The compound was only reported to be antiferromagnetic (AFM), but the value of J was not calculated.

Notations:
Hbz = benzoic acid;
Hoet = 4-ethoxybenzoic acid;
Hcrot = crotonic acid;
DMF = dimethylformamide;
H_2sal = salicylic acid;
bipy = 2, 2′-bipyridyl;
phen = 1, 10-phenanthroline;
bpa = di(2-pyridyl)amine

is the irreducible representation of the rotation group of symmetry). On the basis of Boltzmann distribution, all these states are thermally populated at room temperature due to the weak antiferromagnetic interaction between Gd(III) ions [358]. A good beast-square fit to experimental data was obtained for the exchange parameter $J = -0.16\,\text{cm}^{-1}$ and $g = 2.00$. The agreement factor

$$F = \sum \left[(\chi_{0M})_{\text{obs}} - (\chi_{0M})_{calc}\right]^2 / \sum (\chi_{0M})_{\text{obs}}$$

is 1.9×10^{-9} (χ_{0M} is the molecular static susceptibility per binuclear complex) [355]. This result (a small and negative J value) indicates that the complex is binuclear and undergoes a very weak antiferromagnetic superexchange interaction between Gd(III)

Table 6.14 The isotropic exchange constant J (determined with $g = 2.00$ on the basis of the spin Hamiltonian $\mathcal{H} = -2J\mathbf{S}_{Gd1}\mathbf{S}_{Gd2}$) and Gd(III)–Gd(III) distance in some Gd(III) binuclear compounds [360]

Compound[a]	J, cm^{-1}	$R_{\text{Gd–Gd}}$, Å	Symmetry transformations used to generate atoms
$Gd_2(ClF_2CCOO)_6 \cdot (hypy)_2$	−0.019	3.866(3)	$-x+1, -y+1, -z+1$
$Gd_2(F_3CCOO)_6 \cdot (hypy)_2$	−0.034	3.877(3)	$-x+1, -y+1, -z+1$
$Gd_2(F_2HCCOO)_6 \cdot (hypy)_2$	−0.064	3.880(3)	$-x+1, -y+1, -z+1$
$Gd_2(Cl_2HCOO)_6 \cdot (H_2O)_2(hypy)_2$	−0.011	4.051(2)	$-x+1, -y+1, -z+2$

[a]hypy = 4-hydroxypyridine

ions. The very weak antiferromagnetic behavior of the complex may result mainly from the properties of the bridged ligand [359] and lanthanide (III) ions [357].

The weak antiferromagnetic behavior is characteristic also for other Gd(III) binuclear compounds. In Table 6.14, the experimental data confirming this statement are given.

In [361] a series of six isostructural lanthanide dimers of general formula $[Ln_2(\text{mdea}\,H_2)_2(\text{piv})_6]$, (where *mdea* H_2 is N-methyldiethanolamine, *piv* is pivalate and Ln = La, Ce, Pr, Nd, Sm, and Gd) has been surveyed to gauge the nature of magnetic interactions between the lanthanide centers. The following isostructural lanthanides were synthetized and investigated:

$[La_2(\text{mdea}\,H_2)_2(\text{piv})_6]$ (**I**), $[Ce_2(\text{mdea}\,H_2)_2(\text{piv})_6]$ (**II**),
$[Pr_2(\text{mdea}\,H_2)_2(\text{piv})_6]$ (**III**), $[Nd_2(\text{mdea}\,H_2)_2(\text{piv})_6]$ (**IV**),
$[Sm_2(\text{mdea}\,H_2)_2(\text{piv})_6]$ (**V**) and $[Gd_2(\text{mdea}\,H_2)_2(\text{piv})_6]$ (**VI**).

Of this series of compounds, La_2 (**I**) is a diamagnetic compound, thus its static magnetic susceptibility ($\chi_{0\,\text{La}_2} = \chi_{0\,\text{dia}} = -8.8 \cdot 10^{-4}\,\text{cm}^3\,\text{mol}^{-1}$) was used to correct for the diamagnetic contribution of this system [361]. At room temperature, the experimental $\chi_0 T$ products for crystalline powder samples of (**I**)–(**VI**) are all more or less close to the excepted values (Table 6.15) [361, 362]. When lowering the temperature, the $\chi_0 T$ product decreases down to 1.8 K for all the compounds (**II**)–(**V**), except for Gd_2 (**VI**), where it stays constant until 10 K and then slightly increases to reach $16.8\,\text{cm}^3\,\text{mol}^{-1}$ at 1.8 K.

As it can be seen in Table 6.15, the expected values of the $\chi_0 T$ product for Ln_2 dimers are closer to experimental ones data in comparison with the expected values of $\chi_0 T$ for single Ln ion. It is shown that all compounds in the Table 6.15 contain Ln_2 dimers.

In the case of Gd_2 (**VI**), the slight increase of $\chi_0 T$ product at low temperature indicates dominant ferromagnetic interactions. As Gd(III) has no orbital contribution, it can be considered as an isotropic $S = 7/2$ spin. For Gd_2 (**VI**) $J = +0.0035\,\text{cm}^{-1}$ indicating that Gd–Gd interaction is very weak and the two Gd(III) ions are very weakly coupled ferromagnetically. This value of the constant of isotropic exchange

Table 6.15 Comparison of the $\chi_0 T$ product at room temperature (RT) between the excepted [362] and the experimental values for compounds (**II**)–(**VI**) [361]

Compound	$\chi_0 T$ $(cm^3 K\ mol^{-1})$ theoretical value for each Ln ion at RT [363]	$\chi_0 T$ $(cm^3 K\ mol^{-1})$ expected value for Ln_2 at RT	$\chi_0 T$ $(cm^3 K\ mol^{-1})$ experimental value for Ln_2 at RT
Ce_2 (**II**)	0.80	1.60	1.29
Pr_2 (**III**)	1.60	3.20	2.98
Nd_2 (**IV**)	1.64	3.28	2.93
Sm_2 (**V**)	0.09	0.18	0.46
Gd_2 (**VI**)	7.875	15.75	16.16

interaction was found from the temperature dependence of the static magnetic susceptibility χ_0 (product $\chi_0 T$) by fixing $g = 2.03(1)$ and using the spin Hamiltonian $\mathcal{H} = -2J\mathbf{S}_{Gd1}\mathbf{S}_{Gd2}$, as well as by fitting the magnetization M versus H/T (H is the magnetic field). Both methods lead to the same value of J [361].

The binuclear lanthanide complex series $[Ln_2(valdien)_2(NO_3)_2]$, where Ln = Eu(III), Gd(III), Tb(III), Dy(III) and Ho(III), which exhibits O-bridged Ln(III) ions, have been reported in [364]. The Dy(III) analogue was shown to exhibit a single molecular magnet behavior. It consist in two Dy(III) ions bridged by two phenoxide-oxygen atoms with a Dy(III)–Dy(III) distance of 3.77 Å. The ligand involved in the synthesis of this complex is the Schiff-base ligand, $N1, N3$-bis(3 -methoxysalicylidene)diethylenetriamine (H_2*valdien*), which provides two bridging phenoxide-O atoms acting as super exchange pathways between two Dy(III) metal centers inducting weak antiferromagnetic interactions with the isotropic exchange constant $J_{\mathrm{Dy-Dy}} = -0.21\,\mathrm{cm}^{-1}$ [365]. The Gd(III) analogue, consisting in two Gd(III) ions coupled by exchange interaction into magnetic dimer, exhibits weak antiferromagnetic interactions with $J_{\mathrm{Gd-Gd}} = -0.178(1)\,\mathrm{cm}^{-1}$, if g-factor is fixed at the value $g = 2.00(0)$ [364].

6.2 Combined Time-Reversal Transformation

The invariance of the Hamiltonian with respect to the time-reversal operator is one of the fundamental properties of symmetry of quantum-mechanical systems. Among different equivalent ways to define the time-reversal operator (Sects. 2.3–2.6), in this section it is convenient to use the time-reversal operator in the representation such that the matrix of the unitary operator **U** to be given in the spinor basis. In the case of dimer, in which two paramagnetic ions with spins S_1 and S_2 are coupled with each other by an exchange interaction, the time-reversal operator **T** in the presence of a magnetic field can be written as [61, 366]

$$\mathbf{T} = \mathbf{O}_{\mathbf{A}}\mathbf{U}_1\mathbf{U}_2\mathbf{K}, \tag{6.12}$$

where $\mathbf{U}_1$ and $\mathbf{U}_2$ are unitary operators ($\mathbf{U}_1^+\mathbf{U}_1 = \mathbf{U}_1\mathbf{U}_1^+ = \mathbf{e}^{(S_1)}$, $\mathbf{U}_2^+\mathbf{U}_2 = \mathbf{U}_2\mathbf{U}_2^+ = \mathbf{e}^{(S_2)}$), which are defined by sets of basic spin wave functions $\{|S_1, S_1\rangle, |S_1, S_1 - 1\rangle, \ldots, |S_1, 1 - S_1\rangle, |S_1, -S_1\rangle\}$ and $\{|S_2, S_2\rangle, |S_2, S_2 - 1\rangle, \ldots, |S_2, 1 - S_2\rangle, |S_2, -S_2\rangle\}$; $\mathbf{e}^{(S)}$ is the unit operator given in the spinor basis (the corresponding matrix is unit $(2S + 1) \times (2S + 1)$-matrix), $\mathbf{K}$ is the antilinear operator of complex conjugation ($\mathbf{K}\Psi = \Psi^*$, $\mathbf{K}^2\Psi = \Psi$, $\mathbf{K}^2 = 1$, $\mathbf{K}^{-1} = \mathbf{K}$), $\mathbf{A}$ is the vector potential of the magnetic field and operator $\mathbf{O_A}$ is defined by

$$\mathbf{O_A A} = -\mathbf{A}. \tag{6.13}$$

All matrix elements of the operator $\mathbf{U}$ are equal to zero, with the exception of those disposed on the secondary diagonal and equal in turn to $+1$ and -1. In addition, if the choice of spin basis functions is such that $\langle S, S\,|\,\mathbf{U}\,|\,S, -S\rangle = +1$ (it corresponds to the mentioned above choice of the spinor basis for the matrix of the operator $\mathbf{U}$), then for integer S we have $\langle S, -S\,|\,\mathbf{U}\,|\,S, S\rangle = +1$. The matrix of the operator $\mathbf{U}$ is symmetric and its square equals the identity matrix of dimension $(2S+1)\times(2S+1)$. In the case of half-integer S $\langle S, -S\,|\,\mathbf{U}\,|\,S, S\rangle = -1$ and $\mathbf{U}^2 = -\mathbf{e}^{(S)}$.

Thus, the nonzero matrix elements of the operator $\mathbf{U}$ are

$$\begin{aligned}
\langle S, S\,|\,\mathbf{U}\,|\,S, -S\rangle = \langle S, S-2\,|\,\mathbf{U}\,|\,S, 2-S\rangle = \ldots = \\
= \langle S, 3-S\,|\,\mathbf{U}\,|\,S, S-3\rangle = \langle S, 1-S\,|\,\mathbf{U}\,|\,S, S-1\rangle = 1, \\
\langle S, S-1\,|\,\mathbf{U}\,|\,S, 1-S\rangle = \langle S, S-3\,|\,\mathbf{U}\,|\,S, 3-S\rangle = \ldots = \\
= \langle S, 2-S\,|\,\mathbf{U}\,|\,S, S-2\rangle = \langle S, -S\,|\,\mathbf{U}\,|\,S, S\rangle = -1.
\end{aligned}$$

Particularly, for different values of the spin S of paramagnetic ions in the dimeric cluster ($S = 1/2,\ 1,\ 3/2,\ 2$) we obtain

$$\begin{aligned}
S = 1/2 \\
\{\langle 1/2\,|\,\mathbf{U}\,|-1/2\rangle = 1,\ \langle -1/2\,|\,\mathbf{U}\,|\,1/2\rangle = -1\} \\
S = 1 \\
\{\langle 1\,|\,\mathbf{U}\,|-1\rangle = 1,\ \langle 0\,|\,\mathbf{U}\,|\,0\rangle = -1,\ \langle -1\,|\,\mathbf{U}\,|\,1\rangle = 1\} \\
S = 3/2 \\
\left\{\begin{array}{c} \langle 3/2\,|\,\mathbf{U}\,|-3/2\rangle = 1,\ \langle 1/2\,|\,\mathbf{U}\,|-1/2\rangle = -1, \\ \langle -1/2\,|\,\mathbf{U}\,|\,1/2\rangle = 1,\ \langle -3/2\,|\,\mathbf{U}\,|\,3/2\rangle = -1 \end{array}\right\} \\
S = 2 \\
\left\{\begin{array}{c} \langle 2\,|\,\mathbf{U}\,|-2\rangle = 1,\ \langle 1\,|\,\mathbf{U}\,|-1\rangle = -1,\ \langle 0\,|\,\mathbf{U}\,|\,0\rangle = 1, \\ \langle -1\,|\,\mathbf{U}\,|\,1\rangle = -1,\ \langle -2\,|\,\mathbf{U}\,|\,2\rangle = 1 \end{array}\right\}.
\end{aligned}$$

Let us imagine that under the action of the time-reversal operator on the wave functions of paramagnetic ions of the dimer coupled due to exchange interaction between them, the reversion of signs of the spin projection operators takes place only in one of centers, for example, in the ion with spin S_1 and no changes take place to spin wave functions of the second ion (that is a fortiori not true). Such a transformation could be called a "partial" or "incomplete" time-reversal.

It is clear that under the action of the partial time-reversal operator, the spin Hamiltonian of the isotropic exchange interaction between paramagnetic ions of the magnetic dimer cluster

$$\mathbf{H} = -J\mathbf{S}_1\mathbf{S}_2 \tag{6.14}$$

does not remain invariant, but changes its sign

$$\begin{aligned}\mathbf{T}_1\mathbf{H}\mathbf{T}_1^{-1} &= -J[(\mathbf{U}_1\mathbf{K}S_{1x}\mathbf{K}\mathbf{U}_1^{+})S_{2x} + (\mathbf{U}_1\mathbf{K}S_{1y}\mathbf{K}\mathbf{U}_1^{+})S_{2y}+ \\ &\quad +(\mathbf{U}_1\mathbf{K}S_{1z}\mathbf{K}\mathbf{U}_1^{+})S_{2z}] = J\mathbf{S}_1\mathbf{S}_2 = -\mathbf{H}.\end{aligned} \tag{6.15}$$

The result is not changed if instead of $\mathbf{T}_1$ it is to use the operator $\mathbf{T}_2 = \mathbf{U}_2\mathbf{K}$:

$$\begin{aligned}\mathbf{T}_2\mathbf{H}\mathbf{T}_2^{-1} &= -J[S_{1x}(\mathbf{U}_2\mathbf{K}S_{2x}\mathbf{K}\mathbf{U}_2^{+}) + S_{1y}(\mathbf{U}_2\mathbf{K}S_{2y}\mathbf{K}\mathbf{U}_2^{+})+ \\ &\quad +S_{1z}(\mathbf{U}_2\mathbf{K}S_{2z}\mathbf{K}\mathbf{U}_2^{+})] = J\mathbf{S}_1\mathbf{S}_2 = -\mathbf{H}.\end{aligned} \tag{6.16}$$

If $\mathbf{T}_1$ would be a transformation of symmetry of the spin Hamiltonian, then the change of sign of the Hamiltonian could be described by the change of sign of the exchange parameter J. Since the change of sign of the constant of exchange interaction corresponds to a transition from the exchange interaction of antiferromagnetic type to the exchange interaction of ferromagnetic type and vice versa, this would be a method to obtain ferromagnetic coordination compounds and to liquidate the known disbalance between the number of compounds with antiferromagnetic and ferromagnetic behavior (the number of compounds with antiferromagnetic type of exchange interaction is much more than that with ferromagnetic one). However, this scenario is not possible for two reasons. First, the partial time-reversal operator $\mathbf{T}_1$ is not a transformation of symmetry of the Hamiltonian. Second, the change of sign of the exchange interaction constant J leads to the shift of spin levels of the dimer cluster in opposite directions. For example, if initial positions of spin levels were such that the spin singlet ($S = 0$) was corresponding to the ground state and the spin triplet ($S = 1$) was corresponding to the excited state, then these spin levels would change their places at the change of sign of the constant J. Inversion of spin levels takes place (not the inversion of population of spin levels, as in systems with maser-effects), which is a radical change of the energy spectrum of the dimer. On the other hand, according to the definition of the symmetry transformation, it does not change the physical system. From this, it follows once again that changing the sign of the exchange integral (or equivalent to this transformation of $\mathbf{T}_1$ or $\mathbf{T}_2$) can not be a transformation of symmetry, while the operator $\mathbf{T} = \mathbf{T}_1\mathbf{T}_2$ being a time-reversal operator is a transformation of symmetry ($\mathbf{T}\mathbf{H}\mathbf{T}^{-1} = \mathbf{H}$).

The analysis of the mental experiment described above has shown that there are only two ways for restoration of the time-reversal symmetry $\mathbf{T}_1\mathbf{T}_2\mathbf{H}\mathbf{T}_2^{-1}\mathbf{T}_1^{-1} = \mathbf{H}$: (i) to act on the Hamiltonian $\mathbf{H}$ with operator $\mathbf{T}_2$ in parallel with the operator $\mathbf{T}_1$ (or to act on the Hamiltonian $\mathbf{H}$ with operator $\mathbf{T}_1$ in parallel with the operator $\mathbf{T}_2$

too); (ii) to change the sign of the parameter of exchange interaction J. For changing the sign of the constant J, we introduce the operator $\mathbf{I}_J$:

$$\mathbf{I}_J J = -J, \quad \mathbf{I}_J^2 = 1. \tag{6.17}$$

The spin Hamiltonian $\mathcal{H}$ from (6.14) remains invariant under the simultaneous action of operators $\mathbf{I}_J$ and $\mathbf{T}_1$ on it, that it is the operator $\mathbf{I}_J$ which restores the time-reversal symmetry broken by the action only of the partial time-reversal operator $\mathbf{T}_1$ on $\mathbf{H}$ (see formula (6.15)):

$$\mathbf{T}_1\mathbf{I}_J\mathbf{H}\mathbf{I}_J\mathbf{T}_1^{-1} = -(-J)[(\mathbf{U}_1\mathbf{K}S_{1x}\mathbf{K}\mathbf{U}_1^{+})S_{2x} + (\mathbf{U}_1\mathbf{K}S_{1y}\mathbf{K}\mathbf{U}_1^{+})S_{2y} + \\ +(\mathbf{U}_1\mathbf{K}S_{1z}\mathbf{K}\mathbf{U}_1^{+})S_{2z}] = -J\mathbf{S}_1\mathbf{S}_2 = \mathbf{H}. \tag{6.18}$$

Similarly the simultaneous action of operators $\mathbf{I}_J$ and $\mathbf{T}_2$ on spin Hamiltonian $\mathbf{H}$ preserves its invariance, that is $\mathbf{I}_J$ restores the time-reversal symmetry broken by the action of the partial time-reversal operator $\mathbf{T}_2$ on $\mathbf{H}$ (see formula (6.16)):

$$\mathbf{T}_2\mathbf{I}_J\mathbf{H}\mathbf{I}_J\mathbf{T}_2^{-1} = -(-J)[S_{1x}(\mathbf{U}_2\mathbf{K}S_{2x}\mathbf{K}\mathbf{U}_2^{+}) + S_{1y}(\mathbf{U}_2\mathbf{K}S_{2y}\mathbf{K}\mathbf{U}_2^{+}) + \\ +S_{1z}(\mathbf{U}_2\mathbf{K}S_{2z}\mathbf{K}\mathbf{U}_2^{+})] = -J\mathbf{S}_1\mathbf{S}_2 = \mathbf{H}. \tag{6.19}$$

Thus, the combined operations $\mathbf{T}_1\mathbf{I}_J$ and $\mathbf{T}_2\mathbf{I}_J$ are symmetry operations, which can be called the combined time-reversal operations.

Fulfillment of (i) restores the time-reversal symmetry, but does not lead to new nontrivial physical consequences. Realization of (ii) also restores the time-reversal symmetry, i.e., the invariance of the Hamiltonian with respect to time-reversal operator $\mathbf{T} = \mathbf{T}_1\mathbf{T}_2$. Additionally, at changing the sign of exchange constant J a transition from the exchange interaction of antiferromagnetic type to the exchange interaction of ferromagnetic type takes place.

Thus, at a formal level, the transformation $J \to -J$ and the inversion of spin levels are achieved. It remains only to understand, how to realize such a transformation practically.

6.3 Spin Levels Inversion in Cu(II)–Cu(II) Dimers Caused by Combined Time-Reversal

The magnetic dimer Cu(II)–Cu(II) is characterized by that the every ion of Cu(II), which is a part of this dimer, has an electron configuration $3d^9$ and contains eight electrons with pairwise coupled spins and one electron with an uncompensated spin. This ion appears as a localized electron with the spin $S = 1/2$ in magnetic properties of coordination compounds of the single ion Cu(II), while in the copper binuclear complexes the states are of the total spins $S = 0$ and $S = 1$.

Here it is relevant to note the analogy with the energy band structure of solids, although at first sight there is no reason for such an analogy. Really, the system of two paramagnetic ions, coupled by an exchange interaction is slightly similar to a multiparticle system of N atoms, forming solids with a periodic crystalline lattice, and elementary excitations of electrons and holes with zero and nonzero quasiimpulses in the conduction and valence bands. However, both these systems have a similar behavior with respect to the time-reversal transformation. In the case of solids, the wave function of the hole (its coordinate and spin components) is determined as a result of action of the time-reversal operator on wave function of valence electron taking into account the complex structure of the valence band, if such a structure is known. By analogy with this, in the concrete case of electron configuration $3d^9$ we shall consider that there is one "hole" (which means a "missing electron with unpaired spin in the $3d$-shell"), of which spin wave function is determined as a result of action of the time-reversal operator on the spin wave function of the $3d^1$-electron. The $3d$-hole spin wave functions are

$$\begin{aligned} |1/2,\ 1/2\rangle_h &= -\mathbf{T}\ |1/2,\ -1/2\rangle_e\,, \\ |1/2,\ -1/2\rangle_h &= \mathbf{T}\ |1/2,\ 1/2\rangle_e\,. \end{aligned} \tag{6.20}$$

Thus, the transformation consisting in changing the sign of the exchange interaction constant ($J \rightarrow -J$) is equivalent to the transition from the $3d^9$-electron configuration (which is equivalent to $3d^1$-hole configuration) to the $3d^1$-electron configuration in one ion Cu(II) of exchange-coupled dimer cluster. Such a "compensating" transformation restores the broken time-reversal symmetry under the action of the Hamiltonian (6.14) on operators $\mathbf{T}_1 = \mathbf{U}_1\mathbf{K}$ and $\mathbf{T}_2 = \mathbf{U}_2\mathbf{K}$ separately. The transformation of the symmetry of the Hamiltonian $\mathbf{H}$ from (6.14) is a combined transformation, consisting in application of the operator $\mathbf{T}_1$ or $\mathbf{T}_2$ with a consequent changing the sign of the exchange interaction constant J between electrons of different ions of the dimer. The change of the sign of the constant J is equivalent to the transition from the electron configuration $3d^9$ to the electron configuration $3d^1$ of an ion with spin S_2 in the case the operator $\mathbf{T}_1$ is used. In the case of application of the operator $\mathbf{T}_2$ the transition from the electron configuration $3d^9$ to the electron configuration $3d^1$ is necessary to be done in the ion with spin S_1. In other words, the combined transformation of the symmetry of Hamiltonian $\mathcal{H}$ from (6.14) is determined as "the partial time-reversal operator $\mathbf{T}_1$ (or $\mathbf{T}_2$) + isomorphic replacement of the one of two ions Cu(II) with the electron configuration $3d^9$ by the ion with electron configuration $3d^1$". In this case, the Hamiltonian is invariant with respect to the combined transformation "partial time-reversal + isomorphic substitution", however at such a transformation the isomorphic replacement of one ion by another leads to essential shifts (in opposite directions) of spin levels with $S = 0$ and $S = 1$ of the two-nuclear cluster. Apparently, this is the first case when a generalized symmetry transformation, under the action of which the Hamiltonian remains invariant, describes also significant changes in the structure of the energy spectrum of the quantum-mechanical system.

Let us consider the result of action of the operator $\mathbf{T}_1$ on the spin Hamiltonian (6.14) for the Cu(II)–Cu(II) dimer, in which the ion Cu(II) with electron configuration $3d^9$ and spin $S_2 = 1/2$ is replaced by an ion with the electron configuration $3d^1$. Hypothetically, such an ion may be Cu(X). Although positive copper ions of high charge exist in an eximer laser-produced high density copper plasma [367], such ions are not long-lived and they certainly cannot form any coordination compounds.

The proposed method of transformation of dimeric homonuclear complexes with antiferromagnetic interaction into dimeric heteronuclear complexes with ferromagnetic interaction has an experimental confirmation (see Sect. 6.5).

6.4 Changing the Position of Spin Levels in $3d - 3d$ and $4f - 4f$ Dimer Clusters Caused by Combined Time Reversal

Among the binuclear coordination compounds with ferro- and antiferromagnetic properties, the latter are more widely spread (the number of known ferromagnetic materials is much less that of the antiferromagnetic ones). This non-equivalence of the signs of exchange interaction integrals is not caused by the existence of a supplementary interdiction principle, because the parallel as well as the antiparallel orientations of localized spins are equally possible. Therefore, the difference between the binuclear coordination compounds with ferromagnetic and antiferromagnetic properties may be considered as casual. But it is necessary to note that as a consequence of the Pauli principle the antiparallel orientation of spins is preferable.

In the Sects. 6.2 and 6.3 we have shown that the antiferromagnetic exchange interactions in binuclear compounds can be transformed into ferromagnetic ones and vice versa [366, 368], using the generalized symmetry transformation "the partial time-reversal + isomorphic substitution". Now we will show how to apply this method for transforming the $3d - 3d$ and $4f - 4f$ binuclear compounds, in which the components of dimers are coupled by exchange forces of antiferromagnetic type, into dimers with exchange interactions of ferromagnetic type. In order to change the exchange interaction type, it is necessary make a transition under time-reversal operator from the electron wave functions to the hole states of one ion in the dimer cluster. Thus, for magnetic dimer clusters containing equivalent ions, the partial time-reversal transformation signifies an isomorphic substitution of one of two ions by a "time-reversal" ion. The electron wave functions of such "time-reversed" ion are obtained from the wave functions on the non-time-reversed ion under action of the time-reversal operator. Time-reversed electrons correspond to the holes in the electron shell and vice versa. Additionally, as a result of transforming electrons into holes (or holes into electrons, respectively) a change of the exchange interaction type takes place. As a rule, the exchange interaction of antiferromagnetic type is characteristic for a magnetic dimer containing equivalent ions, regardless of electron and hole structure of dimers. After the partial time-reversal, the distance between "inverted" spin levels changes for heteronuclear cluster as compared with the distance between the spin levels of the initial bihomonuclear cluster.

In the case of $3d$ and $4f$ ions separately forming a dimer cluster, this transformation can be performed by an isomorphic substitution of one of the ions of exchange-coupled dimer, in the following way: one ion with $3d^n$ configuration in the $3d - 3d$ dimer is replaced by an ion with a $3d^{N-n}$ configuration, or one ion with $4f^m$ configuration in the $4f - 4f$ dimer is replaced by an ion with a $4f^{M-m}$ configuration. In these designations, N and M are the numbers of electrons in filled $3d$ and $4f$ shells ($N = 10, M = 14$), respectively. Thus, using the concept of "partial time-reversal + isomorphic substitution" [366, 368], we can perform a controlled synthesis of coordination compounds containing magnetic dimer clusters composed of $3d$ and $4f$ ions with exchange interactions of ferromagnetic or antiferromagnetic types.

The suggested statement stipulates a weak interconfiguration interaction, a weak external perturbations of electron (hole) states of electron shells and possibilities for separating electron and hole states. These conditions are met for $3d$ and $4f$ ions, in spite of different schemes of addition of angular momenta of electrons in these ions. On the basis of this statement, the following classification of exchange interactions of $3d - 3d$ and $4f - 4f$ clusters may be suggested (see Tables 6.16 and 6.17).

Thus, on the basis of the generalized symmetry transformation "partial time-reversal + isomorphic substitution", it was carried out the classification of homo- and heteronuclear $3d - 3d$ dimers, formed of paramagnetic ions of iron group, as well as that of homo- and heteronuclear $4f - 4f$ dimers, formed of paramagnetic ions of rare-earth elements. There were selected in separate classes coordination compounds, containing binuclear clusters with the exchange interaction of antiferro- and ferromagnetic types between electrons of different ions of the dimers. On the basis of this method, every coordination compound containing homobinuclear clusters of ions $3d$ or $4f$ with antiferromagnetic exchange interaction, may be transformed into a compound containing two-component heteronuclear clusters with exchange interaction of ferromagnetic type and vice versa. In this way, it may be eliminated (or at least diminished) the difference between the numbers of coordination compounds, for which the exchange interaction of antiferromagnetic type is realized in magnetic dimers as compared with the number of coordination compounds, for which electrons of different ions of the dimers are coupled with each other by forces of exchange interaction of ferromagnetic type. Moreover, the synthesis of heteronuclear coordinate compounds with ferromagnetic properties are of an independent scientific and practical interest due to the large demand of ferromagnetic materials in micro-, nano- and optoelectronics.

Table 6.16 The homo- and heteronuclear $3d - 3d$ clusters with exchange interactions of antiferro- and ferromagnetic types

Binuclear $3d - 3d$ clusters with exchange interaction of antiferromagnetic type		Binuclear $3d - 3d$ clusters with exchange interaction of ferromagnetic type
Ti(III)–Ti(III)	Cr(III)–V(II)	Ti(III)–Cu(II)
V(IV)–V(IV)	Mn(III)–Mn(III)	V(IV)–Cu(II)
VO(II)–VO(II)	Cr(II)–Cr(II)	VO(II)–Cu(II)
Ti(II)–V(IV)	Mn(III)–Cr(II)	V(III)–Ni(II)
Ti(III)–VO(II)	Fe(II)–Fe(II)	Cr(III)–Co(II)
VO(II)–V(IV)	Co(II)–Co(II)	V(II)–Co(II)
V(III)–V(III)	Ni(II)–Ni(II)	Mn(III)–Fe(II)
Cr(III)–Cr(III)	Cu(II)–Cu(II)	Cr(II)–Fe(II)
V(II)–V(II)	Fe(III)–Fe(III)[a]	Fe(III)–Fe(III)[a]
	Mn(II)–Mn(II)[a]	Mn(II)–Mn(II)[a]
	Fe(III)–Mn(II)[a]	Fe(III)–Mn(II)[a]

[a]The binuclear clusters which are characterized in different compounds by exchange interaction of antiferromagnetic as well as ferromagnetic type

Table 6.17 The homo- and heteronuclear $4f - 4f$ clusters with exchange interaction of antiferro- and ferromagnetic types

Binuclear $4f - 4f$ clusters with exchange interaction of antiferromagnetic type		Binuclear $4f - 4f$ clusters with exchange interaction of ferromagnetic type
Ce(III)–Ce(III)	Tb(III)–Tb(III)	Ce(III)–Yb(III)
Pr(III)–Pr(III)	Dy(III)–Dy(III)	Pr(III)–Tu(III)
Nd(III)–Nd(III)	Ho(III)–Ho(III)	Nd(III)–Er(III)
Pm(III)–Pm(III)	Er(III)–Er(III)	Pm(III)–Ho(III)
Sm(III)–Sm(III)	Tu(III)–Tu(III)	Sm(III)–Dy(III)
Eu(III)–Eu(III)	Yb(III)–Yb(III)	Eu(III)–Tb(III)
	Gd(III)–Gd(III)[a]	Gd(III)–Gd(III)[a]

[a]The binuclear cluster Gd(III)–Gd(III), which is characterized in different compounds by an exchange interaction of antiferromagnetic type, as well as by a ferromagnetic one

6.5 Experimental Evidence of Spin Levels Inversion in Dimer Magnetic Clusters Caused by Combined Time Reversal

The combined symmetry transformation "partial time-reversal + isomorph substitution" contains the operation of isomorphic substitution of atoms (ions), consisting in replacing one ion of the dimer cluster by an ion of the isomorphic atom. Practically, this can be performed very simply, if in the initial homobinuclear compound a half from the total number of ions it substituted by ions of which the electron wave functions coincide with the time reversed electron wave functions of initial ions (the ratio of ions of different types must be 1:1). It is clear that at this substitution of atoms a modification of energetic spectrum of the system will occur. In the case of magnetic homobinuclear clusters an inversion of spin levels will occur. However, if after the isomorphic substitution of atoms the cluster becomes heteronuclear, there will be a not exact inversion of spin levels, although the order of their succession will be inversed. In other words, in this case the transformation of the exchange interaction of antiferromagnetic type into a ferromagnetic one will also occur, but without the conservation of the exchange interaction constant J, which will be less by modulus.

The combined transformation "partial time-reversal + isomorph substitution" discussed in this chapter is of some heuristic interest. However, the generalized symmetry transformation of this type, accompanied by a change of the structure of the energy spectrum of the system, can be taken into account in a serious way only in the case, when there exist reliable data, proving that the modification of the energy spectrum really takes place. In this section, we will show that such data really exist.

The results of research of magnetic properties of complex $CuVO(fsa)_2$ *en* CH_3OH [369, 370] are an experimental proof of the transformation of the $3d^9 - 3d^9$ dimer with antiferromagnetic exchange interaction into a $3d^9 - 3d^1$dimer with a ferromagnetic behavior. As a consequence of the invariance of the spin Hamiltonian (6.14) relative to the combined transformation "partial time-reversal + isomorphic substitution" between Cu(II) and VO(II) in this compound, containing Cu(II)–VO(II) binuclear clusters, it must exist an exchange interaction of the ferromagnetic type. In this particular case one can come to the same conclusion starting from quantum-mechanical reasons, according to which the exchange interaction between ions Cu(II) and VO(II) bears purely a ferromagnetic character due to the orthogonality of orbitals Ψ_{Cu} and Ψ_{VO}.

According to [369], the heterobinuclear complex $CuVO(fsa)_2$ *en* CH_3OH is crystallized in monocline system with the space group of symmetry $P2_1/n$, lattice parameters $a = 11.636(3)$ Å, $b = 13.612(3)$ Å, $c = 12.426(3)$ Å, monocline angle $\beta = 100.8(4)°$ and the coordinating number $z = 4$. The structure consists of heterobinuclear units, in which copper atom is five coordinated with two nitrogen atoms, two phenol oxygens and one oxygen molecule of metanol; the atom of vanadium is also five coordinated by two phenol, two carboxyl oxygens and an oxygen of vanadil group. Atoms of metals and oxygen molecules of metanol and vanadil group are in a mirror σ-plane for two square pyramids CuN_2O_3 and VO_5 [241, 369].

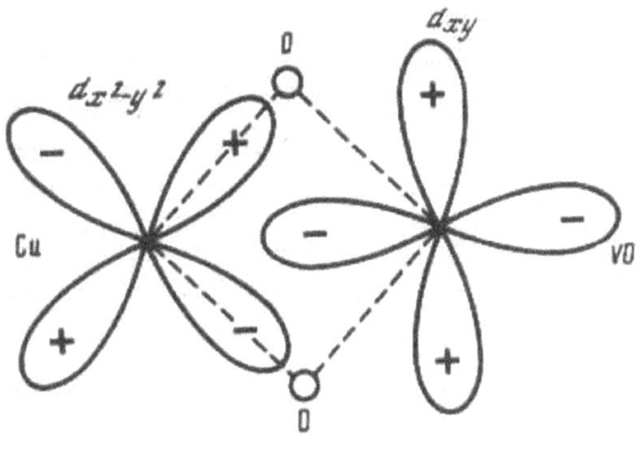

Fig. 6.3 Magnetic orbitals of the dimer Cu(II)–VO(II) [369]

The ion Cu(II) is characterized by a magnetic orbital $d_{x^2-y^2}$, while the ion VO(II) is characterized by the orbital d_{xy} (Fig. 6.3). The exchange interaction between ions of the dimer Cu(II)–VO(II) with spins $S_1 = S_2 = 1/2$ leads to the total spin $S = 0, 1$ ($D_{1/2} \otimes D_{1/2} = D_0 \oplus D_1$, D_S denotes, as earlier, the irreducible representation of the rotation group of symmetry).

On the basis of data on temperature dependence of the static magnetic susceptibility $\chi(T)$ ($T = 18$–300 K), there was found $J = 118\,\mathrm{cm}^{-1}$, $g = 2.017$ [369, 370]. The ground state of the dimer cluster Cu(II)–VO(II) is a spin triplet ($D_1 \in D_{1/2} \otimes D_{1/2}$).

The mechanism of ferromagnetic exchange in the Cu(II)–VO(II) heterobinuclear cluster can be understood on the basis of the approach developed by Kahn and Briat [270]. Since orbitals of copper and vanadil are orthogonal, i.e., $\langle \Psi_{\mathrm{Cu}} | \Psi_{\mathrm{VO}} \rangle = 0$, the integral of exchange interaction is determined by the formula

$$J = \left\langle \Psi_{\mathrm{Cu}}(1)\Psi_{\mathrm{VO}}(2) | r_{12}^{-1} | \Psi_{\mathrm{Cu}}(2)\Psi_{\mathrm{VO}}(1) \right\rangle . \tag{6.21}$$

By examining the structure of the magnetic fragment, one can conclude that orbitals $2p$ give covering σ with $d_{x^2-y^2}$- copper orbital and covering π with d_{xy}-vanadil orbital. The density of covering the orbitals $\rho = \Psi_{\mathrm{Cu}}\Psi_{\mathrm{VO}}$ has two positive extrema on one bridged oxygen along the line Cu-O and two negative extrema, on other oxygen [241]. Since the depth of negative extrema compensates the height of positive extrema, the integral of covering is equal to zero ($\int \rho(1) d\tau(1) = 0$). Just at the same time the two-electron exchange integral $\iint \left[\rho(1)\rho(2)/r_{1/2}\right] d\tau(1) d\tau(2)$, which determines the ferromagnetic contribution, is not zero and is essentially connected with extrema of the covering density ρ.

The singlet–triplet splitting is equal to $118\,\mathrm{cm}^{-1}$ for the dimer Cu(II)–VO(II) with exchange interaction of ferromagnetic type and turns out to be essentially less as compared with similar splitting, equal to $650\,\mathrm{cm}^{-1}$ (see below) for the dimer Cu(II)–Cu(II) with exchange interaction of antiferromagnetic type. This is due to the fact that ions Cu(II) and VO(II) are not equivalent.

The compound $Cu_2(fsa)_2\, en\, CH_3OH$ contains Cu(II)–Cu(II) $(3d^9 - 3d^9)$ antiferromagnetic dimer clusters with $J = -650\,\text{cm}^{-1}$ [371]. The ground state of $3d^9 - 3d^9$ dimer is a spin singlet ($D_0 \in D_{1/2} \otimes D_{1/2}$).

The compound $[Fe_2\,(dhpta)\,(OOCR)\,(H_2O)_2]$, where H_5dhpta = 1, 3 -diamino-2-hydroxypropane- N, N, N', N'-tetraacetic acid, contains Fe(III)–Fe(III) $(3d^5 - 3d^5)$ antiferromagnetic dimer clusters with $J = -14.4, -15.9, -15.1, -15.4$ and $-16.3\,\text{cm}^{-1}$ for $R = C_6H_4OH$, $CH_2CH = CH_2$, $C_6H_4NH_2$, CH_2CH_2Br and $CH = CHCH_3$, respectively [372]. The Hamiltonian $\mathcal{H} = -2J\mathbf{S}_1\mathbf{S}_2$ with $S_1 = S_2 = 5/2$ was used. The ground state of the Fe(III)–Fe(III) cluster is the spin singlet ($D_0 \in D_{5/2} \otimes D_{5/2}$).

The compounds $[Fe_2O\,(TPA)_2\,X]\,(ClO_4)_2 \cdot 2H_2O$, where TPA = tripyridinamine, contain Fe(III)–Fe(III) $(3d^5 - 3d^5)$ antiferromagnetic dimer clusters with $J = -106.6, -102.8, -102.8, -100.8\,\text{cm}^{-1}$ for $X = SO_4$, HPO_4, $HAsO_4$ and MoO_4, respectively [373].

The compounds $Kat_2\,[Fe_2Cl_6O]$ contain Fe(III)–Fe(III) $(3d^5 - 3d^5)$ antiferromagnetic dimer clusters with $J = -116, -117, -112, -117$ and $-127\,\text{cm}^{-1}$ for Kat = $[BzEt_3N]^+$, $[BzPh_3P]^+$, $[Ph_4P]^+$, $[Ph_4As]^+$ and $[HPy]^+$, respectively [374].

The compounds $[Cu_2\,(L_a)_2\,(DMSO)_2] \cdot 2DMSO$ and $[Cu_2\,(L_b)_2\,(DMSO)_2] \cdot 2DMSO$, where $H_2L_{a(b)}$ = N-salicylidene-2-hydroxy-5-chloro (or bromo) benzilamine and DMSO = dimetylsulfoxide, contain Cu(II)–Cu(II) $(3d^9 - 3d^9)$ antiferromagnetic dimer clusters with $J = -223\,\text{cm}^{-1}$ for L_a and $J = -226\,\text{cm}^{-1}$ for L_b [375]. The Hamiltonian $\mathcal{H} = -2J\mathbf{S}_1\mathbf{S}_2$ with $S_1 = S_2 = 1/2$ was used.

The compounds $[Fe_2\,(L_a)_2\,(CH_3COO)_2] \cdot 2THF$ and $[Fe_2\,(L_b)_2\,(CH_3COO)_2] \cdot 2THF$ (for the notation of L_a and L_b see the paragraph on this section; THF is tetrahydrofuran) contain Fe(III)–Fe(III) $(3d^5 - 3d^5)$ ferromagnetic dimer clusters with $J = 2\,\text{cm}^{-1}$ for L_a and $J = 1.6\,\text{cm}^{-1}$ for L_b [375]. The Hamiltonian $\mathcal{H} = -2J\mathbf{S}_1\mathbf{S}_2$ with $S_1 = S_2 = 5/2$ was used. The ground state is 11-fold degenerate ($S = 5$, $D_5 \in D_{5/2} \otimes D_{5/2}$). In the ligand crystalline field the removal of degeneracy takes place. For $3d^5 - 3d^5$ ferromagnetic clusters, the constant of isotropic exchange interaction is significantly smaller than that for the $3d^5 - 3d^5$ antiferromagnetic ones.

As to homobinuclear $3d - 3d$ and $4f - 4f$ magnetic clusters, the *ESR* and static magnetic susceptibility data show that, as a rule, the exchange interaction of the antiferromagnetic type is realized in these clusters especially at relatively large values of the constant J of the isotropic exchange interaction. An exception is the dimer $4f^7 - 4f^7$, for which this interaction may be of antiferro -, as well as of ferromagnetic type, depending on the type of atoms (or group of atoms) which provide an indirect exchange interaction between the paramagnetic ions of the cluster.

Thus the data from Tables 6.1, 6.2, 6.3, 6.4, 6.6, 6.9, 6.11, 6.13, 6.14 show that homobinuclear clusters, excepting the clusters containing ions with half-filled shells, show an exchange interaction of antiferromagnetic type between ions.

The data from Tables 6.11, 6.12 and 6.13 show that the type of exchange interaction in homobinuclear clusters containing ions with half-filled shells depend on the type of bridging. Particularly, there is an exchange interaction of antiferromagnetic type for Gd(III)–Gd(III) $\left(4f^7 - 4f^7\right)$ dimers with A and C bridging and an exchange interaction of ferromagnetic type with B bridging (see Table 6.13).

The method of controlled transformation of homobinuclear compounds with exchange interaction of antiferromagnetic type into heterobinuclear compounds with ferromagnetic behavior using the combined transformation "partial time-reversal + isomorphic substitution" is not limited only to dimer magnetic clusters and can be extended to multinuclear magnetic clusters. This is possible, because in this case the main contribution into exchange interactions is due to the isotropic exchange constants, which are included linearly in the spin Hamiltonian.

Let us consider in detail how the inversion of the spin levels in binuclear coordination compounds occurs, for example, in dimers $3d^9 - 3d^9$ and $3d^1 - 3d^1$. To this end, we perform the diagonalization of the exchange interaction Hamiltonian $\mathbf{H}$ from (6.14) in three cases. In the first case, the matrix of the operator $\mathbf{H}$ is given in the standard spinor basis $\{(\frac{1}{\sqrt{2}})(\alpha_1\beta_2 - \beta_1\alpha_2), \alpha_1\alpha_2, (\frac{1}{\sqrt{2}})(\alpha_1\beta_2 + \beta_1\alpha_2), \beta_1\beta_2\}$ where $\alpha_i(i = 1, 2)$ and $\beta_i(i = 1, 2)$ are the basis spinors belonging to the first and second dimer ions, respectively. The diagonalization of the Hamiltonian (6.14) in this basis leads to a well-known result for singlet, $E^{(s)}$, and triplet, $E^{(t)}$ spin energy levels

$$E^{(s)} = \frac{3}{4}J, \; E^{(t)} = -\tfrac{1}{4}J \tag{6.22}$$

In the second case, the diagonalization of the Hamiltonian $\mathbf{H} = \mathbf{T}_1\mathbf{I}_J\mathbf{H}\mathbf{I}_J\mathbf{T}_1^+$ or $\mathbf{H} = \mathbf{T}_2\mathbf{I}_J\mathbf{H}\mathbf{I}_J\mathbf{T}_2^+$ is carried out in the same basis of spin wave functions. In the third case, the diagonalization of the same Hamiltonian, bordered by operators $\mathbf{T}_1\mathbf{I}_J$ and $\mathbf{I}_J\mathbf{T}_1$ or $\mathbf{T}_2\mathbf{I}_J$ and $\mathbf{I}_J\mathbf{T}_2$, is carried out in the basis of the spin wave functions $\{(\frac{1}{\sqrt{2}})(\alpha_1\beta_{2h} - \beta_1\alpha_{2h}), \alpha_1\alpha_{2h}, (\frac{1}{\sqrt{2}})(\alpha_1\beta_{2h} + \beta_1\alpha_{2h}), \beta_1\beta_{2h}\}$. In this case, the Hamiltonian (6.14) is represented in the form

$$\mathbf{H} = -J\mathbf{S}_1\mathbf{S}_{2h}, \tag{6.23}$$

where $\mathbf{S}_{2h}$ is the spin operator of the hole in the electron $3d$-shell. The values of the energies $E^{(s)}$ and $E^{(t)}$, calculated by the last two methods of diagonalization, are also determined by equations (6.22). The positions of these levels on the energy diagram are determined by the sign of the exchange interaction constant.

Figure 6.4 shows the lowest two energy levels of the dimer $3d^9 - 3d^9$, for which $J < 0$, and the dimer $3d^9 - 3d^1$, for which $J > 0$. For the Hamiltonian (6.14), the first case corresponds to an exchange interaction of the antiferromagnetic type, and the second case corresponds to an exchange interaction of the ferromagnetic type.

As can be seen from Fig. 6.4, the spectrum (b) is the inverted spectrum (a) (replacement of the singlet level by the triplet one and vice versa) and shifted toward high energies by $\frac{1}{2}J$. When the energy spectrum is inverted, the singlet–triplet splitting does not change if it is assumed that the ions with electronic configurations $3d^9$ and $3d^1$ are equivalent. In the case of nonequivalent ions, the singlet–triplet splitting

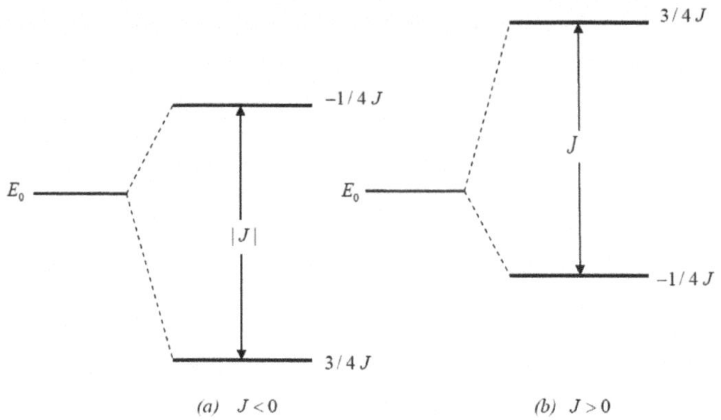

Fig. 6.4 Splitting of a spin quadruplet when the crystal field is neglected in magnetic dimers $3d^9 - 3d^9$ (or $3d^1 - 3d^1$) at $J < 0$ **(a)** and $3d^9 - 3d^1$ (or $3d^1 - 3d^9$) at $J > 0$ **(b)**

before and after the inverting of the spin levels have slightly different values, which is confirmed by experimental data (see Sect. 6.5).

Theorem: *The inversion of the sign of the exchange interaction constant in magnetic binuclear clusters is equivalent to the isomorphic replacement of one of the cluster ions by an ion whose wave functions are time reversed with respect to the wave functions of the unsubstituted ion.*

We shall prove this theorem for homonuclear clusters with an exchange interaction of an antiferromagnetic type containing ions with one electron or one hole in the electron shell, which after isomorphic substitution become heteronuclear clusters with an exchange interaction of the ferromagnetic type. The proof is based on the calculation and comparison among themselves of the following matrix elements:

$$\begin{gathered}
\tfrac{1}{2} < \alpha_1\beta_2 - \beta_1\alpha_2 \mid \mathbf{T}_1\mathbf{I}_J\mathbf{H}\mathbf{I}_J\mathbf{T}_1^+ \mid \alpha_1\beta_2 - \beta_1\alpha_2 >= \tfrac{3}{4}J \\
\tfrac{1}{2} < \alpha_1\beta_2 - \beta_1\alpha_2 \mid \mathbf{T}_2\mathbf{I}_J\mathbf{H}\mathbf{I}_J\mathbf{T}_2^+ \mid \alpha_1\beta_2 - \beta_1\alpha_2 >= \tfrac{3}{4}J \\
-\tfrac{1}{2} < \alpha_1\beta_{2h} - \beta_1\alpha_{2h} \mid J\mathbf{S}_1\mathbf{S}_{2h} \mid \alpha_1\beta_{2h} - \beta_1\alpha_{2h} >= \tfrac{3}{4}J \\
< \alpha_1\alpha_2 \mid \mathbf{T}_1\mathbf{I}_J\mathbf{H}\mathbf{I}_J\mathbf{T}_1^+ \mid \alpha_1\alpha_2 >= -\tfrac{1}{4}J \\
< \alpha_1\alpha_2 \mid \mathbf{T}_2\mathbf{I}_J\mathbf{H}\mathbf{I}_J\mathbf{T}_2^+ \mid \alpha_1\alpha_2 >= -\tfrac{1}{4}J \\
-\tfrac{1}{2} < \alpha_1\beta_{2h} + \beta_1\alpha_{2h} \mid J\mathbf{S}_1\mathbf{S}_{2h} \mid \alpha_1\beta_{2h} + \beta_1\alpha_{2h} >= -\tfrac{1}{4}J \\
< \beta_1\beta_2 \mid \mathbf{T}_1\mathbf{I}_J\mathbf{H}\mathbf{I}_J\mathbf{T}_1^+ \mid \beta_1\beta_2 >= -\tfrac{1}{4}J \\
< \beta_1\beta_2 \mid \mathbf{T}_2\mathbf{I}_J\mathbf{H}\mathbf{I}_J\mathbf{T}_2^+ \mid \beta_1\beta_2 >= -\tfrac{1}{4}J \\
- < \beta_1\beta_{2h} \mid J\mathbf{S}_1\mathbf{S}_{2h} \mid \beta_1\beta_{2h} >= -\tfrac{1}{4}J
\end{gathered} \tag{6.24}$$

As can be seen from (6.24), the operation of combined time reversal ($\mathbf{T}_1\mathbf{I}_J$ or $\mathbf{T}_2\mathbf{I}_J$), containing the operator $\mathbf{I}_J$ of changing the sign of the exchange interaction constant (see (6.17)), is equivalent to the operation of isomorphic substitution of one of the dimer ions by a time-reverse ion. With this substitution, the wave function of the substituting ion is a time-reversed wave function of the substituted ion. In other words, after such isomorphous substitution, the dimer $3d^9 - 3d^9$ is transformed into the dimer $3d^9 - 3d^1$. There are also other alternatives: $3d^9 - 3d^9$ is transformed into $3d^1 - 3d^9$ and $3d^1 - 3d^1$ is transformed to $3d^1 - 3d^9$ or $3d^9 - 3d^1$. The values of matrix elements (6.24) prove that the formal transformation $J \longrightarrow -J$ can be carried out practically by means of this isomorphic substitution of ions in magnetic dimers.

Chapter 7
Is There an Analogy Between Jahn-Teller Effect and an Instability of Spin Populations in Kramers Clusters with Odd Number of Atoms?

The molecules and molecular clusters of high symmetry with orbital degeneracy of the ground state are unstable with respect to a weak structural deformation due to Jahn-Teller effect. In the Sect. 7.1 is discussed the possibility for existence of an analogical effect of instability of spin populations of the trihomonuclear magnetic cluster with respect to a weak structural deformation (the Kahn's instability) in the case when each of three ions of the cluster has the spin $S = 1/2$.

It is shown that the Kahn's instability of spin populations for a triangle of isotropic antiferromagnetically coupled half-integer spins with respect to a weak structural deformation is forbidden. The spin moments associated with spin states of two Kramers doublets for trihomonuclear cluster containing three equivalent ions with half-integer spins situated in the corners of an equilateral triangle ABC have been calculated in Sects. 7.2 and 7.3 for the cases of spins $S_A = S_B = S_C = 1/2$ and $S_A = S_B = S_C = 5/2$, respectively. The full mutual compensation of spin moments corresponding to both components of each Kramers doublet has been found. Such compensation of spin moments is characteristic for all trimer clusters with half-integer values of the spin (Sect. 6.4). It is due to the time-reversal symmetry properties of spin wave functions corresponding to two Kramers doublets of the ground state. For clusters formed by an odd number of atoms with half-integer spins, the instability of spin populations with respect to a weak structural deformation is possible only at breaking the time-reversal symmetry accompanied by its lowering from the complete time-reversal symmetry to the incomplete one.

The incomplete time-reversal operator is an operator under action of which not all three projections of the spin operator **S** are reversed, but only one or only two of them. The conception of incomplete time-reversal differs from the partial time-reversal one used in the Chap. 6 for the case when the signs of all three projections of the spin operator **S** change under the action of the Wigner time-reversal operator. At incomplete time-reversal it take place for only one from two paramagnetic ions of the dimer cluster.

I. I. Geru, *Time-Reversal Symmetry*, Springer Tracts in Modern Physics 281,
https://doi.org/10.1007/978-3-030-01210-6_7

7.1 Kahn's Instability of an Equilateral Spin Trimer $1/2 \otimes 1/2 \otimes 1/2$ Due to a Weak Perturbation

Starting from the concept of spin frustration, introduced by Toulouse relative to the phenomenon of spin glass behavior [376], the term "spin frustration" has been used not only for spin glass systems [377–381], but also in the physics of molecular magnetism to describe polymetallic species, in which a competing interaction leads to a criterion of complexity in the spectrum of low-lying states [382]. It is considered that competing interactions create an orbital degeneracy (or quasi-degeneracy) of the ground state, which can lead to different kinds of instabilities. In the case of discrete molecular species, for which competing spin interactions lead to an orbitally degenerated ground state, with at least one magnetic component, such a situation has been defined as a degenerated frustration [383]. The relation between the spin frustration and the orbital degeneracy (or quasi-degeneracy) is given particularly in the book of Mydosh devoted to spin glasses [381].

The exchange interaction spin Hamiltonian for an ABC triangle (Fig. 7.1a) of isotropic $1/2$ spins antiferromagnetically (AF) coupled can be written as

$$\mathbf{H} = -J\left(\mathbf{S}_A\mathbf{S}_B + \mathbf{S}_B\mathbf{S}_C + \mathbf{S}_C\mathbf{S}_A\right), \tag{7.1}$$

where J is the constant of isotropic exchange interactions.

By introducing the notations

$$\begin{aligned} \mathbf{S}' &= \mathbf{S}_B + \mathbf{S}_C, \\ \mathbf{S} &= \mathbf{S}' + \mathbf{S}_A, \end{aligned} \tag{7.2}$$

the spin states of the system can be written as $\left|S, S'\right\rangle$, where S is one of the values of the cluster total spin ($S_1 = S_2 = 1/2,\ S_3 = 3/2$).

If S_A is half integer, the two spin states $|\,1/2, S_A - 1/2\rangle$ and $|\,1/2, S_A + 1/2\rangle$ have the same lowest energy, which corresponds to an orbitally degenerated ^2E ground state, transformed by irreducible representation of the C_3 symmetry group [383]. According to [383], in the simple case of $S_A = 1/2$ the two components of the degenerated ground state can be defined as $|\,1/2, 0\rangle$ and $|\,1/2, 1\rangle$. There are two Kramers doublets that follows from simple group-theoretic relations

$$D^{(1/2)} \otimes D^{(1/2)} \otimes D^{(1/2)} = (D^{(0)} \oplus D^{(1)}) \otimes D^{(1/2)} = 2D^{(1/2)} \oplus D^{(3/2)}, \tag{7.3}$$

where $D^{(S)}$ is the irreducible representation of the rotations group.

The corresponding spin populations in the corners of the triangle (in $N\mu_B$ units) were obtained as eigenvalues of the gS_Z operator [383], assuming that the Zeeman factor g is equal to 2.00 and ignoring spin delocalization and polarization effects [384]. These eigenvalues for the ABC triangle were found to be equal to 1, 0, 0 and $-1/3$, $2/3$, $2/3$ for the spin states $|\,1/2, 0\rangle$ and $|\,1/2, 1\rangle$ respectively (Fig. 7.1b, c).

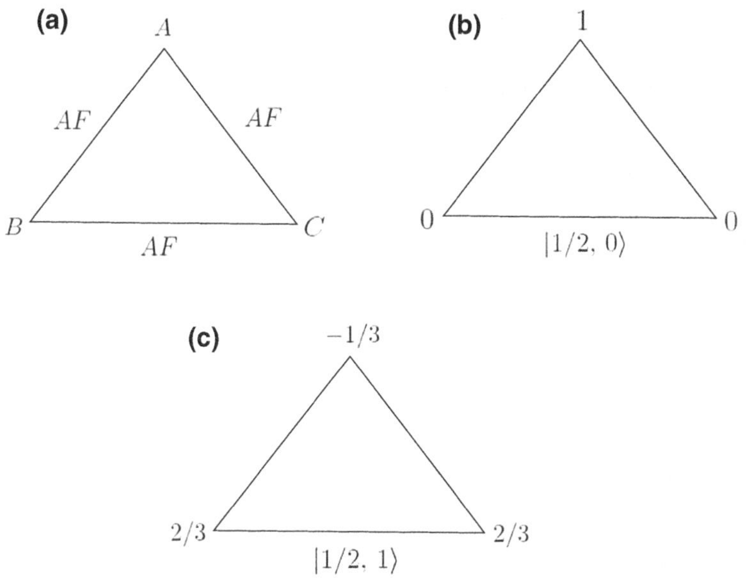

Fig. 7.1 The AF type of exchange interactions in a triangle of isotropic 1/2 spins AF coupled and corresponding spin populations at the corners of the triangle for one component of the ground state Kramers doublets $|1/2, 0\rangle$ and $|1/2, 1\rangle$ (after [383]). **a** the ABC triangle with AF exchange interactions between atoms; **b** the spin populations in the corners of the triangle (in $N\mu_B$ units, N is the number of trihomonuclear clusters and μ_B is the Bohr magneton) for one of the components of the ground state Kramers doublet $|1/2, 0\rangle$; **c** the spin populations in the corners of the triangle (in $N\mu_B$ units) for one of the components of the ground state Kramers doublet $|1/2, 1\rangle$

Here are taken into account only pure spins without orbital contributions, leading to the splitting of the ^2E term in the C_3 symmetry due to spin–orbit coupling [385].

Data presented in Fig. 7.1b, c are not complete, since when these were obtained in [383] it was taken into account only one from the components of each from two Kramers doublets of the ground state. The author of [383] has calculated the variation of spin populations for a triangle of isotropic 1/2 spins AF coupled versus a weak deformation with respect to equilateral topology. In numeric calculations there were used the experimental data concerning the trihomonuclear cation $[Cu_3(pao)_3OH]^{3+}$ [9, 386] with Hpao = pyridine-2-aldehyde-oxime with a perfect threefold symmetry at room temperature. On the basis of these calculations, a conclusion was drawn about the existence of a new effect consisting in the instability of spin populations at the corners A, B and C of the triangle of isotropic 1/2 spins with respect to a weak perturbation transforming the equilateral triangle into an isosceles one. It is supposed that changes of the spin population due to its instability may be experimentally deduced from the polarized neutron diffraction data [387] or magnetic resonance spectroscopy experiments [388].

However, the conclusion about the instability of spin populations is wrong, because in [383] there were taken into account only contributions in spin population due to one component for every of two Kramers doublets of the ground state, but there were not taken into account contributions from the second ("time-reversed") component of Kramers doublets.

In Sects. 7.2 and 7.3 it will be shown that these contributions mutually compensate each other [389, 390] and the indicated spin populations are really stable with respect to weak irreversible replacements of atoms from equilibrium positions.

For trihomonuclear clusters having a Kramers degeneracy of the ground state, the instability of spin populations with respect to a weak deformation takes place only at breaking the time-reversal symmetry (Sect. 7.4).

7.2 Mutual Compensation of Distorted-Induced Spin Polarization in a Trimer $1/2 \otimes 1/2 \otimes 1/2$ Due to Time-Reversal Symmetry

The specificity of the degeneracy of Kramers multiplets, which are splitted in the crystalline field of low symmetry into a set of Kramers doublets, is that the spin wave function φ associated with one component of the doublet is coupled with the spin wave function of the second component $\mathbf{T}\varphi$ by time-reversal operator $\mathbf{T}$ [222]. Therefore, at quantum-mechanical calculations with allowance for Kramers states it is necessary to be careful and confident that there is an non-equivocal correspondence between φ and $\mathbf{T}\varphi$, independently of the origin of Kramers doublets in concrete spin systems.

The Wigner time-reversal operator for a system consisting of N particles with spin $1/2$ is [222]

$$\mathbf{T} = i^N \sigma_{y1}\sigma_{y2}\ldots\sigma_{yN}\mathbf{K}, \tag{7.4}$$

where σ_{yj} $(j = 1, 2, \ldots, N)$ is the imaginary Pauli matrix in the spinor basis $\{|1/2,\ 1/2\rangle,\ |1/2,\ -1/2\rangle\}$ and $\mathbf{K}$ is the operator of complex conjugation.

Such a form of the $\mathbf{T}$ operator is not convenient for calculations in many cases when the spin system consists of particles with spin $S > 1/2$ (for example, magnetic clusters containing ions with high spins). In this case, for spin systems containing subsystems with the summary spin S_i $(i = 1, 2, \ldots, n)$ the time-reversal operator $\mathbf{T}$ is presented as a product of n unitary operators $\mathbf{U}_i$ $(i = 1, 2, \ldots, n;\ \mathbf{U}_i^{+}\mathbf{U}_i = \mathbf{e}^{(S)}$, $\mathbf{e}^{(S)}$ is the unit operator defined in the spinor basis) and the operator $\mathbf{K}$ [220]:

$$\mathbf{T} = \mathbf{U}_1\mathbf{U}_2\ldots\mathbf{U}_n\mathbf{K}, \tag{7.5}$$

where the operator $\mathbf{U}_i$ is defined in the spinor basis

$$\{|S_i,\ S_i\rangle,\ |S_i,\ S_i - 1\rangle,\ \ldots,\ |S_i, 1 -\ S_i\rangle,\ |S_i,\ -S_i\rangle\}.$$

The non-zero matrix elements $(U_i)_{\alpha\beta}$ are equal to $+1$ or -1, as follows:

$$\begin{aligned}(U_i)_{S_i,-S_i} &= (U_i)_{S_i-2,2-S_i} = (U_i)_{S_i-4,4-S_i} = \cdots = \\ &= (U_i)_{2-S_i,S_i-2} = (U_i)_{1-S_i,S_i-1} = 1, \\ (U_i)_{S_i-1,1-S_i} &= (U_i)_{S_i-3,3-S_i} = (U_i)_{S_i-5,5-S_i} = \cdots = \\ &= (U_i)_{3-S_i,S_i-3} = (U_i)_{-S_i,S_i} = -1.\end{aligned} \tag{7.6}$$

All other matrix elements of the operator $\mathbf{U}_i$ are equal to zero (see also Sect. 6.1).

Let us consider again a trimer cluster consisting of atoms with spins $S_A = S_B = S_C = 1/2$. The irreducible representation of the rotation group of symmetry $D^{(1/2)}_{ABC}$ is contained two times in the direct product of representations $D^{(1/2)}_A \otimes D^{(1/2)}_B \otimes D^{(1/2)}_C$. Therefore, there are two Kramers doublets $\{\varphi_1, \varphi_2 = \mathbf{T}\varphi_1\}$ and $\{\varphi_3, \varphi_4 = \mathbf{T}\varphi_3\}$, associated with representations $D^{(1/2)}_{ABC[1]} \in D^{(0)}_{AB} \otimes D^{(1/2)}_C$ and $D^{(1/2)}_{ABC[2]} \in D^{(1)}_{AB} \otimes D^{(1/2)}_C$, respectively.

The single subscript at the irreducible representation of the rotation group $D^{(S)}_p$ $(p = A, B, C)$ indicates, that according to the given irreducible representation, it transforms the spin wave function of ion, localized in the vertex p of the triangle. But if the subscript at the irreducible representation of the rotation group contains notations of two or three vertices of the triangle (for example, $D^{(1)}_{AB}$ and $D^{(1/2)}_{ABC}$), this means that we deal with the Kronecker (direct) product of corresponding irreducible representations. Particularly, in the case of considered examples we have $D^{(1)}_{AB} \in D^{(1/2)}_A \otimes D^{(1/2)}_B$, $D^{(1/2)}_{ABC[1]} \in D^{(0)}_{AB} \otimes D^{(1/2)}_C = D^{(1/2)}_A \otimes D^{(1/2)}_B \otimes D^{(1/2)}_C$ and $D^{(1/2)}_{ABC[2]} \in D^{(1)}_{AB} \otimes D^{(1/2)}_C$, where basic functions of irreducible representations of $D^{(1/2)}_{ABC[1]}$ and $D^{(1/2)}_{ABC[2]}$ of one and the same weight (of the same dimension) are different.

The spin wave functions φ_1 and $\varphi_2 = \mathbf{T}\varphi_1$ are

$$\begin{aligned}\varphi_1 = \frac{1}{\sqrt{2}}\big(&|1/2,\ 1/2\rangle_A\ |1/2,\ -1/2\rangle_B\ |1/2,\ 1/2\rangle_C - \\ &-|1/2,\ -1/2\rangle_A\ |1/2,\ 1/2\rangle_B\ |1/2,\ 1/2\rangle_C\big),\end{aligned} \tag{7.7}$$

$$\varphi_2 = \mathbf{K}\mathbf{U}^{(1/2)}_A \mathbf{U}^{(1/2)}_B \mathbf{U}^{(1/2)}_C \varphi_1 = \mathbf{T}\varphi_1. \tag{7.8}$$

Here the unitary operator $\mathbf{U}^{(1/2)}_p$ $(p = A, B, C)$ is represented by the (2×2)-matrix

$$\mathbf{U}^{(1/2)}_p = \begin{pmatrix} 0 & 1 \\ -1 & 0 \end{pmatrix} \tag{7.9}$$

in the spinor basis $\left\{|1/2,\ 1/2\rangle_p,\ |1/2,\ -1/2\rangle_p\right\}$.

Table 7.1 The spin moments $\langle S_{Zp}\rangle = \langle\varphi_m \mid S_{Zp} \mid \varphi_m\rangle$ $(p = A, B, C)$ associated with the states φ_m $(m = 1 \div 4)$ for a spin trimer with spins $S_p = 1/2$

φ_m	$\langle S_{ZA}\rangle$	$\langle S_{ZB}\rangle$	$\langle S_{ZC}\rangle$
φ_1	0	0	1/2
φ_2	0	0	−1/2
φ_3	1/3	1/3	−1/6
φ_4	−1/3	−1/3	1/6

Taking into account (7.7)–(7.9), the definition of the operator of complex conjugation **K** and using the Clebsh–Gordan coefficients [9], we obtain:

$$\varphi_2 = \frac{1}{\sqrt{2}}\big(|1/2, -1/2\rangle_A \cdot |1/2, 1/2\rangle_B \cdot |1/2, -1/2\rangle_C - \\ - |1/2, 1/2\rangle_A \cdot |1/2, -1/2\rangle_B \cdot |1/2, -1/2\rangle_C\big). \quad (7.10)$$

Two components of the second Kramers doublet correspond to the following spin wave functions

$$\varphi_3 = \frac{1}{\sqrt{6}}\big(2\,|1/2, 1/2\rangle_A\, |1/2, 1/2\rangle_B\, |1/2, -1/2\rangle_C - \\ - |1/2, 1/2\rangle_A\, |1/2, -1/2\rangle_B\, |1/2, 1/2\rangle_C - \\ - |1/2, -1/2\rangle_A\, |1/2, 1/2\rangle_B\, |1/2, 1/2\rangle_C\big), \quad (7.11)$$

$$\varphi_4 = \mathbf{K}\mathbf{U}_A^{(1/2)}\mathbf{U}_B^{(1/2)}\mathbf{U}_C^{(1/2)}\varphi_3 = \mathbf{T}\varphi_3. \quad (7.12)$$

Acting by analogy with the case of finding the function $\varphi_2 = \mathbf{T}\varphi_1$, we will find the next formula for the spin wave function $\varphi_4 = \mathbf{T}\varphi_3$, which is one of the basic functions of irreducible representation of the rotation group $D^{(1/2)}_{ABC[2]} \in D^{(1)}_{AB} \otimes D^{(1/2)}_{C}$,

$$\varphi_4 = \frac{1}{\sqrt{6}}\big(2\,|1/2, -1/2\rangle_A \cdot |1/2, -1/2\rangle_B \cdot |1/2, 1/2\rangle_C - \\ - |1/2, -1/2\rangle_A \cdot |1/2, 1/2\rangle_B \cdot |1/2, -1/2\rangle_C - \\ - |1/2, 1/2\rangle_A \cdot |1/2, -1/2\rangle_B \cdot |1/2, -1/2\rangle_C\big). \quad (7.13)$$

Using the formulas (7.7), (7.10), (7.11) and (7.13), the spin moments $\langle S_{Zp}\rangle = \langle\varphi_m \mid S_{Zp} \mid \varphi_m\rangle$ $(p = A, B, C)$ associated with the states φ_m $(m = 1 \div 4)$ were calculated and presented in the Table 7.1.[1]

[1]The results from Table 7.1 were also obtained in [389] without using the properties of the time-reversal operator, but taking into account the corresponding Clebsh–Gordan coefficients.

For $p = A, B, C$ and $m = 1 \div 4$ it can be shown that

$$\langle S_{Xp} \rangle = \langle \varphi_m \mid S_{Xp} \mid \varphi_m \rangle = 0, \qquad \langle S_{Yp} \rangle = \langle \varphi_m \mid S_{Yp} \mid \varphi_m \rangle = 0. \tag{7.14}$$

As for $\langle S_{Zp} \rangle$, one can see that mean values of operators S_{ZA} and S_{ZB}, calculated using the spin wave function φ_1, are equal to zero, while in accordance to data from [383], which are given in Fig. 7.1b, the spin moments at the corners B and C are equal to zero. Thus, our notations turn into Kahn's ones [383] at substitutions $A \to B$, $B \to C$ and correspondingly $C \to A$.

According to Table 7.1, the spin moment at the corner C is equal to $1/2$ that corresponds to the spin moment equal to 1 at the corner A in Kahn's notations (Fig. 7.1b, calculations are done using the wave function $|1/2, 0\rangle$). This difference of values $1/2$ and 1 is due to different coefficients at spin wave functions φ_1 and $|1/2, 0\rangle$ (the value $1/2$ for the spin moment at the corner C corresponds to the normalized wave function φ_1 from (7.7)).

The mean values of operators S_{ZA}, S_{ZB} and S_{ZC} calculated by means of the spin wave function φ_3 from (7.11) (see Table 7.1) are $1/3$, $1/3$ and $-1/6$ at the corners A, B and C of the triangle. Taking into account the substitutions $A \to B \to C \to A$, this leads to spin moments $-1/3$, $2/3$ and $2/3$ at the corners A, B and C in Kahn's notations (Fig. 7.1b, calculations are done using the wave function $|1/2, 1\rangle$). We see that in this case the given in Table 7.1 data concerning spin moments at the corners A, B and C, calculated using the spin wave function φ_3, and data from [383], obtained using the spin wave function $|1/2, 1\rangle$ (Fig. 7.1c) also, differ by the same constant factor $1/2$. This deviation, as well as the previous one are due to different normalized coefficients at wave functions φ_3 and $|1/2, 1\rangle$. However, as it was already mentioned, the main omission of [383], was to disregard contributions into spin moments at the corners of the triangle caused by spin operators S_{Zp} ($p = A, B, C$) overaged by means of spin wave functions $\varphi_2 = \mathbf{T}\varphi_1$ and $\varphi_4 = \mathbf{T}\varphi_3$.

A comparison of first and second rows, as well as of third and fourth rows in the Table 7.1 shows that any spin moment resulting after averaging the spin operator S_{Zp} ($p = A, B, C$) by means of the wave function of one component of the Kramers doublet will be canceled out by the opposite spin moment resulting after averaging the S_{Zp} operator by means of the wave function of another component of the Kramers doublet. Such a mutually compensation of spin moments is a consequence of the fact that the spin wave functions of the Kramers doublet are time-reversed one with respect to another. The functions φ_1 and φ_2 of one Kramers doublet connected with each other by the relationship $\varphi_2 = \mathbf{T}\varphi_1$, while between functions φ_3 and φ_4 of the another Kramers doublet the relationship $\varphi_4 = \mathbf{T}\varphi_3$ takes place, where $\mathbf{T}$ is the time-reversal operator.

Thus, the instability of spin populations for a triangle of isotropic $1/2$ spins antiferromagnetically coupled with respect to a weak structural deformation is forbidden. This prohibition is due to the time-reversal operator properties, i.e., to the existence of a time-reversal symmetry.

7.3 Mutual Compensation of Distorted-Induced Spin Polarization in a Trimer $5/2 \otimes 5/2 \otimes 5/2$ Due to Time-Reversal Symmetry

The absence of the instability of spin populations with respect to a weak deformation for a triangle of isotropic 1/2 spins antiferromagnetically coupled suggests that such a situation must be also characteristic for trimer clusters containing paramagnetic ions with high half-integer spins. Firstly, in this case the Kronecker product of representations $D^{(S_A)} \otimes D^{(S_B)} \otimes D^{(S_C)}$ $(S_A = S_B = S_C)$ contains only two times the irreducible representation $D^{(1/2)}$, in full analogy with the triangle of isotropic 1/2 spins. In the case of antiferromagnetic exchange interaction between ions of the cluster it must lead to the presence of two Kramer doublets of the ground state, as it occurs for the trihomonuclear cluster of isotropic 1/2 spins. Secondly, matrices of unitary operators $\mathbf{U}_p^{(1/2)}$ and $\mathbf{U}_p^{(S)}$ $(p = A, B, C)$ in the spinor basis have a similar structure. For example, matrices $\mathbf{U}_p^{(3/2)}$ and $\mathbf{U}_p^{(5/2)}$ have the form

$$\mathbf{U}_p^{(3/2)} = \begin{pmatrix} \mathbf{0} & \mathbf{U}_p^{(1/2)} \\ \mathbf{U}_p^{(1/2)} & \mathbf{0} \end{pmatrix}, \quad U_p^{(5/2)} = \begin{pmatrix} \mathbf{0} & \mathbf{0} & \mathbf{U}_p^{(1/2)} \\ \mathbf{0} & \mathbf{U}_p^{(1/2)} & \mathbf{0} \\ \mathbf{U}_p^{(1/2)} & \mathbf{0} & \mathbf{0} \end{pmatrix}, \tag{7.15}$$

where $\mathbf{U}_p^{(1/2)}$ is defined by (7.9) and $\mathbf{0}$ is a the null (2×2)-matrix.

Among coordination compounds with high-spin trihomonuclear clusters there are complex compounds, which contain trinuclear clusters $\{Fe^{3+}\}_3$ with spins $S_p = 5/2$ $(p = A, B, C)$. Particularly, $[Fe_3O(OOCCH_3)_6(H_2O)_3]Cl \cdot 6H_2O$ is one of the first chemical compounds, along with $[Cr_3O(OOCCH_3)_6(H_2O)_3]Cl \cdot 6H_2O$ $(S_p = 3/2,$ $p = A, B, C$), in which there were discovered for the first-time trihomonuclear clusters of ions of metals [370–373, 391–404]. Later, these and other clusters of iron(III) and chromium(III) trihomonuclear compounds were investigated in detail. The following consideration also applies to manganese(II) trihomonuclear compounds $(S_p = 5/2,\ p = A, B, C)$.

Let us calculate the resulting distortion-induced spin polarization in the ground state of a trihomonuclear magnetic cluster with the values of individual spins equal to $S_p = 5/2$ $(p = A, B, C)$. In this case, the ground state of the trinuclear cluster contains two Kramers doublets, as in the case of trihomonuclear clusters with $S_p =$ $1/2$ $(p = A, B, C)$. But the structure of spin wave functions is essentially different in comparison with the structure of φ_m $(m = 1 \div 4)$ wave functions from (7.7), (7.10), (7.11) and (7.13). Using Clebsh–Gordan coefficients for the vector coupling of spins $S_A = S_B = S_C = 5/2$ [9], we obtain two Kramers doublets $\{\psi_1,\ \psi_2 = \mathbf{T}\psi_1\}$ and $\{\psi_3,\ \psi_4 = \mathbf{T}\psi_3\}$ associated with irreducible representations $D_{ABC[3]}^{(1/2)} \in D_{AB}^{(2)} \otimes D_C^{(5/2)}$ and $D_{ABC[4]}^{(1/2)} \in D_{AB}^{(3)} \otimes D_C^{(5/2)}$, respectively.

The spin wave functions ψ_1 and $\psi_2 = \mathbf{T}\psi_1$ are [155]

$$\psi_1 = \frac{1}{2\sqrt{210}} \left\{ \sqrt{10} \left[\sqrt{5} \left(|1/2\rangle_A \, |{-5/2}\rangle_B - |{-5/2}\rangle_A \, |1/2\rangle_B \right) + \right.\right. \tag{7.16}$$

$$+3\left(|-3/2\rangle_A\,|-1/2\rangle_B - |-1/2\rangle_A\,|-3/2\rangle_B\right)\big]\,|5/2\rangle_C +$$
$$+4\left[\sqrt{5}\left(|-5/2\rangle_A\,|3/2\rangle_B - |3/2\rangle_A\,|-5/2\rangle_B\right)+\right.$$
$$\left.+\sqrt{2}\left(|1/2\rangle_A\,|-3/2\rangle_B - |-3/2\rangle_A\,|1/2\rangle_B\right)\right]|3/2\rangle_C +$$
$$+\sqrt{2}\,[5\left(|5/2\rangle_A\,|-5/2\rangle_B - |-5/2\rangle_A\,|5/2\rangle_B\right)+$$
$$+|3/2\rangle_A\,|-3/2\rangle_B - |-3/2\rangle_A\,|3/2\rangle_B +$$
$$+4\left(|1/2\rangle_A\,|-1/2\rangle_B - |-1/2\rangle_A\,|1/2\rangle_B\right)]\,|1/2\rangle_C +$$
$$+2\sqrt{2}\left[\sqrt{5}\left(|-3/2\rangle_A\,|5/2\rangle_B - |5/2\rangle_A\,|-3/2\rangle_B\right)+\right.$$
$$\left.+\sqrt{2}\left(|3/2\rangle_A\,|-1/2\rangle_B - |-1/2\rangle_A\,|3/2\rangle_B\right)\right]|-1/2\rangle_C +$$
$$+\sqrt{2}\left[\sqrt{5}\left(|5/2\rangle_A\,|-1/2\rangle_B - |-1/2\rangle_A\,|5/2\rangle_B\right)+\right.$$
$$\left.+\,3\left(|1/2\rangle_A\,|3/2\rangle_B - |3/2\rangle_A\,|1/2\rangle_B\right)\right]|-3/2\rangle_C \Big\},$$

$$\psi_2 = \mathbf{K}\mathbf{U}_A^{(5/2)}\mathbf{U}_B^{(5/2)}\mathbf{U}_C^{(5/2)}\psi_1 = \mathbf{T}\psi_1. \tag{7.17}$$

In the spinor basis $\{|5/2,\ 5/2\rangle\,,\ |5/2,\ 3/2\rangle\,, ...,\ |5/2,\ -3/2\rangle\,,\ |5/2,\ -5/2\rangle\}$ the unitary matrix $\mathbf{U}_p^{(5/2)}$ $(p = A, B, C)$ is

$$\mathbf{U}_p^{(5/2)} = \begin{pmatrix} 0 & 0 & 0 & 0 & 0 & 1 \\ 0 & 0 & 0 & 0 & -1 & 0 \\ 0 & 0 & 0 & 1 & 0 & 0 \\ 0 & 0 & -1 & 0 & 0 & 0 \\ 0 & 1 & 0 & 0 & 0 & 0 \\ -1 & 0 & 0 & 0 & 0 & 0 \end{pmatrix}. \tag{7.18}$$

Taking into account (7.16)–(7.18), we obtain:

$$\psi_2 = \frac{1}{2\sqrt{210}}\Big\{\sqrt{10}\left[\sqrt{5}\left(|-1/2\rangle_A\,|5/2\rangle_B - |5/2\rangle_A\,|-1/2\rangle_B\right)+\right. \tag{7.19}$$
$$+3\left(|3/2\rangle_A\,|1/2\rangle_B - |1/2\rangle_A\,|3/2\rangle_B\right)\big]\,|-5/2\rangle_C +$$
$$+4\left[\sqrt{5}\left(|5/2\rangle_A\,|-3/2\rangle_B - |3/2\rangle_A\,|5/2\rangle_B\right)+\right.$$
$$\left.+\sqrt{2}\left(|-1/2\rangle_A\,|3/2\rangle_B - |3/2\rangle_A\,|-1/2\rangle_B\right)\right]|-3/2\rangle_C +$$
$$+\sqrt{2}\,[5\left(|-5/2\rangle_A\,|5/2\rangle_B - |5/2\rangle_A\,|-5/2\rangle_B\right)+$$
$$+|-3/2\rangle_A\,|3/2\rangle_B - |3/2\rangle_A\,|-3/2\rangle_B +$$
$$+4\left(|-1/2\rangle_A\,|1/2\rangle_B - |1/2\rangle_A\,|-1/2\rangle_B\right)]\,|-1/2\rangle_C +$$
$$+2\sqrt{2}\left[\sqrt{5}\left(|3/2\rangle_A\,|-5/2\rangle_B - |-5/2\rangle_A\,|3/2\rangle_B\right)+\right.$$
$$\left.+\sqrt{2}\left(|-3/2\rangle_A\,|1/2\rangle_B - |1/2\rangle_A\,|-3/2\rangle_B\right)\right]|1/2\rangle_C +$$
$$+\sqrt{2}\left[\sqrt{5}\left(|-5/2\rangle_A\,|1/2\rangle_B - |1/2\rangle_A\,|-5/2\rangle_B\right)+\right.$$
$$\left.+\,3\left(|-1/2\rangle_A\,|-3/2\rangle_B - |-3/2\rangle_A\,|-1/2\rangle_B\right)\right]|3/2\rangle_C\Big\}.$$

The other two components ψ_3 and $\psi_4 = \mathbf{T}\psi_3$ of the second ground state Kramers doublet, associated with the representation $D^{(1/2)}_{ABC[4]} \in D^{(3)}_{AB} \otimes D^{(5/2)}_{C}$, are

$$\psi_3 = \frac{1}{6\sqrt{35}} \Big\{ \sqrt{5} \big[\left(|-3/2\rangle_A \, |-1/2\rangle_B + |-1/2\rangle_A \, |-3/2\rangle_B \right) - \tag{7.20}$$

$$-\sqrt{5} \left(|-5/2\rangle_A \, |1/2\rangle_B + |1/2\rangle_A \, |-5/2\rangle_B \right) \big] \, |5/2\rangle_C +$$
$$+2 \Big[\sqrt{10} \left(|-5/2\rangle_A \, |3/2\rangle_B + |3/2\rangle_A \, |-5/2\rangle_B \right) +$$
$$+ |-3/2\rangle_A \, |1/2\rangle_B + |1/2\rangle_A \, |-3/2\rangle_B - 2\sqrt{2} \, |-1/2\rangle_A \, |-1/2\rangle_B \Big] \, |3/2\rangle_C +$$
$$+ \left[4 \left(|-1/2\rangle_A \, |1/2\rangle_B + |1/2\rangle_A \, |-1/2\rangle_B \right) - 7 \left(|-3/2\rangle_A \, |3/2\rangle_B + |3/2\rangle_A \, |-3/2\rangle_B \right) - \right.$$
$$\left. -5 \left(|-5/2\rangle_A \, |5/2\rangle_B + |5/2\rangle_A \, |-5/2\rangle_B \right) \right] |1/2\rangle_C +$$
$$+2\sqrt{2} \Big[\sqrt{10} \left(|-3/2\rangle_A \, |5/2\rangle_B + |5/2\rangle_A \, |-3/2\rangle_B \right) +$$
$$+ |-1/2\rangle_A \, |3/2\rangle_B + |3/2\rangle_A \, |-1/2\rangle_B - 2\sqrt{2} \, |1/2\rangle_A \, |1/2\rangle_B \Big] \, |-1/2\rangle_C +$$
$$+5 \big[\left(|1/2\rangle_A \, |3/2\rangle_B + |3/2\rangle_A \, |1/2\rangle_B \right) -$$
$$-\sqrt{5} \left(|-1/2\rangle_A \, |5/2\rangle_B + |5/2\rangle_A \, |-1/2\rangle_B \right) \big] \, |-3/2\rangle_C +$$
$$+2\sqrt{5} \Big[\sqrt{5} \left(|1/2\rangle_A \, |5/2\rangle_B + |5/2\rangle_A \, |1/2\rangle_B \right) - 2\sqrt{2} \, |3/2\rangle_A \, |3/2\rangle_B \Big] \, |-5/2\rangle_C \Big\} ,$$

$$\psi_4 = \mathbf{K} \mathbf{U}^{(5/2)}_A \mathbf{U}^{(5/2)}_B \mathbf{U}^{(5/2)}_C \psi_3 = \mathbf{T}\psi_3. \tag{7.21}$$

Taking into account (7.18), (7.20) and (7.21), we obtain:

$$\psi_4 = \frac{1}{6\sqrt{35}} \Big\{ \sqrt{5} \big[\left(|3/2\rangle_A \, |1/2\rangle_B + |1/2\rangle_A \, |3/2\rangle_B \right) - \tag{7.22}$$

$$-\sqrt{5} \left(|5/2\rangle_A \, |-1/2\rangle_B + |-1/2\rangle_A \, |5/2\rangle_B \right) \big] \, |-5/2\rangle_C +$$
$$+2 \Big[\sqrt{10} \left(|5/2\rangle_A \, |-3/2\rangle_B + |-3/2\rangle_A \, |5/2\rangle_B \right) +$$
$$+ |3/2\rangle_A \, |-1/2\rangle_B + |-1/2\rangle_A \, |3/2\rangle_B - 2\sqrt{2} \, |1/2\rangle_A \, |1/2\rangle_B \Big] \, |-3/2\rangle_C +$$
$$+ \left[4 \left(|1/2\rangle_A \, |-1/2\rangle_B + |-1/2\rangle_A \, |1/2\rangle_B \right) - \right.$$
$$-7 \left(|3/2\rangle_A \, |-3/2\rangle_B + |-3/2\rangle_A \, |3/2\rangle_B \right) -$$
$$\left. -5 \left(|5/2\rangle_A \, |-5/2\rangle_B + |-5/2\rangle_A \, |5/2\rangle_B \right) \right] |-1/2\rangle_C +$$
$$+2\sqrt{2} \Big[\sqrt{10} \left(|3/2\rangle_A \, |-5/2\rangle_B + |-5/2\rangle_A \, |3/2\rangle_B \right) +$$
$$+ |1/2\rangle_A \, |-3/2\rangle_B + |-3/2\rangle_A \, |1/2\rangle_B - 2\sqrt{2} \, |-1/2\rangle_A \, |-1/2\rangle_B \Big] \, |1/2\rangle_C +$$
$$+5 \big[\left(|1/2\rangle_A \, |-3/2\rangle_B + |-3/2\rangle_A \, |-1/2\rangle_B \right) -$$

Table 7.2 The spin moments $\langle S_{Zp}\rangle = \langle \psi_m \mid S_{Zp} \mid \psi_m\rangle$ $(p = A, B, C)$ associated with the states ψ_m $(m = 1 \div 4)$ for a spin trimer with spins $S_p = 5/2$ [155]

ψ_m	$\langle S_{ZA}\rangle$	$\langle S_{ZB}\rangle$	$\langle S_{ZC}\rangle$
ψ_1	−1/3	−1/3	7/6
ψ_2	1/3	1/3	−7/6
ψ_3	2/3	2/3	−5/6
ψ_4	−2/3	−2/3	5/6

$$-\sqrt{5}\left(|1/2\rangle_A\,|{-5/2}\rangle_B + |{-5/2}\rangle_A\,|1/2\rangle_B\right)\Big]\,|3/2\rangle_C +$$
$$+2\sqrt{5}\Big[\sqrt{5}\left(|{-1/2}\rangle_A\,|{-5/2}\rangle_B + |{-5/2}\rangle_A\,|{-1/2}\rangle_B\right) -$$
$$-\;2\sqrt{2}\,|{-3/2}\rangle_A\,|{-3/2}\rangle_B\Big]\,|5/2\rangle_C\Big\}.$$

For compactness, in formulas (7.16), (7.19), (7.20) and (7.22), the product of spin wave functions $|5/2,\,M\rangle_A\left|5/2,\,M'\right\rangle_B\left|5/2,\,M''\right\rangle_C$ is denoted $\mid M\rangle_A\left|\,M'\right\rangle_B\left|\,M''\right\rangle_C$.

The spin moments $\langle S_{Zp}\rangle = \langle \psi_m \mid S_{Zp} \mid \psi_m\rangle$ $(p = A, B, C)$ associated with the states ψ_m $(m = 1 \div 4)$ for the spin trimer with $S_A = S_B = S_C = 5/2$ are given in the Table 7.2.

For $p = A, B, C$ and $m = 1 \div 4$, it can be shown that

$$\langle S_{Xp}\rangle = \langle \psi_m \mid S_{Xp} \mid \psi_m\rangle = 0, \qquad \langle S_{Yp}\rangle = \langle \psi_m \mid S_{Yp} \mid \psi_m\rangle = 0. \tag{7.23}$$

Comparison of the first and second rows, as well as the third and fourth rows in the Table 7.2 in the case of a trimer containing ions with spins $S_p = 5/2$ $(p = A, B, C)$ shows that any spin moment at the corner p of the triangle, resulting after averaging the spin operator S_{Zp} using the spin wave function ψ_1, will be canceled out by the opposite spin moment, resulting after averaging the S_{Zp} operator using the spin wave function ψ_2. Analogically, any spin moment at the corner p of the triangle, resulting after averaging the spin operator S_{Zp} using the spin wave function ψ_3, will be canceled out by the opposite spin moment at the corner p of the triangle, resulting after averaging of S_{Zp} operator using the spin wave function ψ_4. The functions ψ_1 and $\psi_2 = \mathbf{T}\psi_1$ are components of the ground state Kramers doublet, corresponding to the irreducible representation $D^{(1/2)}_{ABC[3]} \in D^{(2)}_{AB} \otimes D^{(5/2)}_{C}$, while the functions ψ_3 and $\psi_4 = \mathbf{T}\psi_3$ are components of the ground state Kramers doublet, corresponding to the irreducible representation $D^{(1/2)}_{ABC[4]} \in D^{(3)}_{AB} \otimes D^{(5/2)}_{C}$. The mutual compensation of spin moments at the corner p $(p = A, B, C)$ of an equilateral triangle with $S_p = 5/2$ (as in the case of the triangle with $S_p = 1/2$) is due to the structure of time-reversal operator **T**. A consequence of this compensation is the stability of spin populations with respect to a weak deformation.

7.4 Distortion-Induced Spin Population Instability of Trimer Homonuclear Kramers Clusters Caused by Time-Reversal Symmetry Violation

As it was shown in Sects. 7.2 and 7.3, the stability of spin populations in trimer clusters with half-integer spin with respect to a weak structural deformation is due to the time-reversal symmetry. Therefore, at violation of the time-reversal symmetry the stability of spin populations should also be broken.

Traditionally, when the effects caused by the time-reversal symmetry violation in many fields of physics are discussed, the sense of the conception "violation of the time-reversal symmetry" is not refined.

Moreover, it is tacitly supposed, that the time-reversal symmetry completely disappears at its violation. However, in the case of space symmetry under the action of an external perturbation it first is lowered gradually as less symmetric atom configurations appear and only in the limiting case it completely disappears (for point groups of symmetry this corresponds to the symmetry group C_1).

The question arises: whether something similar occurs also at time-reversal symmetry, despite the fact that in the non-relativistic quantum mechanics time is separated from space and flows only in one direction? Moreover, even in nuclear physics and physics of elementary particles, in which relativistic effects are essential, it is not revealed the content of the notion "the time-reversal symmetry violation", considering that the complete destruction of time-reversal symmetry takes place too.

In Chap. 8 this problem is discussed on the basis of theoretical-group treatment (discrete non-Abelian and Abelian groups of symmetry) with determination of all incomplete time-reversal operators and their classification. Below, only two of these operators are obtained and used to show that the time-reversal symmetry violation (more exactly, the time-reversal symmetry lowering) leads to the spin populations instability in a trimer cluster with half-integer spin.

Let us consider again the equilateral triangle ABC containing ions with half-integer spins S_p $(p = A, B, C)$ in the corners. As it was shown in [134], even in the case of one particle with spin 1/2 there is a non-Abelian group of symmetry of the eight order $G_8^{(1/2)}$ consisting of operators $i\sigma_y\mathbf{K}$, $-\mathbf{e}$, $-i\sigma_y\mathbf{K}$, $\mathbf{e}$, σ_x, $-\sigma_x$, $\sigma_z\mathbf{K}$ and $-\sigma_z\mathbf{K}$. Here $\mathbf{e}$ is a unit operator defined in the spin space and in the spinor basis set $\{|1/2, 1/2\rangle, |1/2, -1/2\rangle\}$; σ_x, σ_y and σ_z are Pauli operators in the same spinor basis set and $\mathbf{K}$, as earlier, denotes the operator of complex conjugation.

It follows directly from the multiplication table of this group of symmetry (Table 7.3), that for a particle with spin 1/2 the Wigner time-reversal operator $\mathbf{T}$ can be represented as a product of two operators $\mathbf{T}_1$ and $\mathbf{T}_2$ [134]

$$\mathbf{T} = \mathbf{T}_1 \cdot \mathbf{T}_2, \tag{7.24}$$

Table 7.3 Multiplication table for elements of the group $G_8^{(1/2)}$

$G_8^{(1/2)}$	$\pm i\sigma_y \mathbf{K}$	$\mp \mathbf{e}$	$\pm\sigma_x$	$\pm\sigma_z \mathbf{K}$
$i\sigma_y \mathbf{K}$	$\mp \mathbf{e}$	$\mp i\sigma_y K$	$\pm\sigma_z \mathbf{K}$	$\mp\sigma_x$
-**e**	$\mp i\sigma_y \mathbf{K}$	$\pm \mathbf{e}$	$\mp\sigma_x$	$\mp\sigma_z \mathbf{K}$
$-i\sigma_y K$	$\pm \mathbf{e}$	$\pm i\sigma_y \mathbf{K}$	$\mp\sigma_z \mathbf{K}$	$\pm\sigma_x$
e	$\pm i\sigma_y$	$\mp \mathbf{e}$	$\pm\sigma_x$	$\pm\sigma_z \mathbf{K}$
σ_x	$\mp\sigma_z \mathbf{K}$	$\mp\sigma_x$	$\pm \mathbf{e}$	$\mp i\sigma_y \mathbf{K}$
$-\sigma_x$	$\pm\sigma_z \mathbf{K}$	$\pm\sigma_x$	$\mp \mathbf{e}$	$\pm i\sigma_y \mathbf{K}$
$\sigma_z \mathbf{K}$	$\pm\sigma_x$	$\mp\sigma_z \mathbf{K}$	$\pm i\sigma_y \mathbf{K}$	$\pm \mathbf{e}$
$-\sigma_z \mathbf{K}$	$\mp\sigma_x$	$\pm\sigma_z \mathbf{K}$	$\mp i\sigma_y \mathbf{K}$	$\mp \mathbf{e}$

where

$$\mathbf{T}_1 = \sigma_z \mathbf{K}, \tag{7.25}$$

$$\mathbf{T}_2 = \sigma_x. \tag{7.26}$$

The operator $\mathbf{T}$, as well as operators $\mathbf{T}_1$ and $\mathbf{T}_2$ belong to the symmetry group $G_8^{(1/2)}$ ($\mathbf{T}, \mathbf{T}_1, \mathbf{T}_2 \in G_8^{(1/2)}$). Therefore, these operators commute with the corresponding Hamiltonian of the system and the following invariance relationships hold

$$\mathbf{THT}^{-1} = \mathbf{H}, \quad \mathbf{T}_1\mathbf{H}_1\mathbf{T}_1^{-1} = \mathbf{H}_1, \quad \mathbf{T}_2\mathbf{H}_2\mathbf{T}_2^{-1} = \mathbf{H}_2, \tag{7.27}$$

where $\mathbf{H}$, $\mathbf{H}_1$ and $\mathbf{H}_2$ are different Hamiltonians (if the Hamiltonian $\mathbf{H}$ is invariant under time-reversal operator $\mathbf{T}$, then $\mathbf{H}_2$ and $\mathbf{H}_3$, which are invariant under $\mathbf{T}_1$ and $\mathbf{T}_2$, respectively, contain interactions breaking T-symmetry).

The spin-projection operators S_x, S_y and S_z $\left(\mathbf{S} = \frac{1}{2}\sigma\right)$ go into their negatives under the time-reversal [222]. We will refer to the simultaneous change of signs of all operators S_x, S_y and S_z under the action of the operator $\mathbf{T}$ as the complete reversal of motion and the operator $\mathbf{T}$ as the complete time-reversal operator.

On the other hand, taking into account (7.25) and (7.26), we obtain

$$\mathbf{T}_1 S_x \mathbf{T}_1^{-1} = -S_x, \qquad \mathbf{T}_1 S_y \mathbf{T}_1^{-1} = S_y, \qquad \mathbf{T}_1 S_z \mathbf{T}_1^{-1} = S_z. \tag{7.28}$$

$$\mathbf{T}_2 S_x \mathbf{T}_2^{-1} = S_x, \qquad \mathbf{T}_2 S_y \mathbf{T}_2^{-1} = -S_y, \qquad \mathbf{T}_2 S_z \mathbf{T}_2^{-1} = -S_z. \tag{7.29}$$

As follows from (7.28) and (7.29), only S_x changes its sign under the action of the operator $\mathbf{T}_1$ and only S_y and S_z change signs under the action of the operator $\mathbf{T}_2$. Therefore, $\mathbf{T}_1$ and $\mathbf{T}_2$ are "incomplete time-reversal operators".

A similar consideration can be carried out for systems with any high half-integer spins. In particular, for systems with spin $S = 3/2$ the non-Abelian group of symmetry $G_8^{(3/2)}$ is formed from eight operators $\mathbf{U}^{(3/2)}\mathbf{K}$, $-\mathbf{e}^{(3/2)}$, $-\mathbf{U}^{(3/2)}\mathbf{K}$, $\mathbf{e}^{(3/2)}$, $\sigma_x^{(3/2)}$, $-\sigma_x^{(3/2)}$, $\sigma_z^{(3/2)}\mathbf{K}$ and $-\sigma_z^{(3/2)}\mathbf{K}$, where $\mathbf{e}^{(3/2)}$ is the unit operator defined in the spinor basis set $\{|3/2, 3/2\rangle, \ldots, |3/2, -3/2\rangle\}$ and the block matrices $\mathbf{U}^{(3/2)}$, $\sigma_x^{(3/2)}$ and $\sigma_z^{(3/2)}$, in the same basis set, are

$$\mathbf{U}^{(3/2)} = i\begin{pmatrix} \mathbf{0} & \sigma_y \\ \sigma_y & \mathbf{0} \end{pmatrix}, \quad \sigma_x^{(3/2)} = \begin{pmatrix} \mathbf{0} & \sigma_x \\ \sigma_x & \mathbf{0} \end{pmatrix}, \quad \sigma_z^{(3/2)} = \begin{pmatrix} \sigma_z & \mathbf{0} \\ \mathbf{0} & \sigma_z \end{pmatrix}. \tag{7.30}$$

On the basis of the multiplication table of the $G_8^{(1/2)}$ group of symmetry (which is similar to the multiplication table for $G_8^{(3/2)}$ group, taking into account that $\mathbf{U}^{(3/2)} = i\sigma_y^{(3/2)}$) the time-reversal operator $\mathbf{T} = \mathbf{U}^{(3/2)}\mathbf{K}$ for a system with spin $S = 3/2$ can be represented as a product of two operators $\mathbf{T}_1$ and $\mathbf{T}_2$, in analogy with the formula (7.24):

$$\mathbf{T} = \mathbf{U}^{(3/2)}\mathbf{K} = \mathbf{T}_1 \cdot \mathbf{T}_2, \tag{7.31}$$

where

$$\mathbf{T}_1 = \sigma_z^{(3/2)}\mathbf{K}, \tag{7.32}$$

$$\mathbf{T}_2 = \sigma_x^{(3/2)}. \tag{7.33}$$

For systems with the spin $S = 3/2$ there apply the same relationships (7.27), (7.28) and (7.29), which are valid for a particle with the spin $S = 1/2$.

For systems with spin $S = 5/2$ there is a non-Abelian group of symmetry $G_8^{(5/2)}$ formed by operators $\mathbf{U}^{(5/2)}\mathbf{K}$, $-\mathbf{e}^{(5/2)}$, $-\mathbf{U}^{(5/2)}\mathbf{K}$, $\mathbf{e}^{(5/2)}$, $\sigma_x^{(5/2)}$, $-\sigma_x^{(5/2)}$, $\sigma_z^{(5/2)}\mathbf{K}$ and $-\sigma_z^{(5/2)}\mathbf{K}$, where $\mathbf{e}^{(5/2)}$ is the unit operator defined in the spinor basis set $\{|5/2, 5/2\rangle$, ..., $|5/2, -5/2\rangle\}$ and the block matrices $\mathbf{U}^{(5/2)}$, $\sigma_x^{(5/2)}$ and $\sigma_z^{(5/2)}$ in the same basis set are

$$\mathbf{U}^{(5/2)} = i\begin{pmatrix} \mathbf{0} & \mathbf{0} & \sigma_y \\ \mathbf{0} & \sigma_y & \mathbf{0} \\ \sigma_y & \mathbf{0} & \mathbf{0} \end{pmatrix}, \quad \sigma_x^{(5/2)} = \begin{pmatrix} \mathbf{0} & \mathbf{0} & \sigma_x \\ \mathbf{0} & \sigma_x & \mathbf{0} \\ \sigma_x & \mathbf{0} & \mathbf{0} \end{pmatrix}, \quad \sigma_z^{(5/2)} = \begin{pmatrix} \sigma_z & \mathbf{0} & \mathbf{0} \\ \mathbf{0} & \sigma_z & \mathbf{0} \\ \mathbf{0} & \mathbf{0} & \sigma_z \end{pmatrix}. \tag{7.34}$$

The multiplication table for the elements of the $G_8^{(5/2)}$ symmetry group is similar to multiplication tables for the elements of $G_8^{(1/2)}$ and $G_8^{(3/2)}$ non-Abelian groups of symmetry, taking into account that $\mathbf{U}^{(5/2)} = i\sigma_y^{(5/2)}$ (see (7.34)).

In this case, the time-reversal operator **T** can also be represented as a product of operators $\mathbf{T}_1$ and $\mathbf{T}_2$ (see the formulas (7.24)), which commute separately with the Hamiltonian (see the formula (7.27)). These operators are

$$\mathbf{T}_1 = \sigma_z^{(5/2)}\mathbf{K}, \tag{7.35}$$

$$\mathbf{T}_2 = \sigma_x^{(5/2)}. \tag{7.36}$$

The operators $\mathbf{T}_1$ and $\mathbf{T}_2$ satisfy the relationships (7.27), (7.28) and (7.29).

Now we will find out the behavior of distortion-induced spin polarizations at time-reversal symmetry violation with preservation of T_1- or T_2-invariance. In Sect. 7.2, it has been shown that, in the presence of the time-reversal symmetry, the instability with respect to a weak deformation of spin levels populations for a triangle of isotropic $1/2$ spins antiferromagnetically coupled is forbidden.

We will show that such instability can take place at time-reversal symmetry violation. Let us consider, for example, the spin polarization at the corner C of the trimer containing ions with spin $S_p = 1/2$ $(p = A, B, C)$ at time-reversal symmetry violation, when the time-reversal operators $\mathbf{T}_1$, $\mathbf{T}_2$ and **T** are determined by formulas:

$$\mathbf{T}_1 = \sigma_{zA}\sigma_{zB}\sigma_{zC}\mathbf{K}, \tag{7.37}$$

$$\mathbf{T}_2 = \sigma_{xA}\sigma_{xB}\sigma_{xC}, \tag{7.38}$$

$$\mathbf{T} = -i\,\sigma_{yA}\sigma_{yB}\sigma_{yC}\mathbf{K}. \tag{7.39}$$

The results of calculations of matrix elements of the operator S_{zC} for the magnetic trimer $(S_p = 1/2,\ p = A, B, C)$ using spin wave functions φ_1, $\mathbf{T}\varphi_1$, $\mathbf{T}_1\varphi_1$, $\mathbf{T}_2\varphi_1$, φ_3, $\mathbf{T}\varphi_3$, $\mathbf{T}_1\varphi_3$ and $\mathbf{T}_2\varphi_3$ are given for clarity in Tables 7.4, 7.5 and 7.6.

Table 7.4 Mutual compensation of spin moments in the corner C of the triangle ABC of isotropic $1/2$ spins AF coupled at presence of Wigner time-reversal symmetry

Spin wave functions	φ_1	$\mathbf{T}\varphi_1$	φ_3	$\mathbf{T}\varphi_3$
$\langle S_{zC}\rangle$ (in $N\mu_B$ units)	$1/2$	$-1/2$	$-1/6$	$1/6$

Table 7.5 Mutual compensation of spin moments in the corner C of the triangle ABC of isotropic $1/2$ spins AF coupled at lowering the time-reversal symmetry from T (all projections S_x, S_y, S_z of the spin operator **S** are reversed) up to incomplete time-reversal symmetry T_2 (only two projections of the spin operator **S** are reversed)

Spin wave functions	φ_1	$\mathbf{T}_2\varphi_1$	φ_3	$\mathbf{T}_2\varphi_3$
$\langle S_{zC}\rangle$ (in $N\mu_B$ units)	$1/2$	$-1/2$	$-1/6$	$1/6$

Table 7.6 The resulting spin moment (2/3 in $N\mu_B$ units) in the corner C of the triangle ABC of isotropic 1/2 spins AF coupled leading to a spin populations instability with respect to a weak structural deformation due to lowering the time-reversal symmetry from T up to incomplete time-reversal symmetry T_1 (only one projection of the spin operator **S** is reversed)

Spin wave functions	φ_1	$\mathbf{T}_1\varphi_1$	φ_3	$\mathbf{T}_1\varphi_3$
$\langle S_{zC}\rangle$ (in $N\mu_B$ units)	1/2	1/2	−1/6	−1/6

Because the above-discussed symmetries are not realized simultaneously, Tables 7.4, 7.5 and 7.6 clearly show that the spin populations are stable in conditions of time-reversal symmetries T or T_2 and only if the time-reversal symmetry is lowered from T up to T_1 an instability of spin populations with respect to a weak structural deformation arises.

One can see from Table 7.6 that for the cluster with $S_p = 1/2$ at violation of T-invariance (but preservation of T_1-invariance), in the C corner of the triangle there is a spin polarization, which was equal to zero in the presence of the complete time-reversal symmetry (see Sect. 7.2).

Similar results are valid as well for two other corners of the triangle. Moreover, for a triangle containing ions with spins $S_p = 3/2, 5/2$ or $S_p > 5/2$ there are also unstable spin populations at all corners of the triangle at violation of the complete time-reversal symmetry with preservation of the T_1-invariance.

Thus, the theoretical-group treatment allows to find the physical sense of the concept of lowering in time-reversal symmetry. Namely, in the most widespread case of existence of a "the complete time-reversal symmetry", the Hamiltonian is invariant under the action of the operator **T**, whereas at an "incomplete time-reversal transformation", the Hamiltonian is invariant under the action of operators $\mathbf{T}_1$ and $\mathbf{T}_2$ (there are also other opportunities that are examined in Chap. 8). If the Hamiltonian of the system with a half-integer spin is invariant under the action of operator $\mathbf{T}_1$, but is not invariant under the action of the operator **T**, then the Kramers theorem is broken. In this case, an instability of spin populations with respect to a weak structural distortion becomes possible.

According to (7.27), all three operators **T**, $\mathbf{T}_1$ and $\mathbf{T}_2$ commute with the corresponding Hamiltonians as they belong to the same non-Abelian group of symmetry ($G_8^{(1/2)}$, $G_8^{(3/2)}$, $G_8^{(5/2)}$, ... for spin values 1/2, 3/2, 5/2, ..., correspondingly). If the Hamiltonian commutes with $\mathbf{T}_1$ or $\mathbf{T}_2$, but does not commutes with the operator **T**, then there is a violation of the complete time-reversal symmetry, which is lowered up to T_1- or T_2-symmetries. Such a Hamiltonian plays the role of a perturbation operator leading to the violation of the time-reversal symmetry. In particular, under action of a permanent external electric field on trihomonuclear clusters with a Kramers degeneracy of energy levels, the violation of time-reversal symmetry with preservation of T_1-invariance becomes possible. In this case, the spin invariant like $\sum_p(\alpha_{yy}^{(p)} S_y^{(p)} E_y + \alpha_{zz}^{(p)} S_z^{(p)} E_z)$ is allowed. Here $\alpha_{yy}^{(p)}$ and $\alpha_{zz}^{(p)}$ are the constants of interactions between the spin and the electric field, E_y and E_z being y- and z-components of the electric field strength. Note that the Kramers theorem is also broken at violation of T-symmetry with preservation of T_2-invariance and a spin invariant like $\sum_p \alpha_{xx}^{(p)} S_x^{(p)} E_x$ is allowed, while in the case of T-invariance both these spin invariants are forbidden.

Chapter 8
Non-Abelian and Abelian Symmetry Groups Containing Time-Reversal Operators

On the basis of the group-theoretic approach, the existence of six new time-reversal operators is proved, along with the well-known anti-unitary time-reversal operator introduced into quantum mechanics by Wigner. Among the new time-reversal operators, three are anti-unitary and three are unitary. A characteristic feature of the new time-reversal operators is that under their action the signs do not change for all three spin projection operators, but only for two or only for one of them. For this reason, such operators should be called operators of incomplete time reversal, in contrast to the Wigner operator, which in this context is an operator of complete time reversal. Symmetry groups of the eighth and sixteenth orders containing the operators of complete and incomplete time reversal, which are related to systems with a half-integer angular momentum J, are introduced, and they are shown to be non-Abelian groups. Each of these groups contains a fourth-order cyclic subgroup, constructed on the basis of the Wigner time-reversal operator. Moreover, if the symmetry groups contain operators of incomplete time reversal, but do not contain a time-reversal operator, they are Abelian groups. In the case of non-Kramers systems, the corresponding eighth and sixteenth orders groups are Abelian ones. The structure of non-Abelian and, accordingly, Abelian groups is different, but these structures do not change when the magnitude of the angular momentum J varies for both types of groups. This chapter contains the results published by the author [134, 136, 405–409] and their generalization.[1]

[1] In [409], a geometric proof of the Kramers theorem was obtained for the case of a particle with the spin 1/2 based on the $4'm'm$ group of generalized symmetry of a square with neighboring vertices colored in different colors (at the same color of opposite vertices). The group $4'm'm$ is isomorphic to the eighth-order group G_8 containing the time-reversal operator, on the basis of which it would be possible to prove the existence of incomplete time-reversal operators. However, in [409] this was not done.

I. I. Geru, *Time-Reversal Symmetry*, Springer Tracts in Modern Physics 281,
https://doi.org/10.1007/978-3-030-01210-6_8

8.1 Non-Abelian Group of Eighth Order Related to Spin-1/2 Particle

In Chap. 3, it was shown that for the Kramers system the extension of 32 crystallographic point symmetry groups to nontrivial point groups of magnetic symmetry (the number of which is equal to four) can be performed using a fourth-order Abelian group G_4: {$\mathbf{T}$, $\mathbf{T}^2=-e$, $\mathbf{T}^3=-\mathbf{T}$, $\mathbf{T}^4=e$} containing the time-reversal operator $\mathbf{T}$ or, more generally, using the eighth-order non-Abelian group $G_8^{(\frac{1}{2})}$: {$\mathbf{T}$, $\mathbf{T}^2=-e$, $\mathbf{T}^3=-\mathbf{T}$, $\mathbf{T}^4=e$, σ_x, $-\sigma_x$, $\sigma_z\mathbf{K}$, $-\sigma_z\mathbf{K}$}. Here e is a unit (2×2)-matrix, σ_x and σ_z are real Pauli matrices, and $\mathbf{K}$ is the operator of complex conjugation. In both cases, such an extension results in four nontrivial groups of magnetic symmetry, the structure of which depends on the way of expansion of classical point groups.

The group G_4 is isomorphic to the group $4'$ containing four antirotations by 90° around an axis passing through the center of a square, perpendicular to its plane, with neighboring vertices colored in different colors and the opposite ones in a same color.

The group $G_8^{(\frac{1}{2})}$ is isomorphic to the group $4'm'm$ containing all the symmetry transformations of a square with vertices colored as described above (Sect. 2.9, Fig. 2.2). In the Table 8.1, the Kayley table elements of the group $G_8^{(\frac{1}{2})}$ are presented.

According to Table 8.1, the following relation is satisfied:

$$i\sigma_y\mathbf{K} = \sigma_z\mathbf{K}\sigma_x \tag{8.1}$$

Of course, this relation can be obtained directly based on the properties of the Pauli matrices, without the use of the theory of groups. However, the fact that the operators $\mathbf{T}=i\sigma_y\mathbf{K}$, $\sigma_z\mathbf{K}$ and σ_x belong to the same group of symmetry suggests that $\sigma_z\mathbf{K}$ and σ_x together with the time-reversal operator $\mathbf{T}$ commute separately with the corresponding Hamiltonians. However, not all of the spin projection operators $S_\alpha=\frac{1}{2}\sigma_\alpha(\alpha=x,y,z)$ change their sign under the action of the operators $\sigma_z\mathbf{K}$ and σ_x

$$\sigma_z\mathbf{K}S_x\mathbf{K}\sigma_z=-S_x,\quad \sigma_z\mathbf{K}S_y\mathbf{K}\sigma_z=S_y,\quad \sigma_z\mathbf{K}S_z\mathbf{K}\sigma_z=S_z, \tag{8.2}$$

Table 8.1 The Kayley table of the non-Abelian group $G_8^{(\frac{1}{2})}$ [405]

	$i\sigma_y\mathbf{K}$	$-e$	$-i\sigma_y\mathbf{K}$	e	σ_x	$-\sigma_x$	$\sigma_z\mathbf{K}$	$-\sigma_z\mathbf{K}$
$i\sigma_y\mathbf{K}$	$-e$	$-i\sigma_y\mathbf{K}$	e	$i\sigma_y\mathbf{K}$	$\sigma_z\mathbf{K}$	$-\sigma_z\mathbf{K}$	$-\sigma_x$	σ_x
$-e$	$-i\sigma_y\mathbf{K}$	e	$i\sigma_y\mathbf{K}$	$-e$	$-\sigma_x$	σ_x	$-\sigma_z\mathbf{K}$	$\sigma_z\mathbf{K}$
$-i\sigma_y\mathbf{K}$	e	$i\sigma_y\mathbf{K}$	$-e$	$-i\sigma_y\mathbf{K}$	$-\sigma_z\mathbf{K}$	$\sigma_z\mathbf{K}$	σ_x	$-\sigma_x$
e	$i\sigma_y\mathbf{K}$	$-e$	$-i\sigma_y\mathbf{K}$	e	σ_x	$-\sigma_x$	$\sigma_z\mathbf{K}$	$-\sigma_z\mathbf{K}$
σ_x	$-\sigma_z\mathbf{K}$	$-\sigma_x$	$\sigma_z\mathbf{K}$	σ_x	e	$-e$	$-i\sigma_y\mathbf{K}$	$i\sigma_y\mathbf{K}$
$-\sigma_x$	$\sigma_z\mathbf{K}$	σ_x	$-\sigma_z\mathbf{K}$	$-\sigma_x$	$-e$	e	$i\sigma_y\mathbf{K}$	$-i\sigma_y\mathbf{K}$
$\sigma_z\mathbf{K}$	σ_x	$-\sigma_z\mathbf{K}$	$-\sigma_x$	$\sigma_z\mathbf{K}$	$i\sigma_y\mathbf{K}$	$-i\sigma_y\mathbf{K}$	e	$-e$
$-\sigma_z\mathbf{K}$	$-\sigma_x$	$\sigma_z\mathbf{K}$	σ_x	$-\sigma_z\mathbf{K}$	$-i\sigma_y\mathbf{K}$	$i\sigma_y\mathbf{K}$	$-e$	e

$$\sigma_x S_x \sigma_x = S_x, \quad \sigma_x S_y \sigma_x = -S_y, \quad \sigma_x S_z \sigma_x = -S_z, \tag{8.3}$$

Taking into account (8.2) and (8.3), we introduce the notation

$$\mathbf{T}_{1(x)} = \sigma_z \mathbf{K}, \quad \mathbf{T}_{2(yz)} = \sigma_x, \tag{8.4}$$

where the lower indices (x) and (yz) of the operators $\mathbf{T}_{1(x)}$ and $\mathbf{T}_{2(yz)}$ indicate the spin projection operators that change the sign under the action of the corresponding time-reversal operator. Since, according to (8.2)–(8.4), the action of operators $\mathbf{T}_{1(x)}$ and $\mathbf{T}_{2(yz)}$ change the sign of S_x and, respectively, S_y and S_z, but they do not change the sign of all spin projection operators, as it occurs in the case of the operator $\mathbf{T}$, these operators can be referred to as "partial" time-reversal operators. Thus, the Wigner time-reversal operator can be represented as the product of partial time-reversal operators $\mathbf{T}_{1(x)}$ and $\mathbf{T}_{2(yz)}$ [134]:

$$\mathbf{T} = \mathbf{T}_{1(x)} \mathbf{T}_{2(yz)}. \tag{8.5}$$

In this context, $\mathbf{T}$ is a total time-reversal operator. $\mathbf{T}$, $\mathbf{T}_{1(x)}$ and $\mathbf{T}_{2(yz)}$ operators commute with different Hamiltonians, but not with the same Hamiltonian. For example, if $\mathbf{T}$ commutes with $\boldsymbol{H}$, then $\mathbf{T}_{1(x)}$ and $\mathbf{T}_{2(yz)}$ will not commute with this Hamiltonian. Conversely: if $\mathbf{T}_{1(x)}$ or $\mathbf{T}_{2(yz)}$ commute separately with $\mathbf{H}_1$ or $\mathbf{H}_2$, then $\mathbf{T}$ and $\mathbf{T}_{2(yz)}$ or $\mathbf{T}$ and $\mathbf{T}_{1(x)}$ will not be commuting with $\mathbf{H}$. Therefore there is no reason to give preference to the element $\mathbf{T}$ of the group $G_8{}^{(\frac{1}{2})}$ instead of elements $\mathbf{T}_{1(x)}$ and $\mathbf{T}_{2(yz)}$. The basis for the representation of the $\mathbf{T}$ operator as a product of the operators $\mathbf{T}_{1(x)}$ and $\mathbf{T}_{2(yz)}$ (see (8.5)) is a feature of the structure of Kayley table of the group $G_8{}^{(\frac{1}{2})}$ (Table 8.2). However, it follows from the same Kayley table that the operator $\mathbf{T}_{1(x)}$ (the action of which changes only the sign of the operator S_x, whereas the operators S_y and S_z are invariant under such transformation) can be represented as the product of operators $\mathbf{T}$ and $\mathbf{T}_{2(yz)}$. This can use confusion due to the fact that the operators $\mathbf{T}$ and $\mathbf{T}_{2(yz)}$ have higher symmetry as compared with operator $\mathbf{T}_{1(x)}$. In reality, there is no contradiction, since the operators S_y and S_z change the sign twice under the action of the operator $\mathbf{T}\mathbf{T}_{2(yz)}$. As a result, these operators are invariant with respect to $\mathbf{T}\mathbf{T}_{2(yz)}$:

$$\mathbf{T}\mathbf{T}_{2(yz)} S_y \mathbf{T}_{2(yz)}^{-1} \mathbf{T}^{-1} = -\mathbf{T} S_y \mathbf{T}^{-1} = S_y, \tag{8.6}$$

$$\mathbf{T}\mathbf{T}_{2(yz)} S_z \mathbf{T}_{2(yz)}^{-1} \mathbf{T}^{-1} = -\mathbf{T} S_z \mathbf{T}^{-1} = S_z. \tag{8.7}$$

In this case, the sign of the spin projection operator S_x is changed under the action of $\mathbf{T}$ operator, but it does not change under the action of $\mathbf{T}_{2(yz)}$ operator, and as a result we have the relation

$$\mathbf{T}\mathbf{T}_{2(yz)} S_x \mathbf{T}_{2(yz)}^{-1} \mathbf{T}^{-1} = -S_x, \tag{8.8}$$

which is equivalent to the relation

$$\mathbf{T}_{1(x)} S_x \mathbf{T}_{1(x)}^{-1} = -S_x, \tag{8.9}$$

while (8.6) and (8.7) are equivalent to the relations

$$\mathbf{T}_{1(x)} S_y \mathbf{T}_{1(x)}^{-1} = S_y, \quad \mathbf{T}_{1(x)} S_z \mathbf{T}_{1(x)}^{-1} = S_z. \tag{8.10}$$

Thus, using the symmetry properties of the $G_8{}^{(\frac{1}{2})}$ group allows to justify the introduction of incomplete time-reversal operators and to represent the Wigner $\mathbf{T}$ operator of complete time-reversal as a product of the incomplete time-reversal operators $\mathbf{T}_{1(x)}$ and $\mathbf{T}_{2(yz)}$ (see (8.5)). It is easy to understand that, along with $\mathbf{T}_{1(x)}$ and $\mathbf{T}_{2(yz)}$, there must be also other operators of incomplete time-reversal, such as $\mathbf{T}_{1(y)}$, $\mathbf{T}_{2(zx)}$, $\mathbf{T}_{1(z)}$, $\mathbf{T}_{2(xy)}$. However, the existence of these operators does not follow from the symmetry of the non-Abelian group $G_8{}^{(\frac{1}{2})}$. In order to prove the existence of the operators of incomplete time-reversal other than $\mathbf{T}_{1(x)}$ and $\mathbf{T}_{2(yz)}$, it is necessary to build and study other groups of eighth order with elements other than the elements of the $G_8{}^{(\frac{1}{2})}$ group. Such groups may be obtained, for example, on the basis of the non-Abelian group

$$G_8{}^{(\frac{1}{2})} : \{i\sigma_y\mathbf{K}, -e, -i\sigma_y\mathbf{K}, e, \sigma_x, -\sigma_x, \sigma_z\mathbf{K}, -\sigma_z\mathbf{K}\} \tag{8.11}$$

by performing the cyclic permutation of indicies x, y, z. It is easy to see that one can build only two groups of eighth order, different from the $G_8{}^{(\frac{1}{2})}$ (which we denoted by $G_8{}^{(\frac{1}{2},a)}$ and $G_8{}^{(\frac{1}{2},b)}$), under cyclic permutation of indices $x \to z \to y \to x$ and $x \to y \to z \to x$:

$$G_8{}^{(\frac{1}{2},a)} : \{i\sigma_x\mathbf{K}, -e, -i\sigma_x\mathbf{K}, e, \sigma_z, -\sigma_z, \sigma_y\mathbf{K}, -\sigma_y\mathbf{K}\}, \tag{8.12}$$

$$G_8{}^{(\frac{1}{2},b)} : \{i\sigma_z\mathbf{K}, -e, -i\sigma_z\mathbf{K}, e, \sigma_y, -\sigma_y, \sigma_x\mathbf{K}, -\sigma_x\mathbf{K}\}. \tag{8.13}$$

The Kayley tables of the group $G_8{}^{(\frac{1}{2},a)}$ and $G_8{}^{(\frac{1}{2},b)}$ are presented in Tables 8.2 and 8.3.

Let's consider again the non-Abelian symmetry group $G_8{}^{(\frac{1}{2})}$ and represent it, taking into account (8.4), as a group consisting of the unit element e, the operator of full time-reversal $\mathbf{T}$ and two different operators of partial time-reversal $\mathbf{T}_{1(x)}$ and $\mathbf{T}_{2(yz)}$. Thus, each of the elements e, $\mathbf{T}$, $\mathbf{T}_{1(x)}$, $\mathbf{T}_{2(yz)}$ will be in the group twice: once with the coefficient +1, and other time with the coefficient −1:

$$G_8{}^{(\frac{1}{2})} : \{\mathbf{T}, -e, -\mathbf{T}, e, \mathbf{T}_{2(yz)}, -\mathbf{T}_{2(yz)}, \mathbf{T}_{1(x)}, -\mathbf{T}_{1(x)}\}. \tag{8.14}$$

Table 8.2 The Kayley table of the non-Abelian group $G_8^{\left(\frac{1}{2},a\right)}$ (8.12)

	$i\sigma_x\mathbf{K}$	$-e$	$-i\sigma_x\mathbf{K}$	e	σ_z	$-\sigma_z$	$\sigma_y\mathbf{K}$	$-\sigma_y\mathbf{K}$
$i\sigma_x\mathbf{K}$	e	$-i\sigma_x\mathbf{K}$	$-e$	$i\sigma_x\mathbf{K}$	$\sigma_y\mathbf{K}$	$-\sigma_y\mathbf{K}$	σ_z	$-\sigma_z$
$-e$	$-i\sigma_x\mathbf{K}$	e	$i\sigma_x\mathbf{K}$	$-e$	$-\sigma_z$	σ_z	$-\sigma_y\mathbf{K}$	$\sigma_y\mathbf{K}$
$-i\sigma_x\mathbf{K}$	$-e$	$i\sigma_x\mathbf{K}$	e	$-i\sigma_x\mathbf{K}$	$-\sigma_y\mathbf{K}$	$\sigma_y\mathbf{K}$	$-\sigma_z$	σ_z
e	$i\sigma_x\mathbf{K}$	$-e$	$-i\sigma_x\mathbf{K}$	e	σ_z	$-\sigma_z$	$\sigma_y\mathbf{K}$	$-\sigma_y\mathbf{K}$
σ_z	$-\sigma_y\mathbf{K}$	$-\sigma_z$	$\sigma_y\mathbf{K}$	σ_z	e	$-e$	$-i\sigma_x\mathbf{K}$	$i\sigma_x\mathbf{K}$
$-\sigma_z$	$\sigma_y\mathbf{K}$	σ_z	$-\sigma_y\mathbf{K}$	$-\sigma_z$	$-e$	e	$i\sigma_x\mathbf{K}$	$-i\sigma_x\mathbf{K}$
$\sigma_y\mathbf{K}$	$-\sigma_z$	$-\sigma_y\mathbf{K}$	σ_z	$\sigma_y\mathbf{K}$	$i\sigma_x\mathbf{K}$	$-i\sigma_x\mathbf{K}$	$-e$	e
$-\sigma_y\mathbf{K}$	σ_z	$\sigma_y\mathbf{K}$	$-\sigma_z$	$-\sigma_y\mathbf{K}$	$-i\sigma_x\mathbf{K}$	$i\sigma_x\mathbf{K}$	e	$-e$

Table 8.3 The Kayley table of the Abelian group $G_8^{\left(\frac{1}{2},b\right)}$ (8.13)

	$i\sigma_z\mathbf{K}$	$-e$	$-i\sigma_z\mathbf{K}$	e	σ_y	$-\sigma_y$	$\sigma_x\mathbf{K}$	$-\sigma_x\mathbf{K}$
$i\sigma_z\mathbf{K}$	e	$-i\sigma_z\mathbf{K}$	$-e$	$i\sigma_z\mathbf{K}$	$-\sigma_x\mathbf{K}$	$\sigma_x\mathbf{K}$	$-\sigma_y$	σ_y
$-e$	$-i\sigma_z\mathbf{K}$	e	$i\sigma_z\mathbf{K}$	$-e$	$-\sigma_y$	σ_y	$-\sigma_x\mathbf{K}$	$\sigma_x\mathbf{K}$
$-i\sigma_z\mathbf{K}$	$-e$	$i\sigma_z\mathbf{K}$	e	$-i\sigma_z\mathbf{K}$	$\sigma_x\mathbf{K}$	$-\sigma_x\mathbf{K}$	σ_y	$-\sigma_y$
e	$i\sigma_z\mathbf{K}$	$-e$	$-i\sigma_z\mathbf{K}$	e	σ_y	$-\sigma_y$	$\sigma_x\mathbf{K}$	$-\sigma_x\mathbf{K}$
σ_y	$-\sigma_x\mathbf{K}$	$-\sigma_y$	$\sigma_x\mathbf{K}$	σ_y	e	$-e$	$-i\sigma_z\mathbf{K}$	$i\sigma_z\mathbf{K}$
$-\sigma_y$	$\sigma_x\mathbf{K}$	σ_y	$-\sigma_x\mathbf{K}$	$-\sigma_y$	$-e$	e	$i\sigma_z\mathbf{K}$	$-i\sigma_z\mathbf{K}$
$\sigma_x\mathbf{K}$	$-\sigma_y$	$-\sigma_x\mathbf{K}$	σ_y	$\sigma_x\mathbf{K}$	$-i\sigma_z\mathbf{K}$	$i\sigma_z\mathbf{K}$	e	$-e$
$-\sigma_x\mathbf{K}$	σ_y	$\sigma_x\mathbf{K}$	$-\sigma_y$	$-\sigma_x\mathbf{K}$	$i\sigma_z K$	$-i\sigma_z\mathbf{K}$	$-e$	e

Table 8.4 The Kayley table of the non-Abelian group $G_8^{\left(\frac{1}{2}\right)}$, formed by the total and partial time-reversal operators

	$\mathbf{T}$	$-e$	$-\mathbf{T}$	e	$\mathbf{T}_{2(yz)}$	$-\mathbf{T}_{2(yz)}$	$\mathbf{T}_{1(x)}$	$-\mathbf{T}_{1(x)}$
$\mathbf{T}$	$-e$	$-\mathbf{T}$	e	$\mathbf{T}$	$\mathbf{T}_{1(x)}$	$-\mathbf{T}_{1(x)}$	$-\mathbf{T}_{2(yz)}$	$\mathbf{T}_{2(yz)}$
$-e$	$-\mathbf{T}$	e	$\mathbf{T}$	$-e$	$-\mathbf{T}_{2(yz)}$	$\mathbf{T}_{2(yz)}$	$-\mathbf{T}_{1(x)}$	$\mathbf{T}_{1(x)}$
$-\mathbf{T}$	e	$\mathbf{T}$	$-e$	$-\mathbf{T}$	$-\mathbf{T}_{1(x)}$	$\mathbf{T}_{1(x)}$	$\mathbf{T}_{2(yz)}$	$-\mathbf{T}_{2(yz)}$
e	$\mathbf{T}$	$-e$	$-\mathbf{T}$	e	$\mathbf{T}_{2(yz)}$	$-\mathbf{T}_{2(yz)}$	$\mathbf{T}_{1(x)}$	$-\mathbf{T}_{1(x)}$
$\mathbf{T}_{2(yz)}$	$-\mathbf{T}_{1(x)}$	$-\mathbf{T}_{2(yz)}$	$\mathbf{T}_{1(x)}$	$\mathbf{T}_{2(yz)}$	e	$-e$	$-\mathbf{T}$	$\mathbf{T}$
$-\mathbf{T}_{2(yz)}$	$\mathbf{T}_{1(x)}$	$\mathbf{T}_{2(yz)}$	$-\mathbf{T}_{1(x)}$	$-\mathbf{T}_{2(yz)}$	$-e$	e	$\mathbf{T}$	$-\mathbf{T}$
$\mathbf{T}_{1(x)}$	$\mathbf{T}_{2(yz)}$	$-\mathbf{T}_{1(x)}$	$-\mathbf{T}_{2(yz)}$	$\mathbf{T}_{1(x)}$	$\mathbf{T}$	$-\mathbf{T}$	e	$-e$
$-\mathbf{T}_{1(x)}$	$-\mathbf{T}_{2(yz)}$	$\mathbf{T}_{1(x)}$	$\mathbf{T}_{2(yz)}$	$-\mathbf{T}_{1(x)}$	$-\mathbf{T}$	$\mathbf{T}$	$-e$	e

In the Table 8.4 there are given the elements of the Kayley table of the group $G_8^{\left(\frac{1}{2}\right)}$, containing the operators from (8.14).

Both Kayley Tables 8.1 and 8.4 of the group $G_8^{\left(\frac{1}{2}\right)}$ are equivalent. In this case, the definition of the $G_8^{\left(\frac{1}{2}\right)}$ group elements (8.14) through the complete and incomplete time-reversal operators has been possible due to the specific structure of the Kayley

Table 8.1. A similar situation is also typical for groups $G_8{}^{\left(\frac{1}{2},a\right)}$ and $G_8{}^{\left(\frac{1}{2},b\right)}$. As follows from the Table 8.2, the following relationship between the elements $\sigma_y\mathbf{K}$, $i\sigma_x\mathbf{K}$ and σ_z takes place:

$$\sigma_y\mathbf{K} = i\sigma_x\mathbf{K}\sigma_z. \tag{8.15}$$

If we denote

$$\overline{\mathbf{T}} = \sigma_y\mathbf{K},\ \mathbf{T}_{1(z)} = i\sigma_x\mathbf{K},\ \mathbf{T}_{2(xy)} = \sigma_z,\ \overline{\mathbf{T}} = \mathbf{T}_{1(z)}\mathbf{T}_{2(xy)}, \tag{8.16}$$

then the group $G_8{}^{\left(\frac{1}{2},a\right)}$ can be represented as

$$G_8{}^{\left(\frac{1}{2},a\right)} : \{\mathbf{T}_{1(z)}, -e, -\mathbf{T}_{1(z)}, e, \mathbf{T}_{2(xy)}, -\mathbf{T}_{2(xy)}, \overline{\mathbf{T}}, -\overline{\mathbf{T}}\}. \tag{8.17}$$

In the case of the group $G_8{}^{\left(\frac{1}{2},a\right)}$ the operator of complete time-reversal $\overline{\mathbf{T}} = \sigma_y\mathbf{K}$ is different from the operator $\mathbf{T} = i\sigma_y\mathbf{K}$ by an insignificant phase factor $i = exp\left(i\frac{\pi}{2}\right)$. As throughout this book it is used the definition of the time-reversal operator $\mathbf{T} = i\sigma_y\mathbf{K}$ for a particle with the spin $\frac{1}{2}$ [222], we denote the operator $\sigma_y\mathbf{K}$ in (8.16) and (8.17) by $\overline{\mathbf{T}}$ in order to avoid any confusion. Under the action of the operators $\mathbf{T}_{1(z)}$, $\mathbf{T}_{2(xy)}$ and $\overline{\mathbf{T}}$ of the group $G_8{}^{\left(\frac{1}{2},a\right)}$, the spin projection operators S_x,S_y and S_z are transformed as follows:

$$\mathbf{T}_{1(z)}S_x\mathbf{T}_{1(z)}^{-1} = S_x,\quad \mathbf{T}_{1(z)}S_y\mathbf{T}_{1(z)}^{-1} = S_y,\quad \mathbf{T}_{1(z)}S_z\mathbf{T}_{1(z)}^{-1} = -S_z \tag{8.18}$$

$$\mathbf{T}_{2(xy)}S_x\mathbf{T}_{2(xy)}^{-1} = -S_x,\quad \mathbf{T}_{2(xy)}S_y\mathbf{T}_{2(xy)}^{-1} = -S_y,\quad \mathbf{T}_{2(xy)}S_z\mathbf{T}_{2(xy)}^{-1} = S_z \tag{8.19}$$

$$\overline{\mathbf{T}}S_x\overline{\mathbf{T}}^{-1} = -S_x,\quad \overline{\mathbf{T}}S_y\overline{\mathbf{T}}^{-1} = -S_y,\quad \overline{\mathbf{T}}S_z\overline{\mathbf{T}}^{-1} = -S_z \tag{8.20}$$

Taking into consideration (8.16) and (8.18)–(8.20), the Kayley table of the group $G_8{}^{\left(\frac{1}{2},a\right)}$ (8.17) can be expressed in terms of the complete and incomplete time-reversal operators $\overline{\mathbf{T}}$, $\mathbf{T}_{1(z)}$ and $\mathbf{T}_{2(xy)}$ (Table 8.5).

Table 8.5 The Kayley table of the non-Abelian group $G_8{}^{\left(\frac{1}{2},a\right)}$ (8.17)

	$\mathbf{T}_{1(z)}$	$-e$	$-\mathbf{T}_{1(z)}$	e	$\mathbf{T}_{2(xy)}$	$-\mathbf{T}_{2(xy)}$	$\overline{\mathbf{T}}$	$-\overline{\mathbf{T}}$
$\mathbf{T}_{1(z)}$	e	$-\mathbf{T}_{1(z)}$	$-e$	$\mathbf{T}_{1(z)}$	$\overline{\mathbf{T}}$	$-\overline{\mathbf{T}}$	$\mathbf{T}_{2(xy)}$	$-\mathbf{T}_{2(xy)}$
$-e$	$-\mathbf{T}_{1(z)}$	e	$\mathbf{T}_{1(z)}$	$-e$	$-\mathbf{T}_2(xy)$	$\mathbf{T}_{2(xy)}$	$-\overline{\mathbf{T}}$	$\overline{\mathbf{T}}$
$-\mathbf{T}_{1(z)}$	$-e$	$\mathbf{T}_{1(z)}$	e	$-\mathbf{T}_{1(z)}$	$-\overline{\mathbf{T}}$	$\overline{\mathbf{T}}$	$-\mathbf{T}_{2(xy)}$	$\mathbf{T}_{2(xy)}$
e	$\mathbf{T}_{1(z)}$	$-e$	$-\mathbf{T}_{1(z)}$	e	$\mathbf{T}_{2(xy)}$	$-\mathbf{T}_{2(xy)}$	$\overline{\mathbf{T}}$	$-\overline{\mathbf{T}}$
$\mathbf{T}_{2(xy)}$	$-\overline{\mathbf{T}}$	$-\mathbf{T}_{2(xy)}$	$\overline{\mathbf{T}}$	$\mathbf{T}_{2(xy)}$	e	$-e$	$-\mathbf{T}_{1(z)}$	$\mathbf{T}_{1(z)}$
$-\mathbf{T}_{2(xy)}$	$\overline{\mathbf{T}}$	$\mathbf{T}_{2(xy)}$	$-\overline{\mathbf{T}}$	$-\mathbf{T}_{2(xy)}$	$-e$	e	$\mathbf{T}_{1(z)}$	$-\mathbf{T}_{1(z)}$
$\overline{\mathbf{T}}$	$-\mathbf{T}_{2(xy)}$	$-\overline{\mathbf{T}}$	$\mathbf{T}_{2(xy)}$	$\overline{\mathbf{T}}$	$\mathbf{T}_{1(z)}$	$-\mathbf{T}_{1(z)}$	$-e$	e
$-\overline{\mathbf{T}}$	$\mathbf{T}_{2(xy)}$	$\overline{\mathbf{T}}$	$-\mathbf{T}_{2(xy)}$	$-\overline{\mathbf{T}}$	$-\mathbf{T}_{1(z)}$	$\mathbf{T}_{1(z)}$	e	$-e$

For obtaining incomplete time-reversal operators, a similar analysis can be performed for the group $G_8^{(\frac{1}{2},b)}$ (8.13), based on the Kayley table of this group. First of all we note that unlike the groups $G_8^{(\frac{1}{2})}$ and $G_8^{(\frac{1}{2},a)}$, the group $G_8^{(\frac{1}{2},b)}$ (8.13) is Abelian and that among the elements of this group there are not found the complete time-reversal operators $\mathbf{T}$ or $\overline{\mathbf{T}}$.

If we introduce the notation

$$\mathbf{T}_{2(zx)} = \sigma_y, \quad \mathbf{T}_{1(x)} = -i\sigma_z\mathbf{K}, \quad \mathbf{T}_{1(z)} = \sigma_x\mathbf{K} \tag{8.21}$$

then in accordance with the Table 8.3 the following relation exists between the operators $(-i\sigma_z\mathbf{K})$, $\sigma_x\mathbf{K}$ and σ_y:

$$\sigma_y = (-i\sigma_z\mathbf{K})\sigma_x\mathbf{K} \tag{8.22}$$

Using (8.21), this relation can be represented as

$$\mathbf{T}_{2(zx)} = \mathbf{T}_{1(x)}\mathbf{T}_{1(z)}. \tag{8.23}$$

The operators of the spin projections S_x, S_y and S_z transform under incomplete time-reversal operators $\mathbf{T}_{1(x)}$, $\mathbf{T}_{1(z)}$ and $\mathbf{T}_{2(xz)}$ of the group $G_8^{(\frac{1}{2},b)}$, as follows:

$$\mathbf{T}_{1(x)}S_x\mathbf{T}_{1(x)}^{-1} = -S_x, \quad \mathbf{T}_{1(x)}S_y\mathbf{T}_{1(x)}^{-1} = S_y, \quad \mathbf{T}_{1(x)}S_z\mathbf{T}_{1(x)}^{-1} = S_z \tag{8.24}$$

$$\mathbf{T}_{1(z)}S_x\mathbf{T}_{1(z)}^{-1} = S_x, \quad \mathbf{T}_{1(z)}S_y\mathbf{T}_{1(z)}^{-1} = S_y, \quad \mathbf{T}_{1(z)}S_z\mathbf{T}_{1(z)}^{-1} = -S_z \tag{8.25}$$

$$\mathbf{T}_{2(xz)}S_x\mathbf{T}_{2(xz)}^{-1} = -S_x, \quad \mathbf{T}_{2(xz)}S_y\mathbf{T}_{2(xz)}^{-1} = S_y, \quad \mathbf{T}_{2(xz)}S_z\mathbf{T}_{2(xz)}^{-1} = -S_z \tag{8.26}$$

Using (8.21), the group $G_8^{(\frac{1}{2},b)}$ (8.13) can be represented as

$$G_8^{(\frac{1}{2},b)} : \{-\mathbf{T}_{1(x)}, -e, \mathbf{T}_{1(x)}, e, \mathbf{T}_{2(zx)}, -\mathbf{T}_{2(zx)}, \mathbf{T}_{1(z)}, -\mathbf{T}_{1(z)}\} \tag{8.27}$$

The Kayley table of the group $G_8^{(\frac{1}{2},b)}$ (8.27) is presented in the Table 8.6.

As seen from (8.13) and (8.21)–(8.23), as well as from Table 8.3 and 8.6, the group $G_8^{(\frac{1}{2},b)}$ includes only incomplete time-reversal operators and the unit element. In this case, the operator $\mathbf{T}_{2(zx)}$ is the product of the operators $\mathbf{T}_{1(x)}$ and $\mathbf{T}_{1(z)}$ (8.23). The results of the identification of incomplete time-reversal operators based on symmetry properties of non-Abelian groups $G_8^{\frac{1}{2}}$ and $G_8^{(\frac{1}{2},a)}$, and the Abelian group $G_8^{(\frac{1}{2},b)}$ are presented in the Table 8.7.

As already mentioned, the operator $\overline{\mathbf{T}}$ from the Table 8.7 related to the group $G_8^{(\frac{1}{2},a)}$ differs from the Wigner time-reversal operator $\mathbf{T} = i\sigma_y\mathbf{K}$ by the phase factor i. The operator $\mathbf{T}_{1(x)}$ related to the group $G_8^{(\frac{1}{2},b)}$ differs by the factor $-i$ from the operator $\mathbf{T}_{1(x)}$ related to the group $G_8^{(\frac{1}{2})}$. Similarly, the operator $\mathbf{T}_{1(z)}$ related to the group $G_8^{\left(\frac{1}{2},a\right)}$ differs by the factor $-i$ from the operator $\mathbf{T}_{1(z)}$ related to the group $G_8^{\left(\frac{1}{2},b\right)}$. On the basis of the Table 8.7 it can be concluded that the use of the symmetry properties of the groups $G_8^{\left(\frac{1}{2}\right)}$, $G_8^{\left(\frac{1}{2},a\right)}$ and $G_8^{\left(\frac{1}{2},b\right)}$ elements permitted identification of the following incomplete time-reversal operators: $\mathbf{T}_{2(yz)}$,$\mathbf{T}_{2(zx)}$, $\mathbf{T}_{2(xy)}$, $\mathbf{T}_{1(x)}$ and $\mathbf{T}_{1(z)}$. However, this approach doesn't allow identifying the operator $\mathbf{T}_{1(y)}$, and the operators $\mathbf{T}_{1(x)}$ and $\mathbf{T}_{1(z)}$ that are found on the basis of different groups to differ from each other by phase factor. Moreover, it is not proved the possible representation of the operator $\mathbf{T}$ with fixed phase factor as operators $\mathbf{T}_{2(xy)}$ $\mathbf{T}_{1(z)}$ and $\mathbf{T}_{2(zx)}\mathbf{T}_{1(y)}$ as well as the product of operators $\mathbf{T}_{1(x)}$,$\mathbf{T}_{1(y)}$ and $\mathbf{T}_{1(z)}$. An alternative way to determine the operators of incomplete time-reversal, free of the deficiencies noted, will be developed in Sect. 8.2 by extending the $G_8^{\left(\frac{1}{2}\right)}$ group to the non-Abelian group of higher, sixteenth order. Before doing this expansion, we will make a generalization of the $G_8^{\left(\frac{1}{2}\right)}$ group to the case of the same group of eight-order, but related to systems with a high half-integer angular momentum J.

Table 8.6 The Kayley table of the Abelian group $G_8^{\left(\frac{1}{2},b\right)}$ (8.27)

	$-\mathbf{T}_{1(x)}$	$-e$	$\mathbf{T}_{1(x)}$	e	$\mathbf{T}_{2(zx)}$	$-\mathbf{T}_{2(zx)}$	$\mathbf{T}_{1(z)}$	$-\mathbf{T}_{1(z)}$
$-\mathbf{T}_{1(x)}$	e	$\mathbf{T}_{1(x)}$	$-e$	$-\mathbf{T}_{1(x)}$	$-\mathbf{T}_{1(z)}$	$\mathbf{T}_{1(z)}$	$-\mathbf{T}_{2(zx)}$	$\mathbf{T}_{2(zx)}$
$-e$	$\mathbf{T}_{1(x)}$	e	$-\mathbf{T}_{1(x)}$	$-e$	$-\mathbf{T}_{2(zx)}$	$\mathbf{T}_{2(zx)}$	$-\mathbf{T}_{1(z)}$	$\mathbf{T}_{1(z)}$
$\mathbf{T}_{1(x)}$	$-e$	$-\mathbf{T}_{1(x)}$	e	$\mathbf{T}_{1(x)}$	$\mathbf{T}_{1(z)}$	$-\mathbf{T}_{1(z)}$	$\mathbf{T}_{2(zx)}$	$-\mathbf{T}_{2(zx)}$
e	$-\mathbf{T}_{1(x)}$	$-e$	$T_{1(x)}$	e	$\mathbf{T}_{2(zx)}$	$-\mathbf{T}_{2(zx)}$	$\mathbf{T}_{1(z)}$	$-\mathbf{T}_{1(z)}$
$\mathbf{T}_{2(zx)}$	$-\mathbf{T}_{1(z)}$	$-\mathbf{T}_{2(zx)}$	$\mathbf{T}_{1(z)}$	$\mathbf{T}_{2(zx)}$	e	$-e$	$\mathbf{T}_{1(x)}$	$-\mathbf{T}_{1(x)}$
$-\mathbf{T}_{2(zx)}$	$\mathbf{T}_{1(z)}$	$\mathbf{T}_{2(zx)}$	$-\mathbf{T}_{1(z)}$	$-\mathbf{T}_{2(zx)}$	$-e$	e	$-\mathbf{T}_{1(x)}$	$\mathbf{T}_{1(x)}$
$\mathbf{T}_{1(z)}$	$-\mathbf{T}_{2(zx)}$	$-\mathbf{T}_{1(z)}$	$\mathbf{T}_{2(zx)}$	$\mathbf{T}_{1(z)}$	$\mathbf{T}_{1(x)}$	$-\mathbf{T}_{1(x)}$	e	$-e$
$-\mathbf{T}_{1(z)}$	$\mathbf{T}_{2(zx)}$	$\mathbf{T}_{1(z)}$	$-\mathbf{T}_{2(zx)}$	$-\mathbf{T}_{1(z)}$	$-\mathbf{T}_{1(x)}$	$\mathbf{T}_{1(x)}$	$-e$	e

Table 8.7 The operators of incomplete time-reversal related to symmetry properties of groups $G_8^{\left(\frac{1}{2}\right)}$, $G_8^{\left(\frac{1}{2},a\right)}$ and $G_8^{\left(\frac{1}{2},b\right)}$

Group	The operators of incomplete time-reversal
$G_8^{\left(\frac{1}{2}\right)}$	$\mathbf{T}_{2(yz)} = \sigma_x$, $\mathbf{T}_{1(x)} = \sigma_z\mathbf{K}$, $\mathbf{T} = i\sigma_y\mathbf{K} = \mathbf{T}_{1(x)}\mathbf{T}_{2(yz)}$. There are missing the definitions of operators $\mathbf{T}_{1(y)}$ and $\mathbf{T}_{1(z)}$, the representation of the operator $\mathbf{T}_{2(yz)}$ as $\mathbf{T}_{2(yz)} = \mathbf{T}_{1(y)}\mathbf{T}_{1(z)}$, and, accordingly, the representation of the $\mathbf{T}$ operator as $\mathbf{T} = \mathbf{T}_{1(x)}\mathbf{T}_{1(y)}\mathbf{T}_{1(z)}$
$G_8^{\left(\frac{1}{2},a\right)}$	$\mathbf{T}_{2(xy)} = \sigma_z$, $\mathbf{T}_{1(z)} = -i\sigma_x\mathbf{K}$, $\overline{\mathbf{T}} = \sigma_y\mathbf{K} = \mathbf{T}_{1(z)}\mathbf{T}_{2(xy)}$. There are missing the definitions of operators $\mathbf{T}_{1(x)}$ and $\mathbf{T}_{1(y)}$, the representation of the operator $\mathbf{T}_{2(xy)}$ through $\mathbf{T}_{1(x)}$ and $\mathbf{T}_{1(y)}$ and, accordingly, the representation of the $\overline{\mathbf{T}}$ operator through $\mathbf{T}_{1(x)}$, $\mathbf{T}_{1(y)}$, and $\mathbf{T}_{1(z)}$
$G_8^{\left(\frac{1}{2},b\right)}$	The group $G_8^{\left(\frac{1}{2},b\right)}$ is Abelian and does not include the operator of complete time-reversal $\mathbf{T}$. $\mathbf{T}_{2(zx)} = \mathbf{T}_{2(xz)} = \sigma_y$, $\mathbf{T}_{1(z)} = \sigma_x\mathbf{K}$, $\mathbf{T}_{1(x)} = -i\sigma_z\mathbf{K}$, $\mathbf{T}_{2(zx)} = \mathbf{T}_{1(z)}\mathbf{T}_{1(x)} = \mathbf{T}_{1(x)}\mathbf{T}_{1(z)}$. There are missing the representations of the operator $\mathbf{T}$ through $\mathbf{T}_{2(zx)}$ and $\mathbf{T}_{1(y)}$, and, accordingly, through operators $\mathbf{T}_{1(x)}$, $\mathbf{T}_{1(y)}$ and $\mathbf{T}_{1(z)}$

Let us suppose that the system is characterized by angular momentum $J = \frac{3}{2}$. In this case, the group $G_8{}^{(\frac{3}{2})}$ is formed by the following elements:

$$G_8{}^{(\frac{3}{2})} : \{\mathbf{T},\ -e^{(\frac{3}{2})},\ -\mathbf{T},\ e^{(\frac{3}{2})},\ V^{(\frac{3}{2})},\ -V^{(\frac{3}{2})},\ W^{(\frac{3}{2})}\mathbf{K},\ -W^{(\frac{3}{2})}\mathbf{K}\}, \quad (8.28)$$

where $e^{(\frac{3}{2})}$ is a unit (4×4)-matrix, $\mathbf{T}$ is given by:

$$\mathbf{T} = U^{(\frac{3}{2})}\mathbf{K}, \quad (8.29)$$

and $U^{(\frac{3}{2})}$, $V^{(\frac{3}{2})}$, $W^{(\frac{3}{2})}$ represents the following matrices

$$U^{(\frac{3}{2})} = \begin{pmatrix} 0 & i\sigma_y \\ i\sigma_y & 0 \end{pmatrix},\ V^{(\frac{3}{2})} = \begin{pmatrix} 0 & \sigma_x \\ \sigma_x & 0 \end{pmatrix},\ W^{(\frac{3}{2})} = \begin{pmatrix} \sigma_z & 0 \\ 0 & \sigma_z \end{pmatrix}. \quad (8.30)$$

In (8.30) and further in (8.33) and (8.36) 0 denotes the zero 2×2-matrix.

If we consider that $\mathbf{T}$ in the Table 8.1 is the time-reversal operator for a system with the angular momentum $J = \frac{3}{2}$ and replace $e, \sigma_x, i\sigma_y$ and σ_z by $e^{(\frac{3}{2})}, V^{(\frac{3}{2})}, U^{(\frac{3}{2})}$ and $W^{(\frac{3}{2})}$, then we obtain the Kayley table of non-Abelian $G_8{}^{(\frac{3}{2})}$ group, related to a system with the angular momentum $J = \frac{3}{2}$.

Now, let's consider a system with the angular momentum $J = \frac{5}{2}$. In this case, the group $G_8{}^{(\frac{5}{2})}$ is

$$G_8{}^{(\frac{5}{2})} : \{\mathbf{T},\ -e^{(\frac{5}{2})},\ -\mathbf{T},\ e^{(\frac{5}{2})},\ V^{(\frac{5}{2})},\ -V^{(\frac{5}{2})},\ W^{(\frac{5}{2})}\mathbf{K},\ -W^{(\frac{5}{2})}\mathbf{K}\}, \quad (8.31)$$

where $e^{(\frac{5}{2})}$ is a unit 6×6-matrix, and the time-reversal operator $\mathbf{T}$ has the form

$$\mathbf{T} = U^{(\frac{5}{2})}\mathbf{K}. \quad (8.32)$$

In (8.31) and (8.32) $U^{(\frac{5}{2})}$, $V^{(\frac{5}{2})}$ and $W^{(\frac{5}{2})}$ represents the matrix:

$$U^{(\frac{5}{2})} = \begin{pmatrix} 0 & 0 & i\sigma_y \\ 0 & i\sigma_y & 0 \\ i\sigma_y & 0 & 0 \end{pmatrix},\ V^{(\frac{5}{2})} = \begin{pmatrix} 0 & 0 & \sigma_x \\ 0 & \sigma_x & 0 \\ \sigma_x & 0 & 0 \end{pmatrix},\ W^{(\frac{5}{2})} = \begin{pmatrix} \sigma_z & 0 & 0 \\ 0 & \sigma_z & 0 \\ 0 & 0 & \sigma_z \end{pmatrix}. \quad (8.33)$$

If we consider that $\mathbf{T}$ in the Table 8.1 is the time-reversal operator for a system with the angular momentum $J = \frac{5}{2}$ and replace $e, \sigma_x, i\sigma_y$ and σ_z by $e^{(\frac{5}{2})}$, $U^{(\frac{5}{2})}$, $V^{(\frac{5}{2})}$ and $W^{(\frac{5}{2})}$, then we obtain the Kayley table of non-Abelian group $G_8{}^{(\frac{5}{2})}$, related to a system with the angular momentum $J = \frac{5}{2}$.

Now, let $J = \frac{7}{2}$. The non-Abelian group $G_8{}^{(\frac{7}{2})}$ of a system with the angular momentum J contains the following elements:

$$G_8{}^{(\frac{7}{2})} : \{\mathbf{T},\ -e^{(\frac{7}{2})},\ -\mathbf{T},\ e^{(\frac{7}{2})},\ V^{(\frac{7}{2})},\ -V^{(\frac{7}{2})},\ W^{(\frac{7}{2})}\mathbf{K},\ -W^{(\frac{7}{2})}\mathbf{K}\}, \quad (8.34)$$

where $e^{(\frac{7}{2})}$ is a unit (8 × 8)- matrix, and the operator **T** is given by

$$\mathbf{T} = U^{(\frac{7}{2})}\mathbf{K}. \tag{8.35}$$

In (8.34) and (8.35), we use the notation $U^{(\frac{7}{2})}$,$V^{(\frac{7}{2})}$ and $W^{(\frac{7}{2})}$ for the following matrices, respectively:

$$U^{(\frac{7}{2})} = \begin{pmatrix} 0 & 0 & 0 & i\sigma_y \\ 0 & 0 & i\sigma_y & 0 \\ 0 & i\sigma_y & 0 & 0 \\ i\sigma_y & 0 & 0 & 0 \end{pmatrix}, \quad V^{(\frac{7}{2})} = \begin{pmatrix} 0 & 0 & 0 & \sigma_x \\ 0 & 0 & \sigma_x & 0 \\ 0 & \sigma_x & 0 & 0 \\ \sigma_x & 0 & 0 & 0 \end{pmatrix},$$
$$W^{(\frac{7}{2})} = \begin{pmatrix} \sigma_z & 0 & 0 & 0 \\ 0 & \sigma_z & 0 & 0 \\ 0 & 0 & \sigma_z & 0 \\ 0 & 0 & 0 & \sigma_z \end{pmatrix}. \tag{8.36}$$

The form of matrices in (8.36) is similar to that of matrices in (8.33).

If we consider that **T** in the Table 8.1 is the time-reversal operator (8.35) for a system with the angular momentum $J = \frac{7}{2}$ and replace e, σ_x, $i\sigma_y$ and σ_z by $e^{(\frac{7}{2})}$, $U^{(\frac{7}{2})}$,$V^{(\frac{7}{2})}$ and $W^{(\frac{7}{2})}$, then we obtain the Kayley table of non-Abelian group ${G_8}^{(\frac{7}{2})}$, related to a system with the angular momentum $J = \frac{7}{2}$. On the basis of (8.28)–(8.36) we can conclude that for an arbitrary half-integer value of the angular momentum J the non-Abelian group ${G_8}^{(J)}$ has the following structure:

$${G_8}^{(J)} : \{\mathbf{T},\ -e^{(J)},\ -\mathbf{T},\ e^{(J)},\ V^{(J)},\ -V^{(J)},\ W^{(J)}\mathbf{K},\ -W^{(J)}\mathbf{K}\} \tag{8.37}$$

where $e^{(J)}$ is the unit $((2J+1) \times (2J+1))$-matrix, and the complete time-reversal operator **T** is

$$\mathbf{T} = U^{(J)}\mathbf{K}. \tag{8.38}$$

In (8.37) and (8.38) $U^{(J)}$, $V^{(J)}$ and $W^{(J)}$ are the following matrices:

$$U^{(J)} = \begin{pmatrix} 0 & & & i\sigma_y \\ & & i\sigma_y & \\ & i\sigma_y & & \\ & \cdot^{\cdot^{\cdot}} & & 0 \\ i\sigma_y & & & \end{pmatrix}, \quad V^{(J)} = \begin{pmatrix} 0 & & & \sigma_x \\ & & \sigma_x & \\ & \sigma_x & & \\ & \cdot^{\cdot^{\cdot}} & & 0 \\ \sigma_x & & & \end{pmatrix},$$

$$W^{(J)} = \begin{pmatrix} \sigma_z & & & 0 \\ & \sigma_z & & \\ & & \sigma_z & \\ 0 & & & \ddots \\ & & & & \sigma_z \end{pmatrix} \tag{8.39}$$

In (8.39) $i\sigma_y$, σ_x and σ_z in matrices $U^{(J)}$, $V^{(J)}$ and $W^{(J)}$ are found $J + \frac{1}{2}$ times.

If we consider that **T** from Table 8.1 is an operator of complete time-reversal for a system with a half-integer angular momentum J and substitute e, σ_x, $i\sigma_y$ and σ_z by $e^{(J)}$, $U^{(J)}$, $V^{(J)}$ and $W^{(J)}$, than we obtain the Kayley table for the non-Abelian group ${G_8}^{(J)}$ related to a system with the half-integer angular momentum J. Analogically, if in (8.14) and in the Table 8.4 we substitute e by $e^{(J)}$ and define the operator **T** by (8.38), and the operators $\mathbf{T}_{2(yz)}$ and $\mathbf{T}_{1(x)}$ by the formulas

$$\mathbf{T}_{2(yz)} = V^{(J)}, \tag{8.40}$$

$$\mathbf{T}_{1(x)} = W^{(J)}\mathbf{K}, \tag{8.41}$$

where $U^{(J)}$, $V^{(J)}$ and $W^{(J)}$ are defined by (8.39), then we obtain the non-Abelian group ${G_8}^{(J)}$, formed by the complete and incomplete time-reversal operators $\pm\mathbf{T}$, $\pm\mathbf{T}_{2(yz)}$ and $\pm\mathbf{T}_{1(x)}$ (taking into account that for Kramers system $\pm e^{(J)} = \pm\mathbf{T}^{\,4}$), and its Kayley table, for arbitrary half-integer J.

8.2 Extension of the Group ${G_8}^{\left(\frac{1}{2}\right)}$ to Non-Abelian Groups of Sixteenth Order Related to Kramers Systems

We extend the group ${G_8}^{\left(\frac{1}{2}\right)}$ to the group ${G_{16}}^{\left(\frac{1}{2}\right)}$, presenting it as a direct product of groups

$${G_{16}}^{\left(\frac{1}{2}\right)} = {G_8}^{\left(\frac{1}{2}\right)} \otimes G_2, \tag{8.42}$$

where G_2 is a second-order cyclic group formed by the elements **K** and $\mathbf{K}^2 = 1$. The non-Abelian group ${G_{16}}^{\left(\frac{1}{2}\right)}$ consists of the following elements:

$$\begin{aligned} {G_{16}}^{\left(\frac{1}{2}\right)} : \{ & i\sigma_y\mathbf{K}, -e, -i\sigma_y\mathbf{K}, e, \sigma_x, -\sigma_x, \sigma_z\mathbf{K}, -\sigma_z\mathbf{K}, \\ & i\sigma_y, -e\mathbf{K}, -i\sigma_y, e\mathbf{K}, \sigma_x\mathbf{K}, -\sigma_x\mathbf{K}, \sigma_z, -\sigma_z\}. \end{aligned} \tag{8.43}$$

It is easy to create the Kayley table of group ${G_{16}}^{\left(\frac{1}{2}\right)}$, which is conveniently to be represented in the form of four blocks: O_{11}, O_{12}, O_{21} and O_{22} (Tables 8.8, 8.9, 8.10 and 8.11).

The block O_{11} of the Kayley table of ${G_{16}}^{\left(\frac{1}{2}\right)}$ group coincides with the Kayley Table 8.1 of the group ${G_8}^{\left(\frac{1}{2}\right)}$, which has been discussed in the Sect. 8.1.Based on the structure of this Kayley table, it was justified the introduction of incomplete time-reversal operators $\mathbf{T}_{2(yz)}$ and $\mathbf{T}_{1(x)}$ (see (8.4)), and shown that the operator **T** can be represented as a product of operators $\mathbf{T}_{1(x)}$ and $\mathbf{T}_{2(yz)}$ (8.1), (8.4), (8.5).

Based on the structure of the block O_{12} of the Kayley table of the group ${G_{16}}^{\left(\frac{1}{2}\right)}$ it can be set the relation

$$(\sigma_z\mathbf{K})(e\mathbf{K}) = \sigma_z. \tag{8.44}$$

Table 8.8 The Kayley table of the non-Abelian group $G_{16}^{\left(\frac{1}{2}\right)}$ (the block O_{11})

	$i\sigma_y\mathbf{K}$	$-e$	$-i\sigma_y\mathbf{K}$	e	σ_x	$-\sigma_x$	$\sigma_z\mathbf{K}$	$-\sigma_z\mathbf{K}$
$i\sigma_y\mathbf{K}$	$-e$	$-i\sigma_y\mathbf{K}$	e	$i\sigma_y\mathbf{K}$	$\sigma_z\mathbf{K}$	$-\sigma_z\mathbf{K}$	$-\sigma_x$	σ_x
$-e$	$-i\sigma_y\mathbf{K}$	e	$i\sigma_y\mathbf{K}$	$-e$	$-\sigma_x$	σ_x	$-\sigma_z\mathbf{K}$	$\sigma_z\mathbf{K}$
$-i\sigma_y\mathbf{K}$	e	$i\sigma_y\mathbf{K}$	$-e$	$-i\sigma_y\mathbf{K}$	$-\sigma_z\mathbf{K}$	$\sigma_z\mathbf{K}$	σ_x	$-\sigma_x$
e	$i\sigma_y\mathbf{K}$	$-e$	$-i\sigma_y\mathbf{K}$	e	σ_x	$-\sigma_x$	$\sigma_z\mathbf{K}$	$-\sigma_z\mathbf{K}$
σ_x	$-\sigma_z\mathbf{K}$	$-\sigma_x$	$\sigma_z\mathbf{K}$	σ_x	e	$-e$	$-i\sigma_y\mathbf{K}$	$i\sigma_y\mathbf{K}$
$-\sigma_x$	$\sigma_z\mathbf{K}$	σ_x	$-\sigma_z\mathbf{K}$	$-\sigma_x$	$-e$	e	$i\sigma_y\mathbf{K}$	$-i\sigma_y\mathbf{K}$
$\sigma_z\mathbf{K}$	σ_x	$-\sigma_z\mathbf{K}$	$-\sigma_x$	$\sigma_z\mathbf{K}$	$i\sigma_y\mathbf{K}$	$-i\sigma_y\mathbf{K}$	e	$-e$
$-\sigma_z\mathbf{K}$	$-\sigma_x$	$\sigma_z\mathbf{K}$	σ_x	$-\sigma_z\mathbf{K}$	$-i\sigma_y\mathbf{K}$	$i\sigma_y\mathbf{K}$	$-e$	e

Table 8.9 The Kayley table of the non-Abelian group $G_{16}^{\left(\frac{1}{2}\right)}$(the block O_{12})

	$i\sigma_y$	$-e\mathbf{K}$	$-i\sigma_y$	$e\mathbf{K}$	$\sigma_x\mathbf{K}$	$-\sigma_x\mathbf{K}$	σ_z	$-\sigma_z$
$i\sigma_y\mathbf{K}$	$-e\mathbf{K}$	$-i\sigma_y$	$e\mathbf{K}$	$i\sigma_y$	σ_z	$-\sigma_z$	$-\sigma_x\mathbf{K}$	$\sigma_x\mathbf{K}$
$-e$	$-i\sigma_y$	$e\mathbf{K}$	$i\sigma_y$	$-e\mathbf{K}$	$-\sigma_x\mathbf{K}$	$\sigma_x\mathbf{K}$	$-\sigma_z$	σ_z
$-i\sigma_y\mathbf{K}$	$e\mathbf{K}$	$i\sigma_y$	$-e\mathbf{K}$	$-i\sigma_y$	$-\sigma_z$	σ_z	$\sigma_x\mathbf{K}$	$-\sigma_x\mathbf{K}$
e	$i\sigma_y$	$-e\mathbf{K}$	$-i\sigma_y$	$e\mathbf{K}$	$\sigma_x\mathbf{K}$	$-\sigma_x\mathbf{K}$	σ_z	$-\sigma_z$
σ_x	$-\sigma_z$	$-\sigma_x\mathbf{K}$	σ_z	$\sigma_x\mathbf{K}$	$e\mathbf{K}$	$-e\mathbf{K}$	$-i\sigma_y$	$i\sigma_y$
$-\sigma_x$	σ_z	$\sigma_x\mathbf{K}$	$-\sigma_z$	$-\sigma_x K$	$-e\mathbf{K}$	$e\mathbf{K}$	$i\sigma_y$	$-i\sigma_y$
$\sigma_z\mathbf{K}$	$\sigma_x\mathbf{K}$	$-\sigma_z$	$-\sigma_x\mathbf{K}$	σ_z	$i\sigma_y$	$-i\sigma_y$	$e\mathbf{K}$	$-e\mathbf{K}$
$-\sigma_z\mathbf{K}$	$-\sigma_x\mathbf{K}$	σ_z	$\sigma_x\mathbf{K}$	$-\sigma_z$	$-i\sigma_y$	$i\sigma_y$	$-e\mathbf{K}$	$e\mathbf{K}$

Table 8.10 The Kayley table of the non-Abelian group $G_{16}^{\left(\frac{1}{2}\right)}$ (the block O_{21})

	$i\sigma_y\mathbf{K}$	$-e$	$-i\sigma_y\mathbf{K}$	e	σ_x	$-\sigma_x$	$\sigma_z\mathbf{K}$	$-\sigma_z\mathbf{K}$
$i\sigma_y$	$-e\mathbf{K}$	$-i\sigma_y$	$e\mathbf{K}$	$i\sigma_y$	σ_z	$-\sigma_z$	$-\sigma_x\mathbf{K}$	$\sigma_x\mathbf{K}$
$-e\mathbf{K}$	$-i\sigma_y$	$e\mathbf{K}$	$i\sigma_y$	$-e\mathbf{K}$	$-\sigma_x\mathbf{K}$	$\sigma_x\mathbf{K}$	$-\sigma_z$	σ_z
$-i\sigma_y$	$e\mathbf{K}$	$i\sigma_y$	$-e\mathbf{K}$	$-i\sigma_y$	$-\sigma_z$	σ_z	$\sigma_x\mathbf{K}$	$-\sigma_x\mathbf{K}$
$e\mathbf{K}$	$i\sigma_y$	$-e\mathbf{K}$	$-i\sigma_y$	$e\mathbf{K}$	$\sigma_x\mathbf{K}$	$-\sigma_x\mathbf{K}$	σ_z	$-\sigma_z$
$\sigma_x\mathbf{K}$	$-\sigma_z$	$-\sigma_x\mathbf{K}$	σ_z	$\sigma_x\mathbf{K}$	$e\mathbf{K}$	$-e\mathbf{K}$	$-i\sigma_y$	$i\sigma_y$
$-\sigma_x\mathbf{K}$	σ_z	$\sigma_x\mathbf{K}$	$-\sigma_z$	$-\sigma_x\mathbf{K}$	$-e\mathbf{K}$	$e\mathbf{K}$	$i\sigma_y$	$-i\sigma_y$
σ_z	$\sigma_x\mathbf{K}$	$-\sigma_z$	$-\sigma_x\mathbf{K}$	σ_z	$i\sigma_y$	$-i\sigma_y$	$e\mathbf{K}$	$-e\mathbf{K}$
$-\sigma_z$	$-\sigma_x\mathbf{K}$	σ_z	$\sigma_z\mathbf{K}$	$-\sigma_z$	$-i\sigma_y$	$i\sigma_y$	$-e\mathbf{K}$	$e\mathbf{K}$

If the notation $\mathbf{T}_{2(xy)}$ from (8.16) is used and a new notation

$$\mathbf{T}_{1(x)} = \sigma_z\mathbf{K}, \quad \mathbf{T}_{1(y)} = e\mathbf{K} \tag{8.45}$$

Table 8.11 The Kayley table of the non-Abelian group $G_{16}^{(\frac{1}{2})}$ (the block O_{22})

	$i\sigma_y$	$-e\mathbf{K}$	$-i\sigma_y$	$e\mathbf{K}$	$\sigma_x\mathbf{K}$	$-\sigma_x\mathbf{K}$	σ_z	$-\sigma_z$
$i\sigma_y$	$-e$	$-i\sigma_y\mathbf{K}$	e	$i\sigma_y\mathbf{K}$	$\sigma_z\mathbf{K}$	$-\sigma_z\mathbf{K}$	$-\sigma_x$	σ_x
$-e\mathbf{K}$	$-i\sigma_y\mathbf{K}$	e	$i\sigma_y\mathbf{K}$	$-e$	$-\sigma_x$	σ_x	$-\sigma_z\mathbf{K}$	$\sigma_z\mathbf{K}$
$-i\sigma_y$	e	$i\sigma_y\mathbf{K}$	$-e$	$-i\sigma_y\mathbf{K}$	$-\sigma_z\mathbf{K}$	$\sigma_z\mathbf{K}$	σ_x	$-\sigma_x$
$e\mathbf{K}$	$i\sigma_y\mathbf{K}$	$-e$	$-i\sigma_y\mathbf{K}$	e	σ_x	$-\sigma_x$	$\sigma_z\mathbf{K}$	$-\sigma_z\mathbf{K}$
$\sigma_x\mathbf{K}$	$-\sigma_z\mathbf{K}$	$-\sigma_x$	$\sigma_z\mathbf{K}$	σ_x	e	$-e$	$-i\sigma_y\mathbf{K}$	$i\sigma_y\mathbf{K}$
$-\sigma_x\mathbf{K}$	$\sigma_z\mathbf{K}$	σ_x	$-\sigma_z\mathbf{K}$	$-\sigma_x$	$-e$	e	$i\sigma_y\mathbf{K}$	$-i\sigma_y\mathbf{K}$
σ_z	σ_x	$-\sigma_z\mathbf{K}$	$-\sigma_x$	$\sigma_z\mathbf{K}$	$i\sigma_y\mathbf{K}$	$-i\sigma_y\mathbf{K}$	e	$-e$
$-\sigma_z$	$-\sigma_x$	$\sigma_z\mathbf{K}$	σ_x	$-\sigma_z\mathbf{K}$	$-i\sigma_y\mathbf{K}$	$i\sigma_y\mathbf{K}$	$-e$	e

is introduced, then the notation (8.44) can be rewritten as

$$\mathbf{T}_{2(xy)} = \mathbf{T}_{1(x)}\mathbf{T}_{1(y)}. \tag{8.46}$$

It is seen that the operator $\mathbf{T}_{1(x)} = -i\sigma_z\mathbf{K}$ from (8.21), obtained by using the group $G_8^{(\frac{1}{2},b)}$ (8.13), differs by the phase factor $-i$ from the operator $\mathbf{T}_{1(x)}$ from (8.45), defined by using the block O_{12} of the Kayley table of the group $G_{16}^{(\frac{1}{2})}$. However, we note that the difference in phase factors does not affect the result of the action of the $\mathbf{T}_{1(x)}$ operator on the spin projection operators S_x, S_y and S_z.

The structure of the block O_{21} allows writing the relation

$$(e\mathbf{K})(\sigma_z\mathbf{K}) = \sigma_z, \tag{8.47}$$

which coincides with (8.44), since the operators $e\mathbf{K}$ and $\sigma_z\mathbf{K}$ commute. In addition, the rules of multiplication of elements of the $G_{16}^{(\frac{1}{2})}$ group belonging to the block O_{21}, lead to the known relation between the Pauli matrices

$$\sigma_z\sigma_x = i\sigma_y. \tag{8.48}$$

However, it is not surprising this known relationship, confirmed by the Kayley table, but the fact that $i\sigma_y$, σ_z and σ_x are operators of incomplete time-reversal: $\mathbf{T}_{2(zx)} = i\sigma_y$, $\mathbf{T}_{2(xy)} = \sigma_z$, $\mathbf{T}_{2(yz)} = \sigma_x$ and $\mathbf{T}_{2(zx)} = \mathbf{T}_{2(xy)}\mathbf{T}_{2(yz)}$. It should be noted that the operator S_y changes its sign twice under the operator $\mathbf{T}_{2(xy)}\mathbf{T}_{2(yz)}$ (as a result it remains unchanged), while the operators S_x and S_z change the sign only once. As a consequence, the operator $\mathbf{T}_{2(xy)}\mathbf{T}_{2(yz)}$ is equivalent to the operator $\mathbf{T}_{2(zx)}$.

New operator relations containing the incomplete time-reversal operators can also be found based on the structure of the Kayley table of the group $G_{16}^{(\frac{1}{2})}$

$$\mathbf{T} = (i\sigma_y)(e\mathbf{K}) = i\sigma_y\mathbf{K}, \tag{8.49}$$

$$\mathbf{T} = \sigma_z(\sigma_x\mathbf{K}) = i\sigma_y\mathbf{K}. \tag{8.50}$$

According to the rules of multiplication of O_{11} block elements of Kayley table for the group $G_{16}{}^{\left(\frac{1}{2}\right)}$ (which coincide with the rules of multiplication of $G_8{}^{\left(\frac{1}{2}\right)}$ group elements), we have the relations

$$\mathbf{T} = i\sigma_y\mathbf{K} = (\sigma_z\mathbf{K})\sigma_x = \mathbf{T}_{1(x)}\mathbf{T}_{2(yz)}, \tag{8.51}$$

where $T_{1(x)}$ and $T_{2(yz)}$ are defined in (8.4). On the other hand, on the basis of the rules of multiplication of O_{22} block elements we find

$$\mathbf{T}_{2(yz)} = \sigma_x = (e\mathbf{K})(\sigma_x\mathbf{K}) = \mathbf{T}_{1(y)}\mathbf{T}_{1(z)}, \tag{8.52}$$

where

$$\mathbf{T}_{1(y)} = e\mathbf{K},\ \mathbf{T}_{1(z)} = \sigma_x\mathbf{K}. \tag{8.53}$$

Using (8.53), the operator $\mathbf{T}$ of (8.51) can be represented as

$$\mathbf{T} = \mathbf{T}_{1(x)}\mathbf{T}_{2(yz)} = \mathbf{T}_{1(x)}\mathbf{T}_{1(y)}\mathbf{T}_{1(z)}. \tag{8.54}$$

Further, on the basis of the rules of multiplication of elements of the same block we find two expressions for the operator $\mathbf{T}$, different in form but identical in content

$$\begin{aligned}\mathbf{T} &= (i\sigma_y)(e\mathbf{K}) = \mathbf{T}_{2(zx)}\mathbf{T}_{1(y)} = (\sigma_z\mathbf{K})(\sigma_x\mathbf{K})\mathbf{T}_{1(y)} = \\ &= \mathbf{T}_{1(x)}\mathbf{T}_{1(z)}\mathbf{T}_{1(y)} = \mathbf{T}_{1(x)}\mathbf{T}_{1(y)}\mathbf{T}_{1(z)},\end{aligned} \tag{8.55}$$

$$\mathbf{T} = \sigma_z(\sigma_x\mathbf{K}) = \mathbf{T}_{2(xy)}\mathbf{T}_{1(z)}, \tag{8.56}$$

where

$$\mathbf{T}_{2(zx)} = i\sigma_y,\ \mathbf{T}_{2(xy)} = \sigma_z, \tag{8.57}$$

while $\mathbf{T}_{1(y)}$, $\mathbf{T}_{1(z)}$ and $\mathbf{T}_{1(x)}$ are determined by (8.53) and (8.4), respectively. The rules of multiplication of O_{12} block elements of Kayley table for the group $G_{16}{}^{\left(\frac{1}{2}\right)}$ allow the operator $\mathbf{T}_{2(xy)} = \sigma_z$ from (8.57) to be represented as

$$\mathbf{T}_{2(xy)} = (\sigma_z\mathbf{K})(e\mathbf{K}) = \mathbf{T}_{1(x)}\mathbf{T}_{1(y)}. \tag{8.58}$$

Substituting $\mathbf{T}_{2(xy)}$ of (8.58) in (8.56), we find

$$\mathbf{T} = \mathbf{T}_{2(xy)}\mathbf{T}_{1(z)} = \mathbf{T}_{1(x)}\mathbf{T}_{1(y)}\mathbf{T}_{1(z)}. \tag{8.59}$$

It is seen that the relations (8.49) and (8.50) can be represented in the form of (8.55) and (8.56) (or (8.59)), of which right sides contain the operators of incomplete time-reversal. Thus, it was proved the existence of six incomplete time-reversal operators $\mathbf{T}_{1(x)}$, $\mathbf{T}_{1(y)}$, $\mathbf{T}_{1(z)}$, $\mathbf{T}_{2(yz)}$, $\mathbf{T}_{2(zx)}$, and $\mathbf{T}_{2(xy)}$ on the basis of rules of multiplication of the $G_{16}^{\left(\frac{1}{2}\right)}$ group elements.

In quantum mechanics and quantum field theory, T-symmetry is one of the fundamental discrete symmetries. With regard to $T_{1(x)}$-, $T_{1(y)}$-, $\mathbf{T}_{1(z)}$-, $T_{2(yz)}$-, $T_{2(zx)}$-, and $T_{2(xy)}$-symmetry, now their range of application is not determined. Assuming that under some conditions any one of these symmetries takes place, at the same time a violation of T-symmetry will happen. In this case, one should talk about a lowering of the symmetry, not about the violation of the complete time-reversal symmetry, by analogy with the lowering of the spatial symmetry in the presence of an external perturbation. If the first eight elements of the $G_{16}^{\left(\frac{1}{2}\right)}$ group (8.43) are presented in the form of equivalent elements of the $G_8^{\left(\frac{1}{2}\right)}$ group (8.14), and for the remaining elements we take into account that $i\sigma_y = \mathbf{T}_{2(zx)}$, $e\mathbf{K} = \mathbf{T}_{1(y)}$, $\sigma_x\mathbf{K} = \mathbf{T}_{1(z)}$, and $\sigma_z = \mathbf{T}_{2(xy)}$, (8.53) and (8.57), this group can be represented by operators of complete and incomplete time-reversal, taking into account that $e = \mathbf{T}^4$

$$G_{16}^{\left(\frac{1}{2}\right)} : \big\{\mathbf{T},\ -e,\ -\mathbf{T},\ e,\ \mathbf{T}_{2(yz)},\ -\mathbf{T}_{2(yz)},\ \mathbf{T}_{1(x)},\ -\mathbf{T}_{1(x)},\ \mathbf{T}_{2(zx)},\ -\mathbf{T}_{1(y)},\ -\mathbf{T}_{2(zx)},\ \mathbf{T}_{1(y)},\ \mathbf{T}_{1(z)},\ -\mathbf{T}_{1(z)},\ \mathbf{T}_{2(xy)},\ -\mathbf{T}_{2(xy)}\big\}. \tag{8.60}$$

The Kayley table of this group is presented in the Appendix B.

Extension of the $G_{16}^{\left(\frac{1}{2}\right)}$ group to $G_{16}^{(J)}$ at a half-integer angular momentum $J > \frac{1}{2}$ is carried out by analogy with the extension of the $G_8^{\left(\frac{1}{2}\right)}$ group to $G_8^{(J)}$. In this case, the $G_{16}^{(J)}$ group may be represented as direct products of groups $G_8^{(J)}$ and G_2,

$$G_{16}^{(J)} = G_8^{(J)} \otimes G_2, \tag{8.61}$$

where the $G_8^{(J)}$ group is defined in (8.37) and G_2 is a second-order Abelian group, formed of $\mathbf{K}$ and $\mathbf{K}^2 = 1$ elements. Considering this, it is easy to show that the $G_{16}^{(J)}$ group is formed of the following elements

$$G_{16}^{(J)} : \{\mathbf{T},\ -e^{(J)},\ -\mathbf{T},\ e^{(J)},\ V^{(J)},\ -V^{(J)},\ W^{(J)}\mathbf{K},\ -W^{(J)}\mathbf{K},\ U^{(J)},\ -e^{(J)}\mathbf{K},\ -U^{(J)},\ e\mathbf{K},\ V^{(J)}\mathbf{K},\ -V^{(J)}\mathbf{K},\ W^{(J)},\ -W^{(J)}\}, \tag{8.62}$$

where $\mathbf{T}$, $U^{(J)}$, $V^{(J)}$, and $W^{(J)}$ are defined by (8.38) and (8.39). The Kayley table of the $G_{16}^{(J)}$ group (8.62) is presented in the Appendix C.

Using (8.38), (8.40), and (8.41) and introducing the notation

$$\mathbf{T}_{2(zx)} = U^{(J)},\quad \mathbf{T}_{2(xy)} = W^{(J)},\quad \mathbf{T}_{1(y)} = e^{(J)}\mathbf{K},\ \mathbf{T}_{1(z)} = V^{(J)}\mathbf{K} \tag{8.63}$$

and taking into account that $\mathbf{T}^4 = (V^{(J)})^4\mathbf{K}^4 = e^{(J)}$ for Kramers systems, the group $G_{16}{}^{(J)}$ elements can be defined by the operators of complete and incomplete time-reversal

$$G_{16}{}^{(J)} : \{\mathbf{T},\ -e^{(J)},\ -\mathbf{T},\ e^{(J)},\ \mathbf{T}_{2(yz)},\ -\mathbf{T}_{2(yz)},\ \mathbf{T}_{1(x)},\ -\mathbf{T}_{1(x)},\ \mathbf{T}_{2(zx)},\ -\mathbf{T}_{1(y)},\ -\mathbf{T}_{2(zx)},\ \mathbf{T}_{1(y)},\ \mathbf{T}_{1(z)},\ -\mathbf{T}_{1(z)},\ \mathbf{T}_{2(xy)},\ -\mathbf{T}_{2(xy)}\}. \tag{8.64}$$

The Kayley table of the $G_{16}{}^{(J)}$ group (8.64) is presented in the Appendix D.

It should be noted that the Kayley table of the $G_{16}{}^{(J)}$ group (8.64) at half-integer J is identical in form with the Kayley table of the $G_{16}{}^{(\frac{1}{2})}$ group (8.60). This means that the structure of the matrices $U^{(J)}$,$V^{(J)}$ and $W^{(J)}$ from (8.39) at half-integer J has been determined correctly and that there are an infinite number of finite non-Abelian $G_{16}{}^{(J)}$ groups at half-integer $J = \frac{1}{2},\ 1,\ \frac{3}{2},\ldots,\infty$ for which the Kayley tables have the same structure.

With regard to incomplete time-reversal operators, they can be introduced into quantum mechanics for particles with the spin $\frac{1}{2}$ in two different ways: (1) on the basis of Kayley tables of non-Abelian groups $G_8{}^{(\frac{1}{2})}$,$G_8{}^{(\frac{1}{2},a)}$ and $G_8{}^{(\frac{1}{2},b)}$(Tables 8.1, 8.2 and 8.3), and (2) on the basis of the Kayley table of the $G_{16}{}^{(\frac{1}{2})}$ group (Tables 8.8, 8.9, 8.10 and 8.11). As it was shown in Sect. 8.1, the incomplete time-reversal operators obtained by the first method may differ by phase factors. For example, the operator $\mathbf{T}_{1(x)} = -i\sigma_z\mathbf{K}$, found on the basis of the $G_8{}^{(\frac{1}{2},b)}$ group, differs by a phase factor $-i$ from the $\mathbf{T}_{1(x)} = \sigma_z\mathbf{K}$ operator, found on the basis of the group $G_8{}^{(\frac{1}{2})}$. Similarly, the operator $\mathbf{T}_{1(z)} = -i\sigma_x\mathbf{K}$, determined using the $G_8{}^{(\frac{1}{2},a)}$ group, differs by an $-i$ factor from the $\mathbf{T}_{1(z)} = \sigma_x\mathbf{K}$ operator, found on the basis of the $G_8{}^{(\frac{1}{2},b)}$ group. Analogically, the operator $\mathbf{T}_{1(z)} = -i\sigma_x\mathbf{K}$, defined by a group $G_8{}^{(\frac{1}{2},a)}$, differs by a factor $-i$ from the operator $\mathbf{T}_{1(z)} = \sigma_x\mathbf{K}$, found on the basis of the group $G_8{}^{(\frac{1}{2},b)}$. This difference in the phase factors has no effect on the result of action of incomplete time-reversal operators on the spin projection operators and on the wave functions (which are determined themselves up to a phase factor). Nevertheless, it seems that determining the incomplete time-reversal operators is more acceptable by second way, in which their form are uniquely determined.

8.3 Abelian Groups of Eighth and Sixteenth Orders Related to Non-Kramers Systems

The essential difference between non-Kramers and Kramers system is that for the first the Abelian group, built on the basis of the $\mathbf{T}$ and $\mathbf{T}^2$ operators, is a second-order group, while for Kramers systems the lowest order Abelian group is built on the basis of $\mathbf{T}$, $\mathbf{T}^2$, $\mathbf{T}^3$ and $\mathbf{T}^4$ operators and is a fourth-order group.

In the Sect. 3.2, the $G_4{}^{(\frac{1}{2})}: \{\mathbf{T},\ \mathbf{T}^2 = -e,\ \mathbf{T}^3 = -\mathbf{T},\ \mathbf{T}^4 = e\}$ group has been extended to the $G_8{}^{(\frac{1}{2})}$ group on the basis of existence of an isomorphic group $4m'm$ of generalized symmetry of a square with neighboring vertices colored with different colors (at the same time, the opposite vertices must have of the same color). The $4'm'm$ group has eight symmetry elements: four antirotations, $4'_z$, $4'^2_z$, $4'^3_z$, $4'^4_z = e$, two reflections $m_{xy}{}^{(i)}(i = 1, 2)$ in the diagonals of the square and two antireflections $m_x{}^{(i)}(i = 1, 2)$ in the lines passing through the middle of the opposite sides (Fig. 3.1). In the Sect. 8.1, two new groups $G_8{}^{(\frac{1}{2},a)}$ and $G_8{}^{(\frac{1}{2},b)}$, have been obtained on the basis of the $G_8{}^{(\frac{1}{2})}$ group. Based on the Kayley tables of the $G_8{}^{(\frac{1}{2})}$ and $G_8{}^{(\frac{1}{2},a)}$ groups, it has been demonstrated the existence of $\mathbf{T}_{1(x)}$ and $\mathbf{T}_{2(yz)}$, and, respectively, $\mathbf{T}_{1(z)}$ and $\mathbf{T}_{2(xy)}$ operators of incomplete time reversal. The $G_8{}^{(\frac{1}{2},b)}$ group does not contain the time-reversal operator $\mathbf{T}$. However, using this group it has been proven the representation of the $\mathbf{T}_{2(zx)}$ operator as $\mathbf{T}_{2(zx)} = \mathbf{T}_{1(z)}\mathbf{T}_{1(x)}$. It was shown that the regular method of introducing all possible (compatible with the symmetry properties) operators of incomplete time reversal $\mathbf{T}_{2(yz)}$, $\mathbf{T}_{2(zx)}$, $\mathbf{T}_{2(xy)}$, $\mathbf{T}_{1(x)}$, $\mathbf{T}_{1(y)}$ and $\mathbf{T}_{1(z)}$ for a particle with spin $\frac{1}{2}$ is based on the extension of the group $G_8{}^{(\frac{1}{2})}$ to the non-Abelian group $G_{16}{}^{(\frac{1}{2})}$. The Groups $G_8{}^{(\frac{1}{2})}$ and $G_{16}{}^{(\frac{1}{2})}$ were further extended to the groups $G_8{}^{(J)}$ and $G_{16}{}^{(J)}$ at half-integer J.

A similar, but not identical program can also be used to find operators of incomplete time reversal in the case of non-Kramers systems. We begin with the analysis of the simplest case of a non-Kramers system characterized by the angular momentum $J = 1$. It is easy to verify that the following four operators $\mathbf{T} = U^{(1)}\mathbf{K}$, $e^{(1)}$, $W^{(1)}$ and $V^{(1)}\mathbf{K}$ form the group

$$G_4{}^{(1)}: \{\mathbf{T},\ e^{(1)},\ W^{(1)},\ V^{(1)}\mathbf{K}\}, \tag{8.65}$$

where

$$U^{(1)} = \begin{pmatrix} 0 & 0 & 1 \\ 0 & -1 & 0 \\ 1 & 0 & 0 \end{pmatrix},\quad V^{(1)} = \begin{pmatrix} 0 & 0 & 1 \\ 0 & 1 & 0 \\ 1 & 0 & 0 \end{pmatrix},\quad W^{(1)} = \begin{pmatrix} 1 & 0 & 0 \\ 0 & -1 & 0 \\ 0 & 0 & 1 \end{pmatrix} \tag{8.66}$$

and $e^{(1)}$ is the unit 3×3-matrix.

It must be noted that the number of rows and the number of columns of matrices $e^{(1)}$, $U^{(1)}$, $W^{(1)}$ and $V^{(1)}$ is three. This number is odd for any integer J, which makes it impossible to use Kronecker products of operators to extend the group $G_4{}^{(1)}$ to the group $G_4{}^{(J)}$ with an integer J, in contrast to the Kramers systems for which such an extension was carried out. Another significant difference between the groups related to Kramers systems and the groups related to non-Kramers systems is that in the first case these groups are non-Abelian, whereas in the second case they are Abelian groups.

It is easy to verify that the elements of the group $G_4{}^{(1)}$ commute with each other. If, in addition, the extension of the group $G_4{}^{(1)}$ to a group of higher order (for example,

Table 8.12 The Kayley table of the Abelian group ${G_4}^{(1)}$

	$\mathbf{T}$	$e^{(1)}$	$W^{(1)}$	$V^{(1)}\mathbf{K}$
$\mathbf{T}$	$e^{(1)}$	$\mathbf{T}$	$V^{(1)}\mathbf{K}$	$W^{(1)}$
$e^{(1)}$	$\mathbf{T}$	$e^{(1)}$	$W^{(1)}$	$V^{(1)}\mathbf{K}$
$W^{(1)}$	$V^{(1)}\mathbf{K}$	$W^{(1)}$	$e^{(1)}$	$\mathbf{T}$
$V^{(1)}\mathbf{K}$	$W^{(1)}$	$V^{(1)}\mathbf{K}$	$\mathbf{T}$	$e^{(1)}$

Table 8.13 The Kayley table of the Abelian group ${G_4}^{(1)}$ (8.70)

	$\mathbf{T}$	$e^{(1)}$	$\mathbf{T}_{2(xy)}$	$\mathbf{T}_{1(z)}$
$\mathbf{T}$	$e^{(1)}$	$\mathbf{T}$	$\mathbf{T}_{1(z)}$	$\mathbf{T}_{2(xy)}$
$e^{(1)}$	$\mathbf{T}$	$e^{(1)}$	$\mathbf{T}_{2(xy)}$	$\mathbf{T}_{1(z)}$
$\mathbf{T}_{2(xy)}$	$\mathbf{T}_{1(z)}$	$\mathbf{T}_{2(xy)}$	$e^{(1)}$	$\mathbf{T}$
$\mathbf{T}_{1(x)}$	$\mathbf{T}_{2(xy)}$	$\mathbf{T}_{1(z)}$	$\mathbf{T}$	$e^{(1)}$

to the group ${G_8}^{(1)}$) is carried out using some cyclic group, then the extended group will also be Abelian. The Kayley table of the group ${G_4}^{(1)}$ is given in the Table 8.12.

In the Table 8.12, the time-reversal operator $\mathbf{T}$ is represented as $\mathbf{T} = U^{(1)}\mathbf{K}$, and $W^{(1)}$ and $V^{(1)}\mathbf{K}$ are operators of incomplete time reversal

$$W^{(1)}S_x\left(W^{(1)}\right)^{-1} = -S_x, \quad W^{(1)}S_y\left(W^{(1)}\right)^{-1} = -S_y, \quad W^{(1)}S_z\left(W^{(1)}\right)^{-1} = S_z, \tag{8.67}$$

$$\begin{aligned} &V^{(1)}\mathbf{K}S_x(V^{(1)}\mathbf{K})^{-1} = S_x, \qquad V^{(1)}\mathbf{K}S_y(V^{(1)}\mathbf{K})^{-1} = S_y, \\ &V^{(1)}\mathbf{K}S_z(V^{(1)}\mathbf{K})^{-1} = -S_z. \end{aligned} \tag{8.68}$$

On the basis of (8.67) and (8.68) and in accordance with the notations adopted in this book, operators $W^{(1)}$ and $V^{(1)}\mathbf{K}$ should be represented through the operators $\mathbf{T}_{2(xy)}$ and $\mathbf{T}_{1(z)}$:

$$\mathbf{T}_{2(xy)} = W^{(1)}, \quad \mathbf{T}_{1(z)} = V^{(1)}\mathbf{K}. \tag{8.69}$$

In this case, the group ${G_4}^{(1)}$ (8.65) can be represented in the form

$${G_4}^{(1)} : \{\, \mathbf{T},\ e^{(1)},\ \mathbf{T}_{2(xy)},\ \mathbf{T}_{1(z)} \,\}. \tag{8.70}$$

Taking into account that $e^{(1)} = \mathbf{T}^2$, the Kayley table of the group ${G_4}^{(1)}$ can be represented through operators of complete and incomplete time reversal (Table 8.13).

Using the cyclic group $G_2 : \{\mathbf{K},\ \mathbf{K}^2 = 1\}$, the group ${G_4}^{(1)}$can be extended to the group ${G_8}^{(1)} = {G_4}^{(1)} \otimes G_2$:

$${G_8}^{(1)} : \{\, \mathbf{T},\ e^{(1)},\ W^{(1)},\ V^{(1)}\mathbf{K},\ U^{(1)},\ e^{(1)}\mathbf{K},\ W^{(1)}\mathbf{K},\ V^{(1)} \,\}. \tag{8.71}$$

Table 8.14 The Kayley table of the Abelian group ${G_8}^{(1)}$ (8.71)

	$\mathbf{T}$	$e^{(1)}$	$W^{(1)}$	$V^{(1)}\mathbf{K}$	$U^{(1)}$	$e^{(1)}\mathbf{K}$	$W^{(1)}\mathbf{K}$	$V^{(1)}$
$\mathbf{T}$	$e^{(1)}$	$\mathbf{T}$	$V^{(1)}\mathbf{K}$	$W^{(1)}$	$e^{(1)}\mathbf{K}$	$U^{(1)}$	$V^{(1)}$	$W^{(1)}\mathbf{K}$
$e^{(1)}$	$\mathbf{T}$	$e^{(1)}$	$W^{(1)}$	$V^{(1)}\mathbf{K}$	$U^{(1)}$	$e^{(1)}\mathbf{K}$	$W^{(1)}\mathbf{K}$	$V^{(1)}$
$W^{(1)}$	$V^{(1)}\mathbf{K}$	$W^{(1)}$	$e^{(1)}$	$\mathbf{T}$	$V^{(1)}$	$W^{(1)}\mathbf{K}$	$e^{(1)}\mathbf{K}$	$U^{(1)}$
$V^{(1)}\mathbf{K}$	$W^{(1)}$	$V^{(1)}\mathbf{K}$	$\mathbf{T}$	$e^{(1)}$	$W^{(1)}\mathbf{K}$	$V^{(1)}$	$U^{(1)}$	$e^{(1)}\mathbf{K}$
$U^{(1)}$	$e^{(1)}\mathbf{K}$	$U^{(1)}$	$V^{(1)}$	$W^{(1)}\mathbf{K}$	$e^{(1)}$	$\mathbf{T}$	$V^{(1)}\mathbf{K}$	$W^{(1)}$
$e^{(1)}\mathbf{K}$	$U^{(1)}$	$e^{(1)}\mathbf{K}$	$W^{(1)}\mathbf{K}$	$V^{(1)}$	$\mathbf{T}$	$e^{(1)}$	$W^{(1)}$	$V^{(1)}\mathbf{K}$
$W^{(1)}\mathbf{K}$	$V^{(1)}$	$W^{(1)}\mathbf{K}$	$e^{(1)}\mathbf{K}$	$U^{(1)}$	$V^{(1)}\mathbf{K}$	$W^{(1)}$	$e^{(1)}$	$\mathbf{T}$
$V^{(1)}$	$W^{(1)}\mathbf{K}$	$V^{(1)}$	$U^{(1)}$	$e^{(1)}\mathbf{K}$	$W^{(1)}$	$V^{(1)}\mathbf{K}$	$\mathbf{T}$	$e^{(1)}$

The Kayley table of the group ${G_8}^{(1)}$ is presented in Table 8.14.

On the basis of the Kayley Table 8.13 and taking into account that $\mathbf{T} = U^{(1)}\mathbf{K}$, we found the following relations:

$$\mathbf{T} = \mathbf{T}_{2(xy)}\mathbf{T}_{1(z)}, \quad \mathbf{T} = \mathbf{T}_{1(y)}\mathbf{T}_{2(zx)}, \quad \mathbf{T} = \mathbf{T}_{1(x)}\mathbf{T}_{2(yz)}, \tag{8.72}$$

where

$$\mathbf{T}_{2(xy)} = W^{(1)}, \quad \mathbf{T}_{1(z)} = V^{(1)}\mathbf{K}, \quad \mathbf{T}_{1(y)} = e^{(1)}\mathbf{K}, \ \mathbf{T}_{2(zx)} = U^{(1)}, \tag{8.73}$$
$$\mathbf{T}_{1(x)} = W^{(1)}\mathbf{K}, \ \mathbf{T}_{2(yz)} = V^{(1)}.$$

Taking account of (8.73), the group ${G_8}^{(1)}$ (8.71) can be represented through the operators of complete and incomplete time reversal

$${G_8}^{(1)} : \{\, \mathbf{T},\ e^{(1)},\ \mathbf{T}_{2(xy)},\ \mathbf{T}_{1(z)},\ \mathbf{T}_{2(zx)},\ \mathbf{T}_{1(y)},\ \mathbf{T}_{1(x)},\ \mathbf{T}_{2(yz)} \,\}. \tag{8.74}$$

The Kayley table of the Abelian group ${G_8}^{(1)}$ (8.74), expressed in terms of operators of complete and partial time reversal, is presented in the Table 8.15.

The Table 8.15 contains all six operators of incomplete time reversal $\mathbf{T}_{2(yz)}$,$\mathbf{T}_{2(zx)}$, $\mathbf{T}_{2(xy)}$, $\mathbf{T}_{1(x)}$, $\mathbf{T}_{1(y)}$, $\mathbf{T}_{1(z)}$, along with the operator of full time reversal $\mathbf{T}$. Thus, using the Kayley Table 8.14 of the Abelian group ${G_8}^{(1)} = {G_4}^{(1)} \otimes G_2$ from (8.71) there can be found all six operators of incomplete time reversal for the system with the angular momentum $J = 1$.

Another method for determining operators of incomplete time reversal for a system with the angular momentum $J = 1$ is to use the Kayley table of the group $\widetilde{G_8}^{(1)}$ that differs in structure from the group ${G_8}^{(1)}$ (8.71) and is made up of the following elements:

$$\widetilde{G_8}^{(1)} : \{\, e^{(1)},\ -e^{(1)},\ \mathbf{T},\ -\mathbf{T},\ V^{(1)},\ -V^{(1)},\ W^{(1)}\mathbf{K},\ -W^{(1)}\mathbf{K} \,\}, \tag{8.75}$$

Table 8.15 The Kayley table of the Abelian group $G_8{}^{(1)}$ (8.74)

	$\mathbf{T}$	$e^{(1)}$	$\mathbf{T}_{2(xy)}$	$\mathbf{T}_{1(z)}$	$\mathbf{T}_{2(zx)}$	$\mathbf{T}_{1(y)}$	$\mathbf{T}_{1(x)}$	$\mathbf{T}_{2(yz)}$
$\mathbf{T}$	$e^{(1)}$	$\mathbf{T}$	$\mathbf{T}_{1(z)}$	$\mathbf{T}_{2(xy)}$	$\mathbf{T}_{1(y)}$	$\mathbf{T}_{2(zx)}$	$\mathbf{T}_{2(yz)}$	$\mathbf{T}_{1(x)}$
$e^{(1)}$	$\mathbf{T}$	$e^{(1)}$	$\mathbf{T}_{2(xy)}$	$\mathbf{T}_{1(z)}$	$\mathbf{T}_{2(zx)}$	$\mathbf{T}_{1(y)}$	$\mathbf{T}_{1(x)}$	$\mathbf{T}_{2(yz)}$
$\mathbf{T}_{2(xy)}$	$\mathbf{T}_{1(z)}$	$\mathbf{T}_{2(xy)}$	$e^{(1)}$	$\mathbf{T}$	$\mathbf{T}_{2(yz)}$	$\mathbf{T}_{1(x)}$	$\mathbf{T}_{1(y)}$	$\mathbf{T}_{2(zx)}$
$\mathbf{T}_{1(z)}$	$\mathbf{T}_{2(xy)}$	$\mathbf{T}_{1(z)}$	$\mathbf{T}$	$e^{(1)}$	$\mathbf{T}_{1(x)}$	$\mathbf{T}_{2(yz)}$	$\mathbf{T}_{2(zx)}$	$e^{(1)}\mathbf{K}$
$\mathbf{T}_{2(zx)}$	$\mathbf{T}_{1(y)}$	$\mathbf{T}_{2(zx)}$	$\mathbf{T}_{2(yz)}$	$\mathbf{T}_{1(x)}$	$e^{(1)}$	$\mathbf{T}$	$\mathbf{T}_{1(z)}$	$\mathbf{T}_{2(xy)}$
$\mathbf{T}_{1(y)}$	$\mathbf{T}_{2(zx)}$	$\mathbf{T}_{1(y)}$	$\mathbf{T}_{1(x)}$	$\mathbf{T}_{2(yz)}$	$\mathbf{T}$	$e^{(1)}$	$\mathbf{T}_{2(xy)}$	$\mathbf{T}_{1(z)}$
$\mathbf{T}_{1(x)}$	$\mathbf{T}_{2(yz)}$	$\mathbf{T}_{1(x)}$	$\mathbf{T}_{1(y)}$	$\mathbf{T}_{2(zx)}$	$\mathbf{T}_{1(z)}$	$\mathbf{T}_{2(xy)}$	$e^{(1)}$	$\mathbf{T}$
$\mathbf{T}_{2(yz)}$	$\mathbf{T}_{1(x)}$	$\mathbf{T}_{2(yz)}$	$\mathbf{T}_{2(zx)}$	$\mathbf{T}_{1(y)}$	$\mathbf{T}_{2(xy)}$	$\mathbf{T}_{1(z)}$	$\mathbf{T}$	$e^{(1)}$

where $e^{(1)}$ is the unit 3×3-matrix, $U^{(1)}$, $V^{(1)}$, and $W^{(1)}$ are determined by (8.66), and $\mathbf{T} = U^{(1)}\mathbf{K}$. The Kayley table of the group $\widetilde{G}_8{}^{(1)}$ is presented in the Table 8.16.

The group $\widetilde{G}_8{}^{(1)}$ also allows us to find the operators of incomplete time reversal, since on the basis of Table 8.16 there is the relation

$$\mathbf{T} = (W^{(1)}\mathbf{K})V^{(1)} \tag{8.76}$$

that coincides with the third relation from (8.72), where $\mathbf{T} = U^{(1)}\mathbf{K}$ and $\mathbf{T}_{1(x)}$ and $\mathbf{T}_{2(yz)}$ are defined in (8.73). Introducing the operators $\mathbf{T}_{2(yz)}$ and $\mathbf{T}_{1(x)}$, the group $\widetilde{G}_8{}^{(1)}$ (8.75) becomes a group formed from the operators of complete and incomplete time reversal

$$\widetilde{G}_8{}^{(1)} : \{\, e^{(1)},\ -e^{(1)},\ \mathbf{T},\ -\mathbf{T},\ \mathbf{T}_{2(yz)},\ -\mathbf{T}_{2(yz)},\ \mathbf{T}_{1(x)},\ -\mathbf{T}_{1(x)} \,\}. \tag{8.77}$$

Table 8.16 The Kayley table of Abelian group $\widetilde{G}_8^{(1)}$ (8.75)

	$e^{(1)}$	$-e^{(1)}$	$\mathbf{T}$	$-\mathbf{T}$	$V^{(1)}$	$-V^{(1)}$	$W^{(1)}\mathbf{K}$	$-W^{(1)}\mathbf{K}$
$e^{(1)}$	$e^{(1)}$	$-e^{(1)}$	$\mathbf{T}$	$-\mathbf{T}$	$V^{(1)}$	$-V^{(1)}$	$W^{(1)}\mathbf{K}$	$-W^{(1)}\mathbf{K}$
$-e^{(1)}$	$-e^{(1)}$	$e^{(1)}$	$-\mathbf{T}$	$\mathbf{T}$	$-V^{(1)}$	$V^{(1)}$	$-W^{(1)}\mathbf{K}$	$W^{(1)}\mathbf{K}$
$\mathbf{T}$	$\mathbf{T}$	$-\mathbf{T}$	$e^{(1)}$	$-e^{(1)}$	$W^{(1)}\mathbf{K}$	$-W^{(1)}\mathbf{K}$	$V^{(1)}$	$-V^{(1)}$
$-\mathbf{T}$	$-\mathbf{T}$	$\mathbf{T}$	$-e^{(1)}$	$e^{(1)}$	$-W^{(1)}\mathbf{K}$	$W^{(1)}\mathbf{K}$	$-V^{(1)}$	$V^{(1)}$
$V^{(1)}$	$V^{(1)}$	$-V^{(1)}$	$W^{(1)}\mathbf{K}$	$-W^{(1)}\mathbf{K}$	$e^{(1)}$	$-e^{(1)}$	$\mathbf{T}$	$-\mathbf{T}$
$-V^{(1)}$	$-V^{(1)}$	$V^{(1)}$	$-W^{(1)}\mathbf{K}$	$W^{(1)}\mathbf{K}$	$-e^{(1)}$	$e^{(1)}$	$-\mathbf{T}$	$\mathbf{T}$
$W^{(1)}\mathbf{K}$	$W^{(1)}\mathbf{K}$	$-W^{(1)}\mathbf{K}$	$V^{(1)}$	$-V^{(1)}$	$\mathbf{T}$	$-\mathbf{T}$	$e^{(1)}$	$-e^{(1)}$
$-W^{(1)}\mathbf{K}$	$-W^{(1)}\mathbf{K}$	$W^{(1)}\mathbf{K}$	$-V^{(1)}$	$V^{(1)}$	$-\mathbf{T}$	$\mathbf{T}$	$-e^{(1)}$	$e^{(1)}$

Table 8.17 The Kayley table of the Abelian group $\widetilde{G_8}^{(1)}$ (8.77)

	$e^{(1)}$	$-e^{(1)}$	$\mathbf{T}$	$-\mathbf{T}$	$\mathbf{T}_{2(yz)}$	$-\mathbf{T}_{2(yz)}$	$\mathbf{T}_{1(x)}$	$-\mathbf{T}_{1(x)}$
$e^{(1)}$	$e^{(1)}$	$-e^{(1)}$	$\mathbf{T}$	$-\mathbf{T}$	$\mathbf{T}_{2(yz)}$	$-\mathbf{T}_{2(yz)}$	$\mathbf{T}_{1(x)}$	$-\mathbf{T}_{1(x)}$
$-e^{(1)}$	$-e^{(1)}$	$e^{(1)}$	$-\mathbf{T}$	$\mathbf{T}$	$-\mathbf{T}_{2(yz)}$	$\mathbf{T}_{2(yz)}$	$-\mathbf{T}_{1(x)}$	$\mathbf{T}_{1(x)}$
$\mathbf{T}$	$\mathbf{T}$	$-\mathbf{T}$	$e^{(1)}$	$-e^{(1)}$	$\mathbf{T}_{1(x)}$	$-\mathbf{T}_{1(x)}$	$\mathbf{T}_{2(yz)}$	$-\mathbf{T}_{2(yz)}$
$-\mathbf{T}$	$-\mathbf{T}$	$\mathbf{T}$	$-e^{(1)}$	$e^{(1)}$	$-\mathbf{T}_{1(x)}$	$\mathbf{T}_{1(x)}$	$-\mathbf{T}_{2(yz)}$	$\mathbf{T}_{2(yz)}$
$\mathbf{T}_{2(yz)}$	$\mathbf{T}_{2(yz)}$	$-\mathbf{T}_{2(yz)}$	$\mathbf{T}_{1(x)}$	$-\mathbf{T}_{1(x)}$	$e^{(1)}$	$-e^{(1)}$	$\mathbf{T}$	$-\mathbf{T}$
$-\mathbf{T}_{2(yz)}$	$-\mathbf{T}_{2(yz)}$	$\mathbf{T}_{2(yz)}$	$-\mathbf{T}_{1(x)}$	$\mathbf{T}_{1(x)}$	$-e^{(1)}$	$e^{(1)}$	$-\mathbf{T}$	$\mathbf{T}$
$\mathbf{T}_{1(x)}$	$\mathbf{T}_{1(x)}$	$-\mathbf{T}_{1(x)}$	$\mathbf{T}_{2(yz)}$	$-\mathbf{T}_{2(yz)}$	$\mathbf{T}$	$-\mathbf{T}$	$e^{(1)}$	$-e^{(1)}$
$-\mathbf{T}_{1(x)}$	$-\mathbf{T}_{1(x)}$	$\mathbf{T}_{1(x)}$	$-\mathbf{T}_{2(yz)}$	$\mathbf{T}_{2(yz)}$	$-\mathbf{T}$	$\mathbf{T}$	$-e^{(1)}$	$e^{(1)}$

The Kayley table of this group in terms of the operators of complete and incomplete time reversal is presented in the Table 8.17.

The generalization of the groups ${G_8}^{(1)}$ (8.71) and $\widetilde{G_8}^{(1)}$ (8.75) to the case of non-Kramers systems with angular momentum $J > 1$ is carried out by a direct verification of the fulfillment of group postulates for $J = 2$ and $J = 3$. Assuming

$$U^{(2)} = \begin{pmatrix} 0 & & & & 1 \\ & & & -1 & \\ & & 1 & & \\ & -1 & & & \\ 1 & & & & 0 \end{pmatrix}, \quad V^{(2)} = \begin{pmatrix} 0 & & & & 1 \\ & & & 1 & \\ & & 1 & & \\ & 1 & & & \\ 1 & & & & 0 \end{pmatrix}, \tag{8.78}$$

$$W^{(2)} = \begin{pmatrix} 1 & & & & 0 \\ & -1 & & & \\ & & 1 & & \\ & & & -1 & \\ 0 & & & & 1 \end{pmatrix}$$

we can see that if in (8.71) and (8.75) to replace $e^{(1)}$, $U^{(1)}$, $V^{(1)}$ and $W^{(1)}$ by $e^{(2)}$, $U^{(2)}$, $V^{(2)}$ and $W^{(2)}$, then the groups ${G_8}^{(1)}$ and $\widetilde{G_8}^{(1)}$ go over into the groups ${G_8}^{(2)}$ and $\widetilde{G_8}^{(2)}$, respectively. As for the Kayley tables of groups ${G_8}^{(2)}$ and $\widetilde{G_8}^{(2)}$, they will remain the same as for the ${G_8}^{(1)}$ and $\widetilde{G_8}^{(1)}$ groups, taking into account the above replacements.

A direct check of the fulfillment of group postulates in the transition from $e^{(2)}$, $U^{(2)}$, $V^{(2)}$ and $W^{(2)}$ to $e^{(3)}$, $U^{(3)}$, $V^{(3)}$ and $W^{(3)}$, where

$$U^{(3)} = \begin{pmatrix} 0 & & & & 1 \\ & & & -1 & \\ & & 1 & & \\ & \iddots & & & \\ 1 & & & & 0 \end{pmatrix}, \quad V^{(3)} = \begin{pmatrix} 0 & & & & 1 \\ & & & 1 & \\ & & 1 & & \\ & \iddots & & & \\ 1 & & & & 0 \end{pmatrix}, \tag{8.79}$$

$$W^{(3)} = \begin{pmatrix} 1 & & & & \\ & -1 & & & 0 \\ & & 1 & & \\ & & & \ddots & \\ 0 & & & & 1 \end{pmatrix}$$

shows that ${G_8}^{(3)}$ and $\widetilde{G_8}^{(3)}$, as well as ${G_8}^{(2)}$, ${G_8}^{(1)}$ and $\widetilde{G_8}^{(2)}$, $\widetilde{G_8}^{(1)}$ are Abelian symmetry groups. On the basis of the method of mathematical induction, we can conclude that ${G_8}^{(J)}$ and $\widetilde{G_8}^{(J)}$ are eighth-order Abelian symmetry groups for any integer angular momentum J and consist of the following elements:

$${G_8}^{(J)} : \{\mathbf{T},\ e^{(J)},\ W^{(J)},\ V^{(J)}\mathbf{K},\ U^{(J)},\ e^{(J)}\mathbf{K},\ W^{(J)}\mathbf{K},\ V^{(J)}\}, \tag{8.80}$$

$$\widetilde{{G_8}^{(J)}} : \{e^{(J)},\ -e^{(J)},\ \mathbf{T},\ -\mathbf{T},\ V^{(J)},\ -V^{(J)},\ W^{(J)}\mathbf{K},\ -W^{(J)}\mathbf{K}\}, \tag{8.81}$$

where $U^{(J)}$, $V^{(J)}$ and $W^{(J)}$ are the following $(2J+1) \times (2J+1)$-matrices:

$$U^{(J)} = \begin{pmatrix} 0 & & & & 1 \\ & & & -1 & \\ & & 1 & & \\ & \iddots & & & \\ 1 & & & & 0 \end{pmatrix}, \quad V^{(J)} = \begin{pmatrix} 0 & & & & 1 \\ & & & 1 & \\ & & 1 & & \\ & \iddots & & & \\ 1 & & & & 0 \end{pmatrix}, \tag{8.82}$$

$$W^{(J)} = \begin{pmatrix} 1 & & & & \\ & -1 & & & 0 \\ & & 1 & & \\ & & & \ddots & \\ 0 & & & & 1 \end{pmatrix},$$

$\mathbf{T} = U^{(J)}\mathbf{K}$ is the complete time-reversal operator and $e^{(J)}$ is a unit $(2J+1) \times (2J+1)$-matrix.

In this case, each of the matrices $U^{(J)}$, $V^{(J)}$ and $W^{(J)}$ is symmetric with respect to the reflection of its matrix elements at the center of the matrix and has the property

$$(U^{(J)})^2 = e^{(J)},\ (V^{(J)})^2 = e^{(J)},\ (W^{(J)})^2 = e^{(J)}. \tag{8.83}$$

The group $G_8^{(J)}$ can be easily extended to the Abelian group $G_{16}^{(J)} = G_8^{(J)} \otimes G_2$, where the cyclic group G_2 is formed by elements $\mathbf{K}$ and $\mathbf{K}^2 = 1$.

8.4 Peculiarities of the Structure of Eighth- and Sixteenth-Order Non-Abelian Groups

In order to elucidate the structure of the non-Abelian group $G_8{}^{(\frac{1}{2})}$ (8.11), we denote by g_i and g_k the elements of the cyclic subgroup $G_4{}^{(\frac{1}{2})} (g_i, g_k \in G_4{}^{(\frac{1}{2})})$, containing four consecutive antirotations at an angle of 90° about an axis passing through the center of a square with colored vertices, perpendicular to its plane, as shown in Fig. 2.2. Let denote by h_i and h_k two of the four remaining elements of the group $G_8{}^{(\frac{1}{2})}$ that do not belong to the subgroup $G_4{}^{(\frac{1}{2})}$. These four elements constitute the set $H_4{}^{(\frac{1}{2})} (h_i, h_k \in H_4{}^{(\frac{1}{2})})$. In this case, we have

$$g_i g_k \in G_4{}^{(\frac{1}{2})},\ h_i h_k \in G_4{}^{(\frac{1}{2})},\ g_i h_k \in H_4{}^{(\frac{1}{2})},\ h_i g_k \in H_4{}^{(\frac{1}{2})}. \tag{8.84}$$

As seen from (8.84), any product of elements of the set $H_4{}^{(\frac{1}{2})}$ belongs to the subgroup $G_4{}^{(\frac{1}{2})}$.

The indicated properties of the elements of group $G_8{}^{(\frac{1}{2})}$ follow directly from the Kayley table (Table 8.1) of this group. The relation $g_i g_k \in G_4{}^{(\frac{1}{2})}$ from (8.84) follows from the definition of the group $G_4{}^{(\frac{1}{2})}$ containing the elements $i\sigma_y\mathbf{K}$, $(i\sigma_y\mathbf{K})^2 = -e$, $(i\sigma_y\mathbf{K})^3 = -i\sigma_y\mathbf{K}$ and $(i\sigma_y\mathbf{K})^4 = e$ and is trivial. The correctness of the remaining three relations from (8.84) can be verified using Table 8.1:

$$\begin{aligned} &h_i h_k = g_l \in G_4{}^{(\frac{1}{2})} \\ &\sigma_x(-\sigma_x) = -e,\ \sigma_x(\pm\sigma_z\mathbf{K}) = \mp\mathbf{T}, \\ &\sigma_z\mathbf{K}(\pm\sigma_x) = \pm\mathbf{T},\ \sigma_z\mathbf{K}(-\sigma_z\mathbf{K}) = -e. \end{aligned} \tag{8.85}$$

$$\begin{aligned} &g_i h_k = h_l \in H_4{}^{(\frac{1}{2})} \\ &\mathbf{T}(\pm\sigma_x) = \pm\sigma_z\mathbf{K},\ \mathbf{T}(\pm\sigma_z\mathbf{K}) = \mp\sigma_x, \\ &e(\pm\sigma_x) = \pm\sigma_x,\ e(\pm\sigma_z\mathbf{K}) = \pm\sigma_z\mathbf{K}. \end{aligned} \tag{8.86}$$

$$h_i g_k = h_m \in H_4{}^{(\frac{1}{2})}$$
$$\sigma_x(\pm\mathbf{T}) = \mp\sigma_z\mathbf{K},\ \sigma_x(\pm e) = \pm\sigma_x, \tag{8.87}$$
$$(\sigma_z\mathbf{K})(\pm\mathbf{T}) = \pm\sigma_x,\ (\sigma_z\mathbf{K})(\pm e) = \pm\sigma_z\mathbf{K}.$$

Similar relations hold for the group $G_8{}^{(J)}$ (8.37) related to Kramers systems with half-integer angular momenta $J > 1/2$:

$$h_i h_k = g_l \in G_4{}^{(J)}$$
$$V^{(J)}(-V^{(J)}) = -e^{(J)},\ V^{(J)}(\pm W^{(J)}\mathbf{K}) = \mp\mathbf{T}, \tag{8.88}$$
$$W^{(J)}\mathbf{K}(\pm V^{(J)}) = \pm\mathbf{T},\ (W^{(J)}\mathbf{K})(-W^{(J)}\mathbf{K}) = -e^{(J)}.$$

$$g_i h_k = h_l \in H_4{}^{(J)}$$
$$\mathbf{T}(\pm V^{(J)}) = \pm W^{(J)}\mathbf{K},\ \mathbf{T}(\pm W^{(J)}\mathbf{K}) = \mp V^{(J)}, \tag{8.89}$$
$$e^{(J)}(\pm V^{(J)}) = \pm V^{(J)},\ e^{(J)}(\pm W^{(J)}\mathbf{K}) = \pm W^{(J)}\mathbf{K}.$$

$$h_i g_k = h_m \in H_4{}^{(J)}$$
$$V^{(J)}(\pm\mathbf{T}) = \mp W^{(J)}\mathbf{K},\ V^{(J)}(\pm e^{(J)}) = \pm V^{(J)}, \tag{8.90}$$
$$(W^{(J)}\mathbf{K})(\pm\mathbf{T}) = \pm V^{(J)},\ W^{(J)}\mathbf{K}(\pm e^{(J)}) = \pm W^{(J)}\mathbf{K}.$$

As in the case of the group $G_8{}^{(\frac{1}{2})}$ (8.11), the relation $g_i g_k \in G_4{}^{(J)}$ holds, where the fourth-order cyclic group $G_4{}^{(J)}$ is formed by the elements $U^{(J)}\mathbf{K}$, $(U^{(J)}\mathbf{K})^2 = -e^{(J)}$, $(U^{(J)}\mathbf{K})^3 = -U^{(J)}\mathbf{K}$ and $(U^{(J)}\mathbf{K})^4 = e^{(J)}$. In (8.88), (8.89) and (8.90) $\mathbf{T} = U^{(J)}\mathbf{K}$, $e^{(J)}$ is the unit $(2J+1)\times(2J+1)$-matrix, while $U^{(J)}$, $V^{(J)}$ and $W^{(J)}$ are determined by (8.39).

The elements of the group $G_8{}^{(\frac{1}{2},a)}$ (8.12) are characterized by the same features as the elements of the group $G_8{}^{(\frac{1}{2})}$ (8.11):

$$h_i h_k = g_l \in G_4{}^{(\frac{1}{2},a)}$$
$$\sigma_z(-\sigma_z) = -e,\ \sigma_x\mathbf{K}(\pm\sigma_z) = \pm\sigma_y\mathbf{K}, \tag{8.91}$$
$$\sigma_z(\pm i\sigma_x\mathbf{K}) = \mp\sigma_y\mathbf{K},\ (i\sigma_x\mathbf{K})(-i\sigma_x\mathbf{K}) = -e.$$

$$g_i h_k = h_l \in H_4{}^{(\frac{1}{2},a)}$$
$$\sigma_y\mathbf{K}(\pm i\sigma_x\mathbf{K}) = \mp\sigma_z,\ \sigma_y\mathbf{K}(\pm\sigma_z) = \pm i\sigma_x\mathbf{K}, \tag{8.92}$$
$$e(\pm\sigma_z) = \pm\sigma_z,\ e(\pm i\sigma_x\mathbf{K}) = \pm i\sigma_x\mathbf{K}.$$

$$h_i g_k = h_m \in H_4{}^{(\frac{1}{2},a)}$$
$$(i\sigma_x\mathbf{K})(\pm\sigma_y\mathbf{K}) = \pm\sigma_z,\ \sigma_z(\pm e) = \pm\sigma_z, \tag{8.93}$$
$$(\pm i\sigma_x\mathbf{K})e = \pm i\sigma_x\mathbf{K},\ \sigma_z(\pm\sigma_y\mathbf{K}) = \mp i\sigma_x\mathbf{K}.$$

Table 8.18 The Kayley table of the non-Abelian group ${G_8}^{(J,a)}$ at half-integer angular momentum J (8.94)

	$iV^{(J)}\mathbf{K}$	$-e^{(J)}$	$-iV^{(J)}\mathbf{K}$	$e^{(J)}$	$W^{(J)}$	$-W^{(J)}$	$-iU^{(J)}\mathbf{K}$	$iU^{(J)}\mathbf{K}$
$iV^{(J)}\mathbf{K}$	$e^{(J)}$	$-iV^{(J)}\mathbf{K}$	$-e^{(J)}$	$iV(J)\mathbf{K}$	$iU^{(J)}\mathbf{K}$	$-iU^{(J)}\mathbf{K}$	$-W^{(J)}$	$W^{(J)}$
$-e^{(J)}$	$-iV^{(J)}\mathbf{K}$	$e^{(J)}$	$iV^{(J)}\mathbf{K}$	$-e^{(J)}$	$-W^{(J)}$	$W^{(J)}$	$iU^{(J)}\mathbf{K}$	$-iU^{(J)}\mathbf{K}$
$-iV^{(J)}\mathbf{K}$	$-e^{(J)}$	$iV^{(J)}\mathbf{K}$	$e^{(J)}$	$-iV^{(J)}\mathbf{K}$	$-iU^{(J)}\mathbf{K}$	$iU^{(J)}\mathbf{K}$	$W^{(J)}$	$-W^{(J)}$
$e^{(J)}$	$iV^{(J)}\mathbf{K}$	$-e^{(J)}$	$-iV^{(J)}\mathbf{K}$	$e^{(J)}$	$W^{(J)}$	$-W^{(J)}$	$-iU^{(J)}\mathbf{K}$	$iU^{(J)}\mathbf{K}$
$W^{(J)}$	$iU^{(J)}\mathbf{K}$	$-W^{(J)}$	$-iU^{(J)}\mathbf{K}$	$W^{(J)}$	$e^{(J)}$	$-e^{(J)}$	$-iV^{(J)}\mathbf{K}$	$iV^{(J)}\mathbf{K}$
$-W^{(J)}$	$-iU^{(J)}\mathbf{K}$	$W^{(J)}$	$iU^{(J)}\mathbf{K}$	$-W^{(J)}$	$-e^{(J)}$	$e^{(J)}$	$iV^{(J)}\mathbf{K}$	$-iV^{(J)}\mathbf{K}$
$-iU^{(J)}\mathbf{K}$	$-W^{(J)}$	$iU^{(J)}\mathbf{K}$	$W^{(J)}$	$-iU^{(J)}\mathbf{K}$	$-iV^{(J)}\mathbf{K}$	$iV^{(J)}\mathbf{K}$	$e^{(J)}$	$-e^{(J)}$
$iU^{(J)}\mathbf{K}$	$W^{(J)}$	$-iU^{(J)}\mathbf{K}$	$-W^{(J)}$	$iU^{(J)}\mathbf{K}$	$iV^{(J)}\mathbf{K}$	$-iV^{(J)}\mathbf{K}$	$-e^{(J)}$	$e^{(J)}$

As in the case of the group ${G_8}^{(\frac{1}{2})}$, a fourth-order cyclic group ${G_4}^{(\frac{1}{2},a)}(g_i g_k \in {G_4}^{(\frac{1}{2},a)})$ is separated in the group ${G_8}^{(\frac{1}{2},a)}$. The fourth-order cyclic group ${G_4}^{(J,a)}$ is also separated in the group ${G_8}^{(J,a)}$ (8.94). The group ${G_8}^{(J,a)}$, which is a generalization of the group ${G_8}^{(\frac{1}{2},a)}$ to the case of an arbitrary half-integer angular momentum J, is formed by the elements

$$ {G_8}^{(J,a)} : \{iV^{(J)}\mathbf{K},\ -e^{(J)},\ -iV^{(J)}\mathbf{K},\ e^{(J)},\ W^{(J)},\ -W^{(J)},\ -iU^{(J)}\mathbf{K},\ iU^{(J)}\mathbf{K}\}, \tag{8.94} $$

where $e^{(J)}$ is a unit $((2J+1)\times(2J+1))$-matrix and $U^{(J)}$, $V^{(J)}$ and $W^{(J)}$ are determined by (8.39). The Kayley table of this group is presented in the Table 8.18.

On the basis of Table 8.18, one can see that the elements of the group ${G_8}^{(J,a)}$ are characterized by the same features as the elements of the group ${G_8}^{(\frac{1}{2},a)}$

$$ \begin{aligned} & h_i h_k = g_l \in {G_4}^{(J,a)} \\ & W^{(J)}(-W^{(J)}) = -e^{(J)},\ W^{(J)}(\pm iV^{(J)}\mathbf{K}) = \pm iU^{(J)}\mathbf{K}, \\ & iV^{(J)}\mathbf{K}(-iV^{(J)}\mathbf{K}) = -e^{(J)},\ iV^{(J)}\mathbf{K}(\pm W^{(J)}) = \mp iU^{(J)}\mathbf{K}. \end{aligned} \tag{8.95} $$

$$ \begin{aligned} & g_i h_k = h_l \in {H_4}^{(J,a)} \\ & -iU^{(J)}\mathbf{K}(\pm iV^{(J)}\mathbf{K}) = \mp V^{(J)},\ -iU^{(J)}\mathbf{K}(\pm W^{(J)}) = \pm iV^{(J)}\mathbf{K}, \\ & e^{(J)}(\pm W^{(J)}) = \pm W^{(J)},\ e^{(J)}(\pm iV^{(J)}\mathbf{K}) = \pm iV^{(J)}\mathbf{K}. \end{aligned} \tag{8.96} $$

$$ \begin{aligned} & h_i g_k = h_m \in {H_4}^{(J,a)} \\ & W^{(J)}(\mp iU^{(J)}\mathbf{K}) = \mp iV^{(J)}\mathbf{K},\ W^{(J)}(\pm e^{(J)}) = \pm W^{(J)}, \\ & (iV^{(J)}\mathbf{K})(\mp iU^{(J)}\mathbf{K}) = \pm W^{(J)},\ (\pm iV^{(J)}\mathbf{K})e^{(J)} = \pm iV^{(J)}\mathbf{K}. \end{aligned} \tag{8.97} $$

As already noted, the group ${G_8}^{(\frac{1}{2},b)}$ (8.13) does not contain the operator of complete time reversal $\mathbf{T}$ and is an Abelian group.

Table 8.19 The Kayley table of the Abelian group ${G_8}^{(J,b)}$ (8.98) at half-integer angular momentum J

	$iW^{(J)}\mathbf{K}$	$-e^{(J)}$	$-iW^{(J)}\mathbf{K}$	$e^{(J)}$	$-iU^{(J)}$	$iU^{(J)}$	$V^{(J)}\mathbf{K}$	$-V^{(J)}\mathbf{K}$
$iW^{(J)}\mathbf{K}$	$e^{(J)}$	$-iW^{(J)}\mathbf{K}$	$-e^{(J)}$	$iW^{(J)}\mathbf{K}$	$-V^{(J)}\mathbf{K}$	$V^{(J)}\mathbf{K}$	$iU^{(J)}$	$-iU^{(J)}$
$-e^{(J)}$	$-iW^{(J)}\mathbf{K}$	$e^{(J)}$	$iW^{(J)}\mathbf{K}$	$-e^{(J)}$	$iU^{(J)}$	$-iU^{(J)}$	$-V^{(J)}\mathbf{K}$	$V^{(J)}\mathbf{K}$
$-iW^{(J)}\mathbf{K}$	$-e^{(J)}$	$iW^{(J)}\mathbf{K}$	$e^{(J)}$	$-iW^{(J)}\mathbf{K}$	$V^{(J)}\mathbf{K}$	$-V^{(J)}\mathbf{K}$	$-iU^{(J)}$	$iU^{(J)}$
$e^{(J)}$	$iW^{(J)}\mathbf{K}$	$-e^{(J)}$	$-iW^{(J)}\mathbf{K}$	$e^{(J)}$	$-iU^{(J)}$	$iU^{(J)}$	$V^{(J)}\mathbf{K}$	$-V^{(J)}\mathbf{K}$
$-iU^{(J)}$	$V^{(J)}\mathbf{K}$	$iU^{(J)}$	$-V^{(J)}\mathbf{K}$	$-iU^{(J)}$	$-e^{(J)}$	$e^{(J)}$	$-iV^{(J)}\mathbf{K}$	$iV^{(J)}\mathbf{K}$
$iU^{(J)}$	$-V^{(J)}\mathbf{K}$	$-iU^{(J)}$	$V^{(J)}\mathbf{K}$	$iU^{(J)}$	$e^{(J)}$	$-e^{(J)}$	$iV^{(J)}\mathbf{K}$	$-iV^{(J)}\mathbf{K}$
$V^{(J)}\mathbf{K}$	$iU^{(J)}$	$-V^{(J)}\mathbf{K}$	$-iU^{(J)}$	$V^{(J)}\mathbf{K}$	$iW^{(J)}\mathbf{K}$	$-iW^{(J)}\mathbf{K}$	$e^{(J)}$	$-e^{(J)}$
$-V^{(J)}\mathbf{K}$	$-iU^{(J)}$	$V^{(J)}\mathbf{K}$	$iU^{(J)}$	$-V^{(J)}\mathbf{K}$	$-iW^{(J)}\mathbf{K}$	$iW^{(J)}\mathbf{K}$	$-e^{(J)}$	$e^{(J)}$

This group can be easily generalized to the group ${G_8}^{(J,b)}$ at an arbitrary half-integer angular momentum J:

$$ {G_8}^{(J,b)} : \{iW^{(J)}\mathbf{K},\ -e^{(J)},\ -iW^{(J)}\mathbf{K},\ e^{(J)},\ -iU^{(J)},\ iU^{(J)},\ V^{(J)}\mathbf{K},\ -V^{(J)}\mathbf{K}\}, \quad (8.98) $$

where, as before $e^{(J)}$ is a unit $((2J+1)\times(2J+1))$ -matrix and $U^{(J)}$, $V^{(J)}$ and $W^{(J)}$ are defined by (8.39).

The Kayley table of the group ${G_8}^{(J,b)}$ (8.98) is presented in the Table 8.19.

All the groups considered above relating to systems with Kramers degeneracy of energy levels are non-Abelian symmetry groups. Exceptions are the groups ${G_8}^{(\frac{1}{2},b)}$ and ${G_8}^{(J,b)}$, which do not contain the operator of the complete time reversal $\mathbf{T}$, but contain operators of incomplete time reversal. Although they refer to systems with a half-integer angular momentum J, these groups do not contain fourth-order cyclic groups based on the $\mathbf{T}$ operator, related to the Kramers degeneracy of energy levels.

Unlike the systems with a half-integer angular momentum, for an integer J the groups containing the time-reversal operator $\mathbf{T}$ are Abelian symmetry groups. This is due to the fact that such groups can not contain a fourth-order cyclic subgroup, but contain only a second-order cyclic subgroup formed by the operators $\mathbf{T}$ and $\mathbf{T}^2 = e^{(J)}$. Nevertheless, in this case there is also a definite structure of the symmetry groups, which is preserved with increasing the angular momentum ($J > 1$). Moreover, due to the commutation of the elements of these groups, their Kayley tables become more symmetric in comparison with Kayley tables of the groups related to Kramers systems. In particular, the group ${G_8}^{(1)}$ (8.71) contains a fourth-order non-cyclic subgroup formed by the elements $e^{(1)}$, $U^{(1)}\mathbf{K}$, $W^{(1)}$ and $V^{(1)}\mathbf{K}$. As follows from the Table 8.14, two equivalent (4×4)-blocks of the Kayley table of this group consist of only elements of the subgroup ${G_4}^{(1)}$ and are located on the main diagonal of the Kayley table. The elements of the group ${G_8}^{(1)}$ that are not part of the subgroup G_4 form two other equivalent (4×4)-blocks located on the secondary diagonal of the Kayley table. Such a structure of the Kayley table is also characteristic of the group ${G_8}^{(J)}$ for any integer J.

An analogous situation is also characteristic of the group $\widetilde{G_8}^{(J)}$, for which the elements $e^{(J)}$, $-e^{(J)}$, $\mathbf{T}$ and $-\mathbf{T}$ form a cyclic subgroup of $\widetilde{G_4}^{(J)}$, and the remaining four elements $V^{(J)}$, $-V^{(J)}$, $W^{(J)}\mathbf{K}$, and $-W^{(J)}\mathbf{K}$ do not enter into $\widetilde{G_4}^{(J)}$. As can be seen from the Table 8.16 at $J = 1$, in this case also the elements of the subgroup $\widetilde{G_8}^{(J)}$ form two equivalent (4×4)-blocks located on the main diagonal of the Kayley table. The remaining elements of the group $\widetilde{G_8}^{(J)}$ form two other equivalent (4×4)-blocks located on the secondary diagonal of the Kayley table.

In the case of the Kramers systems, the structure embedded in the group $G_8{}^{(J)}$ is preserved even when it is expanded by a second-order cyclic group $G_2 : \{\mathbf{K},\ \mathbf{K}^2 = 1\}$ to the groups $G_{16}^{(J)}$ (8.62) and $G_{16}^{(J)}$ (8.64) (Appendices C and D).

Chapter 9
Factorization of Wigner Time-Reversal Operator and Reduction of Time-Reversal Symmetry

The Wigner time-reversal operator **T** is represented as the product of two or three so-called operators of incomplete time reversal, under the action of which not all the angular momentum projection operators change sign. It is shown that when the symmetry of time reversal is violated (reduced) in systems with Kramers degeneracy of energy levels, a violation of the Kramers theorem occurs, with the exception of one case when such reducing is insufficient to remove the Kramers degeneracy. The commutation and anticommutation relations between operators of incomplete time reversal, as well as between these operators and the operator **T**, are found. It is shown that these relations are different for Kramers and non-Kramers systems. In the two-boson representation of the angular momentum J, it is shown that for an integer J the sum of the occupation numbers of bosons of two types can be negative. This corresponds to the antibosonic states, which can be obtained from boson states under the action of the time-reversal operator.

The existence of six incomplete time-reversal operators, found in Chap. 8, causes six possibilities for reducing the time-reversal symmetry to one of the six incomplete time-reversal symmetries. For a systems consisting of spin-$1/2$ particles, six types of reduced T-symmetry could be restored by using the concept of meta-particles. There are six types of meta-particles related to spin-$1/2$ elementary particle and other six types of meta-particles related to a spin-1 elementary particles (Sect. 9.7). For systems with integer angular momentum the lowering of the T-symmetry to T_2- or T_1-symmetry also is possible. Particularly, this may be relevant for interactions between elementary particles of nuclei through spin-1 gauge vector bosons. In this case the lowered T-symmetry (from T- to T_2- or T_1-symmetry) can be restored by introducing six types of gauge vector meta-bosons.

This chapter is written on the basis of the results obtained and published by the author [134, 136, 431–433, 455, 456].

I. I. Geru, *Time-Reversal Symmetry*, Springer Tracts in Modern Physics 281,
https://doi.org/10.1007/978-3-030-01210-6_9

9.1 Six New Types of Time-Reversal Symmetry Related to Kramers Systems

In Sect. 8.1, a non-Abelian group of the 8th order, $G_8{}^{(\frac{1}{2})}$, was introduced for a particle with spin $\frac{1}{2}$, which contains the time-reversal operator $\mathbf{T}$. This group is isomorphic to the $4'm'm$ group of generalized symmetry of a square with neighboring vertices colored in different colors, provided that opposite vertices are of the same color [134]. Because of the isomorphism of the groups $G_8{}^{(\frac{1}{2})}$ and $4'm'm$, there is a one-to-one correspondence between the operators $\mathbf{T}, \mathbf{T}^2 = -e, \mathbf{T}^3 = -\mathbf{T}, \mathbf{T}^4 = e$ and successive anti-rotations by an angle 90° about the axis passing through the center of the square, perpendicular to its plane $4'_z$, ${4'}^2{}_z$, ${4'}^3{}_z$ and ${4'}^4{}_z$. As for the remaining four elements of the group $G_8{}^{(\frac{1}{2})}$, $(\sigma_x,\ -\sigma_x,\ \sigma_z\mathbf{K},\ -\sigma_z\mathbf{K})$, they correspond to two reflections in the diagonals of the square and two anti-reflections in the lines passing through the centers of opposite sides.

We see that the operators σ_x and $\sigma_z\mathbf{K}$ describe the symmetry and antisymmetry transformations if the values of the wave functions Ψ, $\mathbf{T}\Psi$, $\mathbf{T}^2\Psi$ and $\mathbf{T}^3\Psi$ correspond to the vertices of the square (Fig. 2.2). This is not surprising, considering, for example, that the time-reversal operator $\mathbf{T}$ $\left(\mathbf{T} = i\sigma_y\mathbf{K}\right)$ is defined on the basis of the imaginary Pauli operator σ_y [222]. Moreover, under certain conditions, in the case of a violation of T-symmetry, the operators σ_x and $\sigma_z\mathbf{K}$ can act as operators of incomplete time reversal $\mathbf{T}_{2(yz)}$ and $\mathbf{T}_{1(x)}$ (8.4) and (8.5).

Equations 8.4 and 8.5 were found on the basis of the properties of the Kayley table of the group $G_8{}^{(\frac{1}{2})}$. It turned out that the operator $\mathbf{T}$ can be represented as the product of the operators $\mathbf{T}_{1(x)}$ and $\mathbf{T}_{2(yz)}$. By the concept of factorization of the operator $\mathbf{T}$ we mean such a product of operators that form part of the operator $\mathbf{T}$, which are elements of the same symmetry group.

In the case of Kramers systems, this is the group $G_8{}^{(\frac{1}{2})}$ and its various generalizations, which were fulfilled in Sects. 8.1 and 8.2. The representation of the time-reversal operator $\mathbf{T}$ as the product of two (or three) operators, each of which commutes with the corresponding Hamiltonian, means the existence of a potential possibility of breaking the T-symmetry or, more precisely, its lowering (from T- to $T_{2(yz)}$-symmetry or $T_{1(z)}$-symmetry in the case of the $G_8{}^{(\frac{1}{2})}$ group). This happens if the Hamiltonian is invariant, for example, with respect to the operator $\mathbf{T}_{1(x)}$, but is not invariant under the operators $\mathbf{T}$ and $\mathbf{T}_{2(yz)}$. This situation holds, in particular, for the Hamiltonian $\mathbf{H} = \alpha_{yz}\left(E_yS_y + E_zS_z\right)$, which is invariant under the operator $\mathbf{T}_{1(x)}$, but is not invariant with respect to the operators $\mathbf{T}_{2(yz)}$ and $\mathbf{T}$, which leads to a decrease of the T-symmetry to $T_{1(x)}$- symmetry. In the Hamiltonian $\mathbf{H}$, E_y and E_z are components of the constant electric field strength, and α_{yz} is the spin-electric field coupling constant (see Sect. 7.4).

The nontrivial factorization of the time-reversal operator, found on the basis of the symmetry properties of the group $G_8{}^{(\frac{1}{2})}$ (8.11), is not unique. Using the properties of the group $G_8{}^{(\frac{1}{2},a)}$ (8.12), we found a nontrivial factorization of the time-reversal operator $\overline{\mathbf{T}} = \mathbf{T}_{1(z)}\mathbf{T}_{2(xy)}$, where $\overline{\mathbf{T}} = \sigma_y\mathbf{K}$ (the operator that differs by the factor

i from the Wigner time-reversal operator $\mathbf{T} = i\sigma_y\mathbf{K}$), $\mathbf{T}_{1(z)}$ and $\mathbf{T}_{2(xy)}$ are defined in (8.16).

The symmetry properties of the Abelian group $G_8{}^{(\frac{1}{2},b)}$, which contains the operators of incomplete time reversal, together with the unit element, but does not contain the operator $\mathbf{T}$, allow us to find the nontrivial factorization of the operator of incomplete time reversal $\mathbf{T}_{2(zx)}$ in the form $\mathbf{T}_{2(zx)} = \mathbf{T}_{1(z)}\mathbf{T}_{1(x)} = \mathbf{T}_{1(x)}\mathbf{T}_{1(z)}$ (8.23), where $\mathbf{T}_{2(zx)}$, $\mathbf{T}_{1(z)}$ and $\mathbf{T}_{1(x)}$ are defined in (8.21).

The extension of the group $G_8{}^{(\frac{1}{2})}$ (8.11) to the group $G_{16}{}^{(\frac{1}{2})}$ (8.43) made it possible to prove on the basis of (8.45), (8.52), (8.53) and (8.57) that in the case of a particle with spin $\frac{1}{2}$ there are only six operators of incomplete time reversal:

$$\begin{aligned} &\mathbf{T}_{1(x)} = \sigma_z\mathbf{K},\ \mathbf{T}_{1(y)} = e\mathbf{K},\ \mathbf{T}_{1(z)} = \sigma_x\mathbf{K}, \\ &\mathbf{T}_{2(yz)} = \sigma_x,\ \mathbf{T}_{2(zx)} = i\sigma_y,\ \mathbf{T}_{2(xy)} = \sigma_z, \end{aligned} \tag{9.1}$$

where e is the unit element of the group $G_{16}{}^{(\frac{1}{2})}$.

Using the incomplete time inversion operators from (9.1), the nontrivial factorization of the operator of complete time reversal $\mathbf{T}$ is realized in the following ways:

$$\begin{aligned} &\mathbf{T} = \mathbf{T}_{1(x)}\mathbf{T}_{2(yz)} = \mathbf{T}_{1(x)}\mathbf{T}_{1(y)}\mathbf{T}_{1(z)}, \\ &\mathbf{T} = \mathbf{T}_{2(zx)}\mathbf{T}_{1(y)} = \mathbf{T}_{1(x)}\mathbf{T}_{1(z)}\mathbf{T}_{1(y)} = \mathbf{T}_{1(x)}\mathbf{T}_{1(y)}\mathbf{T}_{1(z)}, \\ &\mathbf{T} = \mathbf{T}_{2(xy)}\mathbf{T}_{1(z)} = \mathbf{T}_{1(x)}\mathbf{T}_{1(y)}\mathbf{T}_{1(z)}. \end{aligned} \tag{9.2}$$

The Wigner's time-reversal operator $\mathbf{T}$ together with the incomplete time reversal operators from (9.1) form a system of seven time-reversal operators in the simple case of a particle of spin $\frac{1}{2}$. In Sect. 8.1, it was shown that if the group $G_8{}^{(\frac{1}{2})}$ is extended to the group $G_8{}^{(J)}$ for a system with half-integer angular momentum J, and the group $G_8{}^{(J)}$ is extended further to the group $G_{16}{}^{(J)}$ (8.62), then it can be proved on the basis of (8.40), (8.41) and (8.63) that in this case there exist the following six operators of incomplete time reversal:

$$\begin{aligned} &\mathbf{T}_{1(x)} = W^{(J)}\mathbf{K},\ \mathbf{T}_{1(y)} = e^{(J)}\mathbf{K},\ \mathbf{T}_{1(z)} = V^{(J)}\mathbf{K}, \\ &\mathbf{T}_{2(yz)} = V^{(J)},\ \mathbf{T}_{2(zx)} = U^{(J)},\ \mathbf{T}_{2(xy)} = W^{(J)}, \end{aligned} \tag{9.3}$$

where $e^{(J)}$ is a unit $((2J+1) \times (2J+1))$-matrix and the matrices $U^{(J)}$, $V^{(J)}$ and $W^{(J)}$ are defined in (8.39).

The representation of the Wigner time-reversal operator $\mathbf{T} = U^{(J)}\mathbf{K}$ (8.38) as a product of two or three operators of incomplete time-reversal from (9.3) is determined by formulas (9.2), which were found for a particle with the spin $\frac{1}{2}$, if expressions for operators of incomplete time reversal from (9.3) are used for the latter in the case of a system with a half-integer angular momentum J.

Thus, in the general case of systems with a half-integer angular momentum J, there are six different operators of incomplete time reversal (9.3), with the help of which four types of nontrivial factorization (9.2) of the time-reversal operator $\mathbf{T}$ can be performed. The representation of the operator $\mathbf{T}$ as a product of operators of incomplete time reversal in the case of systems with any half-integer angular momentum J, as well as in the case of a particle with spin $\frac{1}{2}$, indicates the possibility (but not necessarily) of a violation of T-symmetry in some processes. If T-symmetry is violated, with the realization of one of the three T_2-symmetries or one of the three T_1-symmetries, only its decrease occurs, but not the complete annihilation.

9.2 Violation of Kramers Theorem

The Kramers theorem [69] refers to systems with an odd number of particles with spin $\frac{1}{2}$, such as, for example, an atom with an odd number of electrons. In each such system, placed in an arbitrary electric field, but in the absence of a magnetic field, there is an additional n-fold degeneracy of all energy levels (where n is an integer even number) caused by the time-reversal symmetry.

There are at least three ways to prove the Kramers theorem. Two of them are given in Sects. 2.4 and 2.9, namely, the traditional proof in 2.4 and a geometric proof in 2.9. The third original way of proving this theorem, based on transformations of complex-conjugate basis spinors at the rotation of the real axes of coordinates O_x, O_y and O_z, is given in Heine's monograph [9].

Since the additional degeneracy of the energy levels is due to the time-reversal symmetry, it can disappear when this symmetry decreases. If, however, there is a complete violation of the time-reversal symmetry, when none of the seven time-reversal operators commutes with the corresponding Hamiltonians, then it can be said with certainty that the Kramers degeneracy of energy levels is absent even in systems with a half-integer angular momentum, that is, a violation of the Kramers theorem occurs.

A convenient way to verify the presence or absence of an additional degeneracy of the energy levels caused by the operators of incomplete time reversal $\mathbf{T}_{2(yz)}$, $\mathbf{T}_{2(zx)}$, $\mathbf{T}_{2(xy)}$, $\mathbf{T}_{1(x)}$, $\mathbf{T}_{1(y)}$, and $\mathbf{T}_{1(z)}$ is to find the squares of these operators. Since $\mathbf{T}^2 = -e$ (in the case of spin $S = \frac{1}{2}$) and $\mathbf{T}^2 = -e^{(J)}$ (in the case of semi-integer angular momentum $J > \frac{1}{2}$) for the Kramers systems, the violation of similar relations for the squares of partial time-reversal operators would mean the violation of the Kramers theorem when T-symmetry is decreasing. Let us verify in what cases the Kramers theorem is violated for a particle with spin $\frac{1}{2}$. Based on (9.1) we find

$$\begin{aligned} &\mathbf{T}_{1(x)} = \sigma_z \mathbf{K},\ \mathbf{T}^2_{1(x)} = \sigma_z^2 = e, \\ &\mathbf{T}_{2(yz)} = \sigma_x,\ \mathbf{T}^2_{2(yz)} = \sigma_x^2 = e. \end{aligned} \tag{9.4}$$

Using (9.1), we obtain an analogous result for the operators $\mathbf{T}^2_{1(y)}$, $\mathbf{T}^2_{1(z)}$ and $\mathbf{T}^2_{2(xy)}$

$$\begin{aligned} &\mathbf{T}_{1(y)} = e\mathbf{K},\ \mathbf{T}^2_{1(y)} = e, \\ &\mathbf{T}_{1(z)} = \sigma_x\mathbf{K},\ \mathbf{T}^2_{1(z)} = \sigma_x^2 = e, \\ &\mathbf{T}_{2(xy)} = \sigma_z,\ \mathbf{T}^2_{2(xy)} = e. \end{aligned} \tag{9.5}$$

However, in the case of $T_{2(zx)}$-symmetry ($\mathbf{T}_{2(zx)} = i\sigma_y$, see (9.1), i.e., when T-symmetry decreases to $T_{2(zx)}$-symmetry, the Kramers theorem is not violated:

$$\mathbf{T}_{2(zx)} = i\sigma_y,\ \mathbf{T}^2_{2(zx)} = -e \tag{9.6}$$

In the case of Kramers systems with half-integer angular momentum $J > \frac{1}{2}$, on the basis of (9.3), we obtain

$$\begin{aligned} &\mathbf{T}_{1(x)} = W^{(J)}\mathbf{K},\ \mathbf{T}^2_{1(x)} = e^{(J)},\ \mathbf{T}_{1(y)} = e^{(J)}\mathbf{K},\ \mathbf{T}^2_{1(y)} = e^{(J)}, \\ &\mathbf{T}_{1(z)} = V^{(J)}\mathbf{K},\ \mathbf{T}^2_{1(z)} = e^{(J)},\ \mathbf{T}_{2(yz)} = V^{(J)},\ \mathbf{T}^2_{2(yz)} = e^{(J)}, \\ &\mathbf{T}_{2(xy)} = W^{(J)},\ \mathbf{T}^2_{2(xy)} = e^{(J)},\ \mathbf{T}_{2(zx)} = U^{(J)},\ \mathbf{T}^2_{2(zx)} = -e^{(J)} \end{aligned} \tag{9.7}$$

Thus, for Kramers systems, in five of the six cases of T-symmetry breaking, when this symmetry is reduced to $T_{1(x)}$-, $T_{1(y)}$-, $T_{1(z)}$-, $T_{2(yz)}$- and $T_{2(xy)}$- symmetry, there is also a violation of the Kramers theorem, and the Kramers theorem is not violated only in the case of decreasing T-symmetry to $T_{2(zx)}$-symmetry. This exception is due presumably to the fact that although T-symmetry decreases to $T_{2(zx)}$-symmetry, this decrease is not sufficient to remove the Kramers degeneracy of energy levels. In other words, the remaining time-reversal symmetry ensures the conservation of the Kramers degeneracy of energy levels, which existed in the system in the presence of T-symmetry.

9.3 Six New Types of Time-Reversal Symmetry Related to Non-Kramers Systems

In Sect. 8.3 we introduced and studied the properties of Abelian groups ${G_4}^{(1)}$ (8.65), ${G_8}^{(1)}$ (8.71), $\widetilde{{G_8}^{(1)}}$ (8.75) and $\widetilde{{G_8}^{(J)}}$ (8.81). Based on the structure of Kayley tables of these groups, it was found that in the case of non-Kramers systems there are also six operators of incomplete time reversal, which belong to the groups listed above, along with the Wigner's operator of complete time reversal $\mathbf{T}$. This circumstance makes it possible to call them operators of incomplete time reversal. Just as in the case of Kramers systems, the operator $\mathbf{T}$ can be represented as the product of two or three operators of incomplete time reversal. However, there is a significant difference between the structures of groups related to Kramers systems and non-Kramers ones. For example, the Abelian group ${G_4}^{(1)}$ (8.65) consisting of elements $U^{(1)}\mathbf{K}$, $e^{(1)}$, $W^{(1)}$ and $V^{(1)}\mathbf{K}$ (with symmetric matrices $U^{(1)}$, $V^{(1)}$ and $W^{(1)}$ defined by (8.66) and related to non-Kramers systems with angular momentum $J = 1$, differs in structure from the cyclic group ${G_4}^{(\frac{1}{2})}$ consisting of the elements $\mathbf{T}$, $\mathbf{T}^2 = -e$, $\mathbf{T}^3 = -\mathbf{T}$, $\mathbf{T}^4 = e$. A similar difference exists also between the Abelian group ${G_4}^{(J)}$ formed by the elements $U^{(J)}\mathbf{K}$, $e^{(J)}$, $W^{(J)}$ and $V^{(J)}\mathbf{K}$ and the fourth-order non-Abelian group consisting of the elements $U^{(J)}\mathbf{K}$, $-e^{(J)}$,$-U^{(J)}\mathbf{K}$ and $e^{(J)}$, where J is an integer in the first case and a half-integer in the second. In the first case, $\left(U^{(J)}\right)^2 = e^{(J)}$, and in the second case $\left(U^{(J)}\right)^2 = -e^{(J)}$. This is due to the fact that a fourth-order cyclic group constructed on the basis of the operator $\mathbf{T}$ exists only for Kramers systems, while for non-Kramers systems it is a second-order group.

As for the Abelian group ${G_8}^{(1)} = {G_4}^{(1)} \otimes G_2$ (8.71) and its extension to the group ${G_8}^{(J)}$ (8.80) for the integer angular momentum $J > 1$, for the above reason the structure of these groups also differs from the structure of the non-Abelian groups ${G_8}^{(\frac{1}{2})}$ (8.11) and ${G_8}^{(J)}$ (8.37) for an arbitrary half-integer J. From this point of view, it is of interest to compare the structures of the groups $\widetilde{{G_8}^{(1)}}$ (8.75) and $\widetilde{{G_8}^{(J)}}$ (8.81) for the integer J and, respectively, the groups ${G_8}^{(\frac{1}{2})}$ (8.11) and ${G_8}^{(J)}$ (8.37) for a half-integer J. The groups $\widetilde{{G_8}^{(1)}}$ and $\widetilde{{G_8}^{(J)}}$ are Abelian and their elements coincide in form with the elements of the non-Abelian groups ${G_8}^{(\frac{1}{2})}$ and ${G_8}^{(J)}$, if they are written in terms of the operator $\mathbf{T}$ and operators of incomplete time reversal. The difference lies in the fact that in the first case $\left(U^{(1)}\right)^2 = e^{(1)}$ and $\left(U^{(J)}\right)^2 = e^{(J)}$ and in the second case $\left(U^{(J)}\right)^2 = -e^{(J)}$ ($U^{(J)}$ is the unitary part of the operator $\mathbf{T} = i\sigma_y\mathbf{K} = U\mathbf{K}$) and $\left(U^{(J)}\right)^2 = -e^{(J)}$. This difference leads to the fact that the groups $\widetilde{{G_8}^{(1)}}$ and $\widetilde{{G_8}^{(J)}}$ are Abelian, in contrast to the non-Abelian groups ${G_8}^{(\frac{1}{2})}$ and ${G_8}^{(J)}$ and in accordance with the fact that for systems with integer angular momentum J there is no Kramers degeneracy of the energy levels.

For non-Kramers systems, the nontrivial factorization of the time reversal operator T is carried out in the same way as in the case of the Kramers systems (9.2). Moreover, for non-Kramers systems, each of the six operators of incomplete time reversal has the same form as for the Kramers systems (9.1). The difference consists only in the fact that for an integer J the matrices $U^{(J)}$, $V^{(J)}$, and $W^{(J)}$ are symmetric, in contrast to the case of half-integer J. As already noted, for an integer J we have $\left(U^{(J)}\right)^2 = e^{(J)}$, while for a half-integer J we have $\left(U^{(J)}\right)^2 = -e^{(J)}$. This is an essential difference in behavior of Kramers and non-Kramers systems under the action of operators of complete and incomplete time reversal.

Since in the case of an integer J from the relation $\left(U^{(J)}\right)^2 = e^{(J)}$ it follows that $\mathbf{T}^2 = e^{(J)}$, then all operators $\mathbf{T}_{1(x)}^2$, $\mathbf{T}_{1(y)}^2$, $\mathbf{T}_{1(z)}^2$, $\mathbf{T}_{2(yz)}^2$, $\mathbf{T}_{2(zx)}^2$ and $\mathbf{T}_{2(xy)}^2$ at integer J are unit operators. This corresponds to the absence of a degeneracy of energy levels due to the complete and incomplete time-reversal symmetry as it should be for non-Kramers systems.

9.4 Commutation and Anticommutation Relations for Time-Reversal Operators

In the case of Kramers systems, there are eight nonzero commutators for operators of incomplete time reversal from (9.3)

$$\begin{aligned}
&\left[\mathbf{T}_{1(x)}, \mathbf{T}_{1(z)}\right] = 2\mathbf{T}_{2(zx)}, \ \left[\mathbf{T}_{1(x)}, \mathbf{T}_{2(yz)}\right] = 2\mathbf{T}, \\
&\left[\mathbf{T}_{1(x)}, \mathbf{T}_{2(zx)}\right] = 2\mathbf{T}_{1(z)}, \ \left[\mathbf{T}_{1(z)}, \mathbf{T}_{2(zx)}\right] = -2\mathbf{T}_{1(x)}, \\
&\left[\mathbf{T}_{1(z)}, \mathbf{T}_{2(xy)}\right] = -2\mathbf{T}, \ \left[\mathbf{T}_{2(yz)}, \mathbf{T}_{2(zx)}\right] = -2\mathbf{T}_{2(xy)}, \\
&\left[\mathbf{T}_{2(yz)}, \mathbf{T}_{2(xy)}\right] = -2\mathbf{T}_{2(zx)}, \ \left[\mathbf{T}_{2(zx)}, \mathbf{T}_{2(xy)}\right] = -2\mathbf{T}_{2(yz)}.
\end{aligned} \tag{9.8}$$

Commutators $\left[\mathbf{T}_{1(x)}, \mathbf{T}_{1(y)}\right]$, $\left[\mathbf{T}_{1(x)}, \mathbf{T}_{2(xy)}\right]$, $\left[\mathbf{T}_{1(y)}, \mathbf{T}_{1(z)}\right]$, $\left[\mathbf{T}_{1(y)}, \mathbf{T}_{2(yz)}\right]$, $\left[\mathbf{T}_{1(y)}, \mathbf{T}_{2(zx)}\right]$, $\left[\mathbf{T}_{1(y)}, \mathbf{T}_{2(xy)}\right]$, $\left[\mathbf{T}_{1(z)}, \mathbf{T}_{2(yz)}\right]$ are zero.

Thus, there are eight nonzero commutators (9.8), and seven commutators are zero.

The operators of incomplete time reversal for Kramers systems satisfy the anticommutation relations

$$\begin{aligned}
&\{\mathbf{T}_{1(x)}, \mathbf{T}_{1(y)}\} = 2\mathbf{T}_{2(xy)}, \ \{\mathbf{T}_{1(x)}, \mathbf{T}_{2(xy)}\} = 2\mathbf{T}_{1(y)}, \ \{\mathbf{T}_{1(y)}, \mathbf{T}_{1(z)}\} = 2\mathbf{T}_{2(yz)}, \\
&\{\mathbf{T}_{1(y)}, \mathbf{T}_{2(yz)}\} = 2\mathbf{T}_{1(z)}, \ \{\mathbf{T}_{1(y)}, \mathbf{T}_{2(zx)}\} = 2\mathbf{T}, \\
&\{\mathbf{T}_{1(y)}, \mathbf{T}_{2(xy)}\} = 2\mathbf{T}_{1(x)}, \ \{\mathbf{T}_{1(z)}, \mathbf{T}_{2(yz)}\} = 2\mathbf{T}_{1(y)}.
\end{aligned} \tag{9.9}$$

Thus, there are seven nonzero anticommutators (9.9), and the other eight anticommutators are zero: $\{\mathbf{T}_{1(x)}, \mathbf{T}_{1(z)}\}$, $\{\mathbf{T}_{1(x)}, \mathbf{T}_{2(yz)}\}$, $\{\mathbf{T}_{1(x)}, \mathbf{T}_{2(zx)}\}$, $\{\mathbf{T}_{1(z)}, \mathbf{T}_{2(zx)}\}$, $\{\mathbf{T}_{1(z)}, \mathbf{T}_{2(xy)}\}$, $\{\mathbf{T}_{2(yz)}, \mathbf{T}_{2(zx)}\}$, $\{\mathbf{T}_{2(yz)}, \mathbf{T}_{2(xy)}\}$ and $\{\mathbf{T}_{2(zx)}, \mathbf{T}_{2(xy)}\}$. In addition, there are the following relations between the operator $\mathbf{T}$ and the operators of incomplete time reversal:

$$\begin{aligned}
&\left[\mathbf{T}, \mathbf{T}_{1(x)}\right] = -2\mathbf{T}_{2(yz)}, && \left[\mathbf{T}, \mathbf{T}_{1(z)}\right] = 2\mathbf{T}_{2(xy)}, \\
&\left[\mathbf{T}, \mathbf{T}_{2(yz)}\right] = 2\mathbf{T}_{1(x)}, && \left[\mathbf{T}, \mathbf{T}_{2(xy)}\right] = 2\mathbf{T}_{1(z)}, \\
&\left[\mathbf{T}, \mathbf{T}_{1(y)}\right] = 0, && \left[\mathbf{T}, \mathbf{T}_{2(zx)}\right] = 0.
\end{aligned} \tag{9.10}$$

and the following anticommutation relations of these operators

$$\begin{aligned} &\{\mathbf{T},\mathbf{T}_{1(y)}\}=2\mathbf{T}_{2(zx)},\ \{\mathbf{T},\mathbf{T}_{2(zx)}\}=-2\mathbf{T}_{1(y)},\\ &\{\mathbf{T},\mathbf{T}_{1(x)}\}=\{\mathbf{T},\mathbf{T}_{1(z)}\}=\{\mathbf{T},\mathbf{T}_{2(yz)}\}=\{\mathbf{T},\mathbf{T}_{2(xy)}\}=0. \end{aligned} \tag{9.11}$$

In contrast to the Kramers systems, in the case of systems with an integer angular momentum J, all 21 commutators (15 commutators between six incomplete time-reversal operators and six commutators of the operator $\mathbf{T}$ with operators of incomplete time reversal) are equal to zero, while all 21 anticommutators are nonzero:

$$\begin{aligned} &\{\mathbf{T}_{1(x)},\mathbf{T}_{1(y)}\}=2\mathbf{T}_{2(xy)},\ \{\mathbf{T}_{1(x)},\mathbf{T}_{1(z)}\}=2\mathbf{T}_{2(zx)},\ \{\mathbf{T}_{1(x)},\mathbf{T}_{2(yz)}\}=2\mathbf{T},\\ &\{\mathbf{T}_{1(x)},\mathbf{T}_{2(zx)}\}=2\mathbf{T}_{1(z)},\ \{\mathbf{T}_{1(x)},\mathbf{T}_{2(xy)}\}=2\mathbf{T}_{1(y)},\ \{\mathbf{T}_{1(y)},\mathbf{T}_{1(z)}\}=2\mathbf{T}_{2(yz)},\\ &\{\mathbf{T}_{1(y)},\mathbf{T}_{2(yz)}\}=2\mathbf{T}_{1(z)},\ \{\mathbf{T}_{1(y)},\mathbf{T}_{2(zx)}\}=2\mathbf{T},\ \{\mathbf{T}_{1(y)},\mathbf{T}_{2(xy)}\}=2\mathbf{T}_{1(x)},\\ &\{\mathbf{T}_{1(z)},\mathbf{T}_{2(yz)}\}=2\mathbf{T}_{1(z)},\ \{\mathbf{T}_{1(z)},\mathbf{T}_{2(zx)}\}=2\mathbf{T}_{1(x)},\ \{\mathbf{T}_{1(z)},\mathbf{T}_{2(xy)}\}=2\mathbf{T},\\ &\{\mathbf{T}_{2(yz)},\mathbf{T}_{2(zx)}\}=2\mathbf{T}_{2(xy)},\ \{\mathbf{T}_{2(yz)},\mathbf{T}_{2(xy)}\}=2\mathbf{T}_{2(zx)},\ \{\mathbf{T}_{2(zx)},\mathbf{T}_{2(xy)}\}=2\mathbf{T}_{2(yz)},\\ &\{\mathbf{T},\mathbf{T}_{1(x)}\}=2\mathbf{T}_{1(z)},\ \{\mathbf{T},\mathbf{T}_{1(y)}\}=2\mathbf{T}_{2(zx)},\ \{\mathbf{T},\mathbf{T}_{1(z)}\}=2\mathbf{T}_{2(xy)},\\ &\{\mathbf{T},\mathbf{T}_{2(yz)}\}=2\mathbf{T}_{1(x)},\ \{\mathbf{T},\mathbf{T}_{2(zx)}\}=2\mathbf{T}_{1(y)},\ \{\mathbf{T},\mathbf{T}_{2(xy)}\}=2\mathbf{T}_{1(z)}. \end{aligned} \tag{9.12}$$

The operators of incomplete time reversal in the case of non-Kramers systems are determined by the same formulas (9.3) as for the Kramers systems, with the only significant difference that in this case $U^{(J)}$, $V^{(J)}$ and $W^{(J)}$ are symmetric matrices, where $\left(U^{(J)}\right)^2 = e^{(J)}$ and, respectively, $\mathbf{T}^2 = \left(U^{(J)}\mathbf{K}\right)^2 = e^{(J)}$.

9.5 Unitarity of Spinor Operators in Two-Boson Representation of Angular Momentum and Time-Reversal Symmetry

In a number of cases it turns out to be convenient to use other representations of angular momentum operators along with the traditional representation of these operators in quantum mechanics. In applications, it is not very convenient to represent spin operators through Bose creation and annihilation operators, such as the Holstein–Primakoff [410] and Dyson–Maleev [411, 412] representations, which are used in the quantum theory of magnetism [413] and in the theory of order–disorder phase transitions in ferroelectrics [414]. This is due to the fact that in the Holstein–Primakoff representation the dependence of the operators S_x and S_y is such that the Bose creation and annihilation operators are under the square root. On the other hand, the use of the Dyson–Maleev representation leads to non-Hermitian Hamiltonian. In contrast, from these drawbacks is free the Schwinger representation [415], in which the angular momentum projection operators are expressed through the creation and annihilation operators of two different but correlating Bose fields. Before considering unitary spinor operators in two-boson representation of the angular momentum, it is appropriate to make the following observation. The well-known commutation relation for creation, b^+, and destruction, b, Bose operators

$$b_i b_i^+ - b_i^+ b_i = 1, \tag{9.13}$$

or in the more general form

$$b_{\mathbf{k}\lambda} b^+{}_{\mathbf{k}'\lambda'} - b^+{}_{\mathbf{k}'\lambda'} b_{\mathbf{k}\lambda} = \delta_{\mathbf{k}\mathbf{k}'} \delta_{\lambda\lambda'} \tag{9.14}$$

(where $\mathbf{k}$, $\mathbf{k}'$ and λ, λ' are respectively the wave vector indexes and the indexes of the particles (quasiparticles) polarization), contains an operator on the left side and a c-number on the right-hand side [3, 51, 61, 410–430].

Such an approach is easy to justify, starting from the definition of b_i and $b^+{}_i$ operators [425]:

$$b_i \mid N_i >= \sqrt{N_i} \mid N_i - 1 >, \tag{9.15}$$

$$b_i^+ \mid N_i >= \sqrt{N_i + 1} \mid N_i + 1 >, \tag{9.16}$$

where $\mid N_i >$ is a wave function in the representation of occupation numbers. Using (9.15) and (9.16), we find

$$b_i b_i^+ \mid N_i > = (N_i + 1) \mid N_i >, \tag{9.17}$$

$$b_i^+ b_i \mid N_i > = N_i \mid N_i > . \tag{9.18}$$

Subtracting (9.18) from (9.17), we obtain

$$\left(b_i b_i^+ - b_i^+ b_i\right) \mid N_i >=\mid N_i > \tag{9.19}$$

From this, the commutation relation (9.13) follows. Although the commutation relation for Bose operators in the form (9.13) is applied most often, there is also a justification for this relation for the case when the right-hand side of equation (9.13) is a unit operator in an abstract infinite space. Such an analysis is given in [72], where the commutation relation for the Bose creation and annihilation operators in the quantum state τ is given by the expression

$$b(\tau)b^+(\tau) - b^+(\tau)b(\tau) = I(\tau). \tag{9.20}$$

In (9.20) $I(\tau)$ is a unit infinite matrix. As is known, the operators $b(\tau)$ and $b^+(\tau)$ are also defined in an infinite-dimensional space (the matrices of these operators are given, for example, in [61, 72]). The operator of the number of particles $N(\tau)$ is connected with the operators $b(\tau)$ and $b^+(\tau)$ by the following relations:

$$b^+(\tau)b(\tau) = N(\tau), \; b(\tau)b^+(\tau) = I(\tau) + N(\tau) \tag{9.21}$$

By subtracting the first relation in (9.21) from the second, we obtain the commutation relation (9.20). The relation (9.20) follows from (9.21), but not vice versa. It is important to pay attention to this, since the direct calculation of the commutator (9.20) on the basis of the matrices of the operators $b(\tau)$ and $b^{+}(\tau)$ does not lead to a unit infinite-dimensional matrix. Indeed, it is easy to show that for a matrix corresponding to the operator $b(\tau)b^{+}(\tau) - b^{+}(\tau)b(\tau)$ in the left-hand side of the equation (9.20), all off-diagonal elements are zero, and the diagonal elements satisfy the relations

$$e_{11} = e_{22} = e_{33} = \cdots = e_{n-1,n-1} = 1; \; e_{nn} = -\sum_{i=1}^{n-1} e_{ii} = 1 - n. \quad (9.22)$$

When $n >> 1$, we have $e_{nn} = -n$ and, respectively,

$$\lim_{n \to \infty} e_{nn} = -\infty. \quad (9.23)$$

Thus, the matrix of the operator $b(\tau)b^{+}(\tau) - b^{+}(\tau)b(\tau)$ in an infinite-dimensional basis has the form

$$b(\tau)b^{+}(\tau) - b^{+}(\tau)b(\tau) = \begin{pmatrix} 1 & & & & \\ & 1 & & 0 & \\ & & 1 & & \\ & 0 & & \ddots & \\ & & & & -\infty \end{pmatrix} = I_{\tau}. \quad (9.24)$$

This matrix differs from the unit infinite matrix $I(\tau)$ from (9.20) introduced in [72] by the fact that one (the bottom) diagonal matrix element tends to a negative number infinitely large in absolute value, while each of the remaining matrix elements on the main diagonal are equal to 1. Despite the essential difference between the commutation relation (9.24) and the commutation relations (9.13), (9.14) and (9.20), calculations with their application lead to the same results for the number of bosons and the quantities associated. This is the hidden paradox, connected with the commutation relation for Bose creation and annihilation operators. In the case of fermions, there is no analogous paradox.

In the case of spin $S = \frac{1}{2}$ the transition from the spinor representation of spin operators to Schwinger representation can be realized in a very simple way on the basis of creation and annihilation Bose operators a^{+}, b^{+} and a, b (a and b being commuting operators) by introduction of an unitary spinor operator $U_{\frac{1}{2}} = \begin{pmatrix} a \\ b \end{pmatrix}$. Under action of this operator the following unitary transformation takes place

$$\begin{aligned} U_{\frac{1}{2}} S_z U_{\frac{1}{2}}^{+} &= \tfrac{1}{2}(a^{+}a - b^{+}b), \\ U_{\frac{1}{2}} S_{+} U_{\frac{1}{2}}^{+} &= a^{+}b, \;\; U_{\frac{1}{2}} S_{-} U^{+} = b^{+}a, \end{aligned} \quad (9.25)$$

where $S_{\pm} = S_x \pm i S_y$. The unitary property of the spinor operator $U_{\frac{1}{2}}$ $\left(U^{+}{}_{\frac{1}{2}} U_{\frac{1}{2}} = \mathbf{1}\right)$ leads to relationship

$$a^{+}a + b^{+}b = \mathbf{1}, \tag{9.26}$$

which takes into account that only the self-values of the operator $a^{+}a + b^{+}b$, which satisfy the condition $n_a + n_b = 1$ $\left(n_a =< n_a \mid a^{+}a \mid n_a >, n_b =< n_b \mid b^{+}b \mid n_b >\right)$, and corresponding them self-functions of this operator, are involved in Schwinger representation of the angular momentum. The operator $a^{+}a + b^{+}b$ is an invariant of the Lie group of symmetry $SU(2)$ with respect to rotations taking into consideration that groups $SU(2)$ and R_3 (the group of thee dimensional rotations) are homomorphic [71]. Moreover, this operator commutes with operators S_z, S_+ and S_-

$$\left[a^{+}a + b^{+}b, \hat{S_z}\right] = \left[a^{+}a + b^{+}b, \hat{S_+}\right] = \left[a^{+}a + b^{+}b, \hat{S_-}\right] = 0 \tag{9.27}$$

Therefore the operator $a^{+}a + b^{+}b$ commute also with Zeeman Hamiltonian which depends linearly on S_z, S_+ and S_-.

Thus, the commutation relations (9.27) lead to conservation of the total number of bosons of a- and b-types, i.e. to $n_a + n_b = const$. It is in accordance with relationship (9.26) averaged on boson number of occupation ($n_a + n_b = 1$) with specification that the constant value is equal to unit. The formula (9.26), which is a single consequence of unitarity of the operator $U_{\frac{1}{2}}$ in the case of a particle with spin $S = \frac{1}{2}$, must be transformed into relationship

$$a^{+}a + b^{+}b = 2S \cdot \mathbf{1} \tag{9.28}$$

for any system with spin S, where $\mathbf{1}$ is unit operator represented by unit $(2S + 1) \times (2S + 1)$-matrix. It is a single kinematic condition traditionally related to Schwinger representation of angular momentum. However, excepting (9.28) there are $(2S - 1)$ new kinematic conditions resulting from the unitarity property of the operator U_S, by means of which the transition from the spinor representation of an arbitrary spin S to its two-boson representation is realized [135, 431]. This is the non-Schwinger approach to Schwinger representation of spin operators.

Let us introduce the following unitary spinor operator (M is the self-value of the operator S_z, in the system of units in which $\hbar = 1$ is used):

$$U_s = \begin{bmatrix} [(2S)!]^{-\frac{1}{2}}\, a^{2S} \\ [(2S-1)!]^{-\frac{1}{2}}\, a^{2S-1}b \\ \cdot \\ [(S+M)!\,(S-M)]^{-\frac{1}{2}}\, a^{S+M}b^{S-M} \\ \cdot \\ [(2S-1)!]^{-\frac{1}{2}}\, ab^{2S-1} \\ [(2S)!]^{-\frac{1}{2}}\, b^{2S} \end{bmatrix} \tag{9.29}$$

by means of which the transition from spinor representation of operators S_z, S_+ and S_- to Schwingers one is carried out [135, 220, 431]:

$$\begin{aligned} S_z &= \tfrac{1}{2}\left(a^+a - b^+b\right) O_z^{(S)}, \\ S_+ &= a^+bO_+^{(S)}, \quad S_- = b^+aO_-^{(S)}. \end{aligned} \tag{9.30}$$

Here $O_z^{(S)}$, $O_+^{(S)}$ and $O_-^{(S)}$ are known as operator loads to operators S_z, S_+ and S_- of spin $S = \frac{1}{2}$, which represented some polynomials of $2S-1$ degree in the terms of operator $n = a^+a + b^+b$. According to [135, 220, 431], each from operator leads $O_z^{(S)}$, $O_+^{(S)}$ and $O_-^{(S)}$ is equivalent to unit operator, that is caused by specific structure of unitary spinor operator U_S. Thus, interdependently of the value of spin S the operators S_z, S_+, and S_- in the coupled bosons *(CB)* representation have the form

$$\begin{aligned} S_z &= \tfrac{1}{2}\left(a^+a - b^+b\right), \\ S_+ &= a^+b, \quad S_- = b^+a. \end{aligned} \tag{9.31}$$

Because in *CB* representation the form of operators S_z, S_+, and S_- does not depend on the value of spin S, all of the specific of multilevel spin system is determined by spin wave functions in this representation:

$$\begin{aligned} \mid S, M > &= [(S+M)!\,(S-M)!]^{-\frac{1}{2}} \left(a^+\right)^{S+M} \left(b^+\right)^{S-M} \mid 0 >= \\ &= \mid S+M >_a \mid S-M >_b, \end{aligned} \tag{9.32}$$

where $\mid 0 >$ is the vacuum state ($\mid 0 > = \mid 0 >_a \mid 0 >_b$).

Averaging the operator equation

$$U^+{}_S\left(a^+, b^+\right) U_S\,(a, b) = \mathbf{1}, \tag{9.33}$$

where $U_S\,(a, b)$ is determined by (9.29), by means of wave functions $|S+M>_a$ $|S+M>_b$ that present the set of $2S+1$ functions $\mid 2S >_a \mid 0 >_b$, $|2S-1>_a$ $|1>_b$, $\mid 2S-2 >_a \mid 2 >_b, \ldots, \mid 2 >_a \mid 2S-2 >_b$, $\mid 1 >_a \mid 2S-1 >_b$, $\mid 0 >_a \mid 2S >_b$, we obtain the following algebraic equation of $2S$ degree with respect to variable $n = n_a + n_b$:

$$\begin{aligned} n^{2S} - C_{2S-1}n^{2S-1} + C_{2S-2}n^{2S-2} + \cdots + (-1)^{2S}C_2 n + \\ +(-1)^{2S+1}\left[2\left(S - \frac{1}{2}\right)\right]!n = (2S)! \end{aligned} \tag{9.34}$$

The coefficient from (9.34) for $S = \frac{1}{2}, 1, \frac{3}{2}, 2, \frac{5}{2}, 3, \frac{7}{2}, \ldots$ are

$C_{2S-1} = 0, 1, 3, 6, 10, 15, 21, \ldots$; $C_{2S-2} = 0, 0, 2, 11, 35, 85, 175, \ldots$;

$C_{2S-3} = 0, 0, 0, 6, 50, 225, 735, \ldots$; $C_{2S-4} = 0, 0, 0, 0, 24, 277, 1624, \ldots$;

$C_{2S-5} = 0, 0, 0, 0, 0, 120, 1764, \ldots$; $C_{2S-6} = 0, 0, 0, 0, 0, 0, 720, \ldots$;

The values of coefficients $C_{2S-q} =$ in (9.34) depend on the spin S value. The coefficients C_{2S-1} for angular momenta S and $S + \frac{1}{2}$ satisfy the equality

$$C^{S+\frac{1}{2}}{}_{2S-1} - C^{S}{}_{2S-1} = 2S \tag{9.35}$$

The introduced above spin wave functions (9.32) in CB representation hereinafter will be denoted for simplicity by $|2S\ 0>$, $|2S-1\ 1>$, $|\ 2S-2\ 2>, \cdots, |2\ 2S-2>$, $|1\ 2S-1>$, $|\ 0\ 2S>$.

Lets consider (9.34) and its solutions for some particular cases.

1. $S = \frac{1}{2}$. It was obtained the known result consisting in the existence of a single consequence $n = 1$ from unitarity of the spinor operator $U_{\frac{1}{2}}$.
2. $S = 1$. Equation (9.34) for $S = 1$ have the form

$$n^2 - n - 2 = 0. \tag{9.36}$$

   Two solutions of the (9.36) are $n_1 = 2S = 2$ and $n_2 = -1$. It should be noted that presence of the real positive solution $n = 2S$ of (9.34) is characteristic for any value of the spin S. As for solution $n = -1$ which corresponds to negative numbers of filling of the boson states $(n_a + n_b = -1)$ for integer S, this result will be discussed in the Sect. 9.6.
3. $S = \frac{3}{2}$ The equation

$$n^3 - 3n^2 + 2n - 6 = 0 \tag{9.37}$$

   have the following three solutions: $n_1 = 2S,\ n_{2,3} = \pm i\sqrt{3}$.

Unlike systems with spin $\frac{1}{2}$ and $\frac{3}{2}$ there are one real solution $n_1 = 2S$ and two imaginary complex-conjugate ones (n_2 and n_3). The imaginary solutions of (9.34) at $S = \frac{3}{2}$ and, in more general case, the complex-conjugate solutions of (9.34) at $S > \frac{3}{2}$ also will be discussed in the Sect. 9.6.

By analyzing $2S$ solutions of the (9.34) for a given value of spin S, we can say that the only $x_1 = n_a + n_b = 2S$ solution meets positive occupation numbers of a- and b-bosons $(n_a \geq 0, n_b \geq 0)$. The relationship $n_a + n_b = 2S$ sets the number of bosons of a- and b-types necessary for the implementation of this condition. For a given number of states of bosons of the same type, there must be defined, correlated with it, number of states of bosons of another type. In this context, we can talk about boson–boson correlations.

In the case of a particle with spin $S = \frac{1}{2}$, the Wigner time reversal operator $\mathbf{T}$ in the two-boson representation of the angular momentum has the form

$$\mathbf{T} = \left(a^{+}b - b^{+}a\right)\mathbf{K}, \tag{9.38}$$

where $\mathbf{K}$ is the complex conjugation operator. Similarly, for six operators of incomplete time reversal, in this case we obtain

$$\mathbf{T}_{1(x)} = \left(a^+a - b^+b\right)\mathbf{K},\ \mathbf{T}_{1(y)} = e\mathbf{K},\ \mathbf{T}_{1(z)} = \left(a^+b + b^+a\right)\mathbf{K},$$
$$\mathbf{T}_{2(yz)} = a^+b + b^+a,\ \mathbf{T}_{2(zx)} = a^+b - b^+a,\ \mathbf{T}_{2(xy)} = \left(a^+a - b^+b\right). \tag{9.39}$$

All the results obtained earlier concerning operators of complete and incomplete time reversal using the spinor representation for the spin-1/2 projection operators are also preserved in the case of using the two-boson representation of these operators. In particular, the action of the operator $\mathbf{T} = i\sigma_y\mathbf{K}$ on the spin wave functions $\mid \frac{1}{2}, +\frac{1}{2} >\equiv \begin{pmatrix}1\\0\end{pmatrix}$ and $\mid \frac{1}{2}, -\frac{1}{2} >\equiv \begin{pmatrix}0\\1\end{pmatrix}$ written in the spinor basis leads to the following results:

$$\mathbf{T}\mid \frac{1}{2}, +\frac{1}{2} > = \begin{pmatrix}0 & 1\\-1 & 0\end{pmatrix}\mathbf{K}\begin{pmatrix}1\\0\end{pmatrix} = -\mid \frac{1}{2}, -\frac{1}{2} >, \tag{9.40}$$
$$\mathbf{T}\mid \frac{1}{2}, -\frac{1}{2} > = \begin{pmatrix}0 & 1\\-1 & 0\end{pmatrix}\mathbf{K}\begin{pmatrix}0\\1\end{pmatrix} = \mid \frac{1}{2}, \frac{1}{2} > .$$

It is easy to see that the operator $\mathbf{T}$, written using the two-boson representation of the operator $i\sigma_y = S_+ - S_-$ (see (9.31)), transforms the spin wave functions written in the same representation as follows:

$$\mathbf{T}\mid \frac{1}{2}, +\frac{1}{2} > = \left(a^+b - b^+a\right)\mathbf{K}a^+ \mid 0 >= \tag{9.41}$$
$$= a^+a^+b \mid 0 > -aa^+b^+ \mid 0 >= - \mid 0 >_a\mid 1 >_b= - \mid \frac{1}{2}, -\frac{1}{2} >,$$
$$\mathbf{T}\mid \frac{1}{2}, -\frac{1}{2} > = \left(a^+b - b^+a\right)\mathbf{K}b^+ \mid 0 >= \tag{9.42}$$
$$= a^+bb^+ \mid 0 > -b^+b^+ \mid 0 >= a^+bb^+ \mid 0 >=\mid 1 >_a\mid 0 >_b=\mid \frac{1}{2}, \frac{1}{2} > .$$

In Table 9.1 the matrices of unitary operators $U^{(J)}$ in the two-boson representation are presented, which determine the time-reversal operator $\mathbf{T} = U^{(J)}\mathbf{K}$ for systems with angular momentum J.

As follows from (9.41) and (9.42), the matrix $U^{(\frac{1}{2})}$ that determines the operator $\mathbf{T}$ in the spinor representation for the case of a particle with spin $S = \frac{1}{2}$ ($\mathbf{T} = U^{(\frac{1}{2})}\mathbf{K},\ U^{(\frac{1}{2})} = i\sigma_y,\ U^{(\frac{1}{2})+}U^{(\frac{1}{2})} = U^{(\frac{1}{2})}U^{(\frac{1}{2})+} = e$), coincides in form with the matrix $U^{(\frac{1}{2})}$ written in the two-boson representation (in following basises: $\left\{\mid \frac{1}{2}, \frac{1}{2} >, \mid \frac{1}{2}, -\frac{1}{2} >\right\}$ and $\{\mid 1 >_a\mid 0 >_b, \mid 0 >_a\mid 1 >_b\}$, respectively):

$$U^{(\frac{1}{2})} = \begin{pmatrix}0 & 1\\-1 & 0\end{pmatrix} = \begin{pmatrix}0 & 1\\-1 & 0\end{pmatrix} \tag{9.43}$$

Table 9.1 The matrices of unitary operators $U^{(J)}$ in the two-boson representation of the angular momentum

J	Basis wave functions of angular momentum projection operators in the two-boson representation	The matrices of unitary operators $U^{(J)}$
$\frac{1}{2}$	$\mid 1>_a\mid 0>_b$ $\mid 0>_a\mid 1>_b$	$\begin{pmatrix} 0 & 1 \\ -1 & 0 \end{pmatrix}$
1	$\mid 2>_a\mid 0>_b$ $\mid 1>_a\mid 1>_b$ $\mid 0>_a\mid 2>_b$	$\begin{pmatrix} 0 & 0 & 1 \\ 0 & -1 & 0 \\ 1 & 0 & 0 \end{pmatrix}$
$\frac{3}{2}$	$\mid 3>_a\mid 0>_b,$ $\mid 2>_a\mid 1>_b,$ $\mid 1>_a\mid 2>_b,$ $\mid 0>_a\mid 3>_b$	$\begin{pmatrix} 0 & 0 & 0 & 1 \\ 0 & 0 & -1 & 0 \\ 0 & 1 & 0 & 0 \\ -1 & 0 & 0 & 0 \end{pmatrix}$
2	$\mid 4>_a\mid 0>_b,$ $\mid 3>_a\mid 1>_b$ $\mid 2>_a\mid 2>_b$ $\mid 1>_a\mid 3>_b$ $\mid 0>_a\mid 4>_b$	$\begin{pmatrix} 0 & 0 & 0 & 0 & 1 \\ 0 & 0 & 0 & -1 & 0 \\ 0 & 0 & -1 & 0 & 0 \\ 0 & 1 & 0 & 0 & 0 \\ -1 & 0 & 0 & 0 & 0 \end{pmatrix}$
…	…	…
J	$\mid J>_a\mid 0>_b$ $\mid J-1>_a\mid 1>_b$ $\mid J-2>_a\mid 2>_b,$ …, $\mid 2>_a\mid J-2>_b$ $\mid 1>_a\mid J-1>_b$ $\mid 0>_a\mid J>_b$	$\begin{pmatrix} & & & 1 \\ \mathbf{0} & & -1 & \\ & 1 & & \\ & \cdot^{\cdot^{\cdot}} & & \mathbf{0} \\ \pm 1 & & & \end{pmatrix}$

where sign "+" is selected in the last row when J is integer, and sign "−" is selected when J is half-integer

9.6 Boson–Antiboson Representation of Angular Momentum and Its Correlation with Factorization of Wigner Time-Reversal Operator

Among $2S$ solutions of the (9.34) one is $n_1 = 2S$, other is $n_2 = -1$ (this solutions exist only for integer S), and the remaining $2S - 2$ solutions for integer S and $2S - 1$ solutions for half-integer S are in pairs complex conjugated. The kinematic condition $n_1 = 2S$ leads to simultaneous elimination of the states with filling quantum numbers $n_a > S + M$ and $n_b > S - M$. Since the spin wave function $\mid S, M >$ corresponding to projection M of the spin is a product of boson wave functions $\mid S, M >_a$ and $\mid S, M >_b$ (see (9.32)) there are correlations between bosons of a- and b-types.

More exactly, if $S+M$ a-bosons are excited then for realizing of the spin state $\mid S, M>$ simultaneously $S-M$ b-bosons must be excited too. All other Bose-states are forbidden. Thus only wave functions of such correlated a- and b-bosons must be used for description of the spin state $\mid S, M>$. Because the knowledge of $\mid S, M>$ wave functions ($M=S,\ S-1,\cdots,-S+1,\ -S$) permits to calculate all interested magnetic effects, we can talk about boson–boson correlations in quantum theory of magnetism.

The kinematic condition $n_2=-1$, which corresponds to negative resulting quantum numbers of filling for boson states, leads to boson–antiboson representation of spin operators. This representation was introduced in [432, 433]. Let us consider the unitary transformation of spin projection operators under time-reversal operator $\mathbf{T}$ using obvious forms (9.31) for S_z, S_+ and S_-. In the case of non-Kramers systems ($\mathbf{T}^2=1$) the physical sense of solution $n_2=-1$ of (9.34) consists in existence of antiboson states. In this case, the operator of destruction of the antiboson corresponding to a boson of a-type [135] can be presented as

$$\tilde{a}=\mathbf{T}a\mathbf{T}^+, \tag{9.44}$$

where the designation $\tilde{a}$ does not signifies the transpose operator. It is necessary to mention that, if the kinematic condition $n_2=-1$ takes place, then all other solutions of the (9.34) including kinematic condition $n_1=2S$ that corresponds to boson–boson correlations and Wigner time-reversal invariance of the Hamiltonian are forbidden.

The kinematic condition $n_2=-1$, arising in the case of integer values of the spin S, can be presented as $n_2=n_a+n_{\tilde{a}}=-1$ or $\hbar\omega\left(n_a+\frac{1}{2}\right)=-\hbar\omega\left(n_{\tilde{a}}+\frac{1}{2}\right)$. At $\omega=\omega_a=\omega_{\tilde{a}}$ (ω is the boson frequency) the energy $\varepsilon_{\tilde{a}}$ of an antiboson of $\tilde{a}$-type is negative ($\varepsilon_{\tilde{a}}=-\varepsilon_a, \varepsilon_a=\hbar\omega(n_a+\frac{1}{2})$). Therefore, the wave function of $\tilde{a}$-antiboson coincides with the time-reversed wave function of a-boson. In other words, a^+ is the operator of creation of the boson of a-type and $\widetilde{a}^+$ is the operator of creation of the antiboson corresponding to the time-reversed boson of a-type ($\widetilde{a}^+=\mathbf{T}a^+\mathbf{T}^+$). Thus, at $n_2=-1$, it is impossible to realize the boson–boson (two-boson) representation of spin operators. In this case, the identical representation of spin operators is the boson–antiboson representation. Such a situation with boson–antiboson correlations is characteristic for all solutions of the (9.34) related to integer spin S, excluding the solution $n_1=2S$ corresponding to boson–boson correlations.

In the boson–antiboson representation the spin wave function is

$$\begin{aligned}\mid S, M> &= [(S+M)!(S+M+1)!]^{-\frac{1}{2}}\left(a^+\right)^{S+M}\left(\tilde{a}^+\right)^{S+M+1}\mid 0> = \\ &= \mid S+M>_a\mid S+M+1>_{\tilde{a}}\ .\end{aligned} \tag{9.45}$$

where $\tilde{a}^+$ is the operator of creation of the antiboson, which is time-reversed relatively to the boson of a-type. We see that the spin state $\mid SM>$ from (9.45) is realized using $S+M$ bosons of a-type and correlated with these bosons $S+M+1$ antibosons of $\tilde{a}$-type, $\tilde{a}^+$ being the operator of creation of the antiboson of $\tilde{a}$-type.

The same kinematic condition $n_2 = -1$ can be satisfy using two Bose operators a^+ and $\tilde{b}^+$ ($\tilde{b}^+$ is the operator of creation of antiboson which is time reversed relatively to the boson of b-type). In this case, the spin wave function $\mid SM >$ is

$$\begin{aligned} \mid SM > &= [(S+M)!(S+M+1)!]^{-\frac{1}{2}} \left(a^+\right)^{S+M} \left(\tilde{b}^+\right)^{S+M+1} \mid 0 > = \\ &= \mid S+M >_a \mid S+M+1 >_{\tilde{b}} . \end{aligned} \tag{9.46}$$

According to (9.46), the spin state $\mid SM >$ is realized by means of $S+M$ bosons states of a -type and correlating $S+M+1$ antiboson states of $\tilde{b}$ -type.

In (9.45) and (9.46), the following antiboson states were introduced:

$$\mid S+M+1 >_{\tilde{a}} = \mathbf{T} \mid S+M+1 >_a, \mid S+M+1 >_{\tilde{b}} = \mathbf{T} \mid S+M+1 >_b . \tag{9.47}$$

Let us now consider the action of time-reversal operator $\mathbf{T}$ on spin operators S_z, S_x and S_y from (9.31) when the relationship (9.44) is satisfied

$$\mathbf{T}S_z\mathbf{T}^+ = \frac{1}{2}\left(\mathbf{T}a^+\mathbf{T}^+\mathbf{T}a\mathbf{T}^+ - \mathbf{T}b^+\mathbf{T}^+\mathbf{T}b\mathbf{T}^+\right) = \frac{1}{2}(b^+b - a^+a) = -S_z, \tag{9.48}$$

$$\begin{aligned} \mathbf{T}S_x\mathbf{T}^+ &= \frac{1}{2}\mathbf{T}(a^+b + b^+a)\mathbf{T}^+ = \\ &= \frac{1}{2}\left(\mathbf{T}a^+\mathbf{T}^+\mathbf{T}b\mathbf{T}^+ + \mathbf{T}b^+\mathbf{T}^+\mathbf{T}a\mathbf{T}^+\right) = \frac{1}{2}(b^+a + a^+b) = S_x, \end{aligned} \tag{9.49}$$

$$\begin{aligned} \mathbf{T}S_y\mathbf{T}^+ &= -\frac{1}{2}\mathbf{T}\left[i\left(a^+b - b^+a\right)\right]\mathbf{T}^+ = \\ &= \frac{i}{2}\left(\mathbf{T}a^+\mathbf{T}^+\mathbf{T}b\mathbf{T}^+ - \mathbf{T}b^+\mathbf{T}^+\mathbf{T}a\mathbf{T}^+\right) = \frac{i}{2}\left(b^+a - a^+b\right) = S_y. \end{aligned} \tag{9.50}$$

Equations (9.48)–(9.50) show that, when fulfilling the kinematic condition $n_2 = -1$, a violation of the Wigner time-reversal symmetry takes place. In this case, operator $\mathbf{T}$ is similar to the operator of incomplete time reversal $\mathbf{T}_{1(z)}$ for a particle with spin $S = \frac{1}{2}$ found in the Sect. 8.1 (8.4) on the basis of properties of the non-Abelian symmetry group G_8 (see (8.11)).

The remaining $2S - 2$ solutions of (9.34) for systems with integer spin S, and $2S - 1$ solutions for systems with half-integer S are in pairs complex conjugated. Since these solutions, as it will be shown, also have a physical meaning, they will be taken into account at the end of this Section.

Only for $n_1 = 2S$ the results obtained in the spinor representation and in the two-boson one are the same. Specifically, for $S = 1$ and $n_1 = 2$ energy positions of Zeeman components $\mid 1, 1 >$, $\mid 1, 0 >$ and $\mid 1, -1 >$ in spinor representation are the same as the positions of Zeeman components $\mid 2 >_a \mid 0 >_b$, $\mid 1 >_a \mid 1 >_b$ and $\mid 0 >_a \mid 2 >_b$, respectively, in the two-boson representation. But for $n_2 = -1$ (in the case of spin $S = 1$ there are only two solutions $n_1 = 2$ and $n_2 = -1$) the Zeeman levels corresponding to $\mid 2 >_a \mid 0 >_b$, $\mid 1 >_a \mid 1 >_b$ and $\mid 0 >_a \mid 2 >_b$ states are shifted to high energies by $\left(S + \frac{1}{2}\right) g\mu_B H$ as compared with Zeeman components corresponding to $| 1, 1>$, $| 1, 0>$, and $| 1, -1>$ spin states. Here g is g-factor, μ_B is Bohr magneton and H denotes the magnetic field. At $S > 1$ the high-energy shift of spin levels is accompanied by their broadening due to pairs of complex-conjugated solutions of the (9.34). For $n = n_1, n_2$ this leads to continuous Zeeman energy bands (see below).

As regards T-invariance in Zeeman systems, it takes place if not only the sign of spin projection operators, but also the direction of the magnetic field is reversed. In this case, the infringement of T-invariance can be described in terms of the spin-Hamiltonian depending linearly on magnetic field, and containing the unit operator **1** defined in the space of spin wave functions. The interaction between the total spin S and the magnetic field at infringement of T-invariance is described by the operator

$$\mathbf{H} = g\mu_B H_z S_z + g\mu_B \alpha H_z \cdot \mathbf{1}, \tag{9.51}$$

where α is dimensionless constant which characterizes the interaction between spin S and the magnetic field infringing the **T**-invariance ($H_z = H$). Denoting the first and the second terms in (9.51) by $\mathbf{H}_1$ and $\mathbf{H}_2$, respectively, we see that $\mathbf{H}_2$-operator changes the sign under time-reversal operator

$$\mathbf{O_H U K H_2 K U^+ O_H^+} = -\mathbf{H}_2 \tag{9.52}$$

where, as usual, **U** is the unitary operator defined in the space of spin wave functions, **K** is the operator of complex configuration and $\mathbf{O_H}$ is the operator reversing the direction of the magnetic field

$$\mathbf{O_H H} = -\mathbf{H}, \;\; \mathbf{O_H^2} = 1. \tag{9.53}$$

The spectrum of operator **H** at $\alpha = S + \frac{1}{2}$ calculated in spinor basis or in the Schwinger representation under the kinematic condition $n_1 = 2S$ coincides with the spectrum of the operator $\mathbf{H}_1$ calculated in the Schwinger representation under the kinematic condition $n_2 = -1$. Particularly, for $S = 1$ there are only two solutions of the (9.34): $n_1 = 2$ and $n_2 = -1$. If the solution $n_1 = 2$ is realized then there are three Zeeman levels, $E^{(1)}{}_1 = g\mu_B H$, $E^{(1)}{}_0 = 0$ and $E^{(1)}{}_{-1} = -g\mu_B H$. But if the time-reversal symmetry is violated and the kinematic condition $n_2 = -1$ is realized, then each Zeeman level will be shifted by $\frac{3}{2} g\mu_B H$ to the high-energy range. The both spectra are discrete with the same distance between the neighboring levels.

At $S > 1$ the $2S - 1$ solutions for half-integer S and $2S - 2$ solutions for integer S are in pairs complex-conjugated.The real parts of these solutions give the contribution to high-energy shift of Zeeman levels, while their imaginary parts describe the broadening of the levels at time-reversal symmetry violation [136]. This follows from generalized formula for energy of Zeeman levels

$$\begin{aligned} E^{(1)}_{M,q} &= g\mu_B H \left(M + S - \tfrac{1}{2} n_q^{(S)} \right), \\ M &= S,\; S-1,\; \ldots,\; -S+1,\; -S, \end{aligned} \tag{9.54}$$

Table 9.2 The values $\delta E_s(n)$ and $\gamma_s(n)$ for Zeeman systems with Hamiltonian (9.51)

S	$n_q(q = 1, 2, \ldots, 2S)$	$\delta E_s(n)/g\mu_B H$	$\gamma_s(n)/g\mu_B H$
$\frac{1}{2}$	1	–	–
1	$\begin{cases} 2 \\ -1 \end{cases}$	$\begin{cases} - \\ \frac{3}{2} \end{cases}$	$\begin{cases} - \\ - \end{cases}$
$\frac{3}{2}$	$\begin{cases} 3 \\ +i\sqrt{2} \end{cases}$	$\begin{cases} - \\ \frac{3}{2} \end{cases}$	$\begin{cases} - \\ - \end{cases}$
2	$\begin{cases} 4 \\ -1 \\ \frac{3}{2} \pm i\frac{\sqrt{15}}{2} \end{cases}$	$\begin{cases} - \\ \frac{5}{2} \\ \frac{5}{4} \end{cases}$	$\begin{cases} - \\ - \\ \frac{\sqrt{15}}{4} \end{cases}$
$\frac{5}{2}$	$\begin{cases} 5 \\ 2.915 \pm 2.119i \\ -0.415 \pm 1.295i \end{cases}$	$\begin{cases} - \\ 1.042 \\ 2.707 \end{cases}$	$\begin{cases} - \\ 1.059^* \\ 0.647 \end{cases}$

which was obtained using the representation of spin operators throughout operators $a^+,\ a,\ b^+$, and b, taking into account all solutions $n_q^{(S)}$ $(q = 1,\ 2, \ldots, 2S)$ of (9.34).

In Table 9.2 values $\delta E_s(n)$ and $\gamma_s(n)$ of shifts and broadening of Zeeman levels (in dimensionless units) for high-spin systems $\left(S = \frac{1}{2},\ 1,\ \frac{3}{2},\ 2,\ \frac{5}{2}\right)$ at infringement of T-invariance are presented.

The value 1.059 for $\gamma(\frac{5}{2})/g\mu_B H$ denoted in Table 9.2 by asterisk is larger than the corresponding shift of energy level $\delta E_{\frac{5}{2}}(n)/g\mu_B H = 1.042$. Therefore, in this case the Zeeman spectrum is continuous without forbidden bands.

It is seen in Table 9.2 that values $\delta E_{1/2}(1)$ and $\gamma_{1/2}(1)$ are equal to zero, i.e. for a spin-1/2 particles the time-reversal symmetry violation under magnetic field (second term in (9.51)) is not possible. It is not in contradiction with results concerning six types of the Wigner time-reversal symmetry violation for any free spin-$\frac{1}{2}$ particle obtained on the basis of non-Abelian group $G_{16}^{(1/2)}$ (see Chap. 8, (8.43). In the last case, the types of Hamiltonians are not specified, but there are Hamiltonians, which are not invariant under Wigner time-reversal operator (two examples of such Hamiltonians, including the case of spin-1/2 particles, were discussed in Sect. 7.4). For all other cases, the violation of time-reversal symmetry under second term of (9.51) is possible with high-energy shifts of Zeeman levels and their transformation (excluding the case of $S = 1$) into a broaded levels (bands), which in some cases may overlap.

In conclusion, Table 9.2 contains results for all kinematic conditions that follow from all $2S$ solutions of (9.34). However, it is necessary to remember that, if any of these kinematic conditions is realized (usually, it is the solution $n_1 = 2S$), then the others are forbidden.

9.7 About Restoration of Broken Wigner Time-Reversal Symmetry

It is well known that the reconstruction of the symmetry of spatial inversion when the law of parity conservation in weak interactions is violated [434] occurs, if an additional transformation is performed, referred to as the charge conjugation, with the introduction of the concept of antiparticles [435]. A combined transformation containing spatial inversion and charge conjugation (replacement of particles by antiparticles) was called a combined inversion [436].

Similarly, in order to restore the symmetry of the time reversal in cases where it is violated, "one could postulate that time reversal transforms matter into meta-matter" [437]. Experimentally, the meta-matter was not yet discovered by the time when Wigner monograph was published [437]. The meta-matter has not yet been discovered and, probably, will be discovered in the future, when more powerful accelerators will be built.[1]

At present, there are quite a few publications on the violation of the symmetry of time reversal in meson systems, in light and heavy nuclei, in atoms and molecules, superconductors, in quantum transport, etc. (see Chap. 10). However, these works do not discuss the Wigner heuristic postulate on the existence of meta-particles and meta-matter.

We made an attempt to determine some properties of the meta-particles not yet discovered experimentally. Before presenting the results, it is necessary to make four observations.

1. We consider the meta-particles, with the help of which the broken time-reversal symmetry for spin-1/2 particles is restored, not as a hypothetical elementary meta-matter formations, but as a consequences of the existence of a 16th order non-Abelian group $G_{16}^{(\frac{1}{2})}$ (8.43). The pecularities of meta-particles are determined by the symmetry properties of this group. All six incomplete time-reversal operators are elements of this group $G_{16}^{(\frac{1}{2})}$. For meta-particles related to spin-1 elementary particles, instead of non-Abelian group $G_{16}^{(\frac{1}{2})}$, the analogical role play the Abelian group $G_{16}^{(1)} = G_{8}^{(1)} \otimes G_2 \left(G_{16}^{(J)} \text{ at } J = 1 \right)$.
2. The six incomplete time-reversal operators $\mathbf{T}_{2(yz)}$, $\mathbf{T}_{2(zx)}$, $\mathbf{T}_{2(xy)}$, $\mathbf{T}_{1(x)}$, $\mathbf{T}_{1(y)}$ and $\mathbf{T}_1(z)$ for a spin-1/2 particle ((8.4), (8.53), and (8.57)) found in Chap. 8 are also a group-theoretical consequence of the existence of the non-Abelian group $G_{16}^{(\frac{1}{2})}$.
3. The proved possibility of representing the operator of complete time reversal $\mathbf{T}$ in the form of three products of operators of incomplete time reversal: $\mathbf{T}_{1(x)}\mathbf{T}_{2(yz)}$, $\mathbf{T}_{2(zx)}\mathbf{T}_{1(y)}$, $\mathbf{T}_{2(xy)}\mathbf{T}_{1(z)}$, and one product of three operators of incomplete time reversal $\mathbf{T}_{1(x)}\mathbf{T}_{1(y)}\mathbf{T}_{1(z)}$((8.5), (8.55), (8.56), (8.59)) indicates only the potential

[1] In many books and articles the term "metamaterials" is widely used [438–453], but it has no relation to the concept of meta-matter introduced by Wigner in [437]. Unlike this (as well as unlike the definition of meta-matter introduced in [457], we will use the Wigner conception of meta-particles).

possibility of lowering the symmetry of time reversal, but not the existence of a violation of such symmetry.

4. If the experimental data indicate, for example, that for a spin-1/2 particle, in some interaction, the T-symmetry decreases to $T_{2(yz)}$-symmetry (which will be perceived as a violation of T-symmetry), then to restore the initial symmetry it will be necessary to introduce such an operator, under the action of which the spin projection operator S_x changes its sign. Such an operator is $\mathbf{T}_{1(x)}$. If we now represent the operator $\mathbf{T}$ in the form $\mathbf{T} = \mathbf{T}_{1(x)}\mathbf{T}_{2(yz)}$, then the broken T-symmetry will be restored, since all the operators, S_x, S_y and S_z, change the sign under the action of the operator $\mathbf{T}$. However, such a transformation is trivial. Instead, in this case we postulate the existence of meta-particles whose intrinsic angular momentum differs from the spin $S = \frac{1}{2}$ only in that the operator Σ_x of the projection of this momentum on the direction x has an inverse sign with respect to the operator S_x for a particle with spin $\frac{1}{2}$. The matrices Σ_x, Σ_y and Σ_z of the meta-particle with spin projection[2] $\Sigma = \frac{1}{2}$ have the form (a system of units where $\hbar = 1$ is used):

$$\Sigma_x = -\frac{1}{2}\begin{pmatrix} 0 & 1 \\ 1 & 0 \end{pmatrix}, \quad \Sigma_y = \frac{1}{2}\begin{pmatrix} 0 & -i \\ i & 0 \end{pmatrix}, \quad \Sigma_z = \frac{1}{2}\begin{pmatrix} 1 & 0 \\ 0 & -1 \end{pmatrix} \tag{9.55}$$

We note that in (9.55) the matrices of the operators Σ_x, Σ_y and Σ_z are written in the spinor basis $\{| \frac{1}{2}, \frac{1}{2} >, | \frac{1}{2}, -\frac{1}{2} >\}$.

Here it is necessary to make one refinement, which consists in the fact that the form of the spin projection operators matrices is determined by the requirement that the introduction of meta-particles restores the broken T-symmetry. More specifically, this means the following. As the T-symmetry is reduced to $T_{2(yz)}$-symmetry, the spin projection operators S_x, S_y, and S_z are transformed under the action of the operator $\mathbf{T}_{2(yz)}$ as elements of the non-Abelian group $G_{16}^{(\frac{1}{2})}$ (8.60) as follows:

$$\mathbf{T}_{2(yz)}S_x\mathbf{T}^{-1}{}_{2(yz)} = S_x, \quad \mathbf{T}_{2(yz)}S_y\mathbf{T}^{-1}{}_{2(yz)} = -S_y, \quad \mathbf{T}_{2(yz)}S_z\mathbf{T}^{-1}{}_{2(yz)} = -S_z, \tag{9.56}$$

which is a confirmation of the T-symmetry violation, since the sign of the operator S_x did not change.

If we now postulate the existence of meta-particles with meta-spin $\Sigma = \frac{1}{2}$, then, since, $\Sigma_x = -S_x$, $\Sigma_y = S_y$ and $\Sigma_z = S_z$ the following relations hold:

$$\begin{aligned} \mathbf{T}_{2(yz)}\Sigma_x\mathbf{T}^{-1}{}_{2(yz)} - &= \mathbf{T}_{2(yz)}S_x\mathbf{T}^{-1}{}_{2(yz)} = -S_x, \\ \mathbf{T}_{2(yz)}\Sigma_y\mathbf{T}^{-1}{}_{2(yz)} - &= \mathbf{T}_{2(yz)}S_y\mathbf{T}^{-1}{}_{2(yz)} = -S_y, \\ \mathbf{T}_{2(yz)}\Sigma_z\mathbf{T}^{-1}{}_{2(yz)} - &= \mathbf{T}_{2(yz)}S_z\mathbf{T}^{-1}{}_{2(yz)} = -S_z. \end{aligned} \tag{9.57}$$

[2] Since the matrices Σ_y and Σ_z coincide with S_y and S_z, the sign of the matrix Σ_x differs from that of the matrix S_x, and the commutation relations for the matrices Σ_χ and, accordingly, S_χ ($\chi = x, y, z$) are different, then the own angular momentum of the meta-particle could be not called a spin, but otherwise (for example, meta-spin).

Table 9.3 Relations between spin projection operators $S_x^{(n)}$, $S_y^{(n)}$, $S_z^{(n)}$ of a spin-1/2 particle and meta-spin projection $\Sigma_x^{(n)}$, $\Sigma_y^{(n)}$, $\Sigma_z^{(n)}$ $(n = 1, 2, \ldots, 6)$ of corresponding meta-particles with meta-spins $\Sigma^{(n)} = \frac{1}{2}$

Incomplete time-reversal operators	Correlations between spin projection operators S_x, S_y, S_z of a spin-1/2 particle and corresponding meta-spin projection operators $\Sigma_x^{(n)}$, $\Sigma_y^{(n)}$, $\Sigma_z^{(n)}$ of a meta-spin-1/2 meta-particles		
$\mathbf{T}_{2(yz)}$	$\Sigma_x^{(1)} = -S_x^{(1)}$	$\Sigma_y^{(1)} = S_y^{(1)}$	$\Sigma_z^{(1)} = S_z^{(1)}$
$\mathbf{T}_{2(zx)}$	$\Sigma_x^{(2)} = S_x^{(2)}$	$\Sigma_y^{(2)} = -S_y^{(2)}$	$\Sigma_z^{(2)} = S_z^{(2)}$
$\mathbf{T}_{2(xy)}$	$\Sigma_x^{(3)} = S_x^{(3)}$	$\Sigma_y^{(3)} = S_y^{(3)}$	$\Sigma_z^{(3)} = -S_z^{(3)}$
$\mathbf{T}_{1(x)}$	$\Sigma_x^{(4)} = S_x^{(4)}$	$\Sigma_y^{(4)} = -S_y^{(4)}$	$\Sigma_z^{(4)} = -S_z^{(4)}$
$\mathbf{T}_{1(y)}$	$\Sigma_x^{(5)} = -S_x^{(5)}$	$\Sigma_y^{(5)} = S_y^{(5)}$	$\Sigma_z^{(5)} = -S_z^{(5)}$
$\mathbf{T}_{1(z)}$	$\Sigma_x^{(6)} = -S_x^{(6)}$	$\Sigma_y^{(6)} = -S_y^{(6)}$	$\Sigma_z^{(6)} = S_z^{(6)}$

Table 9.3 shows the possibility to restore the broken T-symmetry at the cost of introducing meta-particles with the meta-spin (9.56). In particular, the reduced T-symmetry to $T_{2(yz)}$-symmetry (see (9.57)) can be restored by introducting meta-particles with meta-spin $\sum = \frac{1}{2}$ and meta-spin projection operators $\sum_x^{(1)}$, $\sum_y^{(1)}$ and $\sum_z^{(1)}$. The operator $\mathbf{T}_{2(yz)}$ is an element of the non-Abelian group $G_8^{(\frac{1}{2})}$, which is a subgroup of $G_{16}^{(\frac{1}{2})}$ group. However, on the basis of only the group $G_8^{(\frac{1}{2})}$, it is impossible to determine all six operators of incomplete time reversal. On the other hand, since the form of the matrices Σ_x, Σ_y and Σ_z also depends on the structure of $G_{16}^{(\frac{1}{2})}$, it can be argued that the combined transformation of "incomplete time reversal + introduction of a meta-particle with spin Σ" restores the symmetry of time reversal. The type of the meta-particle is due to the structure of the group $G_{16}^{(\frac{1}{2})}$.

The existence of the operator $\mathbf{T}_{2(yz)}$ itself is due to the structure of the non-Abelian group.

Table 9.3 presents the relations between the spin projection operators Σ_x, Σ_y, Σ_z of a spin-1/2 particle and the meta-spin projection operators of corresponding meta-particles.

As can be seen from Table 9.3, each of the six operators of incomplete time reversal corresponds to a meta-particle of a certain type, having the meta-spin $\Sigma^{(n)} = \frac{1}{2}$ and meta-spin projection operators $\Sigma_x^{(n)}$, $\Sigma_y^{(n)}$ and $\Sigma_z^{(n)}$ $(n = 1, 2, \ldots, 6)$. In this case, if the performance of each of the incomplete time-reversal operations is accompanied by a transition from a particle to the corresponding meta-particle indicated in Table 9.3, then such a combined transformation is a symmetry operation (combined time-reversal transformation), as a result of which the broken T-symmetry is restored. The price of restoration of the broken T-symmetry is, as already noted, the need to take into account the presence of certain types of meta-particles. Apparently, one could postulate that particles and meta-particles are produced and annihilated in pairs, like as particles and antiparticles. Since there are six operators of incomplete time reversal, the reduced Wigner's T-symmetry can be restored by performing six combined time-reversal transformations involving the six types of meta-particles listed in Table 9.3. The commutation relations for the meta-spin projection operators for the meta-particles listed in the first three rows of Table 9.3 have the form

$$\begin{aligned} \left[\Sigma_y^{(n)}, \Sigma_z^{(n)}\right] &= -i\,\Sigma_x^{(n)}, \\ \left[\Sigma_z^{(n)}, \Sigma_x^{(n)}\right] &= -i\,\Sigma_y^{(n)}, \\ \left[\Sigma_x^{(n)}, \Sigma_y^{(n)}\right] &= -i\,\Sigma_z^{(n)}. \end{aligned} \tag{9.58}$$

Nine commutation relations (9.58) at $n = 1, 2, 3$ differ in sign on the right-hand side from the known commutation relations for the spin projection operators. For the remaining meta-particles ($n = 4, 5, 6$), the other nine commutation relations for the operators $\Sigma_x^{(n)},\ \Sigma_y^{(n)},\ \Sigma_z^{(n)}$ coincide with the commutation relations for the operators $S_x,\ S_y,\ S_z$. However, despite this coincidence, the meta-spin projections operators $\Sigma_x^{(n)},\ \Sigma_y^{(n)}$ and $\Sigma_z{}^{(n)}$ ($n = 4, 5, 6$) are not identical to the particle spin projection operators $S_x,\ S_y$ and S_z. Indeed, it is easy to see that for $n = 4, 5, 6$ the following relations hold

$$\begin{gathered} n = 4 \\ < M \mid \Sigma_x^{(4)} \mid M' >=< M \mid S_x \mid M' >, \\ < M \mid \Sigma_y^{(4)} \mid M' >= - < M \mid S_y \mid M' >,\ < M \mid \Sigma_z^{(4)} \mid M' >= - < M \mid S_z \mid M' >, \end{gathered} \tag{9.59}$$

$$\begin{gathered} n = 5 \\ < M \mid \Sigma_y^{(5)} \mid M' >=< M \mid S_y \mid M' >, \\ < M \mid \Sigma_x^{(5)} \mid M' >= - < M \mid S_x \mid M' >,\ < M \mid \Sigma_z^{(5)} \mid M' >= - < M \mid S_z \mid M' >, \end{gathered} \tag{9.60}$$

$$\begin{gathered} n = 6 \\ < M \mid \Sigma_z^{(6)} \mid M' >=< M \mid S_z \mid M' >, \\ < M \mid \Sigma_x^{(6)} \mid M' >= - < M \mid S_x \mid M' >,\ < M \mid \Sigma_y^{(6)} \mid M' >=< M \mid S_y \mid M' > . \end{gathered} \tag{9.61}$$

The discrepancy of certain matrix elements of the operators $\Sigma_\chi{}^{(n)}$ and S_χ ($\chi = x, y, z$) also occurs when the commutation relations for the spin projection operators of the particle and that for the meta-particle do not coincide ($n = 1, 2, 3$):

$$\begin{gathered} n = 1 \\ < M \mid \Sigma_x^{(1)} \mid M' >= - < M \mid S_x \mid M' >, \\ < M \mid \Sigma_y^{(1)} \mid M' >=< M \mid S_y \mid M' >,\ < M \mid \Sigma_z^{(1)} \mid M' >=< M \mid S_z \mid M' >, \end{gathered} \tag{9.62}$$

$$\begin{gathered} n = 2 \\ < M \mid \Sigma_y^{(2)} \mid M' >= - < M \mid S_y \mid M' >, \\ < M \mid \Sigma_x^{(2)} \mid M' >=< M \mid S_x \mid M' >,\ < M \mid \Sigma_z^{(2)} \mid M' >=< M \mid S_z \mid M' >, \end{gathered} \tag{9.63}$$

$$n = 3$$

$$< M \mid \Sigma_z^{(3)} \mid M' >= - < M \mid S_z \mid M' >, \tag{9.64}$$
$$< M \mid \Sigma_x^{(3)} \mid M' >=< M \mid S_x \mid M' >, < M \mid \Sigma_y^{(3)} \mid M' >=< M \mid S_y \mid M' > .$$

As is known, a two-valued irreducible representation of the rotation group $D^{(1/2)}$, according to which the basis spin wave functions of a particle with spin $\frac{1}{2}$ are transformed, can be found using the method proposed by Weyl [454]. This method is based on the fact that any rotation of space at a fixed coordinate system is characterized by three Euler angles (α, β, γ). It can be represented as a product of three basic rotations: rotation around the Z axis by an angle γ, rotating about the Y axis by an angle β, followed by a second rotation around the Z axis by the angle α. In this case, the Hermitian matrix

$$h \equiv x\sigma_x + y\sigma_y + z\sigma_z \equiv \sigma \mathbf{r} = \begin{pmatrix} z & x - iy \\ x + iy & -z \end{pmatrix} \tag{9.65}$$

is tied to each point of space having coordinates (x, y, z), where σ_x, σ_y and σ_z are the Pauli matrices. Using a unitary 2×2-matrix U with a determinant equal to one $(U^+U = UU^+ = \mathbf{1},\ detU = 1)$, one can construct a Hermitian matrix

$$h' = UhU^+, \tag{9.66}$$

whose trace is equal to one as well as the trace of the matrix h. Consequently, the matrix h' should have the form

$$h' = \begin{pmatrix} z' & x' - iy' \\ x' + iy' & -z' \end{pmatrix} = \sigma \mathbf{r}', \tag{9.67}$$

where $\mathbf{r}' = \mathbf{r}'(x', y', z')$. In this case, the coordinates (x', y', z') are linear combinations of coordinates (x, y, z) with coefficients that depend on the elements of the matrix U.

The matrix U describing the rotation of the space about the Z axis by an angle α has the form:

$$U = \begin{pmatrix} e^{-i\frac{\alpha}{2}} & 0 \\ 0 & e^{i\frac{\alpha}{2}} \end{pmatrix}. \tag{9.68}$$

On the basis of (9.65)–(9.68) we find

$$\begin{pmatrix} e^{-i\frac{\alpha}{2}} & 0 \\ 0 & e^{i\frac{\alpha}{2}} \end{pmatrix} \begin{pmatrix} z & x - iy \\ x + iy & -z \end{pmatrix} \begin{pmatrix} e^{i\frac{\alpha}{2}} & 0 \\ 0 & e^{-i\frac{\alpha}{2}} \end{pmatrix} = \begin{pmatrix} z' & x' - iy' \\ x' + iy' & -z' \end{pmatrix}. \tag{9.69}$$

The solution of the matrix equation (9.69) has the form [454]:

$$\begin{aligned} x' &= x\cos\alpha - y\sin\alpha, \\ y' &= x\sin\alpha + y\cos\alpha, \\ z' &= z. \end{aligned} \tag{9.70}$$

Equation (9.70) show that the vector $\mathbf{r}'$ is obtained from the vector $\mathbf{r}$ as it rotates about the Z axis by an angle α. If, by analogy with (9.65), we determine

$$h^{(n)} = 2\left(x\Sigma_x^{(n)} + y\Sigma_y^{(n)} + z\Sigma_z^{(n)}\right) = 2\mathbf{\Sigma}^{(n)}\mathbf{r},\ n = 1, 2, \ldots, 6, \tag{9.71}$$

and, by analogy with (9.66) and (9.67), determine

$$h'^{(n)} = Uh^{(n)}U^{+} = 2\mathbf{\Sigma}^{(n)}\mathbf{r}', \tag{9.72}$$

then we obtain six matrix equations of the type (9.69), whose solutions, as well as solutions of (9.70), show how the vector $\mathbf{r}'$ is obtained from $\mathbf{r}$ when rotating about the Z axis by an angle α. The obtained formulas of transformation from the coordinates (x, y, z) to the coordinates $(x'.y', z')$ for the rotation of the space about the Z axis by the angle α in six cases of T-symmetry restoration by introducing six types of meta-particles with meta-spins $\Sigma^{(n)} = \frac{1}{2}$ and meta-spin projection operators $\Sigma_x^{(n)}$, $\Sigma_y^{(n)}$ and $\Sigma_y^{(n)} (n = 1, 2, \ldots, 6)$ are presented in Table 9.4.

As can be seen from Table 9.4, in the case of $T_{2(yz)}$-, $T_{2(zx)}$-, $T_{1(x)}$- and $T_{1(y)}$-symmetries, the rotations of the space about the Z axis are performed by the same angle α, but in opposite direction compared to the rotations in the case of $T_{2(xy)}$- and $T_{1(z)}$-symmetries. This can be easily verified if we take into account that $\cos\alpha$ and $\sin\alpha$ are, respectively, even and odd functions of the argument α. In this case, replacing α by $-\alpha$, the equation

$$x' = x\cos\alpha + y\sin\alpha$$

will transform into the equation

$$x' = x\cos\alpha - y\sin\alpha,$$

and the equation

$$y' = -x\sin\alpha + y\cos\alpha$$

will transform into the equation

$$y' = x\sin\alpha + y\cos\alpha.$$

Table 9.4 Transformations of the coordinates (x, y, z) of the vector $\mathbf{r}$ into the coordinates (x', y', z') of the vector $\mathbf{r}'$ when the space rotates about the Z axis by an angle α

Incomplete time-reversal operators	$2\boldsymbol{\Sigma}^{(n)}\mathbf{r}$	The relationship between the coordinates (x', y', z') and (x, y, z) at the rotation of the space about the Z axis by the angle α $(z' = z)$
$\mathbf{T}_{2(yz)}$	$-x\sigma_x + y\sigma_y + z\sigma_z$	$\begin{cases} x' = x\cos\alpha + y\sin\alpha \\ y' = -x\sin\alpha + y\cos\alpha \end{cases}$
$\mathbf{T}_{2(zx)}$	$x\sigma_x - y\sigma_y + z\sigma_z$	$\begin{cases} x' = x\cos\alpha + y\sin\alpha \\ y' = -x\sin\alpha + y\cos\alpha \end{cases}$
$\mathbf{T}_{2(xy)}$	$x\sigma_x + y\sigma_y - z\sigma_z$	$\begin{cases} x' = x\cos\alpha - y\sin\alpha \\ y' = x\sin\alpha + y\cos\alpha \end{cases}$
$\mathbf{T}_{1(x)}$	$x\sigma_x - y\sigma_y - z\sigma_z$	$\begin{cases} x' = x\cos\alpha + y\sin\alpha \\ y' = -x\sin\alpha + y\cos\alpha \end{cases}$
$\mathbf{T}_{1(y)}$	$-x\sigma_x + y\sigma_y - z\sigma_z$	$\begin{cases} x' = x\cos\alpha + y\sin\alpha \\ y' = -x\sin\alpha + y\cos\alpha \end{cases}$
$\mathbf{T}_{1(z)}$	$-x\sigma_x - y\sigma_y + z\sigma_z$	$\begin{cases} x' = x\cos\alpha - y\sin\alpha \\ y' = x\sin\alpha + y\cos\alpha \end{cases}$

The second and fourth of these equations completely agree with the equations for x' and y' from (9.70). We see that a spontaneous change takes place in the direction of rotation of the vector $\mathbf{r}$ around the Z axis (with the same angle α remaining) for interactions lowering the T-symmetry to $T_{2(yz)}$-, $T_{2(zx)}$-, $T_{1(x)}$- and $T_{1(y)}$-symmetries as compared with interactions lowering the T-symmetry to $T_{2(xy)}$- and $T_{1(z)}$-symmetries.

In the case of spin-1 particles, the broken T-symmetry can be restored also by means of corresponding meta-particles. Matrices of meta-spin projection operators for this case are given in Appendix E.

Chapter 10
Time-Reversal Symmetry Violation

The time-reversal symmetry violation is the most mysterious phenomenon that allows explaining the matter–antimatter asymmetry in the present Universe. This is due to the fact that the CP-symmetry violation [458], which causes the asymmetry of matter-antimatter, also leads to a violation of T-symmetry due to the CPT theorem. According to Sakharov [478], this asymmetry could have arisen from CP-violating (and consequently, T-violating) interactions at an early stage of the Big Bang. In the course of further evolution, this led to the present structure of the Universe. The most important motivation for studying the time-reversal symmetry breaking is the fact that its origin is not completely understood.

Sections 10.1–10.6 contain a brief description of the time-reversal symmetry violation based on the results of other authors. In Sects. 10.1–10.3, the time-reversal symmetry breaking in K_L^0 and B^0 meson systems, in atomic nuclei, as well as in atoms and molecules is discussed. As a result of the experiments, a rare (with relative probability $\sim 1.6 \times 10^{-3}$) $\pi^+\pi^-$ mode of decay of a longer lived K_L^0-meson (in comparison with the K_S^0-meson) was found, which indicates a violation of CP- and, respectively, T-symmetry. The violation of CP-symmetry is also characteristic of various decay processes of the B^0-mesons: $B^0 \longrightarrow \pi^+\pi^-$, $B^0 \longrightarrow k^+\pi^-$, $B^0 \longrightarrow k^+k^-$, and others. The violation of P- and T-symmetry in the nuclei causes the presence of an electric dipole moment (EDM) in protons and, correspondingly, in nuclei. In atoms and molecules, P- and T-symmetry breaking leads to atomic and molecular EDM_s, which are enhanced in heavy nuclei ($\mathcal{Z} \geq 50$) (due to the relativistic effects and complex structure of these nuclei) and in paramagnetic molecules.

Section 10.4 is devoted to the discussion of the appearance of localized states with broken time-reversal symmetry and the global T-symmetry violation in the bulk high-T_c superconductors.

Sections 10.5 and 10.6 contain the discussion of time-reversal symmetry breaking in connection with the enhancement of quantum transport and unidirectionality of the time.

I. I. Geru, *Time-Reversal Symmetry*, Springer Tracts in Modern Physics 281,
https://doi.org/10.1007/978-3-030-01210-6_10

In the last, 7th section, a simple method of experimental verification of the violation of T-symmetry in paramagnetic ions and molecules is proposed on the basis of *EPR* spectra.

10.1 Time-Reversal Symmetry Violation in Meson Systems

The fundamental *CPT* theorem (Sect. 2.15) means that the time-reversal symmetry violation occurs not only when T-invariance is violated, but also when the *CP*-invariance is violated. The existence of the *CP*-violation has been observed for the first time in 1964 in the decay of K-mesons (kaons) [109]. Later, starting with 2001, the time-reversal symmetry violation was again detected in neutral B-meson decays in the *BaBar* experiment at SLAC [459–462].

It was shown in [109] that K_2^0-meson decays to two pions with a branching ratio $R = \left(K_2^0 \to \pi^+ + \pi^-\right) / \left(K_2^0 \to \text{all charged modes}\right) = 2.0 \pm 0.4 \cdot 10^{-3}$, where the error is a standard deviation. The presence of a two pion decay mode implies that the state of K_2^0-meson is not a pure eigenstate of CP. Indeed, the neutral K_2^0-meson state can be represented in the form

$$| K_2^0 >= \frac{1}{\sqrt{2}}\left[(| K^0 > - | \overline{K^0} >) + \varepsilon(| K^0 > + | \overline{K^0} >)\right], \tag{10.1}$$

where $| \overline{K^0} >$ is the neutral antimeson state. In this case $| \varepsilon |^2 \cong R_T \tau_1 \tau_2$, where τ_1 and τ_2 are the K_1^0 and K_2^0 average lifetimes and R_T is the branching ratio including decay to neutral pions. Using $R_T = \frac{3}{2}R$ and the above mentioned value for the branching ratio R, it was obtained $| \varepsilon | \cong 2.3 \cdot 10^{-3}$. The states $| K^0 >$ and $| \overline{K^0} >$ are not the eigenstates of the *CP*-operator. Indeed, the operator C replaces $| K^0 >$ by $| \overline{K^0} >$ (and vice versa), and the operator P multiplies each of these states by -1 (their internal parity is -1). As a result, we have

$$CP \mid K^0 >= - \mid \overline{K^0} >, \quad CP \mid \overline{K^0} >= - \mid K^0 > . \tag{10.2}$$

Nevertheless, one can construct from these states the eigenstates of the operator *CP*, denoted by $| K_1^0 >$ and $| K_2^0 >$:

$$\begin{aligned} | K_1^0 > &= \left(\tfrac{1}{\sqrt{2}}\right)\left(| K^0 > - | \overline{K^0} >\right), \\ | K_2^0 > &= \left(\tfrac{1}{\sqrt{2}}\right)\left(| K^0 > + | \overline{K^0} >\right). \end{aligned} \tag{10.3}$$

Taking into account the relations (10.2), it is easy to show that the orthonormal states $| K_1^0 >$ and $| K_2^0 >$ are eigenstates of the operator *PC* with eigenvalues $+1$ and -1:

$$CP \mid K_1^0 >= + \mid K_1^0 > , CP \mid K_2^0 >= - \mid K_2^0 > . \tag{10.4}$$

Turning the equalities (10.3), we can rewrite them in the form

$$\mid K^0 >= \left(\frac{1}{\sqrt{2}}\right)\left(\mid K_1^0 > + \mid K_2^0 >\right), \mid \overline{K^0} >= \left(\tfrac{1}{\sqrt{2}}\right)\left(\mid K_2^0 > - \mid K_1^0 >\right). \tag{10.5}$$

This allows us to give the following interpretation [463]: at the moment of its birth, each neutral K-meson can be in the form of either K_1^0-meson or K_2^0-meson with the probability $\frac{1}{2}$. Since the average lifetime of these particles is very different, the activity associated with the decay of the K_1^0-meson into two π-mesons (π^+ and π^-) is initially approximately 600 times greater than for the decay of the K_2^0-meson. Therefore, firstly, the K_1^0-meson decays into two π-mesons. But after a certain time, which is sufficiently small in comparison with the lifetime of the K_2^0-meson (but exceeding the lifetime of the K_1^0-meson), only the mesons of the decay of the K_2^0 particle are observed, since the K_1^0 mesons have already disappeared as a result of the decay of each of them into two π-mesons. Thus, as a result of experiments carried out in [109], for the first time, a rare mode of $\pi^+ + \pi^-$ decay (with a relative probability of $\sim 1.6 \cdot 10^{-3}$) of the neutral, longer-lived K_2^0 meson was discovered, which after this discovery became known as the K_L^0-meson (respectively, the K_1^0 meson became known as the K_S^0 meson). Although the decay of the K_2^0 meson is very rare, it impugns CP-invariance, since the state $\mid K_1^0 >$ is an eigenstate of the operator CP with the eigenvalue $+1$, while the state $\mid K_2^0 >$ is an eigenstate of the operator CP with the eigenvalue -1 (see (10.4)).

Here it is appropriate to make two comments. One of them is that, in the presence of CP-invariance, each particle has an antiparticle with "CP-conjugated" decay modes [463], and the relative decay probabilities and partial lifetimes for these decay modes are the same as for its corresponding particle. But even at violation of CP-invariance, which is discussed in this section, the basic conclusions concerning antiparticles remain valid in the framework of the CPT theorem, since according to this theorem, in violation of CP-invariance, T-invariance must also be violated for CPT-invariance conservation. The second remark is that the decay of the K_L^0-meson into two π-mesons (π^+ and π^-) with a weight of the order of 10^{-3} indicates that the CP-invariance (and, correspondingly, T-invariance) is only broken "slightly" in the decay of K_L^0-mesons. For comparison, we point out that the violation of P and C invariance is observed in processes of weak interaction with a weight equal to 100%, as evidenced, for example, by the fact that the neutrino exists only in one helicity state [463].

The noted features of the violation of CP- and T-invariances are manifested not only in decays of K-mesons, but also in other types of decay in meson systems

1. The time-dependent CP-violating asymmetries in decay of neutral B-mesons [459]. Authors of [459] have selected events in which one neutral B-meson is fully reconstructed in a final state and the feature of the other neutral B-meson is determined from its decay products. The amplitude of CP-violating asymmetry, which in the Standard Model is proportional to $\sin 2\beta$, was derived from the decay

time distribution in such events. The result $\sin\beta = 0.59 \pm 0.14(stat) \pm 0.05(sust)$ indicated a *CP*-violation in the neutral *B*-meson system.

2. *CP*-violating asymmetries for neutral *B*-meson decays to two-body final states of charged pions and kaons ($B^o \longrightarrow \pi^+\pi^-, K^+\pi^-, K^+K^-$) [460].
3. The direct *CP*-violating asymmetry in the decay of the neutral *B*-meson into a charged kaon and a charged pion ($B^o \longrightarrow K^+\pi^-$). The probability of observing such an asymmetry in the absence of a direct *CP*-violation is $1.3 \cdot 10^{-5}$, corresponding to a 4.2 standard deviation [461].
4. The time-dependent *CP*-violating asymmetries in decays of *B*-mesons to $\eta' K$ mesons [462].

10.2 Time-Reversal Symmetry Violation in Atomic Nuclei

The basis for studying the time-reversal symmetry violation in nuclei is the *CPT* theorem [464, 465], which links the time-reversal symmetry with a combination of particle–antiparticle and parity symmetries (*CP*). This fundamental theoretical result and its experimental evidence [109, 459–462] allow us to expect that *T*-violation should be accompanied by a corresponding *CP*-violation and vice versa.

The evidence of *CP*-violation in the decay of neutral kaons is firmly established. In a number of experiments, the observed effect of *CP*-violation not only in meson systems, but also in atomic nuclei, exceeds the statistical uncertainty by more than an order of magnitude. The best-known mechanisms considered as sources of *CP*-violation are the super weak interaction [462, 466], the Cabibbo–Kobayashi–Maskawa (*CKM*) mixing of quark states [467, 468], the term in the quantum chromodynamics (*QCD*) Lagrangian [469], models involving left- and right-handed gauge bosons [470], and numerous Higgs particles [471, 472] or leptoquarks [473–475]. An indirect evidence of *CP*- or *T*-violation is the excess of matter over the antimatter in the present Universe. If the *CP*- and *T*-invariances were met exactly (that is, if there were not a single violation of these invariants), then all baryons and antibaryons created in the Big Bang would have been annihilated and the Universe would be filled now mostly with energy in the form of massless quanta of radiation [476, 477]. It was suggested by Sakharov [478] that matter-antimatter asymmetry in the present Universe could have arisen due to the time-reversal symmetry violating interaction at an early stage of the Big Bang.

Experimental data and, in particular, lack of the characteristic annihilation radiation suggest that the largest structures in the Universe are composed of regular matter [479]. However, the *CP*-violating interaction, detected in a kaon system [465] and later in the system of neutral *B*-mesons [459–462], is too weak to produce so much matter in the baryon genesis process [480, 481]. For this reason, other unknown mechanisms of breaking the *CP*- and *T*-invariances would also exist. This shows that elucidating the mechanism of violation of *T*- and *CP*-symmetry is a timely problem.

Our task is not to describe in detail the violations of *CP*- and *T*-symmetries in nuclear physics and elementary particle physics, but only a discussion, as far as possible, of some of them. Before doing this, we note that after the discovery of the violation of the parity conservation law in weak interactions [109, 482], Sakharov's paper on the role of violation of *CP*-invariance in the evolution of the Universe was published in 1967 [478]. From that moment, there were all the necessary prerequisites for planning and conducting experiments on the study of time-reversal symmetry breaking in β-decay of polarized nuclei and, primarily, in ^{60}Co nuclei, which were used in the famous experiment [109] on discovering the violation of the parity conservation law. Indeed, in this case the violation of *P*-symmetry was experimentally proved [109], and the data on the structure of the present Universe indicated a significant predominance of the number of particles over the number of antiparticles. Both these facts, taken together, indicated a violation of *CP*-symmetry, which, according to the *CPT* theorem, should lead to a violation of *T*-symmetry. However, the corresponding experimental work was published only many years after the appearance of real opportunities for their planning and implementation (see, for example, [479]).

Further discussion of the time-reversal symmetry breaking in atomic nuclei reduces to the following. An important advantage of the search for time-reversal symmetry violation in nuclei interactions is the possibility of enhancing the *T*-violating observables by many orders of magnitude due to the complex nuclear structure. It could be an enhancement of nuclear electric dipole moments (*EDMs*) or that of time-reversal symmetry violating effects in neutron–nuclei interactions [483, 484]. Taking into account that different models of *CP*-symmetry breaking may contribute differently to a particular *T*- or *CP*-observable, which may have unknown theoretical uncertainties, the time-reversal symmetry violation nuclear effects could be considered valuable complementary experiments for *EDM* measurements.

The comparison of *CP*-odd nucleon-coupling constants that lead to time-reversal symmetry violation effects in neutron–deuteron scattering with constrains on the coupling constants from *EDM* measurements offers the opportunity to estimate the sensitivity of time-reversal symmetry effects in neutron scattering experiments. An obvious way to study the time-reversal symmetry is to compare the scattering process with the same process running backward in time. This leads to the principle of detailed balance (see Sect. 2.12) that is based on the invariance of the scattering matrix under the time-reversal transformation **T**. The detailed balance was tested in cross-sections [485] and in polarization observables [486, 487] in a number of nuclear reactions. The obtained results show that *T*-violating amplitudes are at most 10^{-3} to 10^{-2} of dominating strong interaction amplitudes. New prospects for *T*-violating experiments with strongly interacting systems have appeared as a consequence of discovering the large enhancement factors of 10^3 to 10^6 for parity non-conserving phenomena in the interaction of polarized neutrons with nuclear media [479]. It was argued that a similar amplification mechanism may operate for effects of time-reversal symmetry violation [488, 489].

The presence of nuclear *EDMs* is due to simultaneous breaking of time-reversal and parity symmetries [490]. However, these effects might also be observed as insignificant asymmetries in dynamical processes like nuclear collisions or radiative capture/disintegration reactions. The time-reversal symmetry breaking might be traced when observing specific asymmetries in nuclear reaction cross-sections, involving polarized reaction products. In particular, the transmission of a low-energy polarized neutron beam through the matter is of great interest, because in this case the effects of the time-reversal symmetry violation might be explored at very low (thermal or resonance) neutron energies. The neutron beam may scatter coherently, whereas the presence of T-violating terms in the Hamiltonian of neutron–nucleus interaction will lead to overall neutron spin rotation by an angle proportional to the thickness of the target material.

In [479], the time-reversal symmetry violation in β-decay of polarized ^8Li nuclei was studied. The transverse polarization in a plane perpendicular to the nuclear spin axis has been determined for emitted electrons. Using a sample of ^8Li nuclei with the vector polarization of $\sim$0.11 and a polarimeter with an average analyzing power of $S = -0.10$, the asymmetry of the Mott scattering of decay electrons was measured with an accuracy of $\pm 4 \times 10^{-5}$. From this asymmetry, the transverse spin polarization of electrons has been determined with an accuracy of $\pm 4 \times 10^{-4}$. Also, the amplitude $R = (-0.2 \pm 4.0) \times 10^{-3}$ of the triple correlation between the nuclear spin, momentum, and the electron spin has been obtained. These are the results of the precise measurement of the transverse spin polarization of leptons emitted in weak decays. The time-reversal violating part of the correlation amplitude is $R_{\mathrm{TRV}} = (-0.9 \pm 4) \times 10^{-3}$.

We now discuss the contribution to *EDM* of the nucleus due to Shiff moment (*SM*) and the magnetic quadruple moment (*MQM*), which is the lowest T, P-odd magnetic moment [491]. We note that in *EDM* experiments using nuclei with a valence neutron (for example, ^{199}Hg), the direct valence nucleon contribution is zero and the *SM* is generated primarily by the polarization of the nuclear core by its T, P-odd interaction with the valence neutron [492, 493]. This leads to the fact that *SMs* are sensitive to many-body corrections [492–497]. On the contrary, for the *MQM*, the valence nucleon gives the main contribution and so the result should be less sensitive to many-body corrections. The T, P-odd core polarization contribution to the *MQM* was estimated in [498]. It should be noted that in paramagnetic molecules there is an increase of contributions of SM and *MQM* to T, P-odd effects. This problem will be discussed in the next section.

10.3 Time-Reversal Symmetry Violation in Atoms and Molecules

Atomic and molecular systems are characterized by the presence of electrons in them, in contrast to nuclei, which do not contain electrons, with the exception of nuclei emitting β-particles (electrons). This leads to specific features of the effects of T-symmetry breaking in atoms and molecules consisting, first, in the presence of constant *EDMs* in electrons and, secondly, in the influence of the electronic subsystem on the effects of T-symmetry breaking in nuclei. Atoms and molecules are of interest from the point of view of the time reversal symmetry breaking, because they are composed of both leptons and hadrons and involved in a wide range of fundamental interactions.

After the discovery of the combined charge and parity violation (CP-violation) in K_L^0-meson decay [109] (see Sect. 10.1), the search of *EDMs* of elementary particles has become one of the fundamental problems in physics [499–503]. A permanent *EDM* is induced by the weak interaction that breaks both the spatial symmetry inversion and the time-reversal invariance [504]. The P- and T-violating interaction between electrons and nucleons, which gives rise to atomic and molecular *EDMs* has the form [503]

$$b_{PT}^{e-N} = i\frac{G_F}{\sqrt{2}}\sum_N \Big[C_N^{\mathrm{SP}}\left(\overline{\Psi}_N\Psi_N\right)\left(\overline{\Psi}_e\gamma_5\Psi_e\right) + C_N^{\mathrm{PS}}\left(\overline{\Psi}_N\gamma_5\Psi_N\right)\left(\overline{\Psi}_e\Psi_e\right) + \\ + C_N^T\left(\overline{\Psi}_N\gamma_5\sigma_{\mu\nu}\Psi_N\right)\left(\overline{\Psi}_e\sigma_{\mu\nu}\Psi_e\right)\Big], \tag{10.6}$$

where the sum runs over all nucleons, Ψ_e and Ψ_N under the summation sign are the electron and nucleon wave functions, $\overline{\Psi}_e$ and $\overline{\Psi}_N$ are wave functions of corresponding antiparticles, $G_F \simeq 1.166 \cdot 10^{-5}\,\mathrm{GeV}^{-2}$ is the Fermi weak constant, $2\sigma_{\mu\nu} = i[\gamma_\mu, \gamma_\nu]$, γ_μ, γ_ν and γ_5 are Dirac matrices; C_N^{SP}, C_N^{PS}, and C_N^T are the strength of the scalar–pseudoscalar (SP), pseudoscalar–scalar (PS), and tensor (T) nucleon–nucleon interaction, respectively [505, 506]. For the standard definition of angular wave functions, these interactions produce real matrix elements contributing to mixing the opposite P-states, and give rise to atomic and molecular *EDMs*.

The *EDM* of an atom or molecule in the $\mid a >$ state, $\mathbf{D}(a)$, can arise from the sum of intrinsic *EDMs* of the constituent particles or from the mixing of opposite P-states due to P- and T-odd interaction, $\hat{b}_{PT}$ [502]:

$$\mathbf{D}(a) = 2\sum_n \frac{< a \mid \mathbf{d} \mid n >< n \mid \hat{b}_{PT}^{e-N} \mid a >}{E_a - E_n} \tag{10.7}$$

where $\hat{b}_{PT}^{e-N}$ is the electric dipole operator, determined by (10.7), which lies in the direction of the angular momentum $\mathbf{J}$. The polar vector operator $\mathbf{d}$ is P-odd and T-even, whereas the axial vector operator $\mathbf{J}$ is P-even and T-odd (Fig. 10.1).

An atomic or molecular *EDM* can be generated by several P- and T-violating mechanisms, such as the interaction with the electron *EDM* or the P- and T-odd electron–nucleon and nucleon–nucleon interactions [507]. Different systems vary in sensitivity to various sources depending on their electronic and nuclear structure. In paramagnetic systems (with non-zero angular momentum J) the *EDM* is almost entirely due to the electron *EDM* and P- and T-odd electron-nucleon interactions. However, for diamagnetic systems ($J = 0$) *EDMs* are mostly due to P- and T-violating inter-nuclear forces and interactions described by the effective spin-dependent electron Hamiltonian [503].

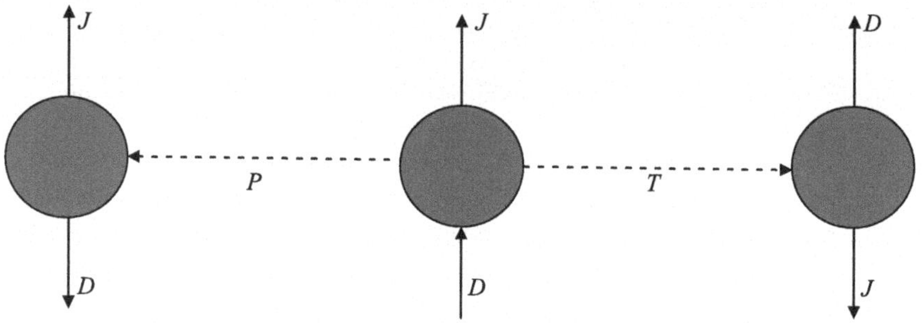

Fig. 10.1 A particle with angular momentum **J** and dipole moment **D** under parity **P** and time reversal **T**. In each of these cases, the result produces a different state with *EDM* either parallel or anti-parallel to **J**

Considerable experimental efforts have been invested to study the atomic *EDMs* induced by proton, neutron and electron *EDMs* and by P, T-odd interactions between them. The available restriction for the electron *EDM*, d_e, was obtained in an atomic *Tl* experiment [508], which established an upper limit of $\mid d_e \mid =< 1.6 \cdot 10^{-27} \mathrm{e} \cdot \mathrm{cm}$, where e is the electron charge. After it was suggested in [509] to use the polar heavy-atom molecules for experimental searching for the proton *EDM*. Such molecules have been considered the most promising objects for such experiments. In heavy atoms, the P- and T-parity non-conservation effects are strongly enhanced due to relativistic and other effect. Particularly, in paramagnetic atoms the enhancement factor for an electron *EDM*, $\frac{d_{\mathrm{atom}}}{d_e}$, is proportional to $\alpha^2 \mathcal{Z}^3 \alpha_D$, where $\alpha \approx \frac{1}{137}$ is the fine structure constant, $\mathcal{Z}$ is the nuclear charge and α_D is the atomic polarizability (the $\alpha^2 \mathcal{Z}^3$ enhancement formula was first derived in [510]). The enhancement factor can be of the order of 100 or greater for highly polarizable heavy atoms ($\mathcal{Z} \geq 50$). Furthermore, the effective intramolecular electric field acting on electrons in polar molecules can be five or more orders of magnitude higher than the maximal field accessible in a laboratory [502].

The statistical sensitivity of experiments to the electron or proton *EDM* depends on some parameters common for all *EDM* experiments. To verify this, it is necessary to apply the Heisenberg uncertainty principle to evaluate the sensitivity of *EDM* measurements. Let suppose the *EDM* of a molecule is measured in an electric field **E**. It is necessary to do not confuse the *EDM* of a polar molecule with the large conventional dipole moment of the molecule, which average to zero in the absence of the electric field in a laboratory coordinate system. In contrast to the latter, the molecular *EDM*, which is vanishingly small, can exist only due to P, T-odd interaction. It is a permanent moment and its direction depends on the sign of the projection of the total electronic angular momentum on the molecular axis [501]. Thus, the interaction energy of the molecular *EDM*, $\boldsymbol{d} = \mathrm{d}\boldsymbol{\lambda}$, (where $\boldsymbol{\lambda}$ is the unit vector along the total angular momentum of the molecule) is $\boldsymbol{d} \cdot \boldsymbol{E}$. The energy difference between the levels corresponding to the opposite directions of the total angular momentum is $2\boldsymbol{d} \cdot \boldsymbol{E}$. If we measure the shift of the energy levels during the time interval τ_0, then the d measurement uncertainty will be $\delta d = \frac{\hbar}{2\tau_0 \boldsymbol{E}\cdot\boldsymbol{\lambda}}$. For such measurements on N uncorrelated molecules, the value of δd will be equal to

$$\delta d = \frac{\hbar}{2\tau_0}\sqrt{N}\boldsymbol{E}\cdot\boldsymbol{\lambda} = \frac{\hbar}{2\tau_0 E_\lambda \sqrt{\frac{\tau \mathrm{d}N}{\mathrm{d}t}}}, \tag{10.8}$$

where $\frac{\mathrm{d}N}{\mathrm{d}t}$ is the counting rate, $E_\lambda = \boldsymbol{E}\cdot\boldsymbol{\lambda}$, and τ is the total measurement time (usually, $\tau >> \tau_0$ and τ_0 is limited by the coherence time of the considered system). Let us now write $d = Gd_e$, where d_e is the value of the electron *EDM* and G is the proportionality coefficient, usually referred to as the enhancement factor. In this case, we have the following final expression:

$$\delta d_e = \frac{\hbar}{2\tau_0 G E_\lambda \sqrt{\tau \frac{\mathrm{d}N}{\mathrm{d}t}}} = \frac{\hbar}{2\tau_0 W \sqrt{\tau \frac{\mathrm{d}N}{\mathrm{d}t}}}, \tag{10.9}$$

where $W = GE_\lambda$ is the effective electric field in the molecule, which can be interpreted as the field that should be applied along the *EDM* of a free electron to have the energy shift $2Wd_e \equiv 2E_\lambda d$ [502]. Of course, for the energy shift caused by the proton *EDM*, a formula analogous to (10.9) is valid if in the latter we replace δd_e by δd_p, and consider G to be the proton enhancement factor.

The energy difference, $2\boldsymbol{d}\cdot\boldsymbol{E}$, is extremely small for completely polarized heavy-atom molecules. Therefore the *EDM* experiment is usually carried out in parallel and antiparallel electric and magnetic ($\boldsymbol{B}$) fields. The interaction energy of the molecular magnetic moment $\boldsymbol{\mu}$, with the magnetic field is much higher than that of the *EDM* with the electric field and energy differences are $2\boldsymbol{\mu}\cdot\boldsymbol{B}+2\boldsymbol{d}\cdot\boldsymbol{E}$ and $2\boldsymbol{\mu}\cdot\boldsymbol{B}-2\boldsymbol{d}\cdot\boldsymbol{E}$ for parallel and antiparallel orientations of the electric and magnetic fields (in practice, the atomic or molecular spin precessions are usually studied instead of direct measurement of the energy shift [511]. When the electric field is reversed, the energy shift, $4\boldsymbol{d}\cdot\boldsymbol{E} = 4d_e W$, indicates the existence of a permanent molecular *EDM* [502].

Now we consider the P, T-odd interaction of the ^{205}Tl nucleus having one unpaired proton with the electromagnetic field of electrons in ^{205}TlF molecule [512]. The Hamiltonian of effective interaction of this electromagnetic field with the *EDM* of the Tl nucleus in TlF can be written in the form [502]

$$\boldsymbol{H}_{\mathrm{eff}} = \left(d^V + d^M\right)\frac{\mathbf{I}}{I}\boldsymbol{\lambda}, \tag{10.10}$$

where $\mathbf{I}$ is the ^{205}Tl nucleus spin operator, $\boldsymbol{\lambda}$ is a unit vector along the Z-axis (pointing from Tl nucleus to the F nucleus), d^V and d^M are the so-called volume and magnetic constants [513]:

$$d^V = 6QX, \tag{10.11}$$

where Q is the magnitude of nuclear Schiff moment (SM),

$$\mathbf{Q} = \frac{e}{6}\left[\frac{3}{5}\sum_n r_n^2 \boldsymbol{r}_n - \frac{1}{Z}\sum_n r_n^2 \sum_{n'} \boldsymbol{r}_{n'}\right], \tag{10.12}$$

where e is the proton charge and $\boldsymbol{r}_n$ is the location of nth proton. X in (10.11) is determined by

$$X = \frac{2\pi}{3}\left[\frac{\partial}{\partial z}\rho_\psi(\boldsymbol{r})\right]_{x,y,z=0}, \tag{10.13}$$

where $\rho_\psi(\boldsymbol{r})$ is the electronic density calculated on the basis of the electron wave function ψ.

The value d^M from (10.10) is

$$d^M = 2\sqrt{2}\,(d_P + d_N)\left(\frac{\mu}{Z} + \frac{1}{2mc}\right)M, \tag{10.14}$$

where d_N is the nuclear EDM arising due to P, T-odd forces acting between the nucleons; μ, m, and Z are the magnetic moment, mass and charge of the Tl nucleus; c is the velocity of light and M is defined as

$$M = \frac{1}{\sqrt{2}} < \psi \mid \sum_i \frac{1}{r_i^3}[\boldsymbol{\alpha}_i \times \boldsymbol{l}_i]_z \mid \psi >, \tag{10.15}$$

where $\boldsymbol{l}_i$ is the orbital momentum operator of the ith electron and $\boldsymbol{\alpha}_i$ are its Dirac matrices. Accounting for $\boldsymbol{H}_{\text{eff}}$ (10.10) leads to a difference in the hyperfine splitting of TlF molecule in parallel and antiparallel electric and magnetic fields. The level shift

$$h\nu = 4\left(d^V + d^M\right)\frac{< \mathbf{I}\boldsymbol{\lambda} >}{I} \tag{10.16}$$

is measured experimentally [514].

In [514], the violation of the time-reversal symmetry in the molecule TlF was studied using a rotationally cold beam from a jet source. It was measured the frequency shift of the thallium nuclear magnetic resonance (NMR) when an external electric field of 29.5 kV/cm was reversed with respect to a magnetic quantization axis. The measured shift value of $(1.4 \pm 2.4) \times 10^{-4}$ Hz is a tenfold improvement in sensitivity over previous measurements. This experiment has reduced the upper limits on the proton and electron electric dipole moments and on other T-violating weak couplings that can be deduced from the frequency shift. The experimental data have been interpreted on the basis of the effective spin Hamiltonian

$$\mathbf{H} = -\mu_{Tl}\boldsymbol{\sigma}_N \cdot \boldsymbol{B}_0 - hd\boldsymbol{\sigma}_N\boldsymbol{\lambda}, \tag{10.17}$$

where the first term is the usual T-conserving hyperfine interaction of the nuclear magnetic dipole moment operator $\mu_{Tl}\boldsymbol{\sigma}_N$ with the internal magnetic field $\boldsymbol{B}_0$ of the molecule and $\boldsymbol{\sigma}_N$ is the Pauli operator related to the nuclear spin operator $\mathbf{I} = \frac{1}{2}\hbar\boldsymbol{\sigma}_N$ ($\hbar = \frac{h}{2\pi}$), i.e., $\boldsymbol{\sigma}_N = \boldsymbol{i}\sigma_{Nx} + \boldsymbol{j}\sigma_{Ny} + \boldsymbol{k}\sigma_{Nz}$, where σ_{Nx}, σ_{Ny}, and σ_{Nz} are Pauli matrices defined in the spinor basis $\{| I = \frac{1}{2}, +\frac{1}{2} >, | I = \frac{1}{2}, -\frac{1}{2} >\}$. The second term in (10.17) describes the P-, T-violating electric dipole interaction between molecular *EDM* and ^{205}Tl nucleus. The unit vector $\boldsymbol{\lambda}$ was defined above, d is the module of the molecular *EDM* (a measure of T-violation in the *TlF* molecule), and $\hbar$ is the Planck constant. In the free space, such an interaction leads to small permanent *EDM*. However, it would be better to detect this interaction by applying a strong electric field $\boldsymbol{E}$, which substantially polarizes the molecule and look for an energy of the form $\boldsymbol{\sigma} \cdot \boldsymbol{E}$. This energy appears as a shift of the *NMR* frequency when $\boldsymbol{E}$ is reversed.

Without going into details of the experiment [502] and its interpretation, we only note that a beam of *TlF* molecules was produced by a hypersonic jet source. The temperature of molecules was sufficiently low so that all they were in the electronic ground state and most of them in the vibrational ground state ($^1\Sigma$, $v = 1$). In this case, a large number of rotational and hyperfine states were occupied, but the experiment was performed using only one particular magnetic hyperfine sublevel of the first excited rotational state with the angular momentum $J = 1$ [502]. This experiment provides the best limits on the proton *EDM* $d_p = (-3.7 \pm 6.3) \times 10^{-23}\, e{\cdot}\text{cm}$ and also places strong limits on the electron–nucleus tensor coupling $C_T = (-1.5 \pm 2.6) \times 10^{-7}$, the nuclear Schiff moment $Q_{Tl} = (2.3 \pm 3.9) \times 10^{-10}\, e{\cdot}fm^3$, and T-odd nucleon-nucleon interactions. Limits derived from this experiment for the electron *EDM*, $d_e = (-2.1 \pm 3.5) \times 10^{-25} e{\cdot}\text{cm}$, are strong, but not as good as those found using paramagnetic atoms (see above in this section). Finally, we note that the second term in the Hamiltonian (10.17) that describes the P-, T-violating internal interaction induces a permanent molecular *EDM*, d_{TlF}, directed along the total angular momentum $\mathbf{F}$ of the molecule. In the $J = 1$ state having $F = 2$, it was found that $d_{TlF} = (-1.7 \pm 2.9) \times 10^{-23}\, e{\cdot}\text{cm}$.

According to Schiff theorem [513, 515], in heavy diamagnetic atoms [516–519] and molecules [514] the *EDM* of the nucleus is entirely screened by electrons. Indeed, here the observable *EDM* is generated by the nuclear Schiff moment Q introduced in the first term of the effective Hamiltonian (10.10) (see also (10.11)–(10.13)). The *SM* is an intra-nuclear charge distribution, generated by T, P-odd interactions within the nucleus, which can introduce an atomic or nuclear *EDM* by polarizing the bound electrons [520, 521]. The *SM* size is $\sim r_N^2 d_N$, where r_N and d_N are the nuclear radius and nuclear *EDM*, respectively. The radius r_N is very small compared to the electron orbital size. Therefore, the atomic *EDM*, d_m, produced by the nuclear *SM* is much smaller than d_N. By contrast, the magnetic interaction between nuclear moments and electrons is not screened. The lowest T-, P-odd magnetic moment is the magnetic quadrupole moment (*MQM*) [522]. It was shown in [520] that in paramagnetic atoms and molecules, as was also noted in [510], the nuclear *MQM* produces a larger *EDM* than *SM* does at the same underlying source of *CP*-violation.

Moreover, the MQM has a collective nature and is significantly enhanced in deformed nuclei (like the case of ordinary electric quadrupole momentum). The time-reversal symmetry violation in paramagnetic molecules induced by nuclear MQM was studied in [522]. It was shown that nuclear T-, P-odd effects are amplified in paramagnetic molecules containing deformed nuclei, where the primary effects arise from the T-, P-odd nuclear MQM. Also, it was argued that *EDMs* measurements in molecules with *MQMs* may, particularly, provide improved limits on the strength of T-, P-odd nuclear forces.

10.4 Time-Reversal Symmetry Violation in Superconductors

After the discovery of ceramic superconductors with high-T_c (HTCS) [523], it became clear that in materials of this class the pairing of electrons with opposite momenta and spin projections is not adequately described by the electron–phonon interaction in the framework of BCS theory [524] developed for conditional metallic superconductors. Considerable efforts have been made to find an alternative mechanism that causes superconductivity in unconventional HTCS, such as $YBa_2Cu_3O_7$ (YBCO) and $Bi_2Sr_2CaCu_2O_8$ (BSCCO), and also in later discovered HTCS MgB_2 [525], iron pnictide compounds and many others. Moreover, the theory of BCS was modified (including both phonon and non-phonon mechanisms), and other additional models were also proposed, such as resonant valence bands, marginal Fermi liquid, etc.

As a result of the performed research, the following was clarified: (a) many of HTSC are quasi-2D and quasi-3D with anisotropic superconducting properties; (b) Fermi surfaces are much more complicated, involving many bands of electronic structure, and (c) on the basis of angle-resolved photoemission spectroscopy data, it was shown the existence of a variety of superconducting gap symmetries (singlet or triplet and s-wave, p-wave, and d-wave) [526]. This variety of superconducting bands symmetries, as it will be seen further, is of a crucial importance for the problem of time-reversal symmetry breaking in unconventional superconductors. According to [525–528], many experimental data show that the pair wave function has a $d_{x^2-y^2}$ character and can be written as

$$\psi(\mathbf{k}) =< c^{+}_{\mathbf{k}\uparrow} c^{+}_{-\mathbf{k}\downarrow} >\propto \cos k_x - \cos k_y, \tag{10.18}$$

where $c^{+}{}_{ks}$ denotes the electron creation operator. The concept of $d_{x^2-y^2}$ wave symmetry pairing is used in various theories. One such theory is based on the idea that the pairing interaction is averaged through the exchange of antiferromagnetic spin fluctuations among quasiparticles [529, 530]. Another theory describes the superconducting states as pairing resulting from the doping of a spin liquid state, so-called resonating valence bonds (RVB) states [531–533]. The third alternative point of view

Table 10.1 Symmetry classification of even-parity pairing states [535]

Γ	$\psi(\mathbf{k})$
A_{1g}	$\cos k_x + \cos k_y$
A_{2g}	$\sin k_x \sin k_y (\cos k_x - \cos k_y)$
B_{1g}	$\cos k_x - \cos k_y$
B_{2g}	$\sin k_x \sin k_y$
E_g	–

assumes the existence of a large continuous symmetry SO(5), which establishes a connection between antiferromagnetic and superconducting order parameters [534].

All oxide high-temperature superconductors consist of CuO_2-superconducting planes and intermediate layers acting as charge reservoirs for removing or adding electrons to the planes. Therefore the discussion of the superconducting state symmetry must be based on the planes lattice structure, which is basically square-like (with a slight orthorombic distortion in some cases) [535]. The pairing states in the lattice are classified according to irreducible representations of the square lattice point group C_{4v} by analogy with the angular momentum classification in rotationally symmetric systems [536].

These representations include both even-(spin singlet) and odd-parity (spin triplet) pairing states. It is possible to distinguish between the two states by some of their magnetic properties. For spin singlet pairing, the spin susceptibility is suppressed in the superconducting state, while it should be only weakly affected for spin triplet states, because the spin freedom degrees in the latter case remain active. Measurements of the Knight shift demonstrated that spin triplet pairing can be ruled out [537]. Thus, it is necessary to restrict even-parity representations, which consist of four one-dimensional (A_{1g}, A_{2g}, B_{1g} and B_{2g}) and a two-dimensional (E_g) representations. The last representation can be ignored, it would imply interlayer pairing. The list of remaining pairing states is given in Table 10.1.

The pair wave functions are given in the momentum space corresponding to real space pairing of particles in the nearest or next nearest sites. The strong Coulomb repulsion among the carriers in the CuO_2-plane forbids them to occupy the same state. Therefore any onsite pairing amplitude in the pair wave function has to vanish [534].

The most symmetric state in the A_{1g} representation cannot be a true s-wave pairing state. It has the form of a so-called extended s-wave with nodes in the first Brillouin zone. There are two pairing wave functions called $d_{x^2-y^2}$-wave and d_{xy}-wave in B_{1g} and, respectively, B_{2g} representations. The state belonging to A_{2g} has a more complicated modal structure and can be termed as g-wave.

From the group theory point of view, all four symmetries are equally good candidates for a superconducting state in the CuO_2-plane. However, only the experiment or a microscopic theory would determine which of them can be realized. A number of experiments including measurements of the Knight shift [537] indicate the presence of low-lying quasiparticle excitations that require nodes in the gap [527]. This requirement is satisfied by all four states. Selection of candidates for the role of a superconducting state can be carried out experimentally. Based on the angle-resolved photoemission spectroscopy (ARPES), locating the nodes of the wave function exclusively along the [110]-direction in BSCCO and YBCO [537], it was shown that A_{2g} and B_{2g} are ruled out, while B_{1g} is the strongest candidate together with A_{1g}.

We now find out when the time-reversal symmetry breaking occurs in superconductors. In particular, let us consider a structure SNS consisting of two superconductors A and B, between which there is a thin layer of a normal metal (the Josephson junction). Suppose, for definiteness, that the superconducting states in materials A and B are transformed with respect to irreducible representations A_{1g} (extended s-wave pairing state) and B_{1g} ($d_{x^2-y^2}$-wave pairing state) of the C_{4v} point group of symmetry. The d-wave nature of the order parameter implies the following three relations concerning the Josephson current I for different, but related angles

$$\begin{aligned} I(-\theta) &= +I(\theta), \\ I\left(\theta \pm \tfrac{\pi}{2}\right) &= -I(\theta), \\ I(\theta \pm \pi) &= +I(\theta) \end{aligned} \tag{10.19}$$

for an arbitrary phase difference $\varphi = \Phi_B - \Phi_A$ [538–541]. In (10.19), θ is the relative angle of one of the main symmetry axes to the interface normal vector. Thus, the critical current I_c may be considered as a θ-dependent function:

$$I_c(\theta) = I_{co} \cos 2\theta, \tag{10.20}$$

which does not reproduce correctly the angular dependence of the coupling strength, but merely gives the proper sign structure.

The sign of $I_c(\theta)$ determines an intrinsic phase shift α of the junction, which is $\alpha = 0$ for $I_c > 0$ and $\alpha = \pi$ for $I_c < 0$ [535].

The presence of the intrinsic phase shift $\alpha = \pi$ can lead to various unusual properties in multiply connected systems. For example, interference phenomena were used to prove the existence of π-phase shifts in some HTSC. The order parameter symmetry in such HTSC was identified as $d_{x^2-y^2}$-wave like [542–548].

The time-reversal symmetry of Josephson junctions is broken if one of the two connected superconductors breaks the time-reversal symmetry. In order for time-reversal symmetry breaking to take place, it is necessary that two wave functions of different symmetries form a linear combination with complex coefficients. In this context, two time-reversal symmetry breaking states are proposed ($s + id$-wave and $d + id$-wave states), both of them with the phase shift $\phi - \phi' = \pm\frac{\pi}{2}$. Here the order parameters are $\eta =\mid \eta \mid e^{i\phi}$ and $\eta' =\mid \eta' \mid= e^{i\phi'}$. The corresponding gap functions in the quasiparticle spectrum

$$\Delta(\mathbf{k}) = \sqrt{\mid \eta\Psi_d(\mathbf{k}) + \eta'\Psi'(\mathbf{k}) \mid^2} \tag{10.21}$$

are nodeless ones in both cases. However, the experimentally observed modes ruled out such time-reversal symmetry breaking states in most of HTSC [535]. Nevertheless, in a number of cases, experiments confirm the existence of time-reversal symmetry breaking superconducting states. It is assumed that the superconductor B violates the time-reversal symmetry described by the two-component order parameter (η_B, η_B'). In this case, the intrinsic phase shift can be obtained using the junction energy [549, 550]

$$E_J = -E_0|\eta_A|\left[\left(|\eta_B| + |\eta_B'|\cos\chi\right)^2 + |\eta_B'|^2\sin^2\chi\right]^{1/2} \cdot \cos(\varphi - \alpha), \qquad (10.22)$$

where

$$\tan\alpha = \frac{|\eta_B'|\sin\chi}{|\eta| + |\eta'|\cos\chi} \qquad (10.23)$$

and $E_0 = I_0\Phi_0/2\pi c$, $\varphi = \Phi_B - \Phi_A$, $\chi = \Phi_B - \Phi_B'$. The time-reversal symmetry is broken for a combination of η and η' if

$$\eta + \eta' \longrightarrow \eta^* + \eta'^* \neq (\eta + \eta')e^{i\gamma}, \qquad (10.24)$$

where γ is an arbitrary phase. Thus, $\eta+\eta'$ and its time-reversed $\eta^*+\eta'^*$ are different, but degenerate states. This is the case if the relative phase $\chi \neq 0, \ \pi$. Note that the phase shift satisfies $\alpha \neq 0, \ \pi$, if the relative phase $\chi \neq 0$, i.e. for time-reversal symmetry breaking states [551, 552].

Let us now discuss the multiple Cooper pair transfer. On a microscopic level, the multiple Cooper pair transfer between two superconductors can be described by Bogolyubov–de Gennes equations or using a quasiclassical approach [553–555]. The interface shall be represented by a simple potential barrier

$$H_{\text{barr}} = Z\delta(x). \qquad (10.25)$$

The current density has the form

$$j(\varphi) = \frac{e}{\pi\hbar}\int_{-\frac{\pi}{2}}^{\frac{\pi}{2}} d\vartheta \frac{1}{\beta}\sum_{\omega_n} \frac{t(\vartheta)\Delta_A\Delta_B(\vartheta-\theta)\sin\varphi}{[\omega_n{}^2 + \Delta_A\Delta_B(\vartheta-\theta)\cos\varphi + E_{nA}E_{nB}]t(\vartheta) + ZE_{nA}E_{nB}}, \qquad (10.26)$$

where $\varphi = \Phi_B - \Phi_A$ and the angle integral over ϑ is restricted, for simplicity, in the basal plane of the d-wave superconductor (the c-axis is always parallel to the interface) [541, 556]. The function $t(\vartheta)$ denotes the tunneling form factor of the junction (by choosing $t(\vartheta) = t_0\cos^2\vartheta$, the perpendicular tunneling is favored). The s-wave gap and the angle-dependent d-wave gap with $\Delta_B(\vartheta) = \Delta_0\cos 2\vartheta$ are denoted by Δ_A and $\Delta_B(\vartheta)$, respectively. The corresponding quasiparticle energies are

$E_n = \left(\omega_n{}^2 + \Delta^2\right)^{\frac{1}{2}}$, where $\omega_n = \frac{\pi(2n+1)}{\beta}$ is the Matsubara frequency ($\beta^{-1} = k_B T$). In the limit $Z >> 1$, it is possible to expand (10.26) in $\Delta_A \Delta_B$ [535]:

$$j(\varphi) = \frac{e}{\pi\hbar} \int_{-\frac{\pi}{2}}^{\frac{\pi}{2}} d\vartheta \frac{1}{\beta} \sum_{\omega_n} \frac{\Delta_A \Delta_B(\vartheta-\theta)t(\vartheta)\sin\varphi}{(\omega_n{}^2+E_{nA}E_{nB})t(\vartheta)+ZE_{nA}E_{nB}} \times \\ \times \left(1 - \frac{\Delta_A \Delta_B(\vartheta-\theta)t(\vartheta)\cos\varphi}{(\omega_n{}^2+E_{nA}E_{nB})t(\vartheta)+ZE_{nA}E_{nB}}\right) + \cdots \tag{10.27}$$

As the temperature at the superconducting side approaches T_c, the first term in (10.27), that is proportional to Δ^2, vanishes more slowly than the second one, proportional to Δ^4. It can be shown that in the specific case of $\theta = \frac{\pi}{4}$ the first term disappears for any temperature so that $\alpha = \pm\frac{\pi}{2}$ minimizes the junction energy for all $T < T_c$. It is a feature specific to the $d_{x^2-y^2}$- wave superconductor. For any other angles there is a critical temperature $\widetilde{T}_c$, below which the phase shift α decreases continuously from 0 to π [541]. Thus, the change occurring at $\widetilde{T}_c$ may be considered as a second order phase transition of the Josephson junction, breaking the time-reversal symmetry.

We consider now the Josephson current, which can be obtained by the derivative of the free energy with respect to φ. The contribution of current at each Fermi momentum $\mathbf{k}_F$ is:

$$J_{\mathbf{k}_F} = \frac{2e\mathbf{k}_F}{\hbar} L_{\mathbf{k}_F} \sum_{n'=0}^{\infty} n' f_{n'}(\nu_{\mathbf{k}_F}) \sin(n'\varphi), \tag{10.28}$$

where $L_{\mathbf{k}_F} = \frac{L}{(\mathbf{k}_F \cdot \mathbf{n})}$ is the traveling distance of the electron or hole in the normal metal, L is the thickness of normal metal layers situated between the superconductors A and B.

It is easily to decompose the current from (10.28) into the perpendicular component, $J_\perp$, passing through the junction (Josephson current) and the component parallel to the metal layer, $J_\parallel$, [557]:

$$\begin{pmatrix} J_\perp \\ J_\parallel \end{pmatrix} = \frac{2ek_F}{\hbar} \int_{-\frac{\pi}{2}}^{\frac{\pi}{2}} \frac{d\nu_{\mathbf{k}_F}}{2\pi} L_{\mathbf{k}_F} \sum_{n'=0}^{\infty} \begin{pmatrix} \cos\nu_{\mathbf{k}_F} \\ \sin\nu_{\mathbf{k}_F} \end{pmatrix} n' f_{n'}(\nu_{\mathbf{k}_F}) \sin(n'\varphi). \tag{10.29}$$

The lowest energy state is obviously given by $J_\perp(\varphi = \varphi_o) = 0$, where

$$\varphi_o = \pm \arccos\left(-\frac{I_1}{2I_2}\right) \text{ for } I_2 < 0 \text{ and } \frac{I_1}{2I_2} < 1. \tag{10.30}$$

The values I_1 and I_2 from (10.30) are determined by the current-phase relation

$$I(\varphi) = I_1 \sin\varphi + I_2 \sin 2\varphi. \tag{10.31}$$

The vanishing of $J_\perp$ implies the overall canceling of all current contributions in different $\mathbf{k}_F$-directions projected on the normal vector of the normal metal layer. However, for $\varphi_o \neq 0,\ \pi$ the parallel current component is not zero. The presence of a finite parallel current component is a manifestation of broken time-reversal symmetry, since there is a degenerate state with reversed current. In the used approach this spontaneous current flows uniformly in the metal layer and in the cases $\theta = \frac{\pi}{4}$ and $T = 0$ the total current is given by [557]

$$I_\parallel = LJ_\parallel\left(\varphi = \frac{\pi}{2}\right) = \frac{ek_F^2}{2\pi^2}. \tag{10.32}$$

This result is valid for two dimensions only, because the motion of quasiparticles along the c-axis has been neglected.

In the continuation of this section, we discuss two effects that seem to be the most promising for the detection of time-reversal symmetry breaking. The first one is connected with the possible fractional flux of quanta [551, 552, 558–560] and the second one is connected with phase slips [561] between two degenerate Josephson junction states.

In general, the phase difference φ is not uniform and can vary over a certain length scale λ_J (Josephson penetration depth) that is the screening length for external magnetic fields in the junction. A Josephson junction can support vortices (flux lines), which contain fluxes

$$\Phi = n\Phi_o, \tag{10.33}$$

where n is an integer positive number and Φ_o is the magnetic flux quantum

$$\Phi_o = \frac{\hbar c}{2e}. \tag{10.34}$$

In (10.34), $\hbar$, c and e are, respectively, the Planck constant, light speed, and the electron charge.

The mentioned vortices are most conveniently described through the spatial variation of φ along the junction. For definiteness, we consider a junction that lies in the (xz)-plane with the direction of the magnetic field along the z-axis and the variation of φ exclusively along x-axis. Using Maxwell's equations and the Josephson current-phase relation in a long inhomogeneous junction, the sine-Gordon equation can be obtained [562]:

$$\frac{\partial^2\varphi}{\partial x^2} = \frac{1}{\lambda_J^2}\sin\varphi, \tag{10.35}$$

where λ_J is the Josephson penetration depth. The constant solutions $\varphi = 2\pi n$ of (10.35) are stable.

Based on (10.35), the half-integer quantization of the magnetic flux can be obtained [538, 563]

$$\Phi = \Phi_o \left(n + \frac{1}{2}\right). \tag{10.36}$$

Along with (10.36), a new flux quantization appears

$$\Phi = \Phi_o \left(n + \frac{\chi}{\pi}\right), \tag{10.37}$$

which is shifted by a junction-specific value and is neither integer, nor half-integer [551, 552, 564]. This is the fractional type quantization.

Following the [555, 560], we present a simple argument for the fact that the observation of anomalous flux quantization (10.37) is uniquely connected with the local violation of the time-reversal symmetry of the superconducting state. Let a flux Φ vortex is located somewhere on the junction. If we apply the time-reversal operation to this system, the vortex flux changes its sign: $\Phi \longrightarrow -\Phi$. These two flux values can only differ by an integer multiple of Φ_o, if the underlying superconducting state is not altered by this operation (apart from inverting the phase):

$$\Phi = -\Phi + \Phi_o n. \tag{10.38}$$

This leads to $\Phi = \Phi_o \frac{n}{2}$, and only integer and half-integer values are allowed to appear. On the other hand, if the superconducting state changes under the time-reversal operation (and as a consequence the time-reversal symmetry is violated), then no constraint is imposed on the difference between Φ and $-\Phi$, and any value of Φ is possible.

An effect known as phase slip, which does not require long inhomogeneous junctions, is related to a kink of φ in time [535]. This effect implies that the phase difference in standard Josephson junctions jumps from one stable state to another by 2π. In the case when the time-reversal symmetry in the junction is violated, the situation may change, because the number of stable states is doubled and a transition between stable junction states includes now differences smaller than 2π. In this case, the transferred flux has a fractional value in the previously defined sense. If we take into account that φ varies in time and yields the voltage

$$V(t) = \left(\frac{\Phi_o}{2\pi}\right) \frac{\partial \varphi}{\partial t}, \tag{10.39}$$

then, under the conditions taken in [535], the voltage signal integrated over time is related to the magnitude of the transferred flux and the total change of φ is

$$\int_{-\infty}^{\infty} V(t) \mathrm{d}t = \frac{\Phi_o}{2\pi} \int_{-\infty}^{\infty} \frac{\partial \varphi}{\partial t} \mathrm{d}t = \frac{\Phi_o}{\pi} \chi, \tag{10.40}$$

i.e. the fractional flux introduced in (10.37) appears again (due to the fractional quantization).

Since the current flow is accompanied by generation of a voltage, energy is absorbed in this process. The barrier between two degenerate minims in the time-reversal breaking state is particularly low and close to the transition in this state. Therefore, oscillating phase slips between two minims can be induced near this transition by microwave radiation. The transition would be accompanied by enhancement of the microwave absorption in the junction [535].

In addition to these two effects, the arrangement of energy levels of quasiparticle bound states can also indicate the violation of time-reversal symmetry. Consideration of the surface of a $d_{x^2-y^2}$- wave superconductor, where states with broken time-reversal symmetry can appear, leads to results similar to those obtained for Josephson junctions, but with some specific features. While two order parameters exist at the interface between two superconductors (one on each side) that can form together a time-reversal symmetry breaking state [558], only one order parameter component is a priori available close to the surface. The pair breaking effect can not only destroy the $d_{x^2-y^2}$- wave order parameter at the surface, but also open the way to an order parameter of different symmetry, which is suppressed in the bulk (such as s-wave or d-wave). This new pairing component can form a complex combination together with the dominant $d_{x^2-y^2}$- wave component [551, 552, 565–567]. Similarly to the interface state, such a state is accompanied by spontaneous currents at the surface. In [535], this state is discussed phenomenologically, based on Ginsburg–Landau theory, and microscopically, using Bogolyubov-de Gennes equations. It was shown that the generated zero-energy state splits into two separate states, one above and other below the Fermi level, leading to lowering the quasiparticle contribution to the free energy. Thus, removal of the large density of states is an important factor contributing to the creation of a time-reversal breaking state.

The splitting of the zero-energy peak also manifests itself in quasiparticle tunneling. The normal metal-superconductor (NS) quasiparticle tunneling spectra made the interface between the metal and semiconductor sufficiently non-transparent [568]. As a consequence, the bound state appears as a resonance in the current-voltage characteristics. Thus, it was suggested that the zero-energy bound state leads to a so-called zero-bias anomaly in the tunneling spectrum. In [565–567, 569, 570], it was suggested that the occurrence of time-reversal symmetry breaking, including splitting of the zero-energy level, should also lead to an observable modification in the tunneling spectrum. This tunneling spectrum modification was indeed identified in the experiments on [110]-oriented NS devices with YBCO and in the zero-bias anomaly splits into a double peak below 4 K [571]. These experiments are a strong proof of the existence of a superconducting state with broken time-reversal symmetry at the surface.

It is easy to show that the spontaneous current is also carried by quasiparticles, similar to the Josephson junction case. Splitting of the zero-energy level leads to an imbalance in the occupation among the electron states with the momentum component parallel and antiparallel to $(1, -1, 0)$. Thus, there exists a finite current along the surface, whose direction depends on which of the two degenerate time-reversal symmetry breaking superconducting states is realized [559, 567]. There were also studied other structures, where spontaneous broken time-reversal states can occur, in particular, the twin boundary in an orthorombically distorted originally tetragonal system [572–574] and the crystal with lattice dislocations [575]. In both cases, a state with spontaneously broken time-reversal symmetry can occur locally.

While in high-T_c semiconductors the time-reversal symmetry breaking is apparently restricted to specific regions in the sample, there are other nonconventional superconductors where the time-reversal symmetry is very likely violated throughout the whole material. As an example, we can indicate the heavy fermion superconductors UPt_3 and $U_{1-x}Th_xBe_{13}(0,017 < x < 0,045)$ [576]. The two consecutive phase transitions identified in these compounds by the temperature dependence of specific heat indicate that these are attributed to the superconductivity involving pairing states of different symmetry. The theory of this double transition is based on multidimensional order parameters $(\eta_1, \eta_2, \ldots)$ whose components have different transition temperatures [577, 578]. The onset of the superconductivity is characterized by the fact that one of the components (denoted by η_1) with the highest critical temperature T_c becomes finite, and one or more other components (denoted by η_2) at the second transition are admixed to form a complex order parameter combination, e.g., $\eta = (\mid \eta_1 \mid, \pm i \mid \eta_2 \mid)$ [535, 550, 579]. Another example is the superconductor Sr_2RuO_4 discovered in [580], which has a single superconducting phase transition leading to an apparently time-reversal symmetry breaking state [535].

Of a particular importance for detecting the broken time-reversal symmetry in these and other compounds is the study of their magnetic properties [550, 578]. As it was noticed earlier, locally time-reversal symmetry breaking states can generate spontaneous currents and field distributions. In order to generate spontaneous currents in a bulk time-reversal symmetry breaking superconductor, it is necessary the order parameter to be not homogeneous As a rule, generated fields do not lead to net magnetic moments and that is why high resolution probes are need for their detection [559, 581, 582]. A suitable experimental method of investigation in this case is the method of spin polarized muons [583, 584]. When such muons are injected into a material, they are trapped quickly in symmetric crystal lattice locations and their spins precess in the local static magnetic field. In the absence of a magnetic order, this field of various origins has, in general, random values and directions. The field distribution can have a Gaussian form

$$D(H) = \frac{1}{\sqrt{2\pi\sigma}} \exp\left(-\frac{H^2}{2\sigma}\right), \tag{10.41}$$

where σ and H are the distribution width and the magnetic field strength. The time dependence of the muon spin polarization along the initial direction is

$$P(t) = \frac{1}{3}\left[1 + 2(1 - \gamma_\mu^2\sigma^2t^2) \exp\left(-\frac{1}{2}\gamma_\mu^2\sigma^2t^2\right)\right], \tag{10.42}$$

where γ_μ is the gyromagnetic ratio of the muon [585]. As it is known, the Gaussian distribution width (10.41) is the measure of the internal field. In a time-reversal symmetry breaking superconductor, the field distribution around each inhomogeneity causes an additional contribution to the width σ. The superconductor contribution to a state of the form $\eta = (\mid \eta_1 \mid, i \mid \eta_2 \mid)$ is $\delta\sigma_{kT}(T) \propto \mid \eta_1(T) \mid\mid \eta_2(T) \mid$, where k and T are the Boltzmann constant and, respectively, the absolute temperature.

For UPt_3 [586] and $U_{1-x}Th_xBe_{13}$ [587], below the second transition at T_c', it was found the approximate form

$$\delta\sigma_{kT}(T) \propto \mid T - T_c' \mid^{\frac{1}{2}} . \tag{10.43}$$

This form is dominated by the second-order parameter η_2, while η_1 is already finite and has a weaker temperature dependence [550, 578]. On the other hand, if the violation of the time-reversal symmetry occurs at the onset of superconductivity, as is probably the case in Sr_2RuO_4, both order parameters have the same temperature dependence and lead to

$$\delta\sigma_{kT}(T) \propto \mid T - T_c \mid \tag{10.44}$$

close to T_c [586].

There are also of interest the results obtained in the BCS formalism concerning the time-reversal symmetry broken (TRSB) state in a multi-band superconductor. For the first time, Moskalenko [588] and Suhl et al. [589], independently of each other, extended the one-band BCS theory to the two-band theory with overlapping the energy band on the Fermi surface. These works were published long before the discovery of high T_c-superconductivity and the synthesis of a number of new superconductors, thus bridging the gap between conventional low T_c superconductors and high T_c cuprates. As it is shown in [526] based on the multi-band BCS theory, frustration between the bands, in three bands BCS superconductors with repulsive inter-band interaction, can lead to an inherently complex gap function, arising out of the phase difference between the bands in the range from 0 to π. The complex conjugate of this state is also a solution: the ground state is degenerate, therefore a TRSB state appears. This state is a result of frustrated repulsive interactions between the three bands that gives rise to a relative phase appearing between the bands. It was shown that the TRSB state only appears in a very small region of the parameter space, where the interaction between bands is of the same order. This allows the frustrated repulsive interaction to form the TRSB state. It was also shown that there is a possible phase transition between the TRSB and conventional BCS states at a finite temperature, which can be experimentally probed.

Let us consider now the case of a $2D$ superconductor with a frustrated checkerboard lattice, which may be considered as a $2D$ projection of the $3D$ corner-sharing lattice of pyrochlore. Using the renormalized mean field theory and the extended Habbard model, it was shown that the $d_{x^2-y^2}$-symmetry state is most stable in a large parameter region for small values of $\mid \frac{t'}{t} \mid$ (t and t' are hopping integrals for the nearest neighbor links and the next nearest neighbor links) [590]. The authors of [590] have found that around $| \frac{t'}{t} |= 1$ a time-reversal symmetry broken state exists. They have shown that $(d + id)$- and $(d + is)$-symmetries states are the most stable.

10.5 Time-Reversal Symmetry Violation and Enhancement of Quantum Transport

After discovery by Wigner [68, 222] of the antiunitary time-reversal symmetry (TRS), this symmetry, as well as various its violations have become the object of many studies in elementary particle physics, in nuclear, atomic and molecular physics, and in solid state physics. However, it cannot be said the same about quantum informatics and, in particular, about quantum transport, communication networks, more effective energy transfer and improved information processing devices.

Ideas about the antiunitary symmetry began to penetrate into these areas of knowledge only in recent years, but already brought encouraging results, consisting, in particular, in the growth of the quantum transport enhancement by time-reversal symmetry breaking [587, 591–596]. Note that the quantum transport enhancement studied in these works is not related with the enhancement of transport due to quantum noise [597, 598], which has been studied in the context of photosynthesis [598, 599]. Here the emphasis is on the time-reversal symmetry breaking of the Hamiltonian dynamics and its influence on the quantum transport.

However it is necessary to make the following remark. The possibility of time-reversal symmetry breaking became a fundamental problem since the creation of quantum mechanics. Until now, the mechanisms of T-violation are not fully understood. In particular, there is the possibility of time-reversal symmetry violation in high-energy physics since some elements of the Cabibbo–Kobayashi–Maskawa (CKM) matrix may have complex values [600]. However, this is related to heavy quark properties and no T-violation is expected to occur in low-energy atomic physics [601]. Search for time-reversal symmetry breaking in low-energy collective modes of an atomic nucleus, as well as for TRS violation in atoms, molecules and solids is a subject of extensive studies, both theoretical and experimental. In contrast to the fundamental violation of T-symmetry, it is relatively easy to find a spontaneous time-reversal symmetry breaking in any quantum circuits. In what follows we will follow the papers [591, 596].

In [596], an approach based on a complex network theory of quantum systems [602–604] has been developed. In this approach the probability transfer is directed by the controlled breaking of time-reversal symmetry that creates a so-called chiral quantum walks [591–595]. Since their recent introduction, continuous time chiral quantum walks have been studied in the context of energy transport in ultra-cold atoms and molecules [592], in non-equilibrium physics [593, 594] and as a method to achieve near perfect state transfer [591, 595]. In [596], conditions of making a Hamiltonian (circuit) time asymmetric have been established in terms of the geometry and edge-weights (gates) of the underlying graph. The time-reversal symmetry breaking is of practical importance because it is equivalent to introducing a biased probability flow in a quantum system. The TRS breaking in a quantum process allows direct state transfer without a biased distribution in initial states, a coupling to an environment or using in situ tunable Hamiltonians [605, 606].

The time-independent quantum walks Hamiltonian is defined by [607–611]:

$$H_{QW} = \sum_{n,m}^{\text{sites}} J_{nm}(| n >< m | + | m >< n |). \tag{10.45}$$

The requirement that hopping weights, J_{nm}, are to be real numbers implies that the induced transitions between two sites are symmetric under time reversal. The time-reversal symmetry can be violated (the hermitian property of the operator being maintained) by appending a complex phase to an edge: $J_{nm} \longrightarrow J_{nm}e^{i\theta_{nm}}$ [612], resulting in a continuous time chiral quantum walk (CQW) Hamiltonian governed by

$$H_{CQW} = \sum_{n,m} J_{nm}e^{i\theta_{nm}} | n >< m | + J_{nm}e^{-i\theta_{nm}} | m >< n | . \tag{10.46}$$

Equation (10.46) can be written in terms of the spin-Pauli matrices:

$$H_{CQW} = \sum_{n,m} J_{nm}\cos(\theta_{nm})(\sigma_n^x\sigma_m^x + \sigma_n^y\sigma_m^y) + \sum_{n,m} J_{nm}\sin(\theta_{nm})(\sigma_n^x\sigma_m^y - \sigma_n^y\sigma_m^x), \tag{10.47}$$

which arise in a variety of physical systems when magnetic fields are considered.

The coherent quantum dynamics and incoherent dynamics within the Markov approximation can be investigated in the chiral quantum walks (CQW) framework. Both types of evolution are included in the Kossakowski–Lindblad equation [613–617]

$$\frac{d}{dt}\rho(t) = \mathcal{L}\{\rho\} = -i[H_{CQW}, \rho] + \sum_k L_k\rho L_k^+ - \frac{1}{2}(L_k^+L_k\rho + \rho L_k^+L_k), \tag{10.48}$$

where $\rho(t)$ is the density operator describing the state of the system at the time t and L_k are Lindblad operators inducing stochastic jumps between quantum states. Using the usual terminology of Markovian processes, the site t can be called a trap if it is coupled to the site s by the Lindblad jump operator, $L_k =| t >< s |$. The site-to-site transfer probability, $P_{n\longrightarrow m}(t) =< n | \rho | m >$, gives the occupancy probability of the site m at the time t at the initial condition $\rho(0) =| n >< n |$.

A quantum switch which violates the time-reversal symmetry and enables directed transport can be introduced and used to create a logic gate.

In this case, $P_{S\longrightarrow E}$ is the occupancy probability of the site E with the particle initially starting from the site S with and without sink. This evolution is time-reversal asymmetric as replacing t by $-t$ results in the particle moving from the site S towards the site F, different from E. When starting at the site E, the particle evolves towards the same site F. By replacing t by $-t$, a particle being initially at the site E evolves towards the initial configuration. Equation (10.48) can be written as follows:

$$\mathcal{L}\{\rho\} = -i[H_{CQW}, \rho] + \sum_{(n,m)} C_{nm} \left(L_{nm}\rho L^{+}_{nm} - \frac{1}{2}\{L^{+}_{nm}L_{nm}, \rho\} \right), \tag{10.49}$$

where the chiral Hamiltonian H_{CQN} is defined in (10.46), and the Lindblad operators are given as $L_{nm} =\mid m >< n \mid$ with $C_{nm} \geq 0$. The transport from the vertex $\mid S >$ to the vertex $\mid E >$ in such a dynamics is characterized by the site-to-site transfer probability, which in the unitary case ($C_{nm} = 0$) is [591]

$$P_{S\longrightarrow E}(t) = Tr\left(e^{-iH_{CQW}t}\rho_S e^{iH_{CQW}t}\rho_E\right) \tag{10.50}$$

with $\rho_S = \mid S >< S \mid$ and $\rho_E = \mid E >< E \mid$, while in the general Markovian case it is

$$\rho_{S\longrightarrow E}(t) = Tr\left(e^{\mathcal{L}t}\{\rho_S\}\rho_E\right). \tag{10.51}$$

In the case of quantum walks, the time-reversal operator T acts as a complex conjugation (with respect to the vertex basis) [591, 618]:

$$T\sum_{v\in V} C_v \mid v >= \sum_{v\in V} C^{*}_v \mid v > . \tag{10.52}$$

The antiunitarity of the T-operator and the relationship $T^2 = e$ (for non-Kramers systems) implies that $T^{+} = T$. The time reversal of a Hamiltonian H is given[1] as $THT^{-1} = THT$. The $H \longrightarrow THT$ action is represented by the replacement $\theta_{mn} \longrightarrow -\theta_{mn}$ in (10.46). The site-to-site transfer probability of $H(P_{S\longrightarrow E}(t))$ and that of $H' = THT$ $(P'_{S\longrightarrow E}(t))$ are related as follows [591]:

$$\begin{aligned} P'_{S\longrightarrow E}(t) &= Tr\left(e^{-i(THT)t}\rho_S e^{i(THT)t}\rho_E\right) = \\ &= Tr\left(Te^{iHt}T\rho_S Te^{-iHt}T\rho_E\right) = \\ &= Tr\left(e^{iHt}T\rho_S Te^{-iHt}T\rho_E T\right) = \\ &= Tr\left(e^{iHt}\rho_S e^{-iHt}\rho_E\right) = P_{S\longrightarrow E}(-t), \end{aligned} \tag{10.53}$$

$$P_{S\longrightarrow E}(-t) = Tr\left(e^{iHt}\rho_S e^{-iHt}\rho_E\right) = Tr\left(e^{-iHt}\rho_E e^{iHt}\rho_S\right) = P_{E\longrightarrow S}(t). \tag{10.54}$$

[1] In this and next sections the Hamiltonian and the time-reversal operator are denoted as H and T instead of **H** and **T**.

A consequence arising from (10.53) and (10.54) is that the transition probabilities in the case of achiral quantum walks are the same at t and $-t$, i.e. $P_{S \longrightarrow E}(t) = P_{S \longrightarrow E}(-t)$. However, $H \neq THT^+$ does not necessarily imply that transition rates are asymmetric in time, because THT^+ might be gauge-equivalent to H (see below). The gauge transformation is simply a local change of basis, i.e., a diagonal unitary transformation U_d:

$$U_d \mid n >= e^{i\alpha_n} \mid n > . \tag{10.55}$$

Note that any unitary basis-change U would induce a transformation of the Lindblad superoperator $\mathcal{L} \longrightarrow \mathcal{L}'$ with

$$\mathcal{L}'\{\rho\} = U\mathcal{L}\{U^+ \rho U\}U^+. \tag{10.56}$$

Using (10.56) and the invariance of localized states under diagonal unitary operator U_d $\left(U_d^+ \rho_v U_d = \rho_v\right)$, the following relationship can be obtained [591]:

$$\begin{aligned} P'_{S \longrightarrow E}(t) &= Tr\left(e^{\mathcal{L}'t}\{\rho_S\}\rho_E\right) = Tr\left(U_d e^{\mathcal{L}t}\{U_d^+ \rho_S U_d\}U_d^+ \rho_E\right) = \\ &= Tr\left(e^{\mathcal{L}t}\{\rho_S\}U_d^+ \rho_E U_d\right) = Tr\left(e^{\mathcal{L}t}\{\rho_S\}\rho_E\right) = P_{S \longrightarrow E}(t), \end{aligned} \tag{10.57}$$

which proves the invariance of the site-to-site probability under the gauge transformation defined by (10.55). Here the cyclicity of the trace and the invariance of the statistical operator $\rho_v (v = S, E)$ under the time-reversal operator $(T \rho_v T = \rho_v)$ have been taken into account. It was also taken into consideration that the antiunitarity of T and $T^2 = e$ implies that $T^{-1} = T^+ = T$. Under diagonal unitary transformations, the quantum walks Hamiltonian parameters transform as

$$\theta_{mn} \longrightarrow \theta_{mn} + \alpha_m - \alpha_n. \tag{10.58}$$

The incoherent part of the Kossakowski–Lindblad (10.49) does not change, since Lindblad operators transform as $L_{nm} \longrightarrow e^{i(\alpha_m - \alpha_n)} L_{nm}$ and phases α_m and α_n cancel in (10.49), because L_{nm} and L_{nm}^+ appear paired.

The approach developed in [591] was used to demonstrate the effect of time-reversal symmetry breaking in some examples, which illustrate the main idea of directionality, suppression, and enhancement of the quantum transport:

1. a unitary quantum switch where the phase, that is, the time-reversal asymmetry parameter controls the direction of quantum transport;
2. the complete suppression of chiral quantum walks on loops with an even number of sites;
3. the exciton transport of energy from antenna to the reaction center in photosynthesis process;

4. randomly generated small-world networks. In the last case, it was shown that by appending time-reversal asymmetric terms to only the network connected to the final site, the speed of the site-to-site transport increases up to 130% [591].

In [596], a quantum circuit version of the time-reversal symmetry theory is developed, which classifies the time-symmetric and time-asymmetric Hamiltonians and circuits in terms of their underlying network elements and geometric structures. These results reveal that many of the typical quantum circuit networks exhibit a time-asymmetry and that the physical effect of time-symmetry breaking plays an essential role in quantum transport science. Using this approach, the most basic time-asymmetric process is identified and it is shown that the fundamental building block can be created and controlled experimentally on the basis of room-temperature liquid-state nuclear magnetic resonance (NMR) of ^{13}C-labeled trichloroethylene dissolved in d-chloroform. It was demonstrated that the time-asymmetry and, as a consequence, the transport probability can be controlled with limited access to the system, namely, by using local z-rotations paired with a naturally emulated time-symmetric evolution, achieving state transfer probabilities approaching unity.

Focusing on a fundamental three-cubit circuit, it was demonstrated experimentally that the time-symmetry breaking can lead to transition probabilities that are enhanced from strictly less than 0.6 toward unity. In thrichloroethylene molecule, C_2HCl_3, the two ^{13}C and one ^1H spins form a 3-qubit register. There is a gate sequence in the single excitation subspace spanned by the computational basis vectors $\{| \ 100 >, \ | \ 010 >, \ | \ 001 >\}$ of a three-qubit system. One of ^{13}C spins is denoted as $C1$(qubit 1), another as $C2$ (qubit 2), and ^1H spin as H (qubit 3). The natural Hamiltonian of this system is

$$H = \sum_{i=1}^{3} \pi \nu_i Z^i + \frac{\pi}{2}\left(J_{13}Z^1Z^3 + J_{23}Z^2Z^3\right) + \frac{\pi}{2}J_{12}\left(X^1X^2 + Y^1Y^2 + Z^1Z^2\right), \tag{10.59}$$

where X^j, Y^j and Z^j are Pauli matrices[2] acting on qubit i, ν_i is the chemical shift of the ith spin and J_{ij} is the scalar coupling strength between spins i and j. As the difference in chemical shifts between $C1$ and $C2$ is not larger (see below) enough to adopt the weak J-coupling approximation [618], these two carbon spins are treated strongly coupled. The parameter of the Hamiltonian (10.59) determined by iteratively fitting the simulated and experimental NMR spectra are presented in Table 10.2.

In Table 10.2 the diagonal elements are the chemical shifts ν_i, and the off-diagonal elements are scalar coupling strengths J_{ij}. T_1 and T_2 are, respectively, the relaxation and dephasing time scales.

The single qubit gates acting on the three-qubit system with the computational basis vectors $\{| \ 100 >, | \ 010 >, | \ 001 >\}$ are

[2] In contrast to quantum informatics, in the literature on magnetism these matrices are denoted by σ_x, σ_y, and σ_z, respectively.

Table 10.2 Parameters of the spin Hamiltonian (10.59) for a three-qubit system in trichloroethylene molecule [596]

(Hz)	$C1$	$C2$	H	$T_1(s)$	$T_2(s)$
$C1$	21784.6	–	–	13.0 ± 0.3	0.45 ± 0.02
$C2$	103.03	20528.0	–	8.9 ± 0.3	1.18 ± 0.02
H	8.52	201.45	4546.9	8.9 ± 0.3	1.7 ± 0.2

$$U^{ij}(\alpha,\theta) = \exp\left\{-i[\cos(\alpha)S^{ij} + \sin(\alpha)A^{ij}]\frac{\theta}{2}\right\}, \tag{10.60}$$

where

$$S^{ij} = X^iX^j + Y^iY^j \; and \; A_{ij} = X^iY^j - Y^iX^j. \tag{10.61}$$

The NMR Hamiltonian (10.59) contains terms like S^{ij} and thus gates with $\alpha = 0$ are naturally implementable. Gates with $\alpha \neq 0$ can be created by additionally applying local z-rotations:

$$U^{ij}(\alpha,\theta) = Z^j(\alpha)U^{ij}(0,\theta)Z^{j+}(\alpha), \tag{10.62}$$

where $Z^j(\alpha) = e^{-i(\frac{\alpha}{2})Z^j}$ is a local z-rotation. Probabilities of transition into one of the computational basis states $\{| \; 100 >, | \; 010 >, | \; 001 >\}$ have been measured experimentally in the same basis by applying the circuit $U(\alpha,\theta)$.

The slice $\alpha = 0$ corresponds to the amplitude and probability time-symmetric case. The slice $\alpha = \frac{\pi}{2}$ corresponds to the maximum probability time-asymmetry in probabilities. Slices corresponding to $\alpha = \pi$ and $\alpha = \frac{3\pi}{2}$ represent a reflection in θ of the former two cases with the effect of changing the direction of time. In the time-symmetric case ($\alpha = 0$), probabilities of transporting the excitation to other spins are always bounded from above by 0.6. However, the time-asymmetry ($\alpha \neq 0$) allows to break this barrier, with transition probabilities approaching unity at the point of maximal time-asymmetry ($\alpha = \frac{\pi}{2}$) [596].

10.6 Time-Reversal Symmetry Violation and Unidirectionality of Time

In quantum mechanics, the evolution in time of a physical system characterized by the state vector $| \; \Psi(\xi,t) >$ that depends on n dynamical variables $\{\xi_1, \xi_2, \ldots, \xi_n\} = \xi$ is considered. For $t > 0$, this evolution is described by the Schrödinger equation. If we replace t by $-t$, then the term containing the first time derivative in the Schrödinger equation changes its sign. However, by performing an additional transformation, introduced by Wigner [222] (Sect. 2), one can easily achieve the reconstruction of the form of the Schrödinger equation, which is satisfied by the state vector $| T\Psi(\xi, t) >$, where T is the time-reversal operator. This leads to the well-known result of the invariance of the Hamiltonian with respect to the operator T ($THT^+ = H$). For T-violating process

$$H \neq THT^+, \tag{10.63}$$

where H is the Hamiltonian for a specific direction of time ($t > 0$) and THT^+ is the version of the Hamiltonian in the reversed direction of time ($t < 0$). In (10.63), T^+ is used instead of T^{-1}, because $T = UK$, where U is a unitary operator ($U^+U = UU^+ = e$) and K is a complex conjugation operator ($K = K^{-1}$).

We can write down the Schrödinger equation for T-violating process, provided the direction of time evolution and the corresponding Hamiltonian are known. By applying the time-reversal operator, we obtain the Schrödinger equation involving the THT^+ version of the Hamiltonian for reversed direction of time. However, we do not have a dynamical equation of motion for the case when the time direction cannot be specified and there is no reason to favor one version of the Hamiltonian over the other [619]. This problem becomes critical when we attempt to describe the Universe as a closed system, because in this case none external clock-like device can be used as a reference for the direction of time. There is no reason to favor one direction of time over the other and, as a consequence, to favor H over THT^+. This major problem has been solved by Vaccaro in a series of recent publications [619–621, 628, 629].

Attempts to understand the implication of the T-violation in the nature of time were also undertaken earlier [622–624], however only Vaccaro showed that it can be clarified by using Feynmans sum [625] to construct a set of all possible paths that the Universe can take over time. The set includes paths that zigzag forwards and backwards through time, corresponding to the Universe evolving according to H in one time direction and according to THT^+ in the reversed time direction. The violation of time-reversal symmetry induces a destructive interference that restricts possible paths in time the Universe can take. The interference eliminates paths meandering forwards and backwards in time and leaves only two main paths corresponding to continuous evolutions forwards and backwards. This analysis solves the problem of modeling the dynamics of T-violation processes in absence of preferred direction of time by incorporating two Hamiltonians, one to forwards evolution and other to backwards one in a single dynamical equation [620]. Further discussion of the Universe evolution and the time-reversal symmetry breaking is based on publications [619–621].

Let the Universe be a closed system. In this case no external clock devices could be used as reference indicators of the direction of time scale and, as a consequence, the evolution of the Universe could be considered independent on the time-scale direction. For convenience, the two time scale directions will be referred to as "forward" and "backward". Let the Universe be in a "origin state" $| \psi_o >$ without reference to the time scale direction. Now, let the Universe evolution in the forward direction over the time interval τ be

$$| \psi_F(\tau) >= U_F(\tau) \mid \psi_o >, \tag{10.64}$$

where $U_F(\tau) = exp(-i\tau H_F)$ is the forward time evolution operator and H_F is the Hamiltonian of the forward time evolution (throughout this section units with $\hbar = 1$ will be used). The backward time evolution operator is

$$U_B(\tau) = TU_F(\tau)T^{-1} = \exp(i\tau H_B), \tag{10.65}$$

where $H_B = TH_AT^{-1}$ is the Hamiltonian of the backward time evolution. The state of the Universe after a time interval τ of backward evolution will be

$$| \psi_B(\tau) >= U_B(\tau) \mid \psi_o > . \tag{10.66}$$

The matrix elements $< \Phi \mid U_F(\tau) \mid \psi_o >$ and $< \Phi \mid U_B(\tau) \mid \psi_o >$ are the probability amplitudes for the Universe in the state $\mid \psi_o >$ to evolve over the time interval τ to $\mid \Phi >$ in two time paths corresponding to forward and, respectively, backward directions. Given that there is no reason to favor one path over the other, Vaccaro [619–621], following Feynman [625], attributes an equal statistical weighting to each of them. In this case, the total probability amplitude for the Universe to evolve from a given state to another is proportional to the sum of probability amplitudes for all possible time paths between the two states. There are only two possible paths in this case and the total amplitude is proportional to

$$< \Phi \mid U_F(\tau) \mid \psi_o > + < \Phi \mid U_B(\tau) \mid \psi_o >=< \Phi \mid [U_F(\tau) + U_B(\tau)] \mid \psi_o > . \tag{10.67}$$

This result is valid for all states $\mid \Phi >$ of the Universe. Therefore the evolution of $\mid \psi_o >$ via both paths, called the symmetric time evolution of the Universe, can be written as

$$| \Psi(\tau) >= [U_F(\tau) + U_B(\tau)] \mid \psi_o > . \tag{10.68}$$

The symmetric time evolution of the Universe in the state $\mid \Psi(\tau) >$ over an additional time interval τ is given by

$$| \Psi(2\tau) >= [U_F(\tau) + U_B(\tau)] \mid \Psi(\tau) >= [U_F(\tau) + U_B(\tau)]^2 \mid \psi_o > . \tag{10.69}$$

By repeating this operation for N such time intervals, one can obtain the following relation for the symmetric time evolution:

$$| \Psi(N\tau) >= [U_F(\tau) + U_B(\tau)]^N \mid \psi_o > . \tag{10.70}$$

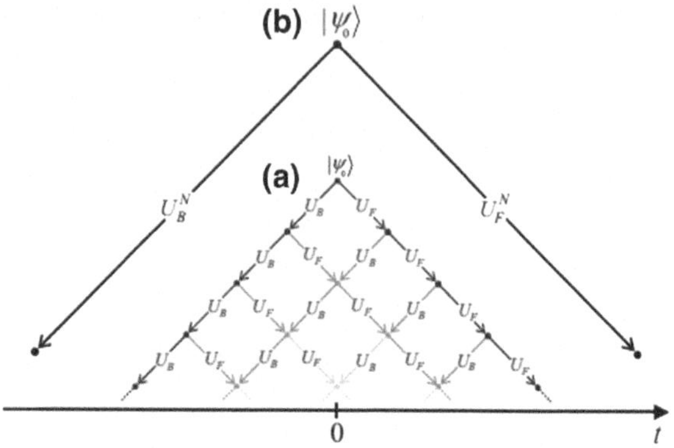

Fig. 10.2 Binary tree representation of the generation of the state $| \Psi(N\tau) >$ from the origin state $| \psi_o >$ according to (10.70) for the case $H_F \neq H_B$ and $N >> 1$: **a** shows an expanded view of the detail near the root node, and **b** shows the whole tree on a much coarser scale [621]

Equation (10.70) can be presented in the form [619]

$$| \Psi(N\tau) >= \sum_{n=0}^{N} S_{N-n,n} \mid \psi_o >, \quad (10.71)$$

where

$$S_{m,n} = U_B(m\tau)U_F(n\tau)\sum_{v=0}^{m} \cdots \sum_{l=0}^{s}\sum_{k=0}^{l} \exp[(v+\cdots+l+k)^2\tau^2[H_F, H_B]] + \quad (10.72)$$
$$+ \exp O(\tau^3).$$

$S_{m,n}$ is a sum containing $\binom{n+m}{n}$ different terms, each of them comprising n factors of $U_F(\tau)$ and m factors of $U_B(\tau)$, where

$$\binom{k}{j} = \frac{k!}{(k-j)!j!}$$

is the binomial coefficient and $[H_F, H_B]$ is the commutator of operators H_F and H_B.

In Fig. 10.2 a graphical interpretation of $| \Psi(n\tau) >$ from (10.70) is represented.

States on Fig. 10.2 are represented as nodes (solid discs) and the unitary evolution by links (arrows) between them. The root node (at the top) represents the origin state $| \psi_o >$ and the leaf nodes (on the bottom row) are components of the state $| \Psi(4\tau) >$.

The expression $< \Phi \mid S_{N-n,n} \mid \psi_o >$ represents the evolution of the Universe from $| \psi_o >$ to $| \Phi >$ over a set of $\binom{N}{n}$ paths through time, where each path comprises a total of n steps in the forward direction and $N-n$ steps in the backward one. The set of paths includes all possible ordering of forward and backward steps. The discussion of consequences of the limit $\tau \to 0$ of infinitely small time steps for a fixed total time suggests that the time interval τ should be a small non-zero number of the order of Planck time, i.e. $\tau \approx 5 \cdot 10^{-44}$ s for a Universe-like system [620, 621].

Returning to equation (10.72), we note that the commutator $[H_F, H_B]$ is an anti-Hermitian one and therefore it has imaginary eigenvalues. Its presence in summations in (10.72) gives rise to interference terms in the basis of eigenstates of $[H_F, H_B]$ of the form [620]

$$\sum_{v=o}^{m} \cdots \sum_{l=0}^{s} \sum_{k=0}^{l} \exp[-(v + \cdots + l + k)\tau^2 i\lambda], \tag{10.73}$$

where λ is an eigenvalue of $i[H_F, H_B]$. This leads to the interference between the set of paths represented by $< \Phi \mid S_{N-n,m} \mid \psi_0 >$.

The estimation of typical values for λ can be done using discussed in the Sect. 10.1 the neutral K meson evolution as a prototypical T-violating process. The phenomenological model of Lee and Wolfenstein [626] and empirical values of Yao et al. [627] give a spectrum of λ values, which have a zero mean value and a standard deviation of $\lambda_{SD} \approx \sqrt{f} \cdot 10^{57}\,\mathrm{s}^{-2}$. Here f is a fraction given by dividing the number of particles associated with T-violating processes by the total number of particles in the Universe (which is assumed to be 10^{80}) [620].

As it is shown in [619], the destructive interference leads to (10.71) being replaced with

$$\mid \Psi(N\tau) >= \left[\sum_{n\approx 0} S_{N-n,n} + \sum_{n \approx N} S_{N-n,n} \right] \mid \psi_0 > \tag{10.74}$$

for the total time $N\tau > 10^{-17}$ s. Ignoring terms of the order of τ gives

$$\mid \Psi(N\tau) >= \{[U_B(\tau)]^N + [U_F(\tau)]^N\} \mid \psi_0 >= [U_B(N\tau)] + U_F(N\tau)] \mid \psi_0 > . \tag{10.75}$$

By presenting the time as $t = N\tau$, we arrive at a key result, the bievolution equation of motion

$$\mid \Psi(t) >= [U_B(t) + U_F(t)] \mid \psi_0 >, \tag{10.76}$$

which is illustrated in Fig. 10.1b. Here the term bievolution refers to the dual evolution generated by two different Hamiltonians. The approximation made at deriving this equation is in the limits $t >> f^{-\frac{1}{2}} 10^{-13}$ s. The destructive interference has eliminated all paths excepting two ones corresponding to either continuous evolution in forward time direction or continuous evolution in backward one.

In [621], the bievolution equation of motion in differential form was derived as follows. By increasing t in (10.76) by a relatively small amount δt we have $\mid \Psi(t + \delta t) >= [U_B(\delta t)U_B(t) + U_F(\delta t)U_F(t)] \mid \psi_0 >$.

As a consequence, the rate of change of the state becomes

$$\frac{\delta \mid \Psi(t) >}{\delta t} = [iH_B U_B(t) - iH_F U_F(t)] \mid \psi_0 > +O(\delta t), \tag{10.77}$$

where $\delta \mid \Psi(t) >=\mid \Psi(t+\delta t) > - \mid \Psi(t) >$. Taking the limit $\delta t \to \tau$ and ignoring the term of the order of τ gives another key result, the Schrödinger equation for bievolution

$$\frac{d \mid \Psi(t) >}{dt} = \frac{d \mid \psi_F(t) >}{dt} - \frac{d \mid \psi_B(t) >}{dt}, \tag{10.78}$$

where $(\frac{d}{dt}) \mid \psi_\mu(t) >= -iH_\mu \mid \psi_\mu(t) >$ and $\mid \psi_\mu(t) >= U_\mu(t) \mid \psi_0 >$ for $\mu = F$ or B.

To the full impact of T-violation process, it is sufficient to compare the above analysis with a Universe without breaking the T-invariance. In this case, $H_F = H_B = H$ and it can be shown [621] that

$$S_{N-n,n} = \exp[i(N-2n)\tau H] \binom{N}{n}, \tag{10.79}$$

which is never zero and therefore all possible paths are included in (10.71). The direction of time is ambiguous as there are no physical reasons to prefer one of the two directions of the time flow corresponding to continuous evolution forwards or backwards direction over the other. The fact that this ambiguity is removed in a Universe with T-violation processes leads to the conclusion that these processes are responsible for the phenomenological unidirectionality of time observed in our Universe.

10.7 Virtual Time-Reversal Method and Its Application to EPR Spectroscopy

In this section we propose a simple experimental method for testing the time-reversal symmetry in systems containing paramagnetic ions or molecules. Before outlining this testing method, it is necessary to make the following remark about the symmetry transformations. To make sure that the space symmetry group of a certain crystal contains certain symmetry elements, it is not necessary to really interchange those crystal atoms that, as a result of these transformations, change to other, but equivalent, locations. For this it is sufficient that such transformations are possible in principle. Similarly, since the angular momentum projection operators reverse their sign under the action of the time-reversal operator, in order to verify the invariance relative to the operator **T** of the Hamiltonian of linear interaction of the spin with a constant magnetic field, it is not necessary to actually change the direction of the magnetic induction vector to the opposite. It is sufficient that such a transformation is possible in principle.

Continuing these analogies, we can conclude that in the case of time reversal symmetry transformation, there is no need to really change the direction of time flow to the opposite one in order to verify the presence of this symmetry transformation. Moreover, due to the law of increasing entropy in irreversible processes, this cannot be done in principle. Nevertheless, one can do what we call a virtual time reversal. The method of virtual time reversal in the application to *EPR* spectroscopy consists in detecting the *EPR* spectrum of paramagnetic gases, liquids and solids in two ways, with a subsequent comparison of the results.

The first way consists of recording the *EPR* spectrum at a slow increase in the induction of the magnetic field in a given interval (B_i, B_f) at a constant rate $(\frac{dB}{\mathrm{d}t} > 0)$. In the second way of detecting the *EPR* spectrum, the induction of the magnetic field B, on the contrary, decreases $(\frac{dB}{\mathrm{d}t} < 0)$ in the interval (B_f, B_i) with the same constant velocity. The essence of a proposed method consists in the rotation of the plane in which the *EPR* spectrum is recorded in the second way (together with the spectrum on it) by an angle of 180° around the c-axis located in the same plane and passing through the middle of the interval (B_f, B_i), perpendicular to the coordinate axis of magnetic fields, and superposition of the rotated spectrum on the original *EPR* spectrum recorded by the first way.

As a result of this rotation, the points B_f and B_i change places,[3] $(B_f, B_i) \longrightarrow (B_i, B_f)$, which seems to correspond to the detection of the rotated *EPR* spectrum by the first method $(\frac{dB}{\mathrm{d}t} > 0)$, but in reality this spectrum was detected by the second method $(\frac{dB}{\mathrm{d}t} < 0)$. There is no contradiction here (the spectrum detected in the conditions of a real experiment cannot change due to the fact that someone turned it to a certain angle) and everything is explained simply.

After the rotation of the spectrum detected by the second way $(\frac{dB}{\mathrm{d}t} < 0)$, by an angle of 180° about the c-axis, a minimum magnetic induction B_i corresponds to the beginning of the magnetic field sweep. In this case, the *EPR* spectrum would be detected at increasing magnetic induction $(dB > 0)$. However, since the *EPR* spectrum was detected experimentally under the opposite condition, $\frac{dB}{\mathrm{d}t} < 0$, to satisfy this condition it is necessary that the inequality $\mathrm{d}t < 0$ holds (for $dB > 0$ and $\mathrm{d}t < 0$, the inequality $\frac{dB}{\mathrm{d}t} < 0$ is satisfied). This is just the virtual inversion of time, just as under these conditions the inequality $dB > 0$ (at $\mathrm{d}t < 0$) means a virtual increase in the induction of the magnetic field.

It is important to note that as a result of the virtual time-reversal operation performed over the *EPR* spectrum, we have obtained such an *EPR* spectrum as it would be at a real time reversal, without changing the direction of the time axis. Since the spin Hamiltonian is invariant under the time-reversal operator $\mathbf{T}$, the spectral positions of the *EPR* lines corresponding to the conditions $\frac{dB}{\mathrm{d}t} > 0$ $(dB > 0, \mathrm{d}t > 0)$ and $\frac{dB}{\mathrm{d}t} < 0$ $(dB > 0, \mathrm{d}t < 0)$ must coincide. On the other hand, since the time-dependent perturbation operator causing quantum *EPR* transitions is also invariant

[3] The intervals (B_f, B_i) and (B_i, B_f) coincide up to the error of the experimental determination of the quantities B_i and B_f.

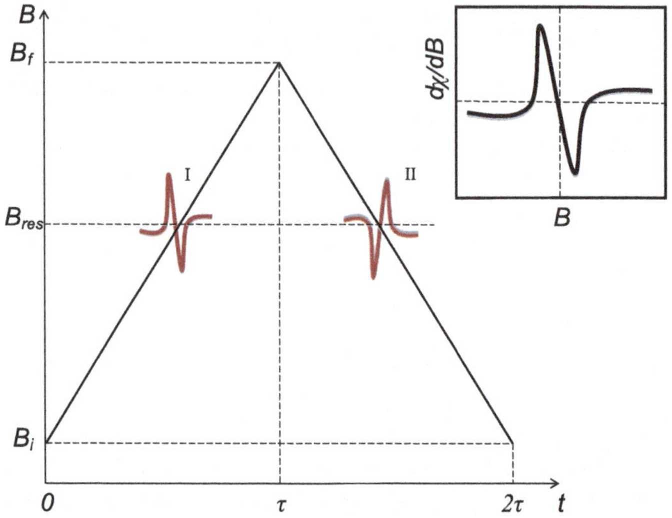

Fig. 10.3 Mirror reflection of the t axis related to the vertical axis passing through point $t = \tau$ (virtual time reversal). *EPR* lines I and II are symmetric related to those axis

with respect to the time-reversal transformation, the *EPR* lines must coincide not only in resonance frequency values, but also in form.

For a supplementary argumentation of the above described virtual time-reversal method, the time dependence of the induction B of magnetic field at detection of the *EPR* spectrum is presented on the Fig. 10.3. The cases of B increasing in the time interval $0 \le t \le \tau$ with subsequent decreasing in the interval $\tau \le t \le 2\tau$ are carried out at conditions of the coincidence of velocities of direct and inverse magnetic field scan. The detected *EPR* signals I and II that have opposite phases are marked by red in Fig. 10.3 (see also the insertion on Fig. 10.3). The mirror reflection of the right side of the isosceles triangle by its height, passing through the point $t = \tau$ orthogonally to the axis t, coincide with the left side. The *EPR* signal I and the mirror reflection of the *EPR* signal II also coincide in the values of resonance fields, as well as in shapes, if the time-reversal symmetry is not broken. It is important to note that the time axis t reverses the direction at mirror reflection. Certainly, changing the direction of the time axis due to the symmetry operation does not lead to a real-time reversal. It is a virtual time reversal.

In the case of breaking time-reversal symmetry, the position and shape of the *EPR* line I do not coincide with those of mirror image of *EPR* line II. As it will be shown below, the detected shift of the *EPR* line with respect to the mirror image of the *EPR* line II is much bigger than the shift due to the time-reversal symmetry violation and has another origin. As for the non-coincidence of shapes of corresponding *EPR* lines, it can be a reliable sign of the time-reversal symmetry violation, if magnetic field inhomogeneities in the resonance cavity of the *EPR* spectrometer are sufficiently small. A supplementary confidence that the non-coincidence of the shapes of discussed *EPR* lines indicates a time-reversal symmetry violation consists in the experimental confirmation of the time-reversal invariance for a standard sample using the same *EPR* spectrometer with the mentioned high-quality resonance cavity.

We demonstrate the application of the proposed method to *EPR* spectroscopy using two examples: the *EPR* spectrum of the reference sample of α, α-diphenyl-β-picrylhydrazyl ($DPPH$) containing free radicals and the *EPR* spectrum of 10^{-3} M aqueous solution of manganese chloride $MnCl_2$.

The $DPPH$, discovered almost 100 years ago (in 1922) [630] and being a traditional standard in *EPR* spectroscopy, is characterized by a single *EPR* line, whose width $\triangle H_{pp}$ lies in the interval (0.15–0.5) mT [631–634], depending on the synthesis technology of this compound. This reference standard sample is used in *EPR* spectroscopy since the discovery of the *EPR* method in 1944 [635, 636].

The experiments were carried out at room temperature on the *EPR* spectrometer SE/X-2544 with detecting the spectra on the computer monitor. The magnetic field induction was measured using the Bruker Teslameter $ER036TM$. In addition, a Hall effect sensor was used to independently measure the magnetic induction and to visualize its time dependence on the monitor at magnetic field sweep in the forward ($\frac{dB}{dt} > 0,\ dB > 0, \mathrm{d}t > 0$) and backward ($\frac{dB}{dt} < 0,\ dB < 0, \mathrm{d}t > 0$) directions.

Two *EPR* spectra of $DPPH$ are presented in Fig. 10.4. One of them (red one) was detected in the usual regime of an increasing magnetic field ($\frac{dB}{dt} > 0,\ dB > 0, \mathrm{d}t > 0$), whereas the second spectrum (blue one) was obtained at a virtual time reversal ($\frac{dB}{dt} < 0,\ dB > 0, \mathrm{d}t < 0$).

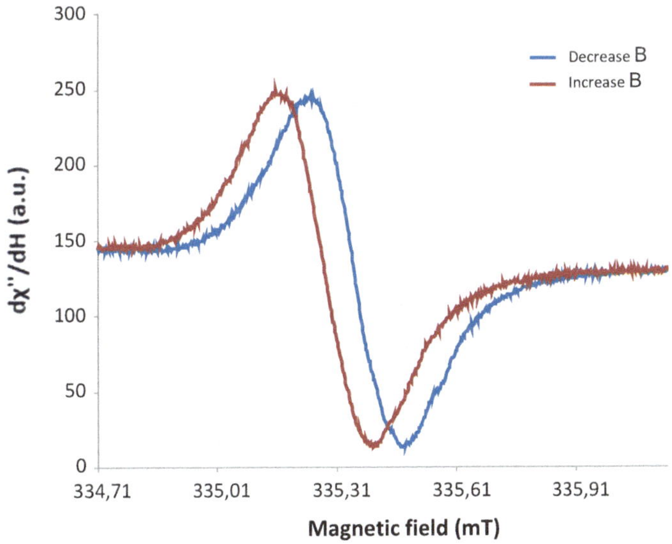

Fig. 10.4 *EPR* spectra of $DPPH$. Red curve : $g = 2.00202 \pm 0.00056$, $\triangle B_{pp} = (0.248 \pm 0.004)$ mT; $\frac{dB}{dt} > 0,\ dB > 0,\ \mathrm{d}t > 0$. Blue curve : $g^T = 2.000660 \pm 0.00060$, $\triangle B^T_{pp} = (0.256 \pm 0.004)$ mT; $\frac{dB}{dt} < 0,\ dB > 0,\ \mathrm{d}t < 0$

In the caption to Fig. 10.4, g^T and ΔB^T_{pp} denote, respectively, the value of the g-factor and the width of the *EPR* line, which was obtained using the method of virtual time reversal.

The *EPR* line corresponding to the virtual time reversal exhibits a shift by $\Delta B_{DPPH} = (0.080 \pm 0.004)\,\text{mT}$ in the direction of large magnetic fields ("blue" shift). It should be noted that the shift ΔB_{DPPH} is much greater than the error in measuring the magnitude of the magnetic induction $|\Delta B| = 0.002\,\text{mT}$.

The shift of the *EPR* spectra (see Fig. 10.4) relative to each other along the magnetic field axis until they coincide shows their good coincidence (Fig. 10.5), as it should be in the presence of time-reversal symmetry. Thus, in the case of *DPPH*, of the two above-mentioned experimental signs of the time-reversal symmetry presence (the coincidence of the resonant frequencies and the coincidence of the *EPR* line shapes in the forward and backward (virtual) time flow), only one is satisfied—the coincidence of the resonance line shapes. However, this does not yet mean that the time-reversal symmetry breaking, since there are other reasons that lead to a discrepancy between the resonance frequencies of the EPR lines, recorded by the conventional method and by the virtual time-reversal method (see above).

The *EPR* spectra of $MnCl_2$ detected at increasing the magnetic field (red curve, $\frac{dB}{dt} > 0, dB > 0, dt > 0$) and using the virtual time-reversal method (blue curve, $\frac{dB}{dt} < 0, dB > 0, dt < 0$) are shown on the Fig. 10.6.

The g-factors g and g^T, whose values are given in the caption to Fig. 10.6, were determined by measuring the magnetic induction corresponding to the center of

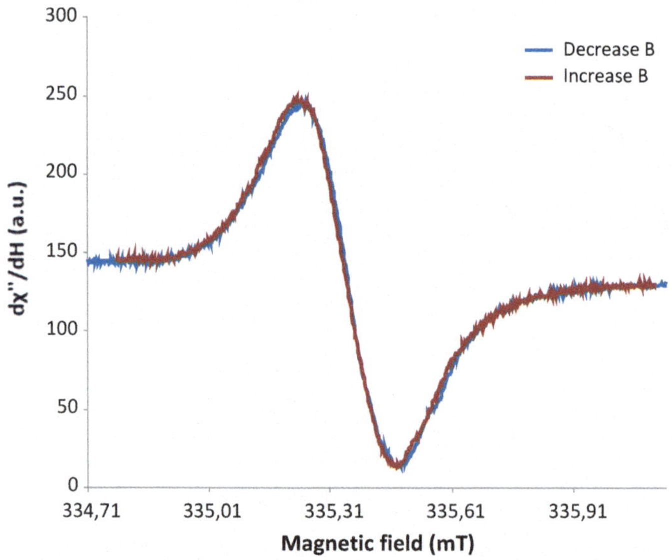

Fig. 10.5 The coincidence of the shapes of *EPR* lines (*DPPH*) at the usual detection of the spectrum (red curve, $\frac{dB}{dt} > 0, dB > 0, dt > 0$) and using the virtual time-reversal method (blue curve, $\frac{dB}{dt} < 0, dB > 0, dt < 0$)

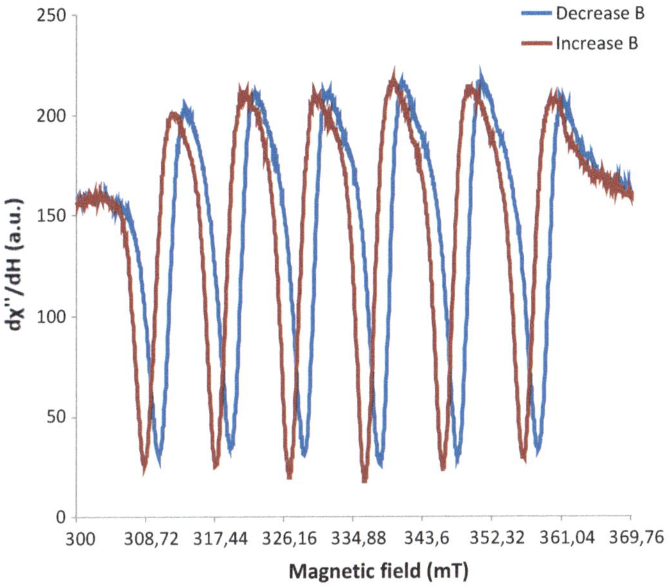

Fig. 10.6 The hyperfine structure of the *EPR* line of $10^{-3}M$ aqueous solution of $MnCl_2$ at usual detecting the spectrum (red curve, $\frac{dB}{dt} > 0,\ dB > 0,\ dt > 0$) and using the virtual time-reversal method (blue curve, $\frac{dB}{dt} < 0,\ dB > 0,\ dt < 0$). Red curve : $a = (9.55 \pm 0.21)\,\text{mT},\ g = 2.0072 \pm 0.0034,\ \triangle B_{pp} = (47.74 \pm 0.38)\,\text{mT}$, blue curve : $a^T = (9.53 \pm 0.21)\,\text{mT},\ g^T = 1.9976 \pm 0.0034,\ \triangle B^T_{pp} = (47.67 \pm 0.16)\,\text{mT}$

gravity of the group of six components of the *EPR* spectrum, characteristic of the contact hyperfine Fermi interaction between the $3d^5$-electrons and the ^{55}Mn nucleus. As seen from Fig. 10.6, both for $MnCl_2$ and *DPPH*, a "blue" shift of the *EPR* spectrum obtained by the virtual time-reversal method is observed with respect to the spectrum detected at increasing magnetic field induction ($\frac{dB}{dt} > 0,\ dB > 0,\ dt > 0$). As in the case of *DPPH*, the magnitude of this shift, $\triangle B_{MnCl_2} = (1.703 \pm 0.223)\,\text{mT}$, is much greater than the magnetic induction measurement error $|\triangle B| = 0.002\,\text{mT}$.

By shifting the *EPR* spectra in Fig. 10.6 relative to each other along the axis of the magnetic fields until they coincide, we can see that, within the experimental accuracy, there is a good agreement between the shapes of all the spectral lines of the hyperfine sextet (Fig. 10.7), as it should be in the presence of time-reversal symmetry.

Let us now proceed to discuss the observed effects. The *EPR* data show that the spin densities in two hydrazyl nitrogen atoms in *DPPH* are large and equally distributed between them. Apparently, partly for this reason, the single *EPR* line of $DPPH^*$ free radical is inhomogeneously broadened, as a result of which the fine structure of this line does not appear. On the other hand, the *DPPH* is a polar molecule having the electric dipole moment $d = 4.92D$. It is known that the parity and the time-reversal violating interactions between electrons and nucleons give rise to atomic and molecular permanent electric dipole moments (*EDMs*) [503], and that

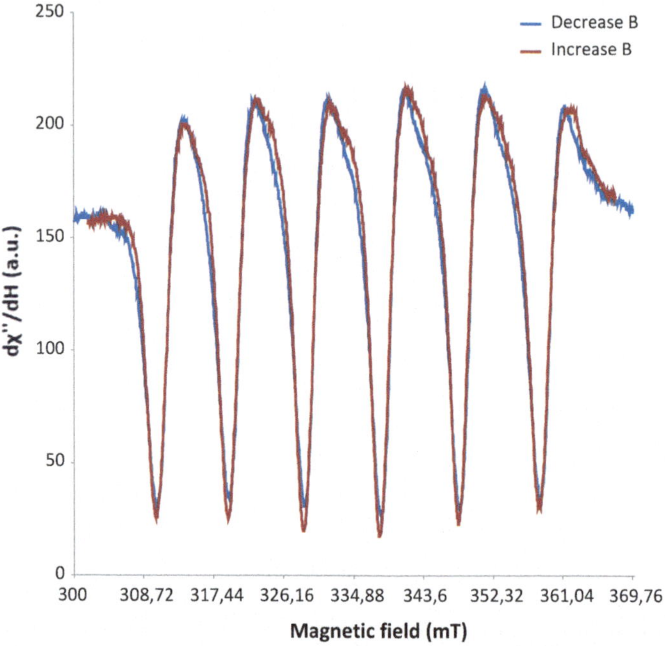

Fig. 10.7 The coincidence of shapes of the lines of the hyperfine structure of the *EPR* spectrum of $10^{-3}M$ aqueous solution of $MnCl_2$ at usual detection of the spectrum (red curve, $\frac{dB}{dt} > 0$, $dB > 0$, $dt > 0$) and using the virtual time-reversal method (blue curve, $\frac{dB}{dt} < 0$, $dB > 0$, $dt < 0$)

the P, T-parity non-conservation effects are strongly enhanced in polar molecules with heavy nuclei [509].

Since the nuclei of N (in the *DPPH* molecule) and ^{55}Mn (in the $\mathrm{MnCl_2}$ molecule) are not heavy (for which $Z \geq 50$), the relativistic effects important for heavy nuclei will not make a significant contribution to the enhancement factor in P, T-parity non-conservation effects. However, another contribution to the enhancement factor, connected with the molecular polarizability, can be significant. The role of paramagnetic atoms in the enhancement factor can also be significant [502, 510]. In the case of paramagnetic atoms (ions), the enhancement factor for an electron *EDM*, $\eta = d_{\text{atom}}/d_e$, is proportional to α_D:

$$\eta = \alpha^2 Z^3 \alpha_D, \tag{10.80}$$

where $\alpha \approx \frac{1}{137}$ is the fine structure constant, Z is the nuclear charge, and α_D is the atomic polarizability. Knowing α_D for N and Mn atoms, we can determine η and, respectively, the atomic *EDMs* of these atoms due to P, T-parity non-conservation effects. However, the observed shifts in the *EPR* lines cannot be explained in this way for the following reason.

Comparison of the frequency (energy) shift due to T, P-odd effect in TlF, $\Delta\nu = -0.13\pm0.22\,\text{mHz}$ [502, 521] ($h\Delta\nu = (-0.54\pm0.91)\cdot10^{-18}\,\text{eV}$), with $\Delta E_{DPPH} = (0.927\pm0.046)\cdot10^{-8}\,\text{eV}$ and $\Delta E_{MnCl_2} = (1.974\pm0.258)\cdot10^{-7}\,\text{eV}$ shows that the observed *EPR* lines shifts are much higher than the shift due to T, P-odd effects and have a different origin.

We note that the resonant values of the magnetic induction B greatly exceed the upper value of B for the hysteresis loop of the *EPR* spectrometer electromagnet. On the other hand, the magnetic field sweep rates at $\frac{dB}{dt} > 0$, $(dB > 0,\ dt > 0)$ and $\frac{dB}{dt} < 0$, $(dB > 0,\ dt < 0)$, which although are close in magnitude, but do not coincide, because the accuracy of the determination B_i and B_f is $\pm0.002\,\text{mT}$. This can lead to an observed shift of the EPR lines.

In the book "*Symmetries and Reflections*" of E. Wigner [437], there is the following remark: "If even the problem is not invariant under the transformation $t' = -t$, nevertheless the transformation containing the time inversion exists. For example, a homogeneous magnetic field has a symmetry element: a reflection in a plane perpendicular to the magnetic field, with a simultaneous change in the sign of t" (E. Wigner, *Etyudy o simmetrii* (Publishing House Mir, Moscow) 1971, p. 275).

In our experiments, the inhomogeneity of the magnetic field is sufficiently small, which allows us to observe the hyperfine structure of the *EPR* spectra. Coincidence of EPR lines shapes detected by standard method and by virtual time-reversal method indicates that the homogeneity of the magnetic field of the electromagnetic wave in the resonance cavity (at the location of the test sample) is sufficiently high in order to no violation of the time-reversal symmetry occurs due to the inhomogeneity of the magnetic field. As for a homogeneous constant magnetic field, in [437] it is not discussed the degree of inhomogeneity (microinhomogeneity) of this field which is allowed in order to a reflection in a plane perpendicular to the magnetic field (with a simultaneous charge in the sign of t) would be the time-reversal symmetry transformation. This is to be determined experimentally.

Note that the virtual time reversal method can be used not only in the study of *EPR* spectra, but also electron-nuclear double resonance (ENDOR) spectra, nuclear magnetic resonance (NMR) spectra and others, if only the equipment used allows detecting spectra in two ways, described in this section.

Appendix A

(See Table A.1)

Table A.1 Matrix of unitary operator U defined by means of basis function operators of the ordinary irreducible representations of the symmetry point groups

Symmetry group	Irreducible representations		Basises [9, 78]	U
	Bethe symbols	Mulliken symbols		
1	2	3	4	5
$\bar{1}$	Γ_1^+	A_g	J_x; J_y; J_z	-1; $+1$; -1
2 and m	Γ_1	AA'	J_z	-1
	Γ_2	BA''	J_x; J_y	-1; $+1$;
222 and $2mm$	Γ_1	AA	J, (222)	$+1$
	Γ_2	B_2B_1	J_y	$+1$
	Γ_3	B_1A_2	J_z	-1
	Γ_4	B_3B_2	J_x	-1
$2/m$	Γ_1^+	A_g	J_z	-1
	Γ_2^+	B_g	J_x; J_y	-1; $+1$
4 and $\bar{4}$	Γ_1	AA	J_z	-1
	Γ_2	BB	$[J_xJ_y]$, (4)	-1
			$J_x^2 - J_y^2$, (4)	$+1$
	Γ_3	$\}E\ E\{$	$\{-J_+,\ J_-\}$	$\begin{bmatrix} 0 & 1 \\ 1 & 0 \end{bmatrix}$
	Γ_4			
422, 4 mm, $\bar{4}$ 2 m	Γ_1	AA	J	$+1$
	Γ_2	A_2A_2	J_z	-1
	Γ_3	B_1B_1	$J_x^2 - J_y^2$, (422)	$+1$
	Γ_4	B_2B_2	$[J_xJ_y]$, (422)	-1
	Γ_5	EE	$\{J_x, J_y\}$	$\begin{bmatrix} -1 & 0 \\ 0 & 1 \end{bmatrix}$

(continued)

I. I. Geru, *Time-Reversal Symmetry*, Springer Tracts in Modern Physics 281,
https://doi.org/10.1007/978-3-030-01210-6

Table A.1 (continued)

Symmetry group	Irreducible representations		Basises [9, 78]	U
	Bethe symbols	Mulliken symbols		
3	Γ_1	A	$J;\ J_z$	$1;\ -1$
	Γ_2	$\}E$	$\{-J_+,\ J_-\}$	$\begin{bmatrix}0&1\\1&0\end{bmatrix}$
	Γ_3			
32 and 3 m	Γ_1	$A_1 A_1$	J	$+1$
	Γ_2	$A_2 A_2$	J_z	-1
	Γ_3	$E E$	$\{J_-,\ -J_+\}$	$\begin{bmatrix}0&1\\1&0\end{bmatrix}$
1	2	3	4	5
6 and $\bar{6}$	Γ_1	$A A'$	J_z, (6); J, (6, $\bar{6}$)	$-1;\ +1$
	Γ_2	$B B'$	–	$+1$
	Γ_3	$\}E_1\ E_1'\{$	$\{-zJ_+,\ zJ_-\}$, ($\bar{6}$)	$\begin{bmatrix}0&1\\1&0\end{bmatrix}$
	Γ_4			
	Γ_5	$\}E_2\ E_2'\{$	$\{-J_+,\ J_-\}$	$\begin{bmatrix}0&1\\1&0\end{bmatrix}$
	Γ_6			
622, 6mm, $\bar{6}m2$	Γ_1	$A_1 A_1 A_1'$	J	1
	Γ_2	$A_2 A_2 A_2'$	J_z	-1
	Γ_3	$B_1 B_2 A_1''$	$i(J_+^3 - J_-^3)$, (622)	$+1$
	Γ_4	$B_2 B_1 A_2''$	$(J_+^3 + J_-^3)$, (622)	-1
	Γ_5	$E_1 E_1 E''$	$\{J_-,\ -J_+\}$	$\begin{bmatrix}0&1\\1&0\end{bmatrix}$
	Γ_6	$E_2 E_2 E'$	$\{J_+^2,\ J_-^2\}$, (622)	$\begin{bmatrix}0&1\\1&0\end{bmatrix}$
			$\{\ iJ_-(J_+^3 - J_-^3),$ $-iJ_+(J_+^3 - J_-^3)\ \}$, (622)	$\begin{bmatrix}0&1\\1&0\end{bmatrix}$
23	Γ_1	A	$J;\ [J_x J_y J_z]$	$+1;\ +1$
	Γ_2	$\}E$	$\frac{1}{\sqrt{2}}(u - v)$	$\begin{bmatrix}1&0\\0&1\end{bmatrix}$
	Γ_3		$\frac{1}{\sqrt{2}}(u + v)$	
	Γ_4	T	$\{J_x,\ J_y,\ J_z\}$	w

(continued)

Table A.1 (continued)

Symmetry group	Irreducible representations		Basises [9, 78]	U
	Bethe symbols	Mulliken symbols		
432 and $\bar{4}3m$	Γ_1	A_1A_1	J	1
	Γ_2	A_2A_2	$[J_xJ_yJ_z]$, (432)	1
	Γ_3	EE	$\{u, v\}$, (432)	$\begin{bmatrix} 1 & 0 \\ 0 & 1 \end{bmatrix}$
	Γ_4	T_1T_1	$\{J_x, J_y, J_z\}$, (432)	w
			$\{J_x^3, J_y^3, J_z^3\}$, (432)	w
			$\{J_x^5, J_y^5, J_z^5\}$, (432)	w
			$\{[J_yJ_z], [J_zJ_x], [J_xJ_y]\}$, ($\bar{4}3m$)	w
			$\{V_x, V_y, V_z\}$, ($\bar{4}3m$)	w
	Γ_5	T_2T_2	$\{J_x, J_y, J_z\}$, ($\bar{4}3m$)	w
			$\{J_x^3, J_y^3, J_z^3\}$, ($\bar{4}3m$)	w
			$\{J_x^5, J_y^5, J_z^5\}$, ($\bar{4}3m$)	w
			$\{[J_yJ_z], [J_zJ_x], [J_xJ_y]\}$, (432)	w
			$\{V_x, V_y, V_z\}$, (432)	w

The symbol J denotes an invariant with respect to proper and improper rotations, $J_\pm = J_x \pm iJ_y$, $u = 3J_z^2 - J(J+1)$, $v = \sqrt{3}(J_x^2 - J_y^2)$, $[J_\alpha J_\beta] = \frac{1}{2}(J_\alpha J_\beta + J_\beta J_\alpha)$, were $\alpha, \beta = y, z;\ z, x$ and x, y; $[J_xJ_yJ_z]=\frac{1}{6}(J_xJ_yJ_z+J_yJ_zJ_x+ J_zJ_xJ_y+ J_xJ_zJ_y+J_zJ_yJ_x+J_yJ_xJ_z)$, $V_x = [J_x(J_y^2 - J_z^2)]$, $V_y = [J_y(J_z^2 - J_x^2)]$, $V_z = [J_z(J_x^2 - J_y^2)]$.

$$w = \begin{bmatrix} -1 & 0 & 0 \\ 0 & 1 & 0 \\ 0 & 0 & -1 \end{bmatrix}.$$

Appendix B

(See Tables B.1, B.2, B.3 and B.4)

Table B.1 The Kayley table of the non-Abelian group $G_{16}^{(\frac{1}{2})}$ in the terms of complete and incomplete time-reversal operators. Block O_{11}

	$\mathbf{T}$	$-e$	$-\mathbf{T}$	e	$\mathbf{T}_{2(yz)}$	$-\mathbf{T}_{2(yz)}$	$\mathbf{T}_{1(x)}$	$-\mathbf{T}_{1(x)}$
$\mathbf{T}$	$-e$	$-\mathbf{T}$	e	$\mathbf{T}$	$\mathbf{T}_{1(x)}$	$-\mathbf{T}_{1(x)}$	$-\mathbf{T}_{2(yz)}$	$\mathbf{T}_{2(yz)}$
$-e$	$-\mathbf{T}$	e	$\mathbf{T}$	$-e$	$-\mathbf{T}_{2(yz)}$	$\mathbf{T}_{2(yz)}$	$-\mathbf{T}_{1(x)}$	$\mathbf{T}_{1(x)}$
$-\mathbf{T}$	e	$\mathbf{T}$	$-e$	$-\mathbf{T}$	$-\mathbf{T}_{1(x)}$	$\mathbf{T}_{1(x)}$	$\mathbf{T}_{2(yz)}$	$-\mathbf{T}_{2(yz)}$
e	$\mathbf{T}$	$-e$	$-\mathbf{T}$	e	$\mathbf{T}_{2(yz)}$	$-\mathbf{T}_{2(yz)}$	$\mathbf{T}_{1(x)}$	$-\mathbf{T}_{1(x)}$
$\mathbf{T}_{2(yz)}$	$-\mathbf{T}_{1(x)}$	$-\mathbf{T}_{2(yz)}$	$\mathbf{T}_{1(x)}$	$\mathbf{T}_{2(yz)}$	e	$-e$	$-\mathbf{T}$	$\mathbf{T}$
$-\mathbf{T}_{2(yz)}$	$\mathbf{T}_{1(x)}$	$\mathbf{T}_{2(yz)}$	$-\mathbf{T}_{1(x)}$	$-\mathbf{T}_{2(yz)}$	$-e$	e	$\mathbf{T}$	$-\mathbf{T}$
$\mathbf{T}_{1(x)}$	$\mathbf{T}_{2(yz)}$	$-\mathbf{T}_{1(x)}$	$-\mathbf{T}_{2(yz)}$	$\mathbf{T}_{1(x)}$	$\mathbf{T}$	$-\mathbf{T}$	e	$-e$
$-\mathbf{T}_{1(x)}$	$-\mathbf{T}_{2(yz)}$	$\mathbf{T}_{1(x)}$	$\mathbf{T}_{2(yz)}$	$-\mathbf{T}_{1(x)}$	$-\mathbf{T}$	$\mathbf{T}$	$-e$	e

Table B.2 The Kayley table of the non-Abelian group $G_{16}^{(\frac{1}{2})}$ in the terms of complete and incomplete time-reversal operators. Block O_{12}

	$\mathbf{T}_{2(zx)}$	$-\mathbf{T}_{1(y)}$	$-\mathbf{T}_{2(zx)}$	$\mathbf{T}_{1(y)}$	$\mathbf{T}_{1(z)}$	$-\mathbf{T}_{1(z)}$	$\mathbf{T}_{2(xy)}$	$-\mathbf{T}_{2(xy)}$
$\mathbf{T}$	$-\mathbf{T}_{1(y)}$	$-\mathbf{T}_{2(zx)}$	$\mathbf{T}_{1(y)}$	$\mathbf{T}_{2(zx)}$	$\mathbf{T}_{2(xy)}$	$-\mathbf{T}_{2(xy)}$	$-\mathbf{T}_{1(z)}$	$\mathbf{T}_{1(z)}$
$-e$	$-\mathbf{T}_{2(zx)}$	$\mathbf{T}_{1(y)}$	$\mathbf{T}_{2(zx)}$	$-\mathbf{T}_{1(y)}$	$-\mathbf{T}_{1(z)}$	$\mathbf{T}_{1(z)}$	$-\mathbf{T}_{2(xy)}$	$\mathbf{T}_{2(xy)}$
$-\mathbf{T}$	$\mathbf{T}_{1(y)}$	$\mathbf{T}_{2(zx)}$	$-\mathbf{T}_{1(y)}$	$-\mathbf{T}_{2(zx)}$	$-\mathbf{T}_{2(xy)}$	$\mathbf{T}_{2(xy)}$	$\mathbf{T}_{1(z)}$	$-\mathbf{T}_{1(z)}$
e	$\mathbf{T}_{2(zx)}$	$-\mathbf{T}_{1(y)}$	$-\mathbf{T}_{2(zx)}$	$\mathbf{T}_{1(y)}$	$\mathbf{T}_{1(z)}$	$-\mathbf{T}_{1(z)}$	$\mathbf{T}_{2(xy)}$	$-\mathbf{T}_{2(xy)}$
$\mathbf{T}_{2(yz)}$	$-\mathbf{T}_{2(xy)}$	$-\mathbf{T}_{1(z)}$	$\mathbf{T}_{2(xy)}$	$\mathbf{T}_{1(z)}$	$\mathbf{T}_{1(y)}$	$-\mathbf{T}_{1(y)}$	$-\mathbf{T}_{2(zx)}$	$\mathbf{T}_{2(zx)}$
$-\mathbf{T}_{2(yz)}$	$\mathbf{T}_{2(xy)}$	$\mathbf{T}_{1(z)}$	$-\mathbf{T}_{2(xy)}$	$-\mathbf{T}_{1(z)}$	$-\mathbf{T}_{1(y)}$	$\mathbf{T}_{1(y)}$	$\mathbf{T}_{2(zx)}$	$-\mathbf{T}_{2(zx)}$
$\mathbf{T}_{1(x)}$	$\mathbf{T}_{1(z)}$	$-\mathbf{T}_{2(xy)}$	$-\mathbf{T}_{1(z)}$	$\mathbf{T}_{2(xy)}$	$\mathbf{T}_{2(zx)}$	$-\mathbf{T}_{2(zx)}$	$\mathbf{T}_{1(y)}$	$-\mathbf{T}_{1(y)}$
$-\mathbf{T}_{1(x)}$	$-\mathbf{T}_{1(z)}$	$\mathbf{T}_{2(xy)}$	$\mathbf{T}_{1(z)}$	$-\mathbf{T}_{2(xy)}$	$-\mathbf{T}_{2(zx)}$	$\mathbf{T}_{2(zx)}$	$-\mathbf{T}_{1(y)}$	$\mathbf{T}_{1(y)}$

I. I. Geru, *Time-Reversal Symmetry*, Springer Tracts in Modern Physics 281,
https://doi.org/10.1007/978-3-030-01210-6

Table B.3 The Kayley table of the non-Abelian group $G_{16}^{(\frac{1}{2})}$ in the terms of complete and incomplete time-reversal operators. Block O_{21}

	$\mathbf{T}$	$-e$	$-\mathbf{T}$	e	$\mathbf{T}_{2(yz)}$	$-\mathbf{T}_{2(yz)}$	$\mathbf{T}_{1(x)}$	$-\mathbf{T}_{1(x)}$
$\mathbf{T}_{2(zx)}$	$-\mathbf{T}_{1(y)}$	$-\mathbf{T}_{2(zx)}$	$\mathbf{T}_{1(y)}$	$\mathbf{T}_{2(zx)}$	$\mathbf{T}_{2(xy)}$	$-\mathbf{T}_{2(xy)}$	$-\mathbf{T}_{1(z)}$	$\mathbf{T}_{1(z)}$
$-\mathbf{T}_{1(y)}$	$-\mathbf{T}_{2(zx)}$	$\mathbf{T}_{1(y)}$	$\mathbf{T}_{2(zx)}$	$-\mathbf{T}_{1(y)}$	$-\mathbf{T}_{1(z)}$	$\mathbf{T}_{1(z)}$	$-\mathbf{T}_{2(xy)}$	$\mathbf{T}_{2(xy)}$
$-\mathbf{T}_{2(zx)}$	$\mathbf{T}_{1(y)}$	$\mathbf{T}_{2(zx)}$	$-\mathbf{T}_{1(y)}$	$-\mathbf{T}_{2(zx)}$	$-\mathbf{T}_{2(xy)}$	$\mathbf{T}_{2(xy)}$	$\mathbf{T}_{1(z)}$	$-\mathbf{T}_{1(z)}$
$\mathbf{T}_{1(y)}$	$\mathbf{T}_{2(zx)}$	$-\mathbf{T}_{1(y)}$	$-\mathbf{T}_{2(zx)}$	$\mathbf{T}_{1(y)}$	$\mathbf{T}_{1(z)}$	$-\mathbf{T}_{1(z)}$	$\mathbf{T}_{2(xy)}$	$-\mathbf{T}_{2(xy)}$
$\mathbf{T}_{1(z)}$	$-\mathbf{T}_{2(xy)}$	$-\mathbf{T}_{1(z)}$	$\mathbf{T}_{2(xy)}$	$\mathbf{T}_{1(z)}$	$\mathbf{T}_{1(y)}$	$-\mathbf{T}_{1(y)}$	$-\mathbf{T}_{2(zx)}$	$\mathbf{T}_{2(zx)}$
$-\mathbf{T}_{1(z)}$	$\mathbf{T}_{2(xy)}$	$\mathbf{T}_{1(z)}$	$-\mathbf{T}_{2(xy)}$	$-\mathbf{T}_{1(z)}$	$-\mathbf{T}_{1(y)}$	$\mathbf{T}_{1(y)}$	$\mathbf{T}_{2(zx)}$	$-\mathbf{T}_{2(zx)}$
$\mathbf{T}_{2(xy)}$	$\mathbf{T}_{1(z)}$	$-\mathbf{T}_{2(xy)}$	$-\mathbf{T}_{1(z)}$	$\mathbf{T}_{2(xy)}$	$\mathbf{T}_{2(zx)}$	$-\mathbf{T}_{2(zx)}$	$\mathbf{T}_{1(y)}$	$-\mathbf{T}_{1(y)}$
$-\mathbf{T}_{2(xy)}$	$-\mathbf{T}_{1(z)}$	$\mathbf{T}_{2(xy)}$	$\mathbf{T}_{1(z)}$	$-\mathbf{T}_{2(xy)}$	$-\mathbf{T}_{2(zx)}$	$\mathbf{T}_{2(zx)}$	$-\mathbf{T}_{1(y)}$	$\mathbf{T}_{1(y)}$

Table B.4 The Kayley table of the non-Abelian group $G_{16}^{(\frac{1}{2})}$ in the terms of complete and incomplete time-reversal operators. Block O_{22}

	$\mathbf{T}_{2(zx)}$	$-\mathbf{T}_{1(y)}$	$-\mathbf{T}_{2(zx)}$	$\mathbf{T}_{1(y)}$	$\mathbf{T}_{1(z)}$	$-\mathbf{T}_{1(z)}$	$\mathbf{T}_{2(xy)}$	$-\mathbf{T}_{2(xy)}$
$\mathbf{T}_{2(zx)}$	$-e$	$-\mathbf{T}$	e	$\mathbf{T}$	$\mathbf{T}_{1(x)}$	$-\mathbf{T}_{1(x)}$	$-\mathbf{T}_{2(yz)}$	$\mathbf{T}_{2(yz)}$
$-\mathbf{T}_{1(y)}$	$-\mathbf{T}$	e	$\mathbf{T}$	$-e$	$-\mathbf{T}_{2(yz)}$	$\mathbf{T}_{2(yz)}$	$-\mathbf{T}_{1(x)}$	$\mathbf{T}_{1(x)}$
$-\mathbf{T}_{2(zx)}$	e	$\mathbf{T}$	$-e$	$-\mathbf{T}$	$-\mathbf{T}_{1(x)}$	$\mathbf{T}_{1(x)}$	$\mathbf{T}_{2(yz)}$	$-\mathbf{T}_{2(yz)}$
$\mathbf{T}_{1(y)}$	$\mathbf{T}$	$-e$	$-\mathbf{T}$	e	$\mathbf{T}_{2(yz)}$	$-\mathbf{T}_{2(yz)}$	$\mathbf{T}_{1(x)}$	$-\mathbf{T}_{1(x)}$
$\mathbf{T}_{1(z)}$	$-\mathbf{T}_{1(x)}$	$-\mathbf{T}_{2(yz)}$	$\mathbf{T}_{1(x)}$	$\mathbf{T}_{2(yz)}$	e	$-e$	$-\mathbf{T}$	$\mathbf{T}$
$-\mathbf{T}_{1(z)}$	$\mathbf{T}_{1(x)}$	$\mathbf{T}_{2(yz)}$	$-\mathbf{T}_{1(x)}$	$-\mathbf{T}_{2(yz)}$	$-e$	e	$\mathbf{T}$	$-\mathbf{T}$
$\mathbf{T}_{2(xy)}$	$\mathbf{T}_{2(yz)}$	$-\mathbf{T}_{1(x)}$	$-\mathbf{T}_{2(yz)}$	$\mathbf{T}_{1(x)}$	$\mathbf{T}$	$-\mathbf{T}$	e	$-e$
$-\mathbf{T}_{2(xy)}$	$-\mathbf{T}_{2(yz)}$	$\mathbf{T}_{1(x)}$	$\mathbf{T}_{2(yz)}$	$-\mathbf{T}_{1(x)}$	$-\mathbf{T}$	$\mathbf{T}$	$-e$	e

Appendix C

(See Tables C.1, C.2, C.3 and C.4)

Table C.1 The Kayley table of non-Abelian group $G_{16}^{(J)}$ at arbitrary half-integer J. The matrices $U^{(J)}$, $V^{(J)}$ and $W^{(J)}$ are defined by (8.39). Block O_{11}

	$\boldsymbol{T}$	$-e^{(J)}$	$-\boldsymbol{T}$	$e^{(J)}$	$V^{(J)}$	$-V^{(J)}$	$W^{(J)}\mathbf{K}$	$-W^{(J)}\mathbf{K}$
$\boldsymbol{T}$	$-e^{(J)}$	$-\boldsymbol{T}$	$e^{(J)}$	$\boldsymbol{T}$	$W^{(J)}\mathbf{K}$	$-W^{(J)}\mathbf{K}$	$-V^{(J)}$	$V^{(J)}$
$-e^{(J)}$	$-\boldsymbol{T}$	$e^{(J)}$	$\boldsymbol{T}$	$-e^{(J)}$	$-V^{(J)}$	$V^{(J)}$	$-W^{(J)}\mathbf{K}$	$W^{(J)}\mathbf{K}$
$-\boldsymbol{T}$	$e^{(J)}$	$\boldsymbol{T}$	$-e^{(J)}$	$-\boldsymbol{T}$	$-W^{(J)}\mathbf{K}$	$W^{(J)}\mathbf{K}$	$V^{(J)}$	$-V^{(J)}$
$e^{(J)}$	$\boldsymbol{T}$	$-e^{(J)}$	$-\boldsymbol{T}$	$e^{(J)}$	$V^{(J)}$	$-V^{(J)}$	$W^{(J)}\mathbf{K}$	$-W^{(J)}\mathbf{K}$
$V^{(J)}$	$-W^{(J)}\mathbf{K}$	$-V^{(J)}$	$W^{(J)}\mathbf{K}$	$V^{(J)}$	$e^{(J)}$	$-e^{(J)}$	$-\boldsymbol{T}$	$\boldsymbol{T}$
$-V^{(J)}$	$W^{(J)}\mathbf{K}$	$V^{(J)}$	$-W^{(J)}\mathbf{K}$	$-V^{(J)}$	$-e^{(J)}$	$e^{(J)}$	$\boldsymbol{T}$	$-\boldsymbol{T}$
$W^{(J)}\mathbf{K}$	$V^{(J)}$	$-W^{(J)}\mathbf{K}$	$-V^{(J)}$	$W^{(J)}\mathbf{K}$	$\boldsymbol{T}$	$-\boldsymbol{T}$	$e^{(J)}$	$-e^{(J)}$
$-W^{(J)}\mathbf{K}$	$-V^{(J)}$	$W^{(J)}\mathbf{K}$	$V^{(J)}$	$-W^{(J)}\mathbf{K}$	$-\boldsymbol{T}$	$\boldsymbol{T}$	$-e^{(J)}$	$e^{(J)}$

Table C.2 The Kayley table of non-Abelian group $G_{16}^{(J)}$ at arbitrary half-integer J. The matrices $U^{(J)}$, $V^{(J)}$ and $W^{(J)}$ are defined by (8.39). Block O_{12}

	$U^{(J)}$	$-e^{(J)}\mathbf{K}$	$-U^{(J)}$	$e^{(J)}\mathbf{K}$	$V^{(J)}\mathbf{K}$	$-V^{(J)}\mathbf{K}$	$W^{(J)}$	$-W^{(J)}$
$\boldsymbol{T}$	$-e^{(J)}\mathbf{K}$	$-U^{(J)}$	$e^{(J)}\mathbf{K}$	$U^{(J)}$	$W^{(J)}$	$-W^{(J)}$	$-V^{(J)}\mathbf{K}$	$V^{(J)}\mathbf{K}$
$-e^{(J)}$	$-U^{(J)}$	$e^{(J)}\mathbf{K}$	$U^{(J)}$	$-e^{(J)}\mathbf{K}$	$-V^{(J)}\mathbf{K}$	$V^{(J)}\mathbf{K}$	$-W^{(J)}$	$W^{(J)}$
$-\boldsymbol{T}$	$e^{(J)}\mathbf{K}$	$U^{(J)}$	$-e^{(J)}\mathbf{K}$	$-U^{(J)}$	$-W^{(J)}$	$W^{(J)}$	$V^{(J)}\mathbf{K}$	$-V^{(J)}\mathbf{K}$
$e^{(J)}$	$U^{(J)}$	$-e^{(J)}\mathbf{K}$	$-U^{(J)}$	$e^{(J)}\mathbf{K}$	$V^{(J)}\mathbf{K}$	$-V^{(J)}\mathbf{K}$	$W^{(J)}$	$-W(J)$
$V^{(J)}$	$-W^{(J)}$	$-V^{(J)}\mathbf{K}$	$W^{(J)}$	$V^{(J)}\mathbf{K}$	$e^{(J)}\mathbf{K}$	$-e^{(J)}\mathbf{K}$	$-U^{(J)}$	$U^{(J)}$
$-V^{(J)}$	$W^{(J)}$	$V^{(J)}\mathbf{K}$	$-W^{(J)}$	$-V^{(J)}\mathbf{K}$	$-e^{J}\mathbf{K}$	$e^{(J)}\mathbf{K}$	$U^{(J)}$	$-U^{(J)}$
$W^{(J)}\mathbf{K}$	$V^{(J)}\mathbf{K}$	$-W^{(J)}$	$-V^{(J)}\mathbf{K}$	$W^{(J)}$	$U^{(J)}$	$-U^{(J)}$	$e^{(J)}\mathbf{K}$	$-e^{(J)}\mathbf{K}$
$-W^{(J)}\mathbf{K}$	$-V^{(J)}\mathbf{K}$	$W^{(J)}$	$V^{(J)}\mathbf{K}$	$-W^{(J)}$	$-U^{(J)}$	$U^{(J)}$	$-e^{(J)}\mathbf{K}$	$e^{(J)}\mathbf{K}$

I. I. Geru, *Time-Reversal Symmetry*, Springer Tracts in Modern Physics 281,
https://doi.org/10.1007/978-3-030-01210-6

Table C.3 The Kayley table of non-Abelian group $G_{16}^{(J)}$ at arbitrary half-integer J. The matrices $U^{(J)}$, $V^{(J)}$ and $W^{(J)}$ are defined by (8.39). Block O_{21}

	$\boldsymbol{T}$	$-e$	$-\boldsymbol{T}$	e	$V^{(J)}$	$-V^{(J)}$	$W^{(J)}\mathbf{K}$	$-W^{(J)}\mathbf{K}$
$U^{(J)}$	$-e^{(J)}\mathbf{K}$	$-U^{(J)}$	$e^{(J)}\mathbf{K}$	$U^{(J)}$	$W^{(J)}$	$-W^{(J)}$	$-V^{(J)}\mathbf{K}$	$V^{(J)}\mathbf{K}$
$-e^{(J)}\mathbf{K}$	$-U^{(J)}$	$e^{(J)}\mathbf{K}$	$U^{(J)}$	$-e^{(J)}\mathbf{K}$	$-V^{(J)}\mathbf{K}$	$V^{(J)}\mathbf{K}$	$-W^{(J)}$	$W^{(J)}$
$-U^{(J)}$	$e^{(J)}\mathbf{K}$	$U^{(J)}$	$-e^{(J)}\mathbf{K}$	$-U^{(J)}$	$-W^{(J)}$	$W^{(J)}$	$V^{(J)}\mathbf{K}$	$-V^{(J)}\mathbf{K}$
$e^{(J)}\mathbf{K}$	$U^{(J)}$	$-e^{(J)}\mathbf{K}$	$-U^{(J)}$	$e^{(J)}\mathbf{K}$	$V^{(J)}\mathbf{K}$	$-V^{(J)}\mathbf{K}$	$W^{(J)}$	$-W^{(J)}$
$V^{(J)}\mathbf{K}$	$-W^{(J)}$	$-V^{(J)}\mathbf{K}$	$W^{(J)}$	$V^{(J)}\mathbf{K}$	$e^{(J)}\mathbf{K}$	$-e^{(J)}\mathbf{K}$	$-U^{(J)}$	$U^{(J)}$
$-V^{(J)}\mathbf{K}$	$W^{(J)}$	$V^{(J)}\mathbf{K}$	$-W^{(J)}$	$-V^{(J)}\mathbf{K}$	$-e^{(J)}\mathbf{K}$	$e^{(J)}\mathbf{K}$	$U^{(J)}$	$-U^{(J)}$
$W^{(J)}$	$V^{(J)}\mathbf{K}$	$-W^{(J)}$	$-V^{(J)}\mathbf{K}$	$W^{(J)}$	$U^{(J)}$	$-U^{(J)}$	$e^{(J)}\mathbf{K}$	$-e^{(J)}\mathbf{K}$
$-W^{(J)}$	$-V^{(J)}\mathbf{K}$	$W^{(J)}$	$V^{(J)}\mathbf{K}$	$-W^{(J)}$	$-U^{(J)}$	$U^{(J)}$	$-e^{(J)}\mathbf{K}$	$e^{(J)}\mathbf{K}$

Table C.4 The Kayley table of non-Abelian group $G_{16}^{(J)}$ at arbitrary half-integer J. The matrices $U^{(J)}$, $V^{(J)}$ and $W^{(J)}$ are defined by (8.39). Block O_{22}

	$U^{(J)}$	$-e^{(J)}\mathbf{K}$	$-U^{(J)}$	$e^{(J)}\mathbf{K}$	$V^{(J)}\mathbf{K}$	$-V^{(J)}\mathbf{K}$	$W^{(J)}$	$-W^{(J)}$
$U^{(J)}$	$-e^{(J)}$	$-\boldsymbol{T}$	$e^{(J)}$	$\boldsymbol{T}$	$W^{(J)}\mathbf{K}$	$-W^{(J)}\mathbf{K}$	$-V^{(J)}$	$V^{(J)}$
$-e^{(J)}\mathbf{K}$	$-\boldsymbol{T}$	$e^{(J)}$	$\boldsymbol{T}$	$-e$	$-V^{(J)}$	$V^{(J)}$	$-W^{(J)}\mathbf{K}$	$W^{(J)}\mathbf{K}$
$-U^{(J)}$	$e^{(J)}$	$\boldsymbol{T}$	$-e^{(J)}$	$-\boldsymbol{T}$	$-W^{(J)}\mathbf{K}$	$W^{(J)}\mathbf{K}$	$V^{(J)}$	$-V^{(J)}$
$e^{(J)}\mathbf{K}$	$\boldsymbol{T}$	$-e^{(J)}$	$-\boldsymbol{T}$	$e^{(J)}$	$V^{(J)}$	$-V^{(J)}$	$W^{(J)}\mathbf{K}$	$-W^{(J)}\mathbf{K}$
$V^{(J)}\mathbf{K}$	$-W^{(J)}\mathbf{K}$	$-V^{(J)}$	$W^{(J)}\mathbf{K}$	$V^{(J)}$	$e^{(J)}$	$-e^{(J)}$	$-\boldsymbol{T}$	$\boldsymbol{T}$
$-V^{(J)}\mathbf{K}$	$W^{(J)}\mathbf{K}$	$V^{(J)}$	$-W^{(J)}\mathbf{K}$	$-V^{(J)}$	$-e^{(J)}$	$e^{(J)}$	$\boldsymbol{T}$	$-\boldsymbol{T}$
$W^{(J)}$	$V^{(J)}$	$-W^{(J)}\mathbf{K}$	$-V^{(J)}$	$W^{(J)}\mathbf{K}$	$\boldsymbol{T}$	$-\boldsymbol{T}$	$e^{(J)}$	$-e^{(J)}$
$-W^{(J)}$	$-V^{(J)}$	$W^{(J)}\mathbf{K}$	$V^{(J)}$	$-W^{(J)}\mathbf{K}$	$-\boldsymbol{T}$	$\boldsymbol{T}$	$-e^{(J)}$	$e^{(J)}$

Appendix D

(See Tables D.1, D.2, D.3 and D.4)

Table D.1 The Kayley table of non-Abelian group $G_{16}^{(J)}$ at arbitrary half-integer J in terms of complete and incomplete time-reversal operators. The operators $\mathbf{T} = U^{(J)}\mathbf{K}, \mathbf{T}_{1(x)} = W^{(J)}$ $\mathbf{K}$, $\mathbf{T}_{1(y)} = e^{(J)}\mathbf{K}$, $\mathbf{T}_{1(z)} = V^{(J)}\mathbf{K}$, $T_{2(yz)} = V^{(J)}$, $\mathbf{T}_{2(zx)}=U^{(J)}$ and $\mathbf{T}_{2(xy)}=W^{(J)}$ are determined by $U^{(J)}$, $V^{(J)}$ and $W^{(J)}$ from (8.39); $\mathbf{K}$ is the operator of complex conjugation. Block O_{11}

	$\mathbf{T}$	$-e^{(J)}$	$-\mathbf{T}$	$e^{(J)}$	$\mathbf{T}_{2(yz)}$	$-\mathbf{T}_{2(yz)}$	$\mathbf{T}_{1(x)}$	$-\mathbf{T}_{1(x)}$
$\mathbf{T}$	$-e^{(J)}$	$-\mathbf{T}$	$e^{(J)}$	$\mathbf{T}$	$\mathbf{T}_{1(x)}$	$-\mathbf{T}_{1(x)}$	$-\mathbf{T}_{2(yz)}$	$\mathbf{T}_{2(yz)}$
$-e^{(J)}$	$-\mathbf{T}$	$e^{(J)}$	$\mathbf{T}$	$-e^{(J)}$	$-\mathbf{T}_{2(yz)}$	$\mathbf{T}_{2(yz)}$	$-\mathbf{T}_{1(x)}$	$\mathbf{T}_{1(x)}$
$-\mathbf{T}$	$e^{(J)}$	$\mathbf{T}$	$-e^{(J)}$	$-\mathbf{T}$	$-\mathbf{T}_{1(x)}$	$\mathbf{T}_{1(x)}$	$\mathbf{T}_{2(yz)}$	$-\mathbf{T}_{2(yz)}$
$e^{(J)}$	$\mathbf{T}$	$-e^{(J)}$	$-\mathbf{T}$	$e^{(J)}$	$\mathbf{T}_{2(yz)}$	$-\mathbf{T}_{2(yz)}$	$\mathbf{T}_{1(x)}$	$-\mathbf{T}_{1(x)}$
$\mathbf{T}_{2(yz)}$	$-\mathbf{T}_{1(x)}$	$-\mathbf{T}_{2(yz)}$	$\mathbf{T}_{1(x)}$	$\mathbf{T}_{2(yz)}$	$e^{(J)}$	$-e^{(J)}$	$-\mathbf{T}$	$\mathbf{T}$
$-\mathbf{T}_{2(yz)}$	$\mathbf{T}_{1(x)}$	$\mathbf{T}_{2(yz)}$	$-\mathbf{T}_{1(x)}$	$-\mathbf{T}_{2(yz)}$	$-e^{(J)}$	$e^{(J)}$	$\mathbf{T}$	$-\mathbf{T}$
$\mathbf{T}_{1(x)}$	$\mathbf{T}_{2(yz)}$	$-\mathbf{T}_{1(x)}$	$-\mathbf{T}_{2(yz)}$	$\mathbf{T}_{1(x)}$	$\mathbf{T}$	$-\mathbf{T}$	$e^{(J)}$	$-e^{(J)}$
$-T_{1(x)}$	$-\mathbf{T}_{2(yz)}$	$\mathbf{T}_{1(x)}$	$\mathbf{T}_{2(yz)}$	$-\mathbf{T}_{1(x)}$	$-\mathbf{T}$	$\mathbf{T}$	$-e^{(J)}$	$e^{(J)}$

Table D.2 The Kayley table of non-Abelian group $G_{16}^{(J)}$ at arbitrary half-integer J in terms of complete and incomplete time-reversal operators. The operators $\mathbf{T} = U^{(J)}\mathbf{K}, \mathbf{T}_{1(x)} = W^{(J)}\mathbf{K}$, $\mathbf{T}_{1(y)} = e^{(J)}\mathbf{K}$, $\mathbf{T}_{1(z)} = V^{(J)}\mathbf{K}$, $T_{2(yz)} = V^{(J)}$, $\mathbf{T}_{2(zx)} = U^{(J)}$ and $\mathbf{T}_{2(xy)} = W^{(J)}$ are determined by $U^{(J)}$, $V^{(J)}$ and $W^{(J)}$ from (8.39); $\mathbf{K}$ is the operator of complex conjugation. Block O_{12}

	$\mathbf{T}_{2(zx)}$	$-\mathbf{T}_{1(y)}$	$-\mathbf{T}_{2(zx)}$	$\mathbf{T}_{1(y)}$	$\mathbf{T}_{1(z)}$	$-\mathbf{T}_{1(z)}$	$\mathbf{T}_{2(xy)}$	$-\mathbf{T}_{2(xy)}$
$\mathbf{T}$	$-\mathbf{T}_{1(y)}$	$-\mathbf{T}_{2(zx)}$	$\mathbf{T}_{1(y)}$	$\mathbf{T}_{2(zx)}$	$\mathbf{T}_{2(xy)}$	$-\mathbf{T}_{2(xy)}$	$-\mathbf{T}_{1(z)}$	$\mathbf{T}_{1(z)}$
$-e^{(J)}$	$-\mathbf{T}_{2(zx)}$	$\mathbf{T}_{1(y)}$	$\mathbf{T}_{2(zx)}$	$-\mathbf{T}_{1(y)}$	$-\mathbf{T}_{1(z)}$	$\mathbf{T}_{1(z)}$	$-\mathbf{T}_{2(xy)}$	$\mathbf{T}_{2(xy)}$
$-\mathbf{T}$	$\mathbf{T}_{1(y)}$	$\mathbf{T}_{2(zx)}$	$-\mathbf{T}_{1(y)}$	$-\mathbf{T}_{2(zx)}$	$-\mathbf{T}_{2(xy)}$	$\mathbf{T}_{2(xy)}$	$\mathbf{T}_{1(z)}$	$-\mathbf{T}_{1(z)}$
$e^{(J)}$	$\mathbf{T}_{2(zx)}$	$-\mathbf{T}_{1(y)}$	$-\mathbf{T}_{2(zx)}$	$\mathbf{T}_{1(y)}$	$\mathbf{T}_{1(z)}$	$-\mathbf{T}_{1(z)}$	$\mathbf{T}_{2(xy)}$	$-\mathbf{T}_{2(xy)}$
$V^{(J)}$	$-\mathbf{T}_{2(xy)}$	$-\mathbf{T}_{1(z)}$	$\mathbf{T}_{2(xy)}$	$\mathbf{T}_{1(z)}$	$\mathbf{T}_{1(y)}$	$-\mathbf{T}_{1(y)}$	$-\mathbf{T}_{2(zx)}$	$\mathbf{T}_{2(zx)}$
$-V^{(J)}$	$\mathbf{T}_{2(xy)}$	$\mathbf{T}_{1(z)}$	$-\mathbf{T}_{2(xy)}$	$-\mathbf{T}_{1(z)}$	$-\mathbf{T}_{1(y)}$	$\mathbf{T}_{1(y)}$	$\mathbf{T}_{2(zx)}$	$-\mathbf{T}_{2(zx)}$
$W^{(J)}\mathbf{K}$	$\mathbf{T}_{1(z)}$	$-\mathbf{T}_{2(xy)}$	$-\mathbf{T}_{1(z)}$	$\mathbf{T}_{2(xy)}$	$\mathbf{T}_{2(zx)}$	$-\mathbf{T}_{2(zx)}$	$\mathbf{T}_{1(y)}$	$-\mathbf{T}_{1(y)}$
$-W^{(J)}\mathbf{K}$	$-\mathbf{T}_{1(z)}$	$\mathbf{T}_{2(xy)}$	$\mathbf{T}_{1(z)}$	$-\mathbf{T}_{2(xy)}$	$-\mathbf{T}_{2(zx)}$	$\mathbf{T}_{2(zx)}$	$-\mathbf{T}_{1(y)}$	$\mathbf{T}_{1(y)}$

I. I. Geru, *Time-Reversal Symmetry*, Springer Tracts in Modern Physics 281,
https://doi.org/10.1007/978-3-030-01210-6

Table D.3 The Kayley table of non-Abelian group $G_{16}^{(J)}$ at arbitrary half-integer J in terms of complete and incomplete time-reversal operators. The operators $\mathbf{T} = U^{(J)}\mathbf{K}$, $\mathbf{T}_{1(x)} = W^{(J)}\mathbf{K}$, $\mathbf{T}_{1(y)} = e^{(J)}\mathbf{K}$, $\mathbf{T}_{1(z)} = V^{(J)}\mathbf{K}$, $T_{2(yz)} = V^{(J)}$, $\mathbf{T}_{2(zx)} = U^{(J)}$ and $\mathbf{T}_{2(xy)} = W^{(J)}$ are determined by $U^{(J)}$, $V^{(J)}$ and $W^{(J)}$ from (8.39); $\mathbf{K}$ is the operator of complex conjugation. Block O_{21}

	$\mathbf{T}$	$-e^{(J)}$	$-\mathbf{T}$	$e^{(J)}$	$\mathbf{T}_{2(yz)}$	$-\mathbf{T}_{2(yz)}$	$\mathbf{T}_{1(x)}$	$-\mathbf{T}_{1(x)}$
$\mathbf{T}_{2(zx)}$	$-\mathbf{T}_{1(y)}$	$-\mathbf{T}_{2(zx)}$	$\mathbf{T}_{1(y)}$	$\mathbf{T}_{2(zx)}$	$\mathbf{T}_{2(xy)}$	$-\mathbf{T}_{2(xy)}$	$-\mathbf{T}_{1(z)}$	$\mathbf{T}_{1(z)}$
$-\mathbf{T}_{1(y)}$	$-\mathbf{T}_{2(zx)}$	$\mathbf{T}_{1(y)}$	$\mathbf{T}_{2(zx)}$	$-\mathbf{T}_{1(y)}$	$-\mathbf{T}_{1(z)}$	$\mathbf{T}_{1(z)}$	$-\mathbf{T}_{2(xy)}$	$\mathbf{T}_{2(xy)}$
$-\mathbf{T}_{2(zx)}$	$\mathbf{T}_{1(y)}$	$\mathbf{T}_{2(zx)}$	$-\mathbf{T}_{1(y)}$	$-\mathbf{T}_{2(zx)}$	$-\mathbf{T}_{2(xy)}$	$\mathbf{T}_{2(xy)}$	$\mathbf{T}_{1(z)}$	$-\mathbf{T}_{1(z)}$
$\mathbf{T}_{1(y)}$	$\mathbf{T}_{2(zx)}$	$-\mathbf{T}_{1(y)}$	$-\mathbf{T}_{2(zx)}$	$\mathbf{T}_{1(y)}$	$\mathbf{T}_{1(z)}$	$-\mathbf{T}_{1(z)}$	$\mathbf{T}_{2(xy)}$	$-\mathbf{T}_{2(xy)}$
$\mathbf{T}_{1(z)}$	$-\mathbf{T}_{2(xy)}$	$-\mathbf{T}_{1(z)}$	$\mathbf{T}_{2(xy)}$	$\mathbf{T}_{1(z)}$	$\mathbf{T}_{1(y)}$	$-\mathbf{T}_{1(y)}$	$-\mathbf{T}_{2(zx)}$	$\mathbf{T}_{2(zx)}$
$-\mathbf{T}_{1(z)}$	$\mathbf{T}_{2(xy)}$	$\mathbf{T}_{1(z)}$	$-\mathbf{T}_{2(xy)}$	$-\mathbf{T}_{1(z)}$	$-\mathbf{T}_{1(y)}$	$\mathbf{T}_{1(y)}$	$\mathbf{T}_{2(zx)}$	$-\mathbf{T}_{2(zx)}$
$\mathbf{T}_{2(xy)}$	$\mathbf{T}_{1(z)}$	$-\mathbf{T}_{2(xy)}$	$-\mathbf{T}_{1(z)}$	$\mathbf{T}_{2(xy)}$	$\mathbf{T}_{2(zx)}$	$-\mathbf{T}_{2(zx)}$	$\mathbf{T}_{1(y)}$	$-\mathbf{T}_{1(y)}$
$-\mathbf{T}_{2(xy)}$	$-\mathbf{T}_{1(z)}$	$\mathbf{T}_{2(xy)}$	$\mathbf{T}_{1(z)}$	$-\mathbf{T}_{2(xy)}$	$-\mathbf{T}_{2(zx)}$	$\mathbf{T}_{2(zx)}$	$-\mathbf{T}_{1(y)}$	$\mathbf{T}_{1(y)}$

Table D.4 The Kayley table of non-Abelian group $G_{16}^{(J)}$ at arbitrary half-integer J in terms of complete and incomplete time-reversal operators. The operators $\mathbf{T} = U^{(J)}\mathbf{K}$, $\mathbf{T}_{1(x)} = W^{(J)}\mathbf{K}$, $\mathbf{T}_{1(y)} = e^{(J)}\mathbf{K}$, $\mathbf{T}_{1(z)} = V^{(J)}\mathbf{K}$, $T_{2(yz)} = V^{(J)}$, $\mathbf{T}_{2(zx)} = U^{(J)}$ and $\mathbf{T}_{2(xy)} = W^{(J)}$ are determined by $U^{(J)}$, $V^{(J)}$ and $W^{(J)}$ from (8.39); $\mathbf{K}$ is the operator of complex conjugation. Block O_{22}

	$\mathbf{T}_{2(zx)}$	$-\mathbf{T}_{1(y)}$	$-\mathbf{T}_{2(zx)}$	$\mathbf{T}_{1(y)}$	$\mathbf{T}_{1(z)}$	$-\mathbf{T}_{1(z)}$	$\mathbf{T}_{2(xy)}$	$-\mathbf{T}_{2(xy)}$
$\mathbf{T}_{2(zx)}$	$-e^{(J)}$	$-\mathbf{T}$	$e^{(J)}$	$\mathbf{T}$	$\mathbf{T}_{1(x)}$	$-\mathbf{T}_{1(x)}$	$-\mathbf{T}_{2(yz)}$	$\mathbf{T}_{2(yz)}$
$-\mathbf{T}_{1(y)}$	$-\mathbf{T}$	$e^{(J)}$	$\mathbf{T}$	$-e^{(J)}$	$-\mathbf{T}_{2(yz)}$	$\mathbf{T}_{2(yz)}$	$-\mathbf{T}_{1(x)}$	$\mathbf{T}_{1(x)}$
$-\mathbf{T}_{2(zx)}$	$e^{(J)}$	$\mathbf{T}$	$-e^{(J)}$	$-\mathbf{T}$	$-\mathbf{T}_{1(x)}$	$\mathbf{T}_{1(x)}$	$\mathbf{T}_{2(yz)}$	$-\mathbf{T}_{2(yz)}$
$\mathbf{T}_{1(y)}$	$\mathbf{T}$	$-e^{(J)}$	$-\mathbf{T}$	$e^{(J)}$	$\mathbf{T}_{2(yz)}$	$-\mathbf{T}_{2(yz)}$	$\mathbf{T}_{1(x)}$	$-\mathbf{T}_{1(x)}$
$\mathbf{T}_{1(z)}$	$-\mathbf{T}_{1(x)}$	$-\mathbf{T}_{2(yz)}$	$\mathbf{T}_{1(x)}$	$\mathbf{T}_{2(yz)}$	$e^{(J)}$	$-e^{(J)}$	$-\mathbf{T}$	$\mathbf{T}$
$-\mathbf{T}_{1(z)}$	$\mathbf{T}_{1(x)}$	$\mathbf{T}_{2(yz)}$	$-\mathbf{T}_{1(x)}$	$-\mathbf{T}_{2(yz)}$	$-e^{(J)}$	$e^{(J)}$	$\mathbf{T}$	$-\mathbf{T}$
$\mathbf{T}_{2(xy)}$	$\mathbf{T}_{2(yz)}$	$-\mathbf{T}_{1(x)}$	$-\mathbf{T}_{2(yz)}$	$\mathbf{T}_{1(x)}$	$\mathbf{T}$	$-\mathbf{T}$	$e^{(J)}$	$-e^{(J)}$
$-\mathbf{T}_{2(xy)}$	$-\mathbf{T}_{2(yz)}$	$\mathbf{T}_{1(x)}$	$\mathbf{T}_{2(yz)}$	$-\mathbf{T}_{1(x)}$	$-\mathbf{T}$	$\mathbf{T}$	$-e^{(J)}$	$e^{(J)}$

Appendix E

Matrices of meta-spin projection operators for six types of meta-particles with meta-spin $\Sigma^{(n)} = 1$ $(n = 1,\ 2,\ \ldots, 6)$.

Time-reversal symmetry restoration from $T_{2(yz)}$ to T $(n = 1)$

$$\Sigma_x^{2(yz)} = \frac{1}{\sqrt{2}}\begin{pmatrix} 0 & -1 & 0 \\ -1 & 0 & -1 \\ 0 & -1 & 0 \end{pmatrix},\ \Sigma_y^{2(yz)} = \frac{1}{\sqrt{2}}\begin{pmatrix} 0 & -i & 0 \\ i & 0 & -i \\ 0 & i & 0 \end{pmatrix},\ \Sigma_z^{2(yz)} = \begin{pmatrix} 1 & 0 & 0 \\ 0 & 0 & 0 \\ 0 & 0 & -1 \end{pmatrix}$$

Time-reversal symmetry restoration from $T_{2(zx)}$ to T $(n = 2)$

$$\Sigma_x^{2(zx)} = \frac{1}{\sqrt{2}}\begin{pmatrix} 0 & 1 & 0 \\ 1 & 0 & 1 \\ 0 & 1 & 0 \end{pmatrix},\ \Sigma_y^{2(zx)} = \frac{1}{\sqrt{2}}\begin{pmatrix} 0 & i & 0 \\ -i & 0 & i \\ 0 & -i & 0 \end{pmatrix},\ \Sigma_z^{2(zx)} = \begin{pmatrix} 1 & 0 & 0 \\ 0 & 0 & 0 \\ 0 & 0 & -1 \end{pmatrix}$$

Time-reversal symmetry restoration from $T_{2(xy)}$ to T $(n = 3)$

$$\Sigma_x^{2(xy)} = \frac{1}{\sqrt{2}}\begin{pmatrix} 0 & 1 & 0 \\ 1 & 0 & 1 \\ 0 & 1 & 0 \end{pmatrix},\ \Sigma_y^{2(xy)} = \frac{1}{\sqrt{2}}\begin{pmatrix} 0 & -i & 0 \\ i & 0 & -i \\ 0 & i & 0 \end{pmatrix},\ \Sigma_z^{2(xy)} = \begin{pmatrix} -1 & 0 & 0 \\ 0 & 0 & 0 \\ 0 & 0 & 1 \end{pmatrix}$$

Time-reversal symmetry restoration from $T_{1(x)}$ to T $(n = 4)$

I. I. Geru, *Time-Reversal Symmetry*, Springer Tracts in Modern Physics 281,
https://doi.org/10.1007/978-3-030-01210-6

$$\Sigma_x^{1(x)} = \frac{1}{\sqrt{2}}\begin{pmatrix} 0 & 1 & 0 \\ 1 & 0 & 1 \\ 0 & 1 & 0 \end{pmatrix}, \ \Sigma_y^{1(x)} = \frac{1}{\sqrt{2}}\begin{pmatrix} 0 & i & 0 \\ -i & 0 & i \\ 0 & -i & 0 \end{pmatrix}, \ \Sigma_z^{1(x)} = \begin{pmatrix} -1 & 0 & 0 \\ 0 & 0 & 0 \\ 0 & 0 & 1 \end{pmatrix}$$

Time-reversal symmetry restoration from $T_{1(y)}$ to T $(n = 5)$

$$\Sigma_x^{1(y)} = \frac{1}{\sqrt{2}}\begin{pmatrix} 0 & -1 & 0 \\ -1 & 0 & -1 \\ 0 & -1 & 0 \end{pmatrix}, \ \Sigma_y^{1(y)} = \frac{1}{\sqrt{2}}\begin{pmatrix} 0 & -i & 0 \\ i & 0 & -i \\ 0 & i & 0 \end{pmatrix}, \ \Sigma_z^{1(y)} = \begin{pmatrix} -1 & 0 & 0 \\ 0 & 0 & 0 \\ 0 & 0 & 1 \end{pmatrix}$$

Time-reversal symmetry restoration from $T_{1(z)}$ to T $(n = 6)$

$$\Sigma_x^{1(z)} = \frac{1}{\sqrt{2}}\begin{pmatrix} 0 & -1 & 0 \\ -1 & 0 & -1 \\ 0 & -1 & 0 \end{pmatrix}, \ \Sigma_y^{1(z)} = \frac{1}{\sqrt{2}}\begin{pmatrix} 0 & i & 0 \\ -i & 0 & i \\ 0 & -i & 0 \end{pmatrix}, \ \Sigma_z^{1(z)} = \begin{pmatrix} 1 & 0 & 0 \\ 0 & 0 & 0 \\ 0 & 0 & -1 \end{pmatrix}$$

References

1. P. Davies, *About Time: Einstein's Unfinished Revolution* (Simon & Schuster, 1996), p. 31. ISBN 978-0684818221
2. R.G. Sachs, *The Physics of Time Reversal* (The University of Chicago Press, Chicago, 1987), 309 pp
3. L.D. Landau, E.M. Lifshitz, *Quantum Mechanics, vol. 3, of a Course of Theoretical Physics* (Pergamon Press, Oxford, 1965)
4. B.V. Medvedev, *Foundations of Theoretical Physics* (Moskow, Izd. Nauka, 1977), p. 496
5. I.V. Saveliev, *Basics of Theoretical Physics*, vol. 1 (Izd. Nauka, Moscow, 1975), 416 pp
6. E. Nöether, Invariante Variationsprobleme. Nachr. Acad. Wiss. Goettingen, Math.-Phys. K1.2, 235 (1918)
7. G.Ya. Lyubarsky, *Groups Theory and Its Application in Physics* (Publishing House Phys.-Math. Literature, Moscow, 1958), 354 pp
8. J.P. Elliott, P.G. Dawber, *Symmetry in Physics*, vols. 1, 2 (The Macmillan Press Ltd, London, 1979)
9. V. Heine, *Group Theory in Quantum Mechanics* (Pergamon Press, London, 1960)
10. R.S. Khamitova, Dokl. Acad. Nauk SSSR **248**, 798 (1979)
11. R.I. Anderson, N.H. Ibragimov, *Lie-Bäcklund Transformations in Applications* (SIAM, Philadelphia, 1979)
12. N.Kh. Ibragimov, Matemat. Sbornik **109**, 229–253 (1979); Math. USSR - Sb. **37**, 205–226 (1980)
13. N.Kh. Ibragimov, Teor. Mat. Fiz. **1**, 350 (1969)
14. J.S. Ames, F.D. Murnaghan, *Theoretical Mechanics* (Ginn, Boston, 1929)
15. K. Gottfrield, Quantum Mechanics, Vol.1: Fundamentals (Benjamin Cummings Publishing Company, Inc., 1966), p. 314; Copyright 1966, 1974, 1989 by Perseus Publishing
16. P. Painlevé, Comptes Rendus **139**, 1170–1174 (1904)
17. A. Einstein, *Collection of Scientific Works in Four Volumes*, vol. 1, ed. by I.E. Tamm, Ya.A. Smorodinskii, B.G. Kuznetsov (Nauka, Moscow, 1965), pp. 7–35, 65–114
18. V.I. Smirnov, *Course of Higher Mathematics*, vol. III, Part. 1, 5th edn. (Publishing House Techn.-Theoret. Literat., Moscow, 1953), 339 pp
19. A.I. Ahiezer, V.B. Berestetskii, *Quantum Electrodynamics* (Nauka, Techn.-Theoret. Leterat., Moscow, 1969), 623 pp
20. J. Loshmidt, Wien Ber. **73**, 139 (1876); **75**, 67 (1877)
21. L. Boltzmann, Wien Ber. **66**, 275 (1872)
22. J.M. Gibbs, *Elementary Principles in Statistical Mechanics* (Charles Scribner's Sons, New York, 1902)

I. I. Geru, *Time-Reversal Symmetry*, Springer Tracts in Modern Physics 281,
https://doi.org/10.1007/978-3-030-01210-6

23. J. Kumicak, X. de Hemptinne, Physica D. **112**, 258–274 (1988)
24. S.G. Brush (ed.), *Kinetic Theory, vol. 2, Irreversible Processes* (Pergamon Press, Oxford, 1966)
25. J.S.W. Lamb, J.A.G. Roberts, Time-reversal symmetry: a survey. Phys. D **112**, 1–39 (1998)
26. W.M. Haddad, V.S. Chellaboina, S.G. Nersesov, Nonlinear Anal.: Real World Appl. **9**, 250–271 (2008)
27. C. Carathéodory, Math. Ann. **67**, 335–386 (1909)
28. C. Carathéodory, Über die Bestimmung de Energie und der absoluten Temperatur mit Hiffe von reversiblen Prozessen, Sitzungsber. preuβ. Akad. Wiss., Math. Phys. **K1**, 39–47 (1925)
29. P. Bridgman, *The Nature of Thermodynamics (Harvard University Press, Cambridge, 1941)* (Printed Peter Smith, Gloucester, 1969)
30. J. Uffink, Bluff your way in the second law of thermodynamics. Stud. Hist. Phil. Mod. Phys. **32**, 305–394 (2001)
31. M. Planck, Über die Begrundung des zweiten Hauptsatzes der Thermodynamik, Sitzungsber. preuβ. Akad. Wisse., Math. Phys. **K1**, 453–463 (1926)
32. H. Reichenbach, *The Direction of Time* (University of California Press, Berkley, 1971)
33. A. Grunbaum, The anistropy of time, in *The Nature of Time*, ed. by T. Gold (Cornell University Press, Ithaca, New York, 1967)
34. J. Earman, Irreversibility and temporal assymmetry. J. Phiolos. **64**, 543–549 (1967)
35. P. Kroes, *Time: Its Structure and Role in Physical Theories* (D. Riedel, Dordrecht, 1985)
36. P. Horwich, *Assymmetries in Time* (MIT Press, Cambridge, 1987)
37. A. Eddington, *The Nature of Physical World* (J.M. Dent & Sons, London, 1935)
38. I. Prigogine, *From Being to Becoming* (W.H. Freeman, San Francisco, 1980)
39. Ch. Dellago, H.A. Posch, Phys. D **112**, 241–249 (1998)
40. G. Gallavotti, Phys. D **112**, 250–257 (1998)
41. W.G. Hoover, Phys. D **112**, 225–240 (1998)
42. I. Prigogine, Int. J. Bifurc. Chaos **05**, 3 (1995). https://doi.org/10.1142/S0218127495000028
43. I. Prigogine, I. Stengers, *Order Out of Chaos: Man's New Dialogue with Nature* (Fontana Press, London, 1985)
44. P.D. Sisterna, Found. Phys. Lett. **13**(3), 205–220 (2000)
45. S.W. Hawking, Phys. Rev. D **32**, 2489 (1985)
46. D.N. Page, Phys. Rev. D **32**, 2496 (1985)
47. S.W. Hawking, R. Laflamme, G.M. Lyons, Phys. Rev. D **47**, 5342 (1993)
48. J. Barbour, *The End of Time* (Oxford University Press, Oxford, 2000)
49. J. Barbour, The emergence of time and its arrow from timelessness, in *Physical Origins of Time Asymmetry*, ed. by J.J. Habliwell, J. Perez-Mercader, W.H. Zurek (Cambridge University Press, Cambridge, 1994), p. 311
50. L. Onsager, Phys. Rev. **37**, 405–426 (1931)
51. V.G. Levich, V.A. Yu.A. Vdovin, Myamlin, *Course of Theoretical Physics*, vol. II (Nauka, Moskva, 1971), 936 pp
52. H. Poincaré, Acta Math. **13**, 1–270 (1890)
53. K. Huang, *Statistical Mechanics*, 2nd edn. (Wiley, New York, 1987), 512 pp
54. P.A.M. Dirac, *The Principles of Quantum Mechanics*, 4th edn. (Clarendon Press, Oxford, 1958)
55. W. Pauli, Die Allgemeinen Prinzipien der Wellenmechanik, in Handbuch der Physics, Bd. V, 1 (Berlin, 1958), p. 69
56. J. Neumann, Mathematische Grundlagen der Quantenmechanik (Berlin, 1932)
57. E. Fermi, *Notes on Quantum Mechanics* (The University of Chicago Press, Chicago, 1995), 191 pp
58. D.J. Griffiths, *Introduction to Quantum Mechanics* (Prentice Hall Inc, Upper Saddle River, 1995)
59. H.A. Bethe, *Intermediate Quantum Mechanics* (W.A. Benjamin Inc, New York, 1964)
60. V.A. Fock, *Nachala kvantovoi mekaniki* (Nauka, Moskva, 1976), 376 pp
61. A.S. Davydov, *Kvantovaya mekanika* (Nauka, Moskva, 1973), 703 pp

62. D.I. Blokhintsev, *Osnovy kvantovoi mehaniki* (Gos. Izd., Vyshaya shkola, Moskva, 1961), 512 pp
63. V.V. Balashov, V.K. Dolinov, *Kurs kvantovoi mekaniki* (Publishing House of Moscow University, Moskva, 1982), 279 pp
64. I.V. Savel'ev, Osnovy teoreticheskoi fiziki, vol. 2, Kvantovaya mekanika (Nauka, Moskva, 1977), 351 pp
65. L.D. Landau, Z. Phys. **45**, 430 (1927)
66. J. von Neumann, Gottinger Nachrichten **1**, 245–272 (1927)
67. W.H. Luissel, *Radiation and Noise in Quantum Electronics* (McGrow-Hill Book Company, New York, 1964)
68. E.P. Wigner, Nachr. Akad. Wiss. Göttingen. Math. Phys. **31**, K1., 546 (1932)
69. H.A. Kramers, Koninkl. Ned. Akad. Wetenschap. Proc. **33**, 959 (1930)
70. F. Haake, *Quantum Signatures of Chaos*. Springer Series in Sinergetics, 3rd edn. (Springer, Berlin, 2010). https://doi.org/10.1007/978-3-642-05428-0-2
71. B.F. Bayman, *Some Lectures on Groups and Their Applications to Spectroscopy* (Gos. Izdat. Fiziko-Matemat. Literat., Moskva, 1961), 226 pp
72. F.A. Kaempffer, *Concepts in Quantum Mechanics* (Academic Press, New York, 1965)
73. G.F. Koster, J.O. Dimmock, R.G. Wheller, H. Statz, *Properties of the Thirti-Two Point Groups* (MIT Press, Cambridge, 1963)
74. G. Frobenius, I. Schur, Berl. Ber. **186** (1906)
75. D.F. Johnston, Proc. R. Soc. A **243**, 546 (1958)
76. H.-W. Streitwolf, *Gruppentheorie in Der Festkörperphysik Akademische Verlagsgesellschaft* (Geest & Portig K.-G, Leipzik, 1967)
77. R.S. Knox, A. Gold, *Summetry in the Solid State* (W.A. Benjamin Inc, New York, 1964)
78. G.L. Bir, G.E. Pikus, *Symmetry and Deformation Effects in Semiconductors* (Nauka, Moscow, 1972), 583 pp
79. K. Herring, Influence of the time-reversal symmetry on energy bands of crystals. Phys. Rev. **52**, 361 (1937)
80. I. Schur, J. Reine Angew. Math. **127**, 20–50 (1904); **132**, 85 (1907); **139**, 155 (1911)
81. M. Hamermesh, Group Theory and its Applications to Physical Problems (Mir, Moscow, 1967), 587 pp. [in Russian]
82. Ch. Ketris, I. Reiner, *Theory of Representations of Finite Groups and Associative Algebres* (Nauka, Moscow, 1969)
83. R.J. Elliott, Phys. Rev. **96**, 280 (1954)
84. V.A. Chaldyshev, N.V. Kudryavtsev, G.F. Karavaev, News of universities. Physics **2**, 46 (1963)
85. I.I. Geru, Doklady AN SSSR **268**, 1392–1394 (1983)
86. O. Klein, Z. Phys. **37**, 895 (1926)
87. V. Fock, Z. Phys. **38**, 242 (1926); **39**, 226 (1926)
88. W. Gordon, Z. Phys. **40**, 117 (1926)
89. C. Itzykson, J.-B. Zuber, *Quantum Field Theory* (McGraw-Hill Book Company, New York, 1980)
90. H. Yamamoto, *Personal communication*, January 22, 2016
91. J.L. Birman, *Theory of Crystal Space Groups and Infra-Red and Raman Lattice Processes of Insulating Crystals* (Springer, Berlin, 1974)
92. M. Born, K. Huang, *Dinamical Theory of Crystal Lattices* (University Press, Oxford, 1954)
93. F. Block, Zsech. Phys. **52**, 555 (1928)
94. J.L. Birman, *Group Theory Methods and Techniques with Applications to Physics* (Wiley, New York, 1977)
95. G. Leilfried, *Microscopical Theory of Mechanical Properties of Crystals* (Handbuch der Physik, Bd. VII/1, S. 104, Springer, Berlin, 1955)
96. V. Frei, Czech. J. Phys. **16**, 207 (1966)
97. J.P. Elliott, P.G. Dawber, *Symmetry in Physics, vol. 2: Further Applications* (The Macmillan Press Ltd., London, 1979), 410 pp. [in Russian]

98. R.F. Streater, A.S. Wightman, *PCT, Spin and Statistics and All That* (W.A. Benjamin, Inc., New York, 1964), 251 pp. [in Russian]
99. A.I. Ahiezer, V.B. Berestetskii, *Quantum Electrodynamics* (Nauka, Moscow, 1969), 623 pp. [in Russian]
100. N.N. Bogoliubov, D.V. Shirkov, *Quantum Fields* (Nauka, Moscow, 1980), 319 pp. [in Russian]
101. N.N. Bogoliubov, A.A. Logunov, I.T. Todorov, Bases of the Axiomatic Approach to Quantum Field Theory (Nauka, Moscow, 1969), 424 pp
102. R. Jost, General theory of quantized fields, in *Proceedings of the Summer Seminar*. Lectures in Applied Mathematics, ed. by M. Kac (Boulder, Colorado, 1960)
103. V. Luc, *Physique Subatomique: Noyaux et Particules*, vol. 2, Developpements (Hermann, Paris, 1982)
104. F.M. Jeager, *The Principle of Symmetry* (Elzevir, Amsterdam, 1920)
105. C.S. Wy, E. Ambler, R.W. Hayward et al., Phys. Rev. **105**, 1413 (1957)
106. T.D. Lee, C.N. Yang, Phys. Rev. **104**, 254 (1956)
107. 60. G.-B. Zhao, Y. Wang, J.-Q. Xia et al., J., Cosmol. Astropart. Phys. (JCAP) 07 **032**, 1–12 (2015), arXiv:1504.04507, https://doi.org/10.1088/1475-7516/2015/07/032
108. Y. Fukuda et al., (Super-Komiokande Collaboration), Evidence for oscillation on atmospheric neutrinos. Phys. Rev. Lett. **81**, 1562 (1998)
109. J.H. Christenson, J.M. Cronin, V.L. Fitch et al., Phys. Rev. Lett. **13**, 138 (1964)
110. G. Lüders, Kgl. Danske Vidensk. Selsk. Mat-Fys. Medd. **28**, N5 (1954)
111. G. Lüders, Ann. Phys. **2**, 1 (1957)
112. W. Pauli, *Niels Bohr and the Development of Physics* (Pergamon, London, 1955)
113. H. Heesh, Z. Krist **71**, 95 (1929)
114. B.A. Tavger and V.M. Zaitsev, Zh. Experim. Teor. Fiz. (U.S.S.R.) **30**, 564 (1956); Soviet. Phys. JETP **3**, 430 (1956)
115. J.P. Elliott, P.G. Dawber, *Symmetry in Physics, vol. 1: Principles and Simple Applications* (The Macmillan Press Ltd, London, 1979), 364 pp. [Russian translation]
116. J.P. Elliott, P.G. Dawber, *Symmetry in Physics, vol. 2: Further Applications* (The Macmillan Press Ltd, London, 1979), 410 pp. [Russian translation]
117. M. Hamermesh, *Group Theory and Its Applications to Physical Problems* (Addison-Wesley Publishing Company Inc, Reading, 1964), 587 pp
118. W. Opechowski, R. Guccione, Magnetic symmetry, in *Magnetism, vol. II, Part A, Statistical Models, Magnetic Symmetry, Hyperfine Interactions, and Metals*, ed. by G.T. Rado, H. Suhl (Academic Press, New York, 1965), pp. 105–165
119. V.L. Indenbom, Kristallografiya **4**, 619 (1959); Soviet Phys. Cryst. (English Transl.) **4**, 578 (1959)
120. A. Niggli, Z. Krist. **111**, 288 (1959)
121. International Tables for X-Ray Crystallography, vol. 1, ed. by F.M.H. Norman, K. Lonsdale (Kynoch Press, New York, 1969)
122. J.O. Dimmock, R.G. Weeler, Phys. Chem. Solids **23**, 729 (1962)
123. J.O. Dimmock, R.G. Weeler, Phys. Rev. **127**, 391 (1962)
124. J.O. Dimmock, J. Math. Phys. **4**, 1307 (1963)
125. I.S. Zheludev, *Symmetry and Its Applications* (Atomizdat, Moscow, 1976), 286 pp
126. V.A. Koptsik, I.N. Kotsev, Magnetic (spin) and magneto-electric point groups of P-symmetry. Preprint/Joint Inst. Nuclear Research, P4-8466, Dubna (1974), 19 pp
127. C.J. Bradley, A.P. Cracknell, *The Mathematical Theory of Symmetry in Solids* (Oxford University Press, Oxford, 1972)
128. A.V. Schubnikov, N.V. Belov, *Coloured Symmetry* (Pergamon Press, London, 1964)
129. N.B. Belov, N.N. Neronova, T.S. Smirnova, Tr. Inst. Kristablogr. Akad. Nauk SSSR **11**, 33–67 (1955)
130. A.V. Schubnikov, *Selected Works on Crystallography* (Nauka, Moscow, 1975), pp. 72–91. [in Russian]
131. A.V. Schubnikov, Perspectives of the development of the theory of symmetry, in *Crystallography (Proceedings of the Fedorov Scientific Session, 1949)* ("Metallur gizdat", Moscow, 1951), pp. 33–47 (in Russian)

132. B.A. Tavger, Krystallografiya **3**, 340 (1958); Soviet Phys. Cryst. (English Transl.) **3**, 341 (1958)
133. B.A. Tavger, Krystallografiya **5**, 677 (1960); Soviet Phys. Cryst. (English Transl.) **5**, 646 (1961)
134. I.I. Geru, Phys. Solid State **44**, 1496–1499. Translated from. Fiz. Tverd. Tela **44**, 1432–1434 (2002)
135. I. Geru, D. Suter, *Resonance Effects of Excitons and Electrons, Basics and Applications* (Spinger, Berlin, 2013), 283 pp
136. I. Geru, Time-reversal symmetry, Boson-Boson and Boson-Antiboson correlations in spin systems, in *Progrese în Fizica Teoretică*, ed. by Co-ordinators V. Cantser, F. Paladi, D. Nica (CEP USM, Chisinau, 2016), pp. 22–95. (in English)
137. A.V. Schubnikov, *Simmetria i antisimmetriya konechnyh figur* (Izd-vo AN SSSR, Moskva, 1951)
138. A.V. Schubnikov, Antisimmetriya konechnyh figur (Tr. Inst. Kristallogr. AN SSSR, N 10, 1954), pp. 3–5
139. A.V. Schubnikov, Kristallografiya **3**, 263–268 (1958)
140. A.V. Schubnikov, Kristallografiya **6**, 490–495 (1961)
141. A.V. Schubnikov, Kristallografiya **7**, 3–6 (1962)
142. A.V. Schubnikov, V.A. Koptsik, *Simmetriya v nauke i iscusstve* (Nauka, Moskva, 1972)
143. N.V. Belov, Kristallografiya **1**, 621–625 (1956)
144. N.V. Belov, E.N. Belova, Kristallografiya **2**, 21–22 (1957)
145. N.V. Belov, E.N. Belova, T.N. Tarhova, Kristallografiya **3**, 618–620 (1958)
146. N.V. Belov, T.N. Tarhova, Kristallografiya **1**, 4–13 (1956)
147. N.V. Belov, T.N. Tarhova, Kristallografiya **1**, 619–620 (1956)
148. A.M. Zamorzaev, Kristallografiya **2**, 15–20 (1957); Soviet Phys. Cryst. (English Transl.) **3**, 401 (1958)
149. A.M. Zamorzaev, Kristallografiya **3**, 399–404 (1958); Soviet Phys. Cryst. (English Transl.) **3**, 401 (1958)
150. A.M. Zamorzaev, Kristallografiya **7**, 813–821 (1962); Soviet Phys. Cryst. (English Transl.) **7**, 661 (1963)
151. A.M. Zamorzaev, Kristallografiya **14**, 195–200 (1969)
152. H.W. Kroto, Carbon **30**, 1139–1141 (1992)
153. M.S. Dresselhaus, G. Dresselhaus, P.C. Eklund, *Science of Fullerenes and Carbon Nanotubes* (Academic Press, New York, 1996)
154. H.A. Jahn, E. Teller, Proc. R. Soc. A **161**, 220 (1937)
155. I.I. Geru, J. Nanoelectron. Optoelectron. **6**, 381–392 (2011)
156. L.A. Welo, Philos. Mag. **6**(36), 481–509 (1928)
157. K. Kambe, J. Phys. Soc. Jpn. **5**(1), 48 (1950)
158. A. Abragam, J. Horowitz, J. Yvon, J. Phys. Rad. **13**, 489 (1952)
159. J. Wucher, H.M. Gijsman, Physica **20**(1–6), 361 (1954)
160. J. Wucher, J.D. Wasscher, Physica **20**(7–12), 721 (1954)
161. B.N. Figgis, G.B. Robertson, Nature **205**, 694 (1964)
162. S.C. Chang, G.A. Jeffrey, Acta Crystollogr. B. **26**(6), 673 (1970)
163. M. Sorai, M. Tachiki, M. Suga, S. Seki, J. Phys. Soc. Jpn. **30**(3), 750 (1971)
164. J.F. Duncan, C.R. Kanekar, K.F. Mok, J. Chem. Soc. A, 480–482 (1969)
165. A.Earnshaw, B.N. Figgis, J. Lewis, J. Chem. Soc. A, 1656–1663 (1966)
166. J.F. Duncan, R.M. Golding, K.F. Mok, J. Inorg. Nucl. Chem. **28**(4), 1114 (1966)
167. K.J. Schenk, H.U. Güdel, Inorg. Chem. **21**(6), 2253 (1982)
168. G.J. Long, W.T. Robinson, W.P. Tappmeyer, D.L. Bridges, J. Chem. Soc. Dalton Trans. (6), 573 (1973)
169. C.T. Dziobkowski, J.T. Wrobleski, D.B. Brown, Inorg. Chem. **20**(3), 671 (1981)
170. C.E. Anson, J.P. Bourke, R.D. Cannon, U.A. Jayasooriya, M. Molinier, A.K. Powell, Inorg. Chem. **36**(6), 1265 (1997)
171. J. Ferguson, H.U. Güder, Chem. Phys. Lett. **17**(4), 547 (1972)

172. H.U. Güdel, F. Furrer, Mol. Phys. **33**(5), 1335 (1977)
173. H.U. Güdel, U. Hauser, A. Furrer, Inorg. Chem. **18**(10), 2730 (1979)
174. U.A. Jayasooriya, R.D. Cannon, R.P. White, G.J. Kearley, Angew. Chem. Int. Ed. Engl. **28**(7), 930 (1989)
175. R.D. Cannon, U.A. Jayasooriya, R. Wu, S.K. arapKoske, J.A. Stride, O.F. Nielsen, R.P. White, G.J. Kearley, D. Summerfield, J. Am. Chem. Soc. **116**(26), 11869 (1994)
176. M.A. Kiskin, I.G. Fomina, A.A. Sidorov, G.G. Alexandrov, O. Yu. Proshenkina, Zh. V. Dobrokhotova, V.N. Ikorskii, Yu. G. Shvedenkov, V.M. Novotortser, I.L. Eremenkov, I.I. Moiseev, Russian Chem. Bull. Int. Edn. **53**(11), 2508 (2004)
177. O. Kahn, *Molecular Magnetism* (Wiley-VCH, Wienheim, 1993)
178. R. Beckett, B.F. Hoskins, J. Chem. Soc., Dalton Trans. (3), 291 (1972)
179. R.J. Butcher, C.J. O'Connor, E. Sinn, Inorg. Chem. **20**(2), 537 (1981)
180. F.B. Hulsbergen, R.W.M. ten Hoedt, G.C. Verschoor, J. Reedijk, A.L. Spek, J. Chem. Soc., Dalton Trans. (3), 539–545 (1983)
181. J.P. Costes, F. Dahan, J.P. Laurent, Inorg. Chem. **25**(4), 413 (1986)
182. K.D. Karlin, Q. Gan, A. Farooq, S. Liu, J. Zubieta, Inorg. Chim. Acta. **165**(1), 37 (1989)
183. M. Angaroni, G.A. Ardizzoia, T. Beringhelli, G.L. Monica, D. Gatteschi, N. Masciocchi, M. Moret, J. Chem. Soc. Dalton Trans. (11), 3305 (1990)
184. Y. Agnus, R. Louis, B. Metz, C. Boudon, J.P. Gisselbrecht, M. Gross, Inorg. Chem. **30**(16), 3155 (1991)
185. P. Chaudhuri, M. Winter, B.P.C. Della Vedova, E. Bill, A. Trautwein, S. Gehring, P. Fleischhauer, B. Nuber, J. Weiss, J. Weiss., Inorg. Chem. **30**(9), 2148 (1991)
186. S. Meenakumari, A.R. Chakravarty, J. Chem. Soc. Dalton Trans. (18), 2749 (1992)
187. D. Christodoulou, C. George, L.K. Keefer, J. Chem. Soc. Chem. Commun. (11), 937 (1993)
188. P.J. van Koningsbruggen, J.W. van Hal, R.A.G. de Graaff, J.G. Haasnoot, J. Reedijk, J. Chem. Soc. Dalton Trans. (14), 2163 (1993)
189. K.D. Karlin, Q.F. Gan, A. Farooq, S. Liu, J. Zubieta, Inorg. Chem. **29**(14), 2549 (1990)
190. P. Chaudhuri, I. Karpenstein, M. Winter, C. Butzlaff, E. Bill, A.X. Trautwein, U. Florke, H.-J. Haupt, J. Chem. Soc. Chem. Commun. (4), 321 (1992)
191. D.E. Fenton, in *Perspectives in Coordination Chemistry*, ed. by A.F. Williams, C. Floriani, A.E. Merbach (VCH Verlagsgesellschaft and Helvetica Chimica Acta Publishers, Weinheim, Basel, 1992), p. 203
192. D.E. Fenton, H. Okawa, J. Chem. Soc. Dalton Trans. (9), 1349 (1993)
193. H. Adams, N.A. Bailey, M.J.S. Dwyer, D.E. Fenton, P.C. Hellier, P.D. Hempstead, J.M. Latour, J. Chem. Soc. Dalton Trans. (8), 1207 (1993)
194. S. Meenakumari, S.K. Tiwary, A.R. Chakravarty, Inorg. Chem. **33**(10), 2085 (1994)
195. L. Spiccia, B.Graham, M.T.W. Hearn, G. Lazarev, B. Moubaraki, K.S. Murray, E.R.T. Tiekink, J. Chem. Soc. Dalton Trans. (21), 4089 (1997)
196. I. Dzyaloshinsky, J. Phys. Chem. Solids. **4**(4), 241 (1958)
197. T. Moriya, Phys. Rev. **120**(1), 91 (1960)
198. J. Padilla, D. Gatteschi, P. Chaudhuri, Inorg. Chim. Acta. **260**(2), 217 (1997)
199. V.A. Gaponenko, M.V. Eremin, Yu. V. Yablokov. Fiz. Tverd. Tela (Russ) **15**, 1336 (1973)
200. Yu.V. Rakitin, T.A. Zhemchuzhnikova, V.V. Zelentsov, Inorg. Chim. Acta. **23**, 145 (1977)
201. S. Ferrer, F. Lloret, I. Bertomeu, G. Alzuet, J. Borrás, S. García-Granda, M. Liu-González, J.G. Haasnoot, Inorg. Chem. **41**(22), 5821 (2002)
202. U. Kortz, S. Nellutla, A.C. Stowe, N.S. Dalal, J. van Tol, B.S. Bassil, Inorg. Chem. **43**(1), 144 (2004)
203. U. Kortz, N.K. Al-Kassem, M.G. Savelieff, N.A. Al Kadi, M. Sadakane, Inorg. Chem. **40**(18), 4742–4749 (2001)
204. R. Clerac, F.A. Cotton, K.R. Dunbar, E.A. Hillard, M.A. Petrukhina, B.W. Smucker, C. R. Acad. Sci. Paris, Chimie/Chem. **4**, 315 (2001)
205. B. Cage, F.A. Cotton, N.S. Dalal, E.A. Hillard, B. Rakvin, C.M. Ramsey, J. Am. Chem. Soc. **125**(18), 5270 (2003)

206. M. Luban, F. Borsa, S. Bud'ko, P. Canfield, S. Jun, J.K. Jung, P. Kögerler, D. Mentrup, A. Müller, R. Modler, D. Procissi, B.J. Suh, M. Torikachvili, Phys. Rev. B. **66**(5), 054407 (2002)
207. J.T. Haraldsen, T. Barnes, J.L. Musfeldt, Phys. Rev. B. **71**(6), 064403 (2005)
208. A. Müller, J. Döring, Angew. Chem. Int. Ed. Enlg. **27**(12), 1721 (1988)
209. A.L. Barra, D. Gatteschi, L. Pardi, A. Müller, J. Döring, J. Am. Chem. Soc. **114**(22), 8509 (1992)
210. D. Gatteschi, L. Pardi, A.L. Barra, A. Müller, J. Döring, Nature **354**, 463–464 (1991)
211. G. Chaboussant, R. Basler, A. Sieber, S.T. Ochsenbein, A. Desmedt, R.E. Lechner, M.T.F. Telling, P. Kögerler, A. Müller, H.-U. Güdel, Europhys. Lett. **59**(2), 291 (2002)
212. G. Chaboussant, S.T. Ochsenbein, A. Sieber, H.-U. Güdel, H. Mutka, A. Müller, A. Barbara, Europhys. Lett. **66**(3), 423 (2004)
213. H. Lypez-Sandoval, R. Contreras, A. Escuer, R. Vicente, S. Bernies, H. Noth, G.J. Leigh, N. Barba-Behrens, J. Chem. Soc., Dalton Trans. (13), 2648 (2002)
214. B.S. Bassil, S. Nellutla, U. Kortz, A.C. Stowe, J. van Tol, N.S. Dalal, B. Keita, L. Nadjo, Inorg. Chem. **44**(8), 2659 (2005)
215. J.R. Galan-Riasearos, C.J. Gomez-Garcia, J.J. Borras-Almenar, E. Coronado, Adv. Rater. **6**, 221 (1996)
216. A. Abragam, B. Bleaney, *Electron Paramagnetic Resonance of Transitions Ions* (Oxford University Press, Oxford, 1970)
217. H. Andres, J.M. Clemente-Juan, M. Aebersold, H.-U. Güdel, E. Coronado, H. Büttner, G. Kearly, J. Melero, R. Burriel, J. Am. Chem. Soc. **121**(43), 10028 (1999)
218. H. Andres, J.M. Clemente-Juan, R. Basler, M. Aebersold, H.-U. Güdel, J.J. Borrás-Almenar, A. Gaita, E. Coronado, H. Büttner, S. Janssen, Inorg. Chem. **40**(8), 1943 (2001)
219. J.M. Clemente-Juan, E. Coronado, A. Gaita-Ariño, C. Giménez-Saiz, G. Chaboussant, H.-U. Güdel, R. Burriel, H. Mutka, Chem. Eur. J. **8**(24), 5701 (2002)
220. I.I. Geru, *Nizkochastotnye Rezonancy Ekcitonov i Primesnyh Tsentrov* (Stiintsa, Kishinev, 1976), 194 pp
221. I.I. Geru, Time inversion symmetry and quasi-energy state degeneracy. Studia Univ. Babeş-Bolyai, Physica, XXXV, **1**, pp. 83–91 (1990)
222. E. Wigner, *Group Theory and Its Applications to the Quantum Mechanics of Atomic Spectra* (New York, 1959)
223. V.V. Balashov, V.C. Dolinov, *Kurs kvantovoi mechaniki* (Izdat. Moskovsk. universit, Moskva, 1982), 279 pp
224. W. Heitler, *The Quantum Theory of Radiation* (Courier Corporation, 1954), 430 pp
225. V.I. Ritus, Shift and energy-level spliting by electromagnetic wave field. JETF **51**, N5(11), 1544–1549, (1966)
226. Ya.B. Zel'dovich, Quasi-energy of quantum system being esposed to periodic action. JETF **51**, N5(12), 1492–1495 (1966)
227. Ya.B. Zel'dovich, UFN **110**, 139 (1973)
228. M.G. Floquet, Ann. Ecole Norm. Sup. **12**, 47–89 (1883)
229. H. Sambe, Phys. Rev. A **7**(6), 2203–2213 (1973)
230. I.I. Geru, About time-reversal operator for high spin systems, in *Quantum Theory of Multiparticle Systems*, ed. by V.A. Moskalenko (Red. Izd. Otdel AN MSSR, 1970), pp. 66–80
231. H.J. Lipkin, *Quantum Mechanics* (North Holland Publishing Company, 1973), 588 pp
232. I.I. Geru, Time inversion for systems with quasi-energy spectrum, in *Physics of Semiconductors and Dielectrics* (Stiintsa, Kishinev, 1982), pp. 54–59 [in Russian]
233. J. Amiel, Comptes rendus (CR) **207**, 1097 (1938)
234. B.C. Guha, Proc. R. Soc. **206A**, 353 (1951)
235. B. Bleaney, K.D. Bowers, Proc. R. Soc. **214A**, 451–465 (1952)
236. B. Bleaney, K.D. Bowers, Phil. Mag. **43**, 372 (1952)
237. H. Abe, J. Shimada, Phys. Rev. **90**, 316 (1953)
238. H. Abe, J. Shimada, Phys. Soc. Jpn. **12**, 1255 (1957)
239. J.H. Van Niekers, F.R.L. Schoening, Acta Crystallogr. **6**, 227 (1953)

240. A. Abraham, B. Bleany, *Electron Paramagnetic Resonance of Transition Ions* (Clarendon Press, Oxford, 1970), 651 pp. [Russian translation]
241. Yu. V. Yablokov, V.K. Voronkova, L.V. Mosina, *Paramagnetic Resonance of Exchainge Clasters* (Nauka, Moscow, 1988), 180 pp
242. H. Abe, J. Phys. Soc. Jpn. **13**, 118 (1957)
243. H. Abe, H. Shirai, J. Phys. Soc. Jpn. **16**, 118 (1961)
244. I.E. Dzyaloshinsky, JETF **32**, 1547–1562 (1957)
245. T. Moryia, Phys. Rev. **120**, 91–99 (1960)
246. B.S. Tsukerblat, M.I. Belinsky, *Magnetochemistry and Radiospectroscopy of Exchange Clusters* (Stiintsa, Kishinev, 1983), 280 pp
247. V.Ya. Mitrofanov, A.E. Nikiforov, V.I. Cherepanov, *Spectroscopy of Exchange Coupled Complexes in Ionic Cristals* (Nauka, Moscow, 1985), 142 pp
248. J. Owen, E.A. Harris, Pair spectra and exchange interaction, in *Electron Paramagnetic Resonance* (Plenum, New York, 1972) pp. 427–492
249. J.M. Baker, Rep. Progr. Phys. **34**, 109–173 (1971)
250. H.A. Kramers, Physica **1**, 182–192 (1934)
251. G.M. Pratt, Phys. Rev. **97**, 926–932 (1955)
252. J.B. Goudenough, A.L. Loeb, Phys. Rev. **98**, 391–408 (1955)
253. P.W. Anderson, Phys. Rev. **115**, 2–13 (1955)
254. P.W. Anderson, Solid State Phys. **14**, 99–214 (1963)
255. K.I. Gondaira, I.A. Tanabe, J. Phys. Soc. Jpn. **21**, 1527–1548 (1966)
256. N.L. Huang, R. Orbah, Phys. Rev. **154**, 487–492 (1967)
257. N.N. Bogolyubov, *Izbrannye trudy*, vol. 2 (Naukova Dumka, Kiev, 1970), 522 pp
258. K.W.H. Stewens, Phys. Status Solidi (b) **97**, 51–62 (1970)
259. N. Fushikami, Y. Tanabe, J. Phys. Soc. Jpn. **45**, 1559–1566 (1978)
260. N. Fushikami, Y. Tanabe, J. Phys. Soc. Jpn. **47**, 505–511 (1979)
261. V.A. Gubanov, A.I. Lihtenstein, A.V. Postnikov, *Magnetism and Chemical Bond in Crystals* (Nauka, Moscow, 1985), p. 245
262. S.V. Vonsovsky, B.V. Karpenko, in *Problems of Theoretical Physics* (Nauka, Moscow, 1967), p. 204
263. J. Goudenough, *Magnetism and Chemical Bond* (Interscience Publishers, Wiley, New York, 1963), p. 393
264. I. Veltrusky, Czech. J. Phys. B **25**, 101 (1975)
265. V.A. Sapozhnikov, A.E. Nikiforov, V. Yu. Mitrofanov, A.N. Meni, Doklady AN SSSR **221**, 577 (1975)
266. M.V. Eremin, Yu. V. Rakitin, Phys. Status Solidi (b) **97**, 51–62 (1980). https://doi.org/10.1002/pssb.2220970103
267. M.V. Eremin, Fiz. Tverd. Tela **22**, 3143–3148 (1980)
268. P.J. Hay, J.C. Thibeaut, R. Hofman, J. Am. Chem. Soc. **97**, 4884–4889 (1975)
269. R.D. Harcourt, G.E. Martin, J. Chem. Soc. Faraday Trans. Part II **73**, 1–14 (1977)
270. O. Kahn, B. Briat, J. Chem. Soc. Faraday Trans. Part II **72**, 268–281 (1976)
271. O. Kahn, B. Briat, J. Chem. Soc. Faraday Trans. Part II **72**, 1441–1446 (1976)
272. Yu.V. Rakitin, V.T. Kalinnikov, Sovremennaya Magnetohimia (Nauka, Moskva, 1994), 276 pp
273. L. Noodleman, J. Chem. Phys. **74**, 5737–5743 (1981)
274. I. Mayer, S.A. Angelov, J. Quant. Chem. **18**(3), 783–796 (1980)
275. G. Van Valkeren, W.W. Schmidt, R. Block, Physica B. **97**, 315–337 (1979)
276. P. de Loth, P. Cassoux, J.P. Marlieu, J. Am. Chem. Soc. **103**, 4007–4016 (1981)
277. M.F. Charlot, M. Verdagner, Y. Yournaux, P. de Loth, Inorg. Chem. **23**, 3802–3808 (1984)
278. M.V. Eremin, Theory of exchange interaction of magnetic ions in dielectrics, in *Spectroscopy of Crystals* (Nauka, Leningrad, 1985), pp. 150–171
279. M.V. Eremin, Fiz. Tverd. Tela **24**, 423–432 (1982)
280. M.V. Eremin, Fiz. Tverd. Tela **24**, 3216–3222 (1982)
281. M.V. Eremin, Fiz. Tverd. Tela **24**, 1981–1987 (1982)

282. M.V. Eremin, Fiz. Tverd. Tela **25**, 1754–1760 (1983)
283. M. Fukuchi, Progr. Theor. Phys. **25**, 939–955 (1961)
284. N.N. Bogolyubov, S.V. Tyablikov, JETF **19**, 256 (1961)
285. P. Lemoine, B. Viossat, G. Morgant et al., J. Inorg. Biochem. **89**, 18–28 (2002)
286. P. Lemoine, D. Nguyen-Huy, B. Viossat et al., Acta Crystallogr. C **55**, 2068–2070 (1999)
287. P. Lemoine, A. Mazurier, I. Billy et al., Z. Krystallogr.-New Cryst. Struct. **215**, 523–525 (2000)
288. M.T. Garland, J.Y. Le Marouille, E. Spodine, Acta Crystallogr. C **41**, 855–858 (1985)
289. X. Solans, L. Ruiz-Ramirez, L. Gasque et al., Acta Ctrystallogr. C **43**, 428–430 (1987)
290. M.T. Garland, D. Grandjean, E. Spodine et al., Acta Crystallogr. C **43**, 1910–1912 (1987)
291. M.T. Garland, J.Y. Le Marouille, E. Spodine et al., Acta Crystallogr. C **42**, 1518–1520 (1986)
292. E. Dubler, U.K. Haring, K.H. Scheller et al., Inorg. Chem. **23**, 3785–3792 (1984)
293. P. Štarha, Z. Trávniček, R. Herchel et al., J. Inorg. Biochem. **103**, 432–440 (2009)
294. G.A. Van Albada, I. Mutikainen, U. Turpeinen et al., Polyhedron **25**, 3278–3284 (2006)
295. P. de Meester, A.C. Skopski, J. Chem. Soc. A. **13**, 2167–2169 (1971)
296. D. Sonnenfroh, R.W. Kreilick, Inorg. Chem. **19**, 1259–1262 (1980)
297. A. Terzis, A.L. Beauchamp, R. Rivest, Inorg. Chem. **12**, 1166–1170 (1973)
298. J.M. González-Pérez, C. Alarcórn-Payer, A. Castiñeiras et al., Inorg. Chem. **45**, 877–882 (2006)
299. C. Mealli, F. Zanobini, J. Chem. Soc. Chem. Commun. **2**, 97–98 (1982)
300. K. Emerson, A. Emad, R.W. Brookes et al., Inorg. Chem. **12**, 978–981 (1973)
301. M. Maloň, Z. Trávniček, M. Maryško et al., Inorg. Chim. Acta **323**, 119–129 (2001)
302. M. Maloň, Z. Trávniček, M. Maryško et al., Trans. Met. Chem. **27**, 580–586 (2002)
303. Z. Trávniček, M. Maloň, Z. Šindelář et al., J. Inorg. Biochem. **84**, 23–32 (2001)
304. A. Karaliota, O. Kresti, C. Tzougraki, J. Inorg. Biochem. **84**, 33–37 (2001)
305. D.P. Graddon, J. Inorg. Nucl. Chem. **17**, 222 (1961)
306. A.B.P. Lever, in, *Inorganic Electronic Spectroscopy*, 2nd edn. (Elsevier, Amsterdam, 1984), pp. 553–572, 636–638
307. E. Kokot, R.L. Martin, Inorg. Chem. **3**, 1306 (1964)
308. L. Dubicki, R.L. Martin, Inorg. Chem. **5**, 2203 (1966)
309. M. Melnic, Coord. Chem. Rev. **36**, 1 (1981)
310. M. Melnic, Coord. Chem. Rev. **42**, 259 (1982)
311. C.C. Hadjikostas, G.A. Katsoulos, M.P. Sigalas et al., Inorg. Chim. Acta **167**, 165 (1990)
312. J.R. Watson, C. Shur, C. Trap, Inorg. Chem. **7**, 469 (1968)
313. Yu.M. Chumakov, V.I. Tsapkov, I.G. Filippova et al., Crystallogr. Rep. **53**, 619–625 (2008)
314. N.M.S. Cusado, R.A. Isaacson, R. Calvo, J. Inorg. Biochem. **84**, 201–206 (2001)
315. E.N. Nfor, E.A. Eno, J.N. Foba-Tendo et al., Zhurnal Struktur. Himii **53**, 1141–1147 (2012)
316. E.L. Isaeva, M.H. Schasutdinova, N.N. Bukov et al., Zhurnal Struktur. Himii **52**, 1062–1064 (2011)
317. R.M. White, *Quantum Theory of Magnetism* (McGraw-Hill Book Company, New York, 1970)
318. G.E. Pake, *Paramagnetic Resonance* (W.A. Benjamin Inc, New York, 1962)
319. M. Julve, M. Verdaguer, M.F. Charlot, O. Kahn, Inorg. Chim. Acta **82**, 5–12 (1984)
320. S. Kremer, Inorg. Chem. **24**, 887–890 (1985)
321. P.D. Krasicky, J.C. Scott, R.H. Silsbee, A.L. Ritter, J. Phys. Chem. Solids **39**, 991–998 (1978)
322. M.Y. Okamura, B.M. Hoffman, J. Chem. Phys. **51**, 3128 (1969)
323. M. Julve, O. Kahn, Inorg. Chim. Acta **76**, L39–L41 (1983)
324. A. Syamal, E.F. Carey, L.J. Theoriot, Inorg. Chem. **12**, 245–248 (1973)
325. B. Jezowska-Trzebiatowska, J. Jezierzka, Chem. Phys. Lett. **34**, 237–239 (1975)
326. D.M. Duggan, D.N. Hendrickson, Inorg. Chem. **14**, 1944–1956 (1975)
327. E.J. Laskowski, D.N. Hendrickson, Inorg. Chem. **17**, 457–470 (1978)
328. L.E. Sweet, L.E. Roy, F. Meng, T. Hughbanks, J. Am. Chem. Soc. **128**, 10193–10201 (2006)
329. W. BingWo, J. ShangDa, W. XiuTeng, Sci. China Ser. B: Chem. **52**, 1739–1758 (2009)
330. O. Kahn, *Molecular Magnetism*, 1st edn. (VCH, Weinheim, 1993)
331. A. Rohde, S.T. Hatscher, W. Urland, J. Alloys Compds. **374**, 137–141 (2004)

332. S.T. Hatscher, W. Urland, Angew. Chem. Int. Ed. **42**, 2862 (2003)
333. S. Lui, L. Gelmini, S.J. Retting, J. Am. Chem. Soc. **114**, 6081 (1992)
334. W. Plass, G. Fries, Z. Anorg, Allg. Chem. **623**, 1205 (1997)
335. P. Guerriero, S. Tamburini, P.A. Vigato, Inorg. Chim. Acta **189**, 19 (1991)
336. J.-P. Costes, F. Dahan, A. Dupuis et al., Inorg. Chem. **27**, 153 (1998)
337. F. Avecilla, C. Platas-Iglesias, R. Rodriguez-Cortinas et al., J. Chem. Soc., Dalton Trans. 4658 (2002)
338. J.-P. Costes, J.M. Clemente-Juan, F. Dahan et al., Angew. Chem. Int. Ed. **41**, 323 (2002)
339. H. Hou, G. Li, L. Li et al., Inorg. Chem. **42**, 428 (2003)
340. J.-P. Costes, A. Dupuis, J.-P. Laurent, Inorg. Chem. Acta **268**, 125 (1998)
341. A. Panagiotopoulos, T.F. Zafiropoulos, S.P. Perlepes et al., Inorg. Chem. **34**, 4918 (1995)
342. D. John, W. Urlad, Eur. J. Inorg. Chem. 4486–4489 (2005). https://doi.org/10.1002/ejic.200500734
343. D. John, W. Urlad, Z. Anorg, Allg. Chem. **631**, 2635 (2005)
344. A. Rohde, W. Urland, J. Alloys Compds. **408–412**, 618–621 (2005)
345. A. Rohde, W. Urland, Z. Anorg, Allg. Chem. **631**, 417 (2005)
346. S.T. Hatscher, W. Urland, Angew. Chem. **115**, 2969 (2003)
347. A. Rohde, W. Urland, Z. Anorg, Allg. Chem. **630**, 2434 (2004)
348. A. Rohde, W. Urland, Synthesis, crystal structure and magnetic behavior of dimeric and tetrameric gadolinium carboxylates with trichloracetic acid. R. Soc. Chem. Dalton Trans. 2974–2978 (2006)
349. S.Y. Niu, J. Jin, X.L. Jin et al., Solid State Sci. **4**, 1103 (2002)
350. A.W.-H. Lam, W.-T. Wong, S. Gao et al., Eur. J. Inorg. Chem. 149 (2003)
351. A. Rizzi, R. Baggio, M.T. Garland et al., Inorg. Chim. Acta **353**, 315 (2003)
352. A.M. Atria, R. Baggio, M.T. Garland et al., Inorg. Chim. Acta **357**, 1997 (2004)
353. A. Panagiotopoulos, T.F. Zafiropoulos, S.P. Perlepes et al., Inorg. Chem. **34**, 4918 (1995)
354. Y.-T. Li, C.-W. Yan, Polish J. Chem. **75**, 1195–1202 (2001)
355. A. Bencini, C. Benelli, A. Caneschi et al., J. Am. Chem. Soc. **107**, 8128 (1985)
356. Y.T. Li, D.Z. Liao, Z.H. Jiang et al., Polyhedron **14**, 2209 (1995)
357. M. Sakamoto, M. Tokagi, T. Ishimori et al., Bull. Chem. Soc. Jpn. **61**, 1613 (1988)
358. O. Kahn, Angew. Chem., Int. Edit. Engl. **24**, 384 (1985)
359. J. Dean, K.J. Maxwell, Mol. Phys. **47**, 551 (1982)
360. D. John, W. Urland, Eur. J. Inorg. Chem. **2006**, 3503–3509 (2006). https://doi.org/10.1002/ejic.200600353
361. G. Abbas, Y. Lan, G. Kostakis et al., Inorg. Chim. Acta **361**, 3494–3499 (2008)
362. C. Benelli, Gatteschi. Chem. Rev. **102**, 2369 (2002)
363. S. Aime, M. Botta, M. Fasano et al., Chem. Soc. Rev. **27**, 19 (1998)
364. P.-H. Lin, T.J. Burchell, Clérac, Angew. Chem. Int. Ed. **47**, 8848 (2008)
365. F. Habib, M. Murugesu, Chem. Soc. Rev. **42**, 3278–3288 (2013)
366. I.I. Geru, Appl. Magn. Reson. **19**, 563–569 (2000)
367. S. Ballanti, P.D. Lazzar, F. Flora et al., Phys. Scr. **51**, 326 (1995)
368. I.I. Geru, Doklady Akademii Nauk SSSR **276**, 1378–1380 (1984)
369. O. Kahn, J. Galy, P. Tola, J. Am. Chem. Soc. **100**, 3931–3933 (1978)
370. O. Kahn, M.F. Charlot, Nouv. J. Chim. **4**, 567–576 (1980)
371. I. Galy, J. Jaud, O. Kahn et al., Inorg. Chim. Acta **36**, 229–236 (1979)
372. M. Kato, Y. Yamada, T. Inogaki et al., Inorg. Chem. **34**, 2645 (1995)
373. R.C. Holz, T.E. Elgreen, L.F. Pearce et al., Inorg. Chem. **32**, 5844–5850 (1993)
374. G. Haselhorst, K. Wieghardt, S. Keller et al., Inorg. Chem. **32**, 520–525 (1993)
375. M. Mikuriya, Y. Kakuta, K. Kawano et al., Jpn. Chem. Lett. **11**, 2031–2034 (1991)
376. G. Toulouse, Commun. Phys. **2**, 115 (1977)
377. G. Toulouse, in *Disordered Sustems and Localization*, eds. by C. Castellani, C. Di Castro, L. Peliti (Springer, Berlin, 1981)
378. D.S. Fisher, G.M. Grinstein, A. Khurana, Phys. Today **41**, 56–67 (1988)
379. A. Pimpinelli, G. Uimin, J. Vilain, J. Phys. Condens. Matter **3**, 4693 (1991)

380. D.L. Stein (ed.), *Spin Glasses in Biology* (World Scientific, Singapore, 1992)
381. J.A. Mydosh, *Spin Glasses: an Experimental Introduction* (Taylor and Francis, London, 1993)
382. J.K. McCuster, E.A. Schmitt, D.N. Hendrickson, in *Magnetic Molecular Materials*. NATO ASI Series, eds. Gatteschi D., Kahn O., Miller J.S. and Palacio F. (Kluwer, Dordrecht, 1991) p. 297
383. O. Kahn, Chem. Phys. Lett. **265**, 109–114 (1997)
384. O. Kahn, Comments Cond. Mat. Phys. **17**, 39 (1994)
385. B.S. Tsukerblat, B.Y. Kuyavskaya, M.I. Belinskii, A.V. Ablov, Theoret. Chim. Acta **38**, 131 (1975)
386. R. Beckett, R. Colton, F.F. Hoskins et al., Aust. J. Chem. **22**, 2527 (1969)
387. A. Zheludev, V. Barone, M. Bonnet et al., J. Am. Chem. Soc. **116**, 2019 (1994)
388. N.M. Altherton, *Principles of Electron Spin Resonance* (Ellis Horwood-Orentice Hall, London, 1993)
389. D. Dadi, M.-H. Whangbo, J. Chem. Phys. **121**, 672–680 (2004)
390. I.I. Geru, *Abstracts International Conference "Modern Development of Magnetic Resonance"* (Kazan, Russian Federation, 2007), pp. 99–101
391. R.D. Cannon, A. Upali, A. Jayassooriya, F.E. Sowrey, C. Tilford, A. Little, P.J. Bourke, R.D. Rogers, J.H. Vincest, G.J. Kearley, Inorg. Chem. **37**, 5675–5677 (1998)
392. R.D. Cannon, R.P. White, Progr. Inorg. Chem. **36**, 195 (1988)
393. H.U. Güdel, J. Chem. Phys. **82**, 2510 (1985)
394. A. Figuerola, V. Tangoulis, J. Ribas, H. Harti, I. Bruüdgam, M. Maestro, C. Diaz, Inorg. Chem. **46**(26), 11017–11024 (2007)
395. U.A. Jayasooriya, R.D. Cannon, R.P. White et al., J. Chem. Phys. **98**, 9303 (1993)
396. J.P. Costes, F. Dahan, J.P. Laurent, Inorg. Chem. **25**, 413 (1986)
397. K.D. Karlin, Q. Gan, A. Farooq et al., Inorg. Chim. Acta **165**, 37 (1989)
398. Y. Agnus, R. Louis, B. Metz el al., Inorg. Chem. **30**, 3155 (1991)
399. R.J. Butcher, C.J. O'Connor, E. Sinn, Inorg. Chem. **20**, 537 (1981)
400. F.B. Hulsbergen, R.W.M. Ten Goedt, G.C. Verschoor, J. Chem. Soc. Dalton Trans. **3**, 539 (1983)
401. K.D. Karlin, Q.F. Gan, A. Farooq et al., Inorg. Chem. **29**, 2549 (1990)
402. P. Chaudhuri, M. Winter, B.P.C. Della Vedova et al., Inorg. Chem. **30**, 2148 (1991)
403. H. Adams, N.A. Bailey, M.J.S. Dwyer, J. Chem. Soc. Dalton Trans. **8**, 1207 (1993)
404. S. Meenakumari, S.K. Tiwary, A.R. Chakravarty, Inorg. Chem. **33**, 2085 (1994)
405. I. Geru, Extended group-theoretical treatment of the time inversion symmetry and Kramers degeneracy: T_3-, T_2- and T_1-invariances, in *Proceedings of the XXXI Congress AMPERE Magnetic Resonance and Related Phenomena* (Poznan, Poland, 2000), p. 130
406. I.I. Geru, Rom. J. Phys. **48**, 485–494 (2003)
407. I.I. Geru, Non-Abelian groups of symmetry and behavior of systems with total half-integer spin under time reversal, in *Conference of Physicists of Moldova* (Chishinau, 2007), p. 174
408. I.I. Geru, Non-Abelian groups of symmetry containing time-reversal operator for systems with spin $S = 1/2$ and $S = 3/2$, in *International Conference Physics of Low Dimensional Structures, 27–28 June 2007, Book of Abstracts* (Publishing House of USM, Chisinau, 2007), pp. 69–71
409. I.I. Geru, Time inversion and generalized symmetry of spin systems, in *Abstracts of All-Union Symposium on Symmetry Theory and Its Generalizations* (Kishinev, 1980), pp. 42–44
410. T. Holstein, H. Primakoff, Phys. Rev. **58**, 1098 (1940)
411. F.J. Dyson, Phys. Rev. **102**, 1217 (1956)
412. S.V. Maleev, Zh. Eksp. Teor. Fiz. **33**, 1010 (1957)
413. S.V. Vonsovsky, *Magnetism* (Wiley, New York, 1974), 1256 pp
414. C. Tsallis, J. Phys. **33**(11–12), 1121–1127 (1972)
415. J. Schwinger, On angular momentum, in *Quantum Theory of Angular Momentum*, ed. by L.C. Biedenham, H. Van Dam (Academic Press, New York, 1956), pp. 229–279
416. N.N. Bogoliubov, D.V. Shirkov, *Quantum Fields* (Nauka, Moscow, 1980), p. 319
417. A.S. Davydov, *Solid State Theory* (Nauka, Moscow, 1978), 639 pp

418. S.V. Tyablikov, *Methods in the Quantum Theory of Magnetism* (Plenum Press, New York, 1967), 354 pp
419. R. Jost, *General Theory of Quantified Fields* (Mir, Moscow, 1967), 236 pp
420. H. Haken, *Quantum Field Theory of Solid State* (Nauka, Moscow, 1980), 341 pp
421. D. Pines, *Elementary Excitations in Solids* (W. A. Benjamin Inc, New York, 1963)
422. A. Isihara, *Statistical Physics* (Academic Press, New York, 1971)
423. J.A. Reisland, *The Physics of Phonons* (Wiley LTD, London, 1973)
424. R. Loudon, *The Quantum Theory of Light* (Clarendon Press, Oxford, 1973)
425. N.H. March, W.H. Young, S. Sampanthar, *The Many-Body Problems in Quantum Mechanics* (Cambridge University Press, Cambridge, 1967)
426. V.M. Agranovich, *The Theory of Excitons* (Nauka, Moscow, 1968), 382 pp
427. Sh. Tsitseica, *Quantum Mechanics* (Publishing House of the Romanian Academy, Bucharest, 1984), 635 pp
428. A.A. Maradudin, E.W. Montrall, G.H. Weiss, *Theory of Lattice Dynamics in the Harmonic Approximation* (Academic Press, New York, 1963), 319 pp
429. I.V. Saveliev, *The Basics of Theoretical Physics*, vol. 2. Quantum Mechanics (Nauka, Moscow, 1977), 351 pp
430. E.G. Pestov, G.M. Lapshin, *Quantum Electronics* (Voenizdat, Moscow, 1972), 334 pp
431. I.I. Geru, V. Geru, Bosonization of angular momentum. Rom. J. Phys. **44**, 97–115 (1999)
432. I.I. Geru, Boson-antiboson representation of spin operators, in *The International Seminar Radiospectroscopy of Condensed Matter. Abstracts of Communications* (Kiev, Ukraine, 1998), pp. 13–14
433. I.I. Geru, New approach in time-inversion symmetry. Rom. J. Phys. **48**, 485–494 (2003)
434. C.S. Wu, E. Ambler, R.W. Hayward et al., Phys. Rev. **105**, 1413 (1957)
435. T.D. Lee, R. Oehme, C.N. Yang, Phys. Rev. **106**(2), 340–345 (1957)
436. L.D. Landau, About conservation laws in weak interactions. JETF **32**, 405–406 (1957)
437. E.P. Wigner, *Symmetries and Reflections* (Indiana University Press, Bloomington, 1970)
438. R.V. Craster, S. Guenneau, *Acoustic Metamaterials. Negative Refraction, Imagining, Lensing and Cloaking* (Springer, Dordrecht, 2013), 319 pp
439. B. Banerjee, *An Introduction to Metamaterials and Waves in Composites* (Taylor and Francis Group, London, 2011), 109 pp
440. V.M. Shalaev, Nature Photonic **1**, 41–48 (2007)
441. C.M. Soukoulis, M. Wegener, Nature Photonic **5**, 523–530 (2011)
442. M. Wegener, S. Linden, Phys. Today **63**, 32–36 (2010)
443. H. Chen, C.T. Chan, P. Sheng, Nature Mater. **9**(5), 387–396 (2010)
444. D. Shuring, J.J. Mock, B.J. Justice et al., Science **314**, 977–980 (2006)
445. J. Mei, G. Ma, M. Yang, Nat. Commun. **3**, 756 (2012). https://doi.org/10.1038/ncomms1758
446. Z.G. Nicolaou, A.E. Motter, Nat. Mater. **11**, 608–613 (2012)
447. H.H. Huang, C.T. Sun, G.L. Huang, Int. J. Eng. Sci. **47**, 610–617 (2009)
448. S.H. Lee, C.M. Park, Y.M. Seo et al., J. Phys. Condens. Mater. **21**, 175704 (2009). https://doi.org/10.1088/0953-8984/21/17/175704
449. A.N. Norris, J. Acoust. Soc. Am. **125**, 839–849 (2009)
450. A. Martin, M. Kadic, R. Schittry, Phys. Rev. B **86**, 155116 (2012). https://doi.org/10.1103/physrevb.86
451. N.I. Zheluder, Y.S. Kivshar, Nat. Mater. **11**, 917–924 (2012)
452. S.A. Tretyakov, J. Opt. **19**, 013002 (2017). https://doi.org/10.1088/2040-8986/19/1/013002
453. M. Kadic, T. Bckmann, R. Schittny et al., Rep. Prog. Phys. **76**, 126501 (2013)
454. B.F. Bayman, *Some Lectures on Groups and their Applications to Spectroscopy* (Nordisk Institut for Teoretisk Atomfysik, Copenhagen, 1957), 225 pp
455. I. Geru, *Proceedings of the 30th Congress AMPERE Magnetic Resonance and Related Phenomena* (Poznan, Poland, 2000), p. 130
456. I.I. Geru, New approach in time inversion symmetry, in *The Third International Balkan Workshop on Applied Physics* (Targovishte, Romania, 2002), p. 38

457. K.J. Boström, *Combined Bohm and Everett: Axiomatics for a Standalone Mechanics* (2012), arXiv:1208.5632v4 [quant-ph]
458. T.D. Lee, C.N. Yang, Phys. Rev. **106**, 340 (1957)
459. B. Aubert et al., [The BaBar Collaboration], Phys. Rev. Lett. **87**, 091801 (2001). arXiv:hep-ex/0107013
460. B. Aubert et al., [The BaBar Collaboration], Phys. Rev. Lett. **89**, 281802 (2002). arXiv:hep-ex/0207055
461. B. Aubert et al., [The BaBar Collaboration], Phys. Rev. Lett. **93**, 131801 (2004). arXiv:hep-ex/0407057
462. B. Aubert et al. [The BaBar Collaboration], Phys. Rev. Lett. **94**, 191802 (2005). arXiv:hep-ex/0502017
463. L. Valentin, *Subatomic Physics: Nuclei and Particles, vol. 2. Developments* (Publishing House Mir, Moscow, 1986), 336 pp
464. G. Lüders, Dan. Mat. Fys. Medd. **28**, 5 (1954)
465. W. Pauli, Exclusion principle, Lorentz group and reflection of space-time and charge, in *Niels Bohr and the Development of Physics*, ed. by W. Pauli (Pergamen Press, New York, 1955)
466. L. Wolfenstein, Phys. Rev. Lett. **13**, 562 (1964)
467. M. Kobayashi, M. Maskawa, Prog. Theor. Phys. **49**, 652 (1972)
468. N. Cabibbo, Phys. Rev. Lett. **10**, 531 (1963)
469. G.'t Hoolt, Phys. Rev. Lett. **37**, 8 (1976)
470. R.N. Mohapatra, J.C. Pati, Phys. Rev. D **11**, 566 (1975)
471. S. Weinberg, Phys. Rev. Lett. **36**, 657 (1976)
472. T.D. Lee, Phys. Rev. D **8**, 1226 (1973)
473. W. Buchmübler, R. Rückl, D. Wyler, Phys. Lett. B **191**, 442 (1987)
474. W. Buchmübler, D. Wyler, Phys. Lett. B **177**, 377 (1986)
475. P. Herczeg, in *Fundamental Symmetries in Nuclei and Particles*, ed. by H. Henrikson, P. Vogel (World Scientific, Singapore, 1989), p. 46
476. S. Weingberg, in *The First Three Minutes* (Basic Book Inc., New York, 1977)
477. S. Weingberg, Phys. Rev. Lett. **42**, 850 (1979)
478. A.D. Sakharov, JETP Lett. **5**, 24 (1967)
479. J. Sromicki, M. Allet, K. Bodek et al., Phys. Rev. C **53**, 932–955 (1996)
480. P. Langacker, in *CP Violation. Advanced Series on Directions in High Energy Physics*, vol. 3, ed. By C. Jarlskog (World Scientific, Singapore, 1989), p. 552
481. A.G. Cohen, D.B. Kaplan, A.E. Nelson, Annu. Rev. Nucl. Part. Sci. **43**, 27 (1993)
482. T.D. Lee, C.N. Yang, Phys. Rev. **106**, 340 (1957)
483. M. Pospelov, A. Ritz, Ann. Phys. **318**, 119 (2005)
484. V.P. Gudkov, Phys. Rep. **212**, 77 (1992)
485. E. Blanke et al., Phys. Rev. Lett. **51**, 355 (1983)
486. T.S. Bhatia et al., Phys. Rev. Lett. **48**, 227 (1982)
487. R.A. Hardekopf et al., Phys. Rev. C **25**, 1090 (1982)
488. V.E. Bunakov, Phys. Rev. Lett. **60**, 2250 (1988)
489. J.D. Bowman, in *Intersections Between Particle and Nuclear Physics, AIP Conference Proceedings*, ed. By D.F. Geesaman, No. 150 (AIP, New York, 1986)
490. R. Lazauskas, Y.H. Song, V. Gudkov, Parity and time-reversal violation in $A = 2 - 4$ nuclei. EPJ Web Conf. **66**, 05014 (2014). https://doi.org/10.1051/epjconf/20146605014
491. I.B. Khriplovich, Zh. Eksp. Teor. Fiz. **71**, 51 (1976). [Sov. Phys. JETP **44**, 25 (1976)]
492. V.V. Flambaum, I.B. Khriplovich, O.P. Sushkov, Nucl. Phys. A **449**, 750 (1986)
493. V.V. Flambaum, I.B. Khriplovich, O.P. Sushkov, Phys. Lett. B **162**, 213 (1985)
494. V.F. Dmitriev, R.A. Sen'kov, Phys. At. Nucl. **66**, 1940 (2003)
495. V.F. Dmitriev, R.A. Sen'kov, N. Auerbach, Phys. Rev. C **71**, 035501 (2005)
496. J.H. de Jesus, J. Engel, Phys. Rev. C **72**, 045503 (2005)
497. S. Ban, J. Dobaczewski, J. Engel, A. Shukla, Phys. Rev. C **82**, 015501 (2010)
498. V.F. Dmitriev, V.B. Telitsin, V.V. Flambaum, V.A. Dzuba, Phys. Rev. C **54**, 3305 (1996)
499. J.S.M. Ginges, V.V. Flambaum, Phys. Rep. **397**, 63–154 (2004)

500. J. Erler, J.M. Ramsey-Musolf, Prog. Part. Nucl. Phys. **54**, 351–442 (2005)
501. E.D. Commins, Adv. At. Mol. Opt. Phys. **40**, 1–55 (1998)
502. A.V. Titov, N.S. Mosyagin, A.N. Petrov et al., Progr. Theor. Chem. Phys. **15**, Chapter II, 1–33 (2006)
503. B.M. Roberts, V.A. Dzuba, V.V. Flambaum, Annu. Rev. Nucl. Part. Sci. **65**, 63–86 (2015)
504. L.D. Landau, Sov. Phys.-JETP **5**, 336–337 (1957)
505. J.S.M. Ginger, V.V. Flambaum, Phys. Rep. **397**, 63 (2004)
506. I.B. Khriplovich, S.K. Lamoreaux, *CP Violation without Strangeness* (Springer, Berlin, 1997)
507. P.G.H. Sandars, Phys. Lett. **14**, 194 (1965)
508. B.C. Regan, E.D. Commins, C.J. Schmidt et al., Phys. Rev. Lett. **86**, 2505 (2001)
509. P.G.H. Sandars, Phys. Rev. Lett. **19**, 1396–1398 (1967)
510. V.V. Flambaum, Yad. Fiz. **24**, 383–386 (1976). [Sov. J. Nucl. Phys. **24**, 199 (1976)]
511. I.B. Khriplovich, S.K. Lamoreaux, *CP without Strangeness. The Electric Dipole Moments of Particles, Atoms, and Molecules* (Springer, Berlin, 1997)
512. E.A. Hinds, C.E. Loving, P.G.H. Sandars, Phys. Lett. B **62**, 97–99 (1976)
513. L.I. Schiff, Phys. Rev. **132**, 2194–2200 (1963)
514. D. Cho, K. Sangster, E.A. Hing, Phys. Rev. A **4**, 2783–2799 (1991)
515. M.D. Swallows, T.H. Loftus, W.C. Griffith et al., Phys. Rev. A **87**, 021102 (2013)
516. M.A. Rosenberry, T.E. Chupp, Phys. Rev. Lett. **86**, 22 (2001)
517. R.J. Holt, I. Ahmad, K. Bailey et al., Nucl. Phys. A **844**, 53 (2010)
518. E.R. Tardiff, E.T. Rand, G.C. Ball et al., Hyperfine Int. **225**, 197 (2014)
519. V.V. Flambaum, I.B. Khriplovich, O.P. Suchkov, Sov. Phys.-JETP **60**, 873 (1984). [ZhETF **87**, 1521, 1984]
520. P.G.H. Sandar, Phys. Rev. Lett. **19**, 1396 (1967)
521. V.V. Flambaum, D. DeMilke, M.G. Kozlov, Phys. Rev. Lett. **113**(10), 103003 (2014)
522. I.B. Khriplovich, *Parity Non-conservation in Atomic Phenomena* (Taylor and Francis Ltd, London, 1991), p. 314
523. J.G. Bednorz, K.A. Muller, Z. Phys. B **64** (2), 189–193 (1986)
524. J. Bardeen, L.N. Cooper, J.R. Schrieffer, Phys. Rev. **106**(1), 162–164 (1957). https://doi.org/10.1103/PhysRev106162
525. M. Iavarone, G. Karapetrov, A.E. Koshelev et al., Phys. Rev. Lett. **89**(18), 187002 (2002). https://doi.org/10.1103/PhysRevLett.89.187002
526. B.J. Wilson, M.P. Das, J. Phys.: Condens. Matter **25**, 425702 (2013). https://doi.org/10.1088/0953-8984/25/425702
527. D.J. Scalapino, Phys. Rep. **250**, 329 (1995)
528. J.F. Annet, N.Gondelfend, A.J. Leggett, *Physical Properties of High-Temperature Superconductors V*, ed. by D.M. Ginsberg (World Scientific, Singapore, 1996), 375 pp
529. P. Monthoux, D. Pines, Phys. Rev. Lett. **69**, 961 (1992)
530. T. Moriya, Y. Takahashi, K. Ueda, J. Phys. Soc. Jpn. **59**, 2005 (1990)
531. P.W. Anderson, Science **235**, 1196 (1987)
532. G. Kotliar, J. Liu, Phys. Rev. B **38**, 5142 (1988)
533. E. Dagotto, T.M. Rice, Science **271**, 618 (1996)
534. S.C. Zhang, Science **275**, 1089 (1997)
535. M. Sigrist, Progr. Theor. Phys. **99**(6), 899–929 (1998)
536. M. Sigrist, T.M. Rice, Z. Phys. B **68**, 9 (1987)
537. S.E. Barrett, J.A. Martindale, D.J. Durand et al., Phys. Rev. Lett. **66**, 108 (1991)
538. M. Sigrist, T.M. Rice, J. Phys. Soc. Jpn. **61**, 4283 (1992)
539. M.B. Walker, J. Luettmer-Strathmann, Phys. Rev. B **54**, 588 (1996)
540. Ch. Bruder, A. Van Otterlo, G.T. Zimanyi, Phys. Rev. B **51**, 12904 (1995)
541. S. Yip, Phys. Rev. B **52**, 3087 (1995)
542. D.J. Van Harlingen, Rev. Mod. Phys. **67**, 515 (1995)
543. D.A. Wollman, D.J. Van Harlingen, W.C. Lee et al., Phys. Rev. Lett. **71**, 2134 (1993)
544. D.A. Wollman, D.J. Van Harlingen, J. Giapintzakis et al., Phys. Rev. Lett. **74**, 797 (1995)
545. D. Brawner, H.R. Ott, Phys. Rev. B **50**, 6530 (1994)

546. A. Mathai, Y. Gim, R.C. Black et al., Phys. Rev. Lett. **74**, 4523 (1995)
547. I. Iguchi, Z. Wen, Phys. Rev. B **49**, 12388 (1994)
548. J.H. Miller, Q.Y. Ying, Z.G. Zou et al., Phys. Rev. Lett. **74**, 2347 (1995)
549. V.B. Geshkenbein, A.I. Larkin, Pisma Zh. Eksp. Teor. Fiz. **43**, 306 (1986). [JETP Lett. **43**, 395 (1986)]
550. M. Sigrist, K. Ueda, Rev. Mod. Phys. **63**, 239 (1991)
551. M. Sigrist, D.B. Bailey, H.B. Laughlin, Phys. Rev. Lett. **74**, 3249 (1995)
552. D.B. Bailey, M. Sigrist, R.B. Laughlin, Phys. Rev. B **55**, 15239 (1997)
553. G.B. Arnold, J. Low. Temp. Phys. **59**, 143 (1985)
554. A. Furusaki, M. Tsukada, Solid State Commun. **78**, 299–302 (1991)
555. Y.S. Barash, A.V. Galaktionov, A.D. Zaikin, Phys. Rev. B **52**, 665 (1995)
556. Y. Tanaka, Phys. Rev. Lett. **72**, 3871 (1994)
557. A. Huck, A. van Otterlo, M. Sigrist, Phys. Rev. B **56**, 14163 (1997)
558. K. Kuboki, M. Sigrist, J. Phys. Soc. Jpn. **65**, 361 (1996)
559. M. Sigrist, T.M. Rice, K. Ueda, Phys. Rev. Lett. **63**, 1727 (1989)
560. M. Sigrist, Y.B. Kim, J. Phys. Soc. Jpn. **63**, 4314 (1994)
561. B. Kuklov, M. Sigrist, Int. J. Mod. Phys. **11**, 1113 (1997)
562. M. Tinkham, *Introduction to Superconductivit*, 2nd edn. (McGraw-Hill International, New York, 1996)
563. V.B. Geshkenbein, A.I. Larkin, A. Barone, Phys. Rev. B **36**, 235 (1987)
564. G.E. Volovik, L.P. Gor'kov, Zh. Eksp. Teor. Fiz. **88**, 1412 (1985). [Sov. Phys. JETP **61**, 843 (1985)]
565. M. Matsumoto, H. Shiba, J. Phys. Soc. Jpn. **64**, 3384 (1995)
566. M. Matsumoto, H. Shiba, J. Phys. Soc. Jpn. **64**, 4867 (1995)
567. M. Matsumoto, H. Shiba, J. Phys. Soc. Jpn. **65**, 2194 (1996)
568. G.E. Blonder, M. Tinkham, T.M. Klapwijk, Phys. Rev. B **25**, 4515 (1982)
569. M. Fogelström, D. Rainer, J.A. Sauls, Phys. Rev. Lett. **79**, 281 (1997)
570. J.X. Zhu, C.S. Ting, Phys. Rev. B **57**, 3038 (1998)
571. M. Covinton, M. Aprili, E. Paraoanu et al., Phys. Rev. Lett. **79**, 277 (1997)
572. M. Sigrist, K. Kuboki, P.A. Lee et al., Phys. Rev. B **53**, 2835 (1996)
573. M.E. Zhitomirsky, M.B. Walker, Phys. Rev. Lett. **79**, 1734 (1997); Phys. Rev. B **57**, 8560 (1998)
574. W. Belzig, Ch. Bruder, M. Sigrist, Phys. Rev. Lett. **80**, 4285 (1998)
575. A.B. Kuklov, Phys. Rev. B **52**, R7002–R7005 (1995)
576. R.H. Heffner, M.R. Norman, Comments Cond. Matter Phys. **17**, 361–408 (1996)
577. P. Kumar, P. Wölfle, Phys. Rev. Lett. **59**, 1954 (1987)
578. M. Sigrist, T.M. Rice, Phys. Rev. B **39**, 2200 (1989)
579. J.A. Sauls, Adv. Phys. **43**, 113 (1994)
580. Y. Maeno, H. Hashimoto, K. Yoshida et al., Nature **372**, 532 (1994)
581. V.P. Mineev, Pis'ma Zh. Eksp. Teor. Fiz. **49**, 624 (1989). [JETP Lett. **49**, 719 (1989)]
582. H. Choi, P. Muzikar, Phys. Rev. B **39**, 9664 (1989)
583. A. Schenk, *Muon Spin Rotation Spectroscopy* (Hilger, London, 1985)
584. A. Amato, Rev. Mod. Phys. **69**, 1119 (1997)
585. R. Kubo, T. Toyabe, in *Magnetic Resonance and Relaxation*, ed. by R. Blinc (NorthHolland, Amsterdam, 1967), p. 810
586. G.M. Luke, A. Keren, L.P. Le et al., Phys. Rev. Lett. **71**, 1466 (1993)
587. O. Mülken, A. Blumen, Phys. Rep. **502**, 37–87 (2011)
588. V.A. Moskalenko, Fiz. Met. Metalloved. **8**, 503 (1959)
589. H. Suhl, B.T. Matthias, L.R. Walker, Phys. Rev. Lett. **3**, 552 (1959)
590. H.-X. Huang, Y.-Q. Li, J.-Y. Gan et al., Phys. Rev. B **75**, 195113 (2007)
591. Z. Zimorás, M. Faccin, Z. Kádár, Sci. Rep. **3**, 2361 (2013). https://doi.org/10.1038/srep02361, arXiv:1208.4049
592. P. Xiang, M. Litinskaya, E.A. Shapiro, R.V. Krems, New J. Phys. **15**, 063015 (2013)
593. S. Bedkihal, M. Bandyopadhyay, D. Segal, The Europ. Phys. J. B **86**, 1–18 (2013)

594. D. Manzano, P.I. Hurtado, Symmetry and the thermodynamics of currents in open quantum systems. arXiv preprint arXiv:1310.7370 (2013)
595. S. Cameron, S. Fehrenbach, L. Granger et al., Linear Algebra Appl. **455**, 115–142 (2014)
596. D. Lu, J.D. Biamonte, J. Li et al., *Chiral Quantum Walks* (2014), arXiv:1405.6209v1
597. I. Sinayskiy, A. Marais, F. Petruccione et al., Phys. Rev. Lett. **108**, 020602 (2012)
598. M. Mohseni, P. Rebentrost, S. Lloyd et al., J. Chem. Phys. **129**, 174106 (2008)
599. S. Lloyd, M. Mohseni, A. Shabani et al., (2011), arXiv:1111.4982v1
600. B.R. Holstein, *Weak Interactions in Nuclei*. Princeton Series in Physics, vol. 106 (1989)
601. E. Grodner, J. Phys.: Conf. Ser. **533**, 012040 (2014)
602. A.M. Childs, Phys. Rev. Lett. **102**, 180501 (2009)
603. M. Faccin, T. Johnson, J. Phys. Rev. X **3**, 041007 (2013)
604. O. Mülken, A. Blumen, Phys. Rep. **502**, 37–87 (2011)
605. M.B. Plenio, S.F. Huelga, New J. Phys. **10**, 113019 (2008)
606. A. Sinayskiy, F. Marais, Petruccione et al. Phys. Rev. Lett. **108**, 020602 (2012), arXiv:1401.3298
607. E. Farhi, S. Gutmann, Phys. Rev. A **58**, 915 (1998)
608. A.M. Childs, R. Cleve, E. Deotto et al., *Proceedings of the 35th Annual ACM STOC* (ACM, New York, 2003), pp. 59–68
609. O. Mülken, A. Blumen, Phys. Rep. **502**, 37 (2011)
610. J. Kempe, Contemp. Phys. **44**, 307 (2003)
611. V. Kendon, Math. Struct. Comput. Sci. **17**, 1169 (2006)
612. M. Mohseni, P. Rebentrost, S. Lloyd et al., J. Chem. Phys. **129**, 174106 (2008)
613. E. Harel, G.S. Engel, Proc. Nat. Acad. Sci. **109**, 706 (2012)
614. A. Kossakowski, Rep. Math. Phys **3**, 247 (1972)
615. G. Lindblad, Commun. Math. Phys **48**, 119 (1975)
616. H.-P. Breuer, F. Petruccione, *The Theory of Open Quantum Systems* (Oxford University Press, Oxford, 2002)
617. J.D. Whitfield, C.A. Rodriguez-Rosario, A. Aspuru-Guzik, Phys. Rev. A **81**, 022323 (2010)
618. L.M.K. Vandersypen, I.L. Chuang, Rev. Mod. Phys. **76**, 1037–1069 (2005)
619. J.A. Vaccaro, Found. Phys. **41**, 1569–1596 (2011). https://doi.org/10.1007/s1070-011-9568-x
620. J.A. Vaccaro, J. Phys.: Conf. Ser. **306**, 012057 (2011). https://doi.org/10.1088/1742-6596/306/1/012057
621. J.A. Vaccaro, *Large Scale Physical Effects of T Violation in Mesons* (2009). Preprint arXiv:0911.4528v2
622. H. Price, *Times Arrow and Archimedes Point* (Oxford University Press, New York, 1966)
623. I.D. Novikov, *The River of Time* (Cambridge University Press, Cambridge, 1998)
624. H.D. Zeh, *The Principal Basis for the Direction of Time* (Springer, Berlin, 2007)
625. R.P. Feynman, Rev. Mod. Phys. **20**, 367–387 (1948)
626. T.D. Lee, L. Wolfenstein, Phys. Rev. **138**, B1490–B1496 (1964)
627. W.-M. Yao et al., J. Phys. G: Nucl. Part. Phys. **33**, 666–684 (2006)
628. J.A. Vaccaro, Found. Phys. **45**, 691–706 (2015). https://doi.org/10.1007/s10701-015-9896-3
629. J.A. Vaccaro, Proc. R. Soc. A **472**, 20150670 (2016). https://doi.org/10.1098/rspa.2015.0670
630. S. Goldschmidt, K. Renn, Ber. Dtsch. Chem. Ges. B **55**, 628–643 (1922)
631. S.A. Al'tshuler, B.M. Kozyrev, *Electron Paramagnetic Resonance* (Academic Press, New York, 1964)
632. B.N. Mishra, S.K. Gupta, Rev. Phys. Appliquée **8**, 117–119 (1973)
633. W. Duffy, Jun, D.L. Strandburg, J. Chem. Phys. **46**, 456 (1967)
634. N. Ohigashi, H. Inokuchi, Bull. Chem. Soc. Jpn. **41**, 269 (1968)
635. E. Zavoisky, Paramagnetic absorption in perpendicular and parallel fields for salts, solutions and metals. Ph.D. thesis, Phys. Inst. Acad. Sci. USSR, Moscow (1944)
636. E. Zavoisky, J. Phys. USSR **9**, 245 (1945)

Index

I. I. Geru, *Time-Reversal Symmetry*, Springer Tracts in Modern Physics 281,
https://doi.org/10.1007/978-3-030-01210-6

M

N

The manufacturer's authorised representative in the EU is Springer Nature Customer Service Centre GmbH, Europaplatz 3, 69115 Heidelberg, Germany. If you have any concerns regarding our products, please contact ProductSafety@springernature.com

Printed and bound by CPI Group (UK) Ltd, Croydon, CR0 4YY
30/09/2026
02225652-0013